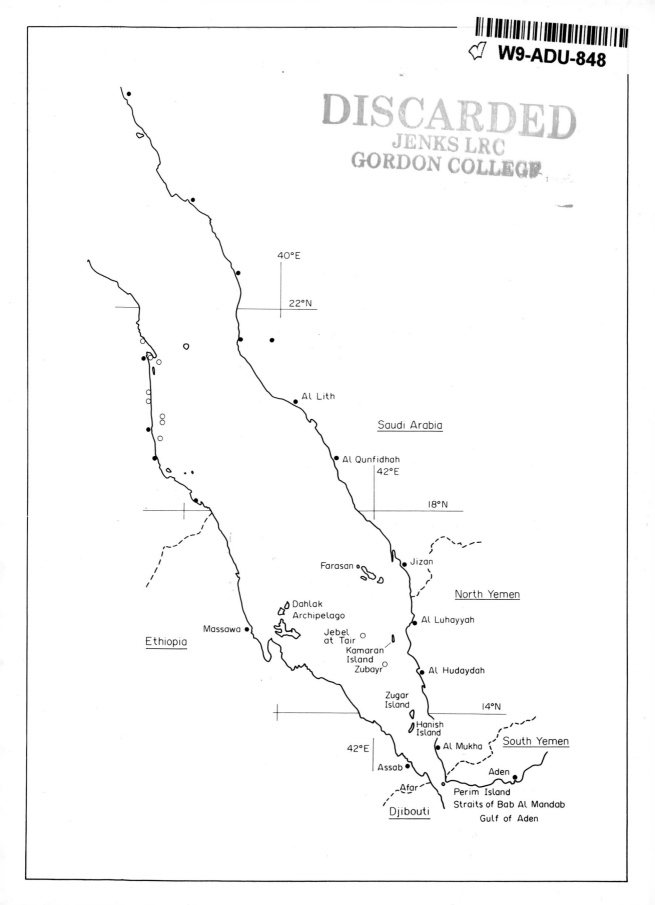

40°E

22°N

Al Lith

Saudi Arabia

Al Qunfidhah
42°E

18°N

Farasan

Jizan

North Yemen

Dahlak
Archipelago

Massawa

Al Luhayyah

Ethiopia

Jebel
at Tair

Kamaran
Island
Zubayr

Al Hudaydah

Zugar
Island

14°N

Hanish
Island

Al Mukha

South Yemen

42°E

Assab

Aden

Afar

Perim Island
Straits of Bab Al Mandab

Djibouti

Gulf of Aden

KEY ENVIRONMENTS

General Editor: J. E. Treherne

RED SEA

The International Union for Conservation of Nature and Natural Resources (IUCN), founded in 1948, is the leading independent international organization concerned with conservation. It is a network of governments, non-governmental organizations, scientists and other specialists dedicated to the conservation and sustainable use of living resources.

The unique role of IUCN is based on its 502 member organizations in 114 countries. The membership includes 57 States, 121 government agencies and virtually all major national and international non-governmental conservation organizations.

Some 2000 experts support the work of IUCN's six Commissions: ecology; education; environmental planning; environmental policy, law and administration; national parks and protected areas; and the survival of species.

The IUCN Secretariat conducts or facilitates IUCN's major functions: monitoring the status of ecosystems and species around the world; developing plans (such as the World Conservation Strategy) for dealing with conservation problems, supporting action arising from these plans by governments or other appropriate organizations, and finding ways and means to implement them. The Secretariat co-ordinates the development, selection and management of the World Wildlife Fund's international conservation projects. IUCN provides the Secretariat for the Ramsar Convention (Convention on Wetlands of International Importance especially as Waterfowl Habitat). It services the CITES convention on trade in endangered species and the World Heritage Site programme of UNESCO.

IUCN, through its network of specialists, is collaborating in the Key Environments Series by providing information, advice on the selection of critical environments, and experts to discuss the relevant issues.

KEY ENVIRONMENTS

RED SEA

Edited by

ALASDAIR J. EDWARDS

University of Newcastle upon Tyne, UK

and

STEPHEN M. HEAD

University of the West Indies, Jamaica

Foreword by

HRH THE DUKE OF EDINBURGH

Published in collaboration with the

INTERNATIONAL UNION FOR CONSERVATION OF
NATURE AND NATURAL RESOURCES

by

PERGAMON PRESS

OXFORD · NEW YORK · BEIJING · FRANKFURT
SÃO PAULO · SYDNEY · TOKYO · TORONTO

U.K.	Pergamon Press, Headington Hill Hall, Oxford OX3 0BW, England
U.S.A.	Pergamon Press, Maxwell House, Fairview Park, Elmsford, New York 10523, U.S.A.
PEOPLE'S REPUBLIC OF CHINA	Pergamon Press, Room 4037, Qianmen Hotel, Beijing, People's Republic of China
FEDERAL REPUBLIC OF GERMANY	Pergamon Press, Hammerweg 6, D-6242 Kronberg, Federal Republic of Germany
BRAZIL	Pergamon Editora, Rua Eça de Queiros, 346, CEP 04011, Paraiso, São Paulo, Brazil
AUSTRALIA	Pergamon Press Australia, P.O. Box 544, Potts Point, N.S.W. 2011, Australia
JAPAN	Pergamon Press, 8th Floor, Matsuoka Central Building, 1-7-1 Nishishinjuku, Shinjuku-ku, Tokyo 160, Japan
CANADA	Pergamon Press Canada, Suite No. 271, 253 College Street, Toronto, Ontario, Canada M5T 1R5

First edition 1987

Library of Congress Cataloging in Publication Data
Red Sea.
(Key environments)
1. Marine ecology — Red Sea. 2. Marine resources conservation — Red Sea. I. Edwards, Alasdair J. II. Head, Stephen M. III. International Union for Conservation of Nature and Natural Resources. IV. Series.
QH93.5.R43R43 1986 333.91′6416′0916533 86-16974

British Library Cataloguing in Publication Data
Red Sea. — (Key environments)
1. Natural history — Red Sea
I. Edwards, Alasdair J. II. Head, Stephen M.
III. International Union for Conservation of Nature and Natural Resources IV. Series
508.3165′33 QH94.3
ISBN 0 08 028873 1

Printed in Great Britain by A. Wheaton & Co. Ltd., Exeter

The general problems of conservation are understood
by most people who take an intelligent interest in the state
of the natural environment. But if adequate measures are to
be taken, there is an urgent need for the problems to be
spelled out in accurate detail.

This series of volumes on "Key Environments" concentrates
attention on those areas of the world of nature that are under
the most severe threat of disturbance and destruction. The
authors expose the stark reality of the situation without
rhetoric or prejudice.

The value of this project is that it provides specialists,
as well as those who have an interest in the conservation of
nature as a whole, with the essential facts without which it is
quite impossible to develop any practical and effective
conservation action.

1984

General Preface

The increasing rates of exploitation and pollution are producing unprecedented environmental changes in all parts of the world. In many cases it is not possible to predict the ultimate consequences of such changes, while in some, environmental destruction has already resulted in ecological disasters.

A major obstacle, which hinders the formulation of rational strategies of conservation and management, is the difficulty in obtaining reliable information. At the present time the results of scientific research in many threatened environments are scattered in various specialist journals, in the reports of expeditions and scientific commissions and in a variety of conference proceedings. It is, thus, frequently difficult even for professional biologists to locate important information. There is consequently an urgent need for scientifically accurate, concise and well-illustrated accounts of major environments which are now or soon will be, under threat. It is this need which these volumes attempt to meet.

The series is produced in collaboration with the International Union for the Conservation of Nature. It aims to identify environments of international ecological importance, to summarize the present knowledge of the flora and fauna, to relate this to recent environmental changes and to suggest where possible, effective management and conservation strategies for the future. The selected environments will be re-examined in subsequent editions to indicate the extent and characteristics of significant changes.

The volume editors and authors are all acknowledged experts who have contributed significantly to the knowledge of their particular environments.

The volumes are aimed at a wide readership, including: academic biologists, environmentalists, conservationists, professional ecologists, some geographers as well as graduate students and informed lay people.

John Treherne

Contents

CHAPTER 1

Introduction

STEPHEN M. HEAD

Department of Zoology, University of The West Indies, Kingston, Jamaica

CONTENTS

1.1. GEOGRAPHICAL SETTING

The Red Sea is a long, narrow body of water separating north-east Africa from the Arabian Peninsula. Its nearly 2000 km of navigable waters connect at the south with the Indian Ocean, and very nearly join the Mediterranean Sea at the north of the Gulf of Suez. The Red Sea has been an important trade route throughout human recorded history, linking the trade goods of India and the Far East with the historical markets of Egypt, the classical world, and Europe. When Ferdinand de Lesseps completed the Suez Canal in 1869, the connection became direct, and now the Red Sea is one of the most important shipping routes in the world.

Seven countries have shorelines on the Red Sea (Fig. 1.1), and they include some of the richest and poorest nations in the world. Some basic statistics about these countries are summarised in Table 1.1. On the western shore lies Egypt to the north, the Sudan borders the central section, and Ethiopia lies to the south. On the eastern shore, the Kingdom of Saudi Arabia occupies the northern and central sections, while the Yemen Arab Republic (North Yemen) borders the southern section. Two other countries, Israel and Jordan, have tiny but strategically important footholds on the Red Sea at the northern tip of the Gulf of Aqaba.

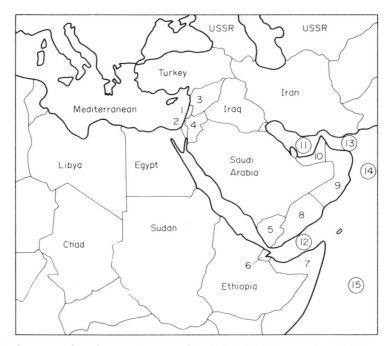

Fig. 1.1. General outline map to show the countries surrounding the Red Sea. Key to numbers: 1, Lebanon. 2, Israel. 3, Syria. 4, Jordan. 5, Yemen Arab Republic (North Yemen). 6, Djibouti. 7, Somalia. 8, People's Democratic Republic of Yemen (South Yemen). 9, Oman. 10, United Arab Emirates. 11, Persian Gulf. 12, Gulf of Aden. 13, Gulf of Oman. 14, Arabian Sea. 15, Indian Ocean.

Through most of the Red Sea region, the spoken language is Arabic, often with major dialectical variations, and so the names of towns and geographical features used in this book have been transliterated into English spelling, following the versions given in *The Times Atlas of the World* (1980). These places are shown on the map on the endpapers, and some of the spelling variants encountered are shown in Table 1.2.

The narrow southern Straits of Bab al Mandab mark the boundary between the Red Sea and the Gulf of Aden. Immediately outside the Red Sea proper, two other countries border the entrance. To the west lies the tiny state of Djibouti, and to the east is the People's Democratic Republic of Yemen (South

TABLE 1.1. Geographical and economic statistics for countries bordering the Red Sea. Largely adapted from data in Year Book (1985).

	Area km^2	Approx. Red Sea shoreline km	Population millions	Trade balance million US$	G.N.P. million US$	Per capita G.N.P. US$
EGYPT	1,002,270	1,386	47.0	−$5,958	$28,160	$600
ETHIOPIA	1,222,896	800	32.0	−$383	$4,530	$141
ISRAEL	20,711	<10	4.2	−$3,455	$20,420	$4,861
JORDAN	97,821	27	3.5	−$2,451	$3,880	$1,108
SAUDI ARABIA	2,151,443	1,740	10.8	+ $38,464	$117,240	$10,855
SUDAN	2,507,857	750	21.1	−$730	$7,390	$350
NORTH YEMEN	195,185	430	5.9	−$748	$910	$433

Yemen), which occupies the southern portion of the Arabian peninsula coast. South Yemen has a small foothold in the Red Sea proper, by occupying the island of Kamaran which lies just off the coast of North Yemen.

The Red Sea lies between 30°N and 12°30′N, and in this considerable latitudinal range one might expect to find a range of climatic conditions. As Frederick Edwards describes in Chapter 3, this is not the case. There are naturally some seasonal wind and air temperature differences between the north and the south, but the whole seaboards of both eastern and western margins are overwhelmingly arid. Rainfall nowhere exceeds about 18 cm per year, and the coastal vegetation is semi-desert. This is not the apparently lifeless sand waste of the Sahara, or the Rub al Khali of central Arabia, but its vegetation cover and productivity are very sparse. Rainfall is very irregular, especially towards the north, where years may intervene between showers. When rain at last falls, the grey-brown desert becomes a carpet of green for a few weeks, but this temporary bounty provides no long-term help for the sparse animal populations of rodents and small antelope which graze the coastal plain. The dearth of rain and vegetation has been the principal reason for the low human population levels anywhere along the Red Sea coast, which even now do not exceed about 20 to the square kilometre except in the vicinity of major towns. Historically, the human population has subsisted by nomadic pastoralism of near-biblical simplicity, artisanal fishing, and in a few favoured ports, by trade, transhipment and the pilgrim business. Nowadays the situation has begun to change very rapidly, at least in those coastal areas where oil-fed economies are investing heavily in plant and building in the coastal zone.

1.2. DIMENSIONS, STRUCTURE AND GEOLOGICAL HISTORY OF THE RED SEA

According to Morcos (1970), the Red Sea is 1932 km long, and averages 280 km in width. Even at its widest, in the south near Massawa, it is only 354 km wide, and this narrows to 29 km at the shallow Straits of Bab al Mandab, a figure which includes the island of Perim, which further restricts the

TABLE 1.2. Alternative place name spellings.

Place	Alternative spelling
Al Hudaydah	Hodeidah
Al Mukha	Mocha
Al Wajh	Wejh
Aqaba	Akaba
Assab	Aseb
Bab al Mandab	Bab el Mandeb
Dahlak Archipelago	Dahlac Archipelago
Dungunab Bay	Dongonab Bay
Elat	Eilat, Eilath
El Kharga	El Karija
Hurghada	Al Ghardaqa, Ghardaqa
Jiddah	Djidda, Jedda, Jeddah, Jidda
Massawa	Mits'iwa
Mecca	Makkah, Meccah
Muhammad Qol	Mohammad Qol
Port Sudan	Bur Sudan
Quseir	Koseir
Riyadh	Ar Riyad
Rub al Khali	Ar Rab'al Khali
Safaga	Bur Safaga
Sinai	Es Sina
Suez	El Suweis
Yanbu al Bahr	Yenbo, Yenbu

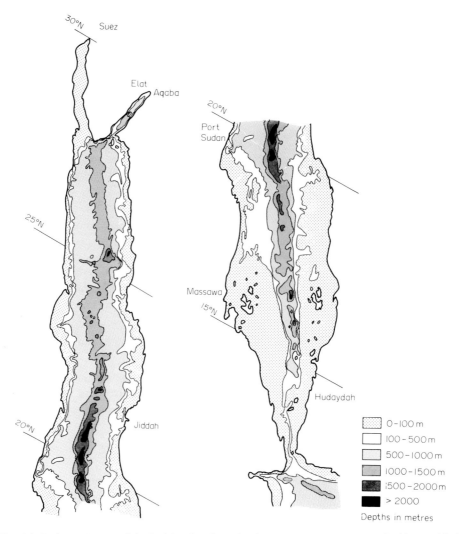

Fig. 1.2. Bathymetric map of the Red Sea, based on the chart in Morcos (1970), considerably simplified.

opening to the Gulf of Aden. The total surface area of the Red Sea is variously estimated between 438 and 450 thousand square kilometres, while the volume is between 215 and 251 thousand cubic kilometres. Morcos considers the average depth to be 491 m, although some other estimates are a little deeper. This compares to an average depth of 3700 m for the World's oceans.

The bathymetry of the Red Sea is shown in Fig. 1.2, and Table 1.3 shows the approximate proportional distribution of depth. The maximum depth recorded in the Red Sea is 2850 m, which is small compared with the great oceans, but large for a body of water of its size. The Red Sea is for example much deeper than the Arabian Gulf. Although in the southern Red Sea the shallow coastal shelf is extensive, in the northern and central Red Sea it may only extend a few kilometres offshore. Morcos (1970) refers to this region as the 'coral reef zone', because over much of the length of the Red Sea, this shelf is covered with actively growing reefs. From the coral reef zone, the sea bottom drops away rapidly in a series of stepped cliffs to the main trough at about 500m depth. The trough gradually

TABLE 1.3. Approximate distribution of surface area by depth for the Red Sea (including northern gulfs). Original, calculated from the bathymetric map in Morcos (1970).

Depth range	Area as proportion of total
0− 50m	23.86%
50− 100m	17.27%
100− 500m	15.71%
500−1000m	29.06%
1000−1500m	11.64%
1500−2000m	2.04%
2000m	0.42%

deepens to about 1000 m near the central axis of the Red Sea, where there may be a further change in topography as the bottom slopes steeply down in the central rift. This central rift or axial trough is about 1500 m deep along much of its length, and contains pits or 'Deeps' where depths may reach over 2500m.

Figure 1.3 shows three representative profiles through the northern (26°N), central (20°N) and southern (14°N) Red Sea. The central region shows the most dramatic topography, with a steep western profile, a well developed main trough or platform at *ca.* 500 m and a plunge to over 2000 m depth in the central rift. To the north the 500 m platform is less well developed and the central rift less prominent. In the south, the shelf is very wide to east and west, and the profile descends steeply to the maximum depth of some 1600 m.

The northern gulfs of Suez and Aqaba present great contrasts in bathymetry. The Gulf of Suez is shallow and flat bottomed basin, with a depth of 55 to 73 m, deepening at the entrance to the Red Sea proper. The Gulf of Aqaba is very steep-sided and deep, reaching maximum depths of over 1800 m near the east coast, although the Gulf is only some 30 km wide. Just as the Red Sea is cut off from the Gulf of Aden by a 100−130 m deep sill about 140 km north of the Straits of Bab al Mandab, the Gulf of Aqaba is separated from the rest of the Red Sea by a sill (the sill of Tiran) at its entrance, about 250−300 m deep (discussed in Chapter 3). The Gulf of Aqaba is in these respects a model, on a very small scale, of the Red Sea itself.

Figure 1.4 shows a hypsometric curve for the Red Sea, including the northern gulfs calculated from the bathymetric chart in Morcos (1970). This curve is obtained by measuring the total area occupied by successive depth bands, and then plotting the cumulative areas against depth. This results in an averaged profile for the Red Sea. The various breaks in topography mentioned above are well demonstrated and it can be seen that about a quarter of the surface area is of 50 m or less depth, thus capable of supporting active coral growth. As Figure 1.2 showed, much of this area is concentrated to the south of the Red Sea. About 14% of the area is deeper than 1000 m and forms part of the axial rift.

The structure and history of the Red Sea are described in Chapter 2 by Colin Braithwaite. By way of introduction a much simplified summary follows here. The Red Sea is part of a system of crustal expansion in which Africa and the Arabian Peninsula are drifting slowly apart, by the same mechanism that created the Atlantic Ocean. The deep axial trough of the Red Sea forms part of a much larger system extending from the Dead Sea to the African Rift Valley, and from Aden eastwards into the Indian Ocean. Associated with this rift system is the uplift of mountain ranges which lie along the length of the Red Sea behind the narrow coastal plain. As the continental masses of Africa and Arabia separate, the deep rift described above is formed, and new crust is continually being injected into this region. The injection of hot magma into the depths of the Red Sea accounts in part for the high temperatures at the bottom of the Red Sea, assisted by the isolation of its deep water from the rest of the Indian Ocean by the sill to the south. It is also responsible for the hot brine pools and valuable metal-rich sediments described in Chapter 4 by Ludwig Karbe.

The Red Sea is a relatively young ocean, and its complex history has encompassed some major environmental upheavals which are discussed in detail by Colin Braithwaite in Chapter 2. It had its origin in crustal sagging which occurred in the Mesozoic era, perhaps 180 million years ago, but only became established as a distinct trough in the Oligocene, about 38 million years ago. During its formation the area was periodically covered by sea but was quite often dry land. A dramatic environmental change occurred in the latter part of the Miocene period (from 25—5 million years ago) when the entire Red Sea basin became a great evaporation pan, forming considerable thicknesses of salt and other evaporite minerals. At this time the Red Sea was separated from the Indian Ocean by a neck of dry land, in the area where the Straits of Bab al Mandab now lie, but was linked across the Isthmus of Suez to the ancient Mediterranean basin (part of the ancient Sea of Tethys). This link was tenuous and periodically the rate of supply of sea water from the Mediterranean was slower than the rate of

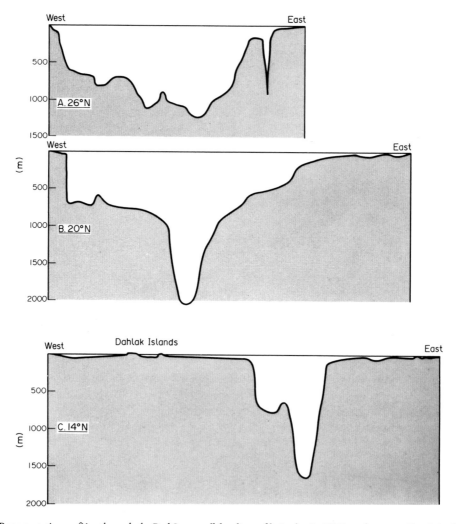

Fig. 1.3. Representative profiles through the Red Sea, parallel to lines of latitude. A. 26°N — from near Quseir in the west to near Al Wajh in the east. B. 20°N — from near Port Sudan in the west to near Al Lith in the east. C. 14°N — from near Massawa in the west to near Al Luhayyah in the east.

evaporation. Fossil evidence points to intermittent refilling, with bands of deep and shallow-water organisms preserved between the evaporite sequences. The Red Sea would thus have been an exceedingly unstable area in a biological sense, with fluctuating depth and salinity inhibiting the development of bottom communities.

The crustal expansion, which had ceased during the Miocene period, started once more about 5 million years ago, creating the present deep axial trough which cuts through the Miocene evaporite sequences. Uplift of the land closed the connection with the Mediterranean, and the Indian Ocean broke through at Bab al Mandab, bringing the first Indo-Pacific faunas, and coral reefs began to grow in shallow waters or on the tops of uplifted blocks.

It seems likely that the fauna of the Red Sea was disturbed once again by depth, salinity and temperature fluctuations during the Ice Ages of the Pleistocene period. Evidence for this can be seen in the changing abundance of fossil planktonic organisms in cores taken from deep sea sediments. Although glaciers never reached the Red Sea, so much water was locked up in the polar ice that at times the sea level fell to over 100 m below its present level. This would have more or less isolated the Red Sea from the Indian Ocean once more, and there is evidence that hypersaline conditions prevailed, like the present Dead Sea. Between glaciations, normal Indian Ocean fauna reappeared, but during glaciations, high salinity, and open water temperatures as low as 13–14°C would have decimated the normal tropical fauna. The last glaciation ended 20 to 15,000 years ago, and the fast-rising sea flooded the basin with warm Indian Ocean water of normal salinity. The sea reached its present level only about 5000 years ago, so present shallow water depositional features, such as coral reefs, are very young. They are however founded upon much older reefs laid down during earlier interglacial periods, and this accounts for their large size and thickness.

It is important to remember the extremely short time in which the tropical communities of the Red Sea have re-established themselves, especially when considering problems of endemism and speciation. For groups such as corals which could not survive the glacial conditions, very little time has elapsed in which the recolonising stocks could evolve differences from their Indian Ocean ancestors.

1.3. INTRODUCTION TO THE ECOLOGY AND INHABITANTS OF THE RED SEA

The Red Sea contains representatives of all the major tropical marine communities except estuaries, which cannot form because it receives no permanent rivers. Figure 1.5 illustrates the principal communities covered in this volume and their zonation. We may start with the pelagic system of the

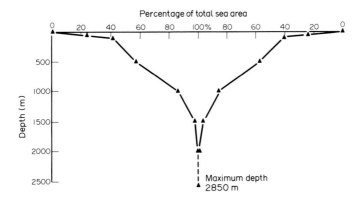

Fig. 1.4. Hypsometric curve for the Red Sea. Original, based on the bathymetric chart summarised in Fig. 1.2. The profile is drawn twice in mirror image to assist visualisation as an average profile.

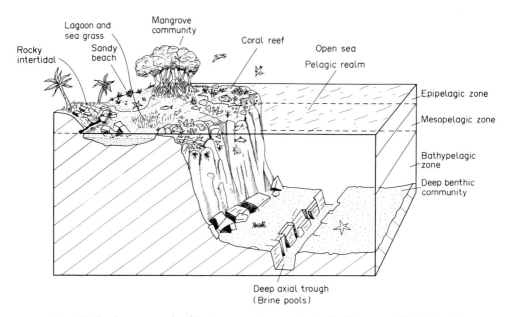

Fig. 1.5. Sketch, not to scale of basic community types in the Red Sea covered in this book.

open sea, dominated by the tiny organisms of the plankton. Planktonic plants or phytoplankton use sunlight, water and inorganic nutrients to create new organic material, and they are eaten by planktonic animals or zooplankton. Predatory zooplankton feed on the herbivores, and are in turn eaten by fish and benthic or bottom-dwelling animals. The zooplankton form the link between primary producers in the sea and the larger animals with which we are more familiar. In the Red Sea, three depth zones can be distinguished in the pelagic environment. The epipelagic zone, from the surface to 100 m depth, is the well-lit region of primary production. Below it lies the mesopelagic zone, to about 800 m depth. This is a region of change, the light extinguishes, the temperature and food supply drops, and oxygen levels may rise and fall. Below 800 m is the bathypelagic zone, sparsely populated because of the very limited food supply, and extending right to the sea bottom. The plankton of the Red Sea is described in Chapter 5 by Horst Weikert, and discussion will be found of the variations in nutrient levels, and production which explain some of the ecological gradients from north to south observed in other Red Sea organisms.

The fall-out of detrital material from the upper layers of the sea forms the food resource for animals living on the deep sea bottom. The unique physical conditions of the deep Red Sea — its high temperature and isolation from the Indian Ocean — make it an exceptional habitat. Low food supply, and high metabolic rates in high ambient temperatures limit the biomass of the deep benthos to very low levels. Chapter 6 by Hjalmar Thiel describes the fauna living between about 500 m and 1000 m depth on the floor of the Red Sea. Dr. Thiel's Chapter, like chapters 4 and 5, is important in summarising the results from some very recent research conducted during environmental impact assessment studies associated with possible mining of deep Red Sea sediments.

Below the depths covered in Chapter 5, lies the deep axial trough of the Red Sea, where new oceanic crust is being injected into the widening rift between Arabia and Africa. Within deep pockets in this rift lie the unique hot brine pools of the Red Sea, and beneath these can be found the economically valuable metal-bearing sediments. Ludwig Karbe's Chapter 4 introduces the extraordinary environmental conditions of the brine pools, which must be among the most extreme habitats on the planet. Dr. Karbe describes the brines and sediments and their limited fauna, and goes on to detail the present pattern of heavy metal concentrations within the water column. He also describes the techniques proposed to recover metal ore from the deep Red Sea, and the precautions being taken to avoid contaminating the productive upper waters with toxic metallic tailings.

So far, we have examined the communities of the deep open sea, but the richest and most exciting communities are those of the shallow coastal waters along the Red Sea margins. Here, in calm, well lit and crystal clear waters is the realm of the coral reefs. Reefs border most of the coasts and islands of the Red Sea, and despite their very recent post-glacial re-establishment, they are among the most attractive of any in the world. Reefs are formed by the secretion of immense quantities of calcium carbonate by living organisms, mainly corals and algae, using energy directly from sunlight or from the oceanic plankton supplied by water currents. The skeletal carbonate becomes consolidated and resists breakdown by wave action, but is complex in outline and full of cavities. This complex substrate is one of the many features which contribute to the overwhelming diversity of animal life on all healthy reefs. Red Sea reefs are visibly dominated by corals and clouds of colourful fish, but a close inspection reveals large numbers of other species from most of the phyla of the animal kingdom. An introduction to the ecology of coral reefs is given by Stephen Head in Chapter 7, which then discusses the coral fauna of the Red Sea and its distribution on reefs in more detail. Reefs are highly zoned communities, and the types and abundance of organisms found at any site depend upon the depth, light intensity and wave energy characteristics of the site. Although one conventionally pictures a hard-substrate or coral dominated assemblage as typical of a reef, in practice much of their surface area is covered with sand. Reef sands support a rich biological community, which in the Red Sea as elsewhere, has received much less study than the fauna of the rocky substrates.

In the shallow lagoons behind coastal or offshore reefs, the sand surface is frequently colonised by dense growths of seagrass. Seagrasses are specialised angiosperms, or 'higher plants', and as this name suggests are rather grass-like in appearance. Seagrass beds may have very high productivity, and they also serve to stabilise sediment, allowing a rich and specialised fauna to develop among them. Seagrasses of the Red Sea are described by Mohammad Wahbeh as part of Chapter 9.

Seagrasses exist from moderate depths (70 m) to the immediate sub-tidal or even mid-tidal position on suitable beaches. Shorewards of the seagrasses lies the rest of the intertidal zone. This region, which is regularly wetted and left to dry by the advancing and receding tides, can be very rich and diverse in highly productive coastal areas with a large tidal range. The environment is of course extreme and unstable, and its inhabitants need special survival adaptations and strategies. The intertidal zone of the Red Sea is particularly harsh and uninviting, with very high solar heating and desiccation stress, and a very low tidal range, especially in the central Red Sea region. Nevertheless, an interesting rocky intertidal fauna can be found, especially to north and south in the Red Sea where the tidal range is more substantial. This community is described by David Jones and Mostapha Ghamrawy in the first part of Chapter 9. These authors also discuss the other Red Sea intertidal communities, the sandy beaches and mangrove swamps or mangals. As in other tropical areas, the beach faunas of the Red Sea tend to be characterised by a diversity of Crustacea in the upper section, with a rich and varied community including worms, molluscs and echinoderms in the lower sections. Mangroves are specially salt-adapted trees which colonise sheltered intertidal sites, often in areas of low salinity near estuaries. Estuaries are absent from the Red Sea, and the mangroves are rather less important there than in other tropical coastal areas. Conditions in the swamps can be extreme, with high water temperature and salinity, low oxygen levels and abundant clogging silt. The Red Sea mangrove fauna is therefore rather impoverished compared with that of other areas.

The coastal communities of mangals, seagrass beds and coral reefs are very interdependent (Ogden and Gladfelter, 1983); some of the principal interactive routes are illustrated in Figure 1.6. Mangroves serve to stabilise the coastline and reduce the impact of heavy sedimentation from the land; they also trap land-originating nutrients. Both sediments and high nutrient input can wreak havoc on inshore coral reefs. The reefs in turn have a vital role in protecting coasts from severe wave and current action, providing the sheltered conditions in which mangrove and seagrass communities can develop. In their turn, seagrasses stabilise reef and lagoon sediments, reducing the damage caused by storms, which otherwise tend to raise sediments, which scour reefs and smother mangroves.

Several plant and animal groups are sufficiently important in the Red Sea to warrant special attention; the detailed coverage of corals, seagrasses and mangroves has already been mentioned. The benthic algae, which are the most important primary producers in many shallow water habitats are described by Diana Walker in Chapter 8. Algae are prominent in some intertidal and lagoonal environments, but the larger species are not conspicuous on reefs, probably as a result of heavy grazing pressure. Bare rock surfaces on reefs are rapidly colonised by thin lawns or turfs of filamentous algae. The standing crop of such lawns may be low, but the growth rates and cropping rates can be very high, so they can be major contributors to reef productivity. Calcareous algae are important reef cementers, and contribute substantially to reef sediments, while filamentous algae compete for space with corals and other sessile animals. If the nutrient balance of a reef is disturbed by enrichment, algae may rapidly come to dominate the substrates with the loss of most of the typical reef fauna including the corals.

Two groups of invertebrates are singled out as of particular interest in the Red Sea and are the subject of special chapters. Michael Mastaller describes the extensive mollusc fauna in Chapter 10, while Andrew Campbell introduces the echinoderms in Chapter 11. Both groups of animals occupy a surprising range of niches, including filter feeders, grazers, detritivores, scavengers and specialist carnivores. Neither group is necessarily conspicuous to a casual observer, but a brief search in cavities, under rocks and in the upper sediment layers reveals a great diversity of molluscs. The echinoderm fauna is less rich, but highly spectacular. With the exception of some rock-burrowing sea urchins and a few starfish and sea cucumbers, the echinoderms are best seen at night, when they emerge from hiding to graze or expand beautiful feeding fans in the plankton-rich night waters. Many other groups of

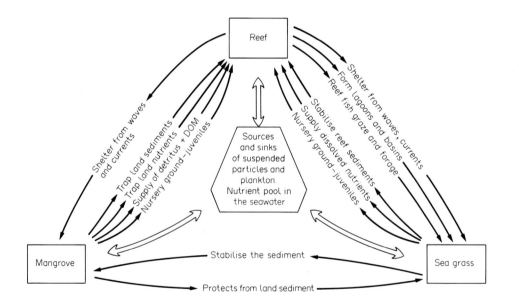

Fig. 1.6. Diagram to summarise principal interactive mechanisms between reefs, seagrass beds and mangroves. Original, based on discussion in Ogden and Gladfelter (1983).

invertebrates are present in the Red Sea, and some, such as the sponges and crustaceans, arguably deserve their own chapters in this book. Space limitations restrict coverage of other invertebrate groups to Chapter 12 by Stephen Head, in which non-coral cnidarians, sponges and crustaceans are fairly extensively covered, with short notes on other taxa.

Fish are extremely abundant and diverse in the Red Sea. They fill all manner of ecological niches, as herbivores, large and small predators, plankton feeders, sediment reworkers and a plethora of highly specialised niches, often as symbiotic partners. Fish are by far the most conspicuous animals other than corals in all shallow water habitats in the Red Sea, and they penetrate, often as bizarre specialists, into very deep water. It is perhaps the colour and movement that reef fish impart to their environment that makes a coral reef so attractive to tourist divers. Their habits and diversity are discussed in Chapter 13 by Rupert Ormond whilst their relationships to Indian Ocean fishes and their invasion of the Eastern Mediterranean through the Suez Canal are described by Alasdair Edwards. Unfortunately, reef fish are very susceptible to overfishing by traditional or non-traditional means, such as spearfishing. In Chapter 17, Stephen Head outlines the present fisheries of the Red Sea and discusses their future prospects. At present most of the area (with the exception of the Gulf of Suez) is probably underfished, a function more of low human population density and poor communications than of deliberate conservation. However, the fisheries yield is unlikely ever to be high, as, except in the southern Red Sea, the area of suitable shallow shelf is too small, and the overall productivity too low.

Certain other vertebrates inhabit the Red Sea, and are important less from their contribution to its ecology than from their global conservation status. Chapter 14 reviews three such groups, the turtles, Dugong and whales. Turtle species are under threat through most of their range, and the Red Sea fauna is reviewed from the taxonomic and conservation stance by Jack Frazier. The Hawksbill turtle is still moderately common, and four other species may occasionally be encountered. Sadly, large heaps of turtle bones on offshore islands testify to a once prolific turtle fauna, exploited to the point of collapse before or during the last century. The Dugong is a large inoffensive mammal which feeds exclusively on seagrasses. Colin Bertram's description of Red Sea Dugongs covers the ecology and human use of this curious sea animal, which is reputed to have given rise to the legend of the mermaid. The restricted shelf area and seagrass development of the Red Sea make it a less than perfect environment for Dugong, and probably only a few hundred are present. Even this small population may be of significance however, since the Dugong is under severe pressure throughout its range. The cetaceans of the Red Sea are described in the final part of Chapter 14 by Peter Evans. Like the other large air breathers, the population numbers and the species list are small compared with other Indian Ocean areas. The isolation, somewhat extreme physical conditions and low productivity of the Red Sea probably contribute to the small number of species living there, but in the cetaceans as with many other groups more detailed surveys may reveal a richer fauna than presently recorded.

Peter Evans summarises a wealth of observations on the sea birds of the Red Sea in Chapter 15. Most of the species of gulls, terns and boobies are still present in good numbers, feeding in the more productive inshore waters and nesting on the offshore cays and small islands plentiful along many areas of the coast. Some of the bird populations are large enough to be of global significance, and with increasing human activity in the Red Sea, breeding populations may be in danger of disturbance.

Human beings are a part of the natural ecology of the Red Sea. We interact with it in several ways, fishing the shallower shelves and diving in the coral reef zone. Many large and small vessels navigate the waters of the Red Sea, carrying pilgrims, trade goods and oil. Accidental and deliberate discharges of oil occur, and oil pollution is becoming a serious problem in the heavily used northern Red Sea and gulfs of Aqaba and Suez. The four closing chapters in this book summarise the role and impact of people in the ecology of the Red Sea; the chapter on fisheries has already been mentioned. Mark Horton has contributed Chapter 16, a fascinating history of human settlement in the area, spanning from the earliest pre-human beginnings,through the stone age, classical, medieval and modern times. This chapter serves to illustrate how central the Red Sea has been in man's history of trade and communication, linking as it does the Mediterranean world and the Indian Ocean. It also shows how limited the scale of human occupation has been along the coast of the Red Sea until recent oil wealth has initiated major coastal construction and settlement in certain areas.

To close the book, Brian Dicks in Chapter 18, and Rupert Ormond in Chapter 19 give accounts covering the pollution, conservation problems and opportunities for management of the Red Sea environment. Until very recently, the Red Sea was probably one of the areas of the world least affected by pollution and other despoliations of man. The situation is changing fast, especially in the oil-polluted Gulf of Suez, and along parts of the Saudi Arabian coast where coastal development is proceeding as fast as anywhere in the world. The picture is not wholly bleak however. As Chapter 19 shows, serious efforts in conservation and environmental protection are now underway in several Red Sea countries, and in many of these, marine parks either exist or have been proposed. In an area as small and enclosed as the Red Sea however, special international measures will have to be agreed to safeguard water quality and prevent general contamination from tanker discharges and a few sources of industrial, mining or sewage pollution.

1.4. BIOGEOGRAPHY OF THE RED SEA

1.4.1. Faunal affinities

The Red Sea was at one time a southern extension of the Mediterranean, and the two seas are today separated by only about but 120 km of low-lying land, but the fauna and flora of the Red Sea is wholly Indian Ocean in affinity and origin. Probably the only other place in the world where such a small distance separates such dissimilar ecosystems would be the Isthmus of Panama which separates the Pacific and Caribbean Sea.

Some details of faunal origins are given in later chapters in this book. In all cases, species in the Red Sea seem to have arrived from the Indian Ocean, and are identical to, or very closely related to species found in the Indian Ocean as a whole. This is to be expected from our knowledge of the recent geological history of the Red Sea described above and in Chapter 2. Although the Red Sea was once connected to the Palaeo-Mediterranean, the great evaporations of the Miocene would have completely eliminated the Mediterranean-derived fauna. As we have seen there is also evidence that even after the colonisation of the Red Sea by Indo-Pacific animals and plants, further disruptions occurred during the Ice Ages, and would have included decimation of warm-temperature species. The modern Red Sea fauna, requiring warm water of 'normal' salinity may have been present for as little as 10–15 thousand years. Given this short period it is hardly surprising that the fauna is closely similar to that of the rest of the Indian Ocean.

In certain oceanographic respects the Red Sea differs significantly from the Indian Ocean, and this may be manifested in some biogeographical peculiarities. As Chapter 3 describes, the Red Sea is very saline, reaching levels of over 40‰ to the north, and as high as 42‰ in the Gulf of Suez. The central Red Sea is also warmer than the north-western Indian Ocean, especially in summer. These differences make the north of the Red Sea a rather extreme environment, that may be excluding some species. Most taxa seem a little less diverse to the north of the Red Sea than in the central or southern portions. As we shall see in Chapter 5, Indian Ocean plankton carried into the Red Sea by water currents tend to die or migrate into cooler deeper water as they are transported north. The decline may also be related to productivity differences, since the productivity of the central and northern Red Sea, out of reach of nutrients from the Indian Ocean, tends to be less than in the south.

One interesting effect of the high ambient surface temperature of Red Sea water, is that many organisms are able to penetrate further north in the Red Sea than in other parts of their range. The mangroves of Sinai are the most northerly in the Indo-Pacific province. Elat and Aqaba, at the tip of the Gulf of Aqaba have flourishing, rich and diverse coral reefs at very nearly 30°N, making these reefs among the northernmost in the world. Only those of southern Japan and Bermuda are appreciably further north, in both cases because of warm water currents.

A further unique characteristic of the Red Sea is the very high temperature of the deep Red Sea water — about 22°C. Deep water organisms are normally adapted to life at temperatures usually less than 5°C. High temperatures encourage high metabolic rates in animals whose body temperature varies with that of the environment, and this increases their food requirements. The low surface productivity of the Red Sea means a very small supply of food to deep water organisms. It is probably as a result of this as well as the shallow sill near Bab al Mandab that few deep water animals from the Indian Ocean live in the deep Red Sea. Many shallow water species which are normally excluded from deep water by their intolerance of cold, appear to penetrate much deeper in the Red Sea than elsewehere (Ekman 1953 — *see also* chapters 6 and 11).

1.4.2. Endemism

The real extent of endemism in the Red Sea fauna and flora is hard to estimate. Levels as high as 70% have been proposed for the crinoids. For the fishes the figure is about 17% (*see* Chapter 13), in algae the figure is about 9%, and in the case of corals is no higher than 8.5%. For reasons mentioned in Chapter 7, the endemism level in corals may be even lower. In general, the recorded proportion of endemism (often based on old taxonomic accounts) seems to be high (Briggs, 1974), averaging perhaps about 20%. For certain taxa it is likely that the endemism estimates may reduce in the light of modern research but for fishes, at least, earlier estimates of about 15% endemism (quoted in Briggs, 1974) have been found to be substantially correct. In any case, proving endemism requires extensive comparative collections and study: a species may appear to be an endemic simply because no-one has yet collected it elsewhere, or because it has never been critically compared with specimens from another area. In this way, high levels of endemism in Crustacea from the deep Red Sea may reflect inadequate sampling in other Indian Ocean sites. In groups such as the corals and molluscs which have received extensive recent study, the estimates of endemism have been reduced.

The semi-isolation of the Red Sea from the main body of the Indian Ocean might lead one to predict high proportions of endemic species, encouraged also by the rather extreme oceanographic conditions. On the other hand, the environmental disturbances of the last Ice Age have allowed little time for the recolonising warm water species to evolve differences from their Indian Ocean ancestral stock. It could however be argued that the extraordinary burst of colonisation that must have occurred at the end of the last Ice Age must in itself have provided a possible mechanism for accelerated evolutionary change, at least at the sub-species level. It is also probable that some taxa are currently speciating in the the Red Sea. One such may be the abundant coral genus *Stylophora*, six species of which have been described from the Red Sea. On the highly diverse Great Barrier Reef and through much of the Pacific, only a single species is known.

Although the Red Sea is usually regarded as an integral part of the Western Indian Ocean Province of the Indo-West Pacific zoogeographic region, Briggs (1974) has suggested that the high recorded endemism levels justify its separation as a zoogeographical province in its own right. To be more confident of this distinction, a thorough modern re-examination of the degree of endemism of the Red Sea fauna would be necessary, a very difficult task, but one with many interesting implications for study of evolutionary rates and mechanisms in the sea.

1.4.3. Lessepsian migration

Although not strictly a Red Sea phenomenon, the migration of species through the Suez Canal is of considerable interest. The canal was built by Ferdinand de Lesseps, and opened in 1869. It lies at sea level and links the Eastern Mediterranean directly with the northernmost tip of the Gulf of Suez. The canal offers a migration route between the two biogeographic areas, and the phenomenon has been termed Lessepsian Migration by Por (1971, 1978) and Por *et al.* (1972), who has studied the problem in great detail.

Currents in the Suez Canal are to the north for most of the year, but tend to reverse and run south from July to October (Morcos 1970). Passive movement of animals could therefore occur in both directions, but while many Red Sea species have penetrated through to the Mediterranean, very few have accomplished the reverse migration. It is likely that for some time after the canal opened, no migration could take place because of the high salinity (68‰) barrier presented by the Bitter Lakes. As canal water flowed through the Bitter Lakes the salinity fell, reaching 52‰ in 1924 when the canal fauna was studied by a British expedition (Fox 1926, 1929). It is now about 41‰, no higher than that in the northern Gulf of Suez. Another migration barrier has been diminished in recent years by the damming of the Nile at Aswan, which has resulted in an increase of salinity to about 31‰ in the Nile Delta, into which the canal opens (Por 1971). The present fauna of the canal includes a special 'Isthmus element' of highly salt tolerant species probably present in the lagoons of the isthmus before the canal was cut. This group was probably the first to colonise the new waterway. Subsequently, the Bitter Lakes have been occupied by a number of Red Sea species including the seagrass *Halophila stipulacea* characteristic of slightly hypersaline coastal inlets, and tolerant of shallow salty and muddy conditions.

Por (1978) recorded 128 species which have made the migration from the Red Sea to the Mediterranean. Although about 53 species have been suggested to have migrated in the opposite direction he considers most of these cases to be dubious. Many of the migrants to the Mediterranean have become very successful, spreading north and west along the Levant Coast as far as Turkey. The fish in particular have been successful as described by Dr. Edwards in Chapter 13. The reasons for the greater migration success of Red Sea stocks can probably be attributed to their pre-adaptation to the saline conditions of the canal and low temperatures of the Eastern Mediterranean by virtue of the high ambient salinities and comparatively low temperatures in the Gulf of Suez. In addition, the fauna of the eastern Mediterranean is rather impoverished, leaving niches and resources free for the taking. It is probable that the dominant northward current has also assisted migration in that direction.

As Por *et al.* (1972) pointed out, even with the reduction of the salinity barrier of the Bitter Lakes, there are still other problems limiting the spread of animals along the canal. The canal is shallow, with a soft muddy bottom, and the water is very turbid from the disturbance of the bottom by shipping. These factors would effectively inhibit the colonisation of the canal by the more exotic Red Sea fauna such as corals and coral-associated animals. It is however worth noting that the closure of the Suez canal following the Six Day War allowed sediments to settle, leading to greatly enhanced colonisation of the canal (Por 1971). This may have given a great, if temporary, boost to Lessepsian migration.

1.5. HISTORY OF EXPLORATION AND RESEARCH

From the earliest recorded history, the Red Sea has been a trade route, although one of fluctuating importance. Trade routes attract naturalists, and this was especially true of the Red Sea, which is the nearest tropical marine area to the classical world and Europe. The constant trade through the region caught in its wake a trickle of natural curios, which eventually found their way into the great natural history collections of the seventeenth to nineteenth centuries. These factors combine to give the Red Sea an unusually significant position in the development of biological sciences as one of the major sources of specimens and descriptions relating to tropical marine systems.

The earliest descriptions of the Red Sea date from about the beginning of the Christian era; the most important are those of Pliny the Elder (23–79 AD) and Strabo (64 BC–21 AD). These accounts give interesting information on the people of the Red Sea and their customs, and also some insight into the natural history of the region. Pliny in particular was a rather credulous writer, and some of his strange animal lore, refined by time and retelling, surfaces in the bizarre fauna of medieval bestiaries (Costello, 1979). By the late classical period, the name 'Red Sea' had somehow become fixed, although its origins are still unclear. Thomson (1948) writes

'How the name "Red" was come by is obscure, and later writers can only make bad guesses about reddish water or mountains, or a fabulous King Erythras or Red; was it perhaps the sea of the red sunrise, as there was a mythical "red" island Erytheia at the sunset'.

The early trade routes tended to avoid the peristent north winds of Suez, and vessels docked at the Egyptian port of Quseir. Trade was brisk under Emperor Augustus, Strabo reports 120 ships working the route to India; later the trade received a boost with the discovery by the 1st century AD captain Hippalus that fast voyages could be made from the Red Sea to India out of sight of land using the monsoon wind. Pliny bemoans the shocking increase in expenditure on decadent eastern luxuries that this trade provoked in Rome (Sykes 1949). The later trade patterns in the Red Sea are the province of Chapter 16 in this book.

Perhaps the first important traveller's account of the medieval or post classical period was that of the great Arab explorer Ibn Battuta of Tangier, whose peregrinations began at the age of twenty in 1325. Ibn Battuta's travels included Spain and West Africa, Sri Lanka, the Maldive Islands, India and Java, with extensive coverage of the near and middle east. Given his remarkable itinerary it is perhaps surprising that Battuta ventured so little on the Red Sea itself, even allowing for his reluctance to take the same route twice. Perhaps his narrow escape from drowning after his first pilgrimage served as a warning; the ship he nearly set sail in from Jiddah to Quseir foundered with the loss of all the passengers (Gibb, 1962). Other early accounts of the Red Sea region include that of Joseph Pitts, who had the appalling misfortune in 1678 at the age of 16, to be captured by Algerian pirates and sold into slavery. After passing through various hands, and becoming (outwardly) a Muslim, Pitts became the first Englishman to complete the Haj pilgrimage and enter Mecca. Pitts gives a brief account of his voyage from Suez to Rabigh and Jiddah. At Mecca he was given his freedom, and after further adventures returned at last to England in 1695 (Foster 1949). He published in 1731 an account of the *Religion and Manners of the Mahometans*, for many years the definitive source of information on the subject until the audacious *Pilgrimage to Al-Madinah and Meccah* of Sir Richard Burton in 1853, published as a best-seller in 1855.

Early writers on travel in the Red Sea such as Pitts, Daniel (who ended up in Al Mukha (Mocha) in 1700 after an abortive attempt to carry despatches to India) and even Burton lay much stress upon the difficulties of navigation in the Red Sea 'infamous region of rocks reefs and shoals'. It was customary for vessels to navigate by day and to anchor in natural harbours of the coast or in the lee of an offshore reef at night. On occasions, this could limit the safe sailing period to as little as five hours. To pass beyond a known anchorage late in the day would be to risk being caught after dusk in reef-dotted waters with no anchorage in sight. The principal impression given by these early travellers of their feelings for the Red Sea is one of distrust and frustration.

Scheer and Pillai (1983) give an excellent resumé of the scientific exploration of the Red Sea, from the viewpoint of corals and reefs, and this topic is touched upon by other authors later in this book. We may regard the first 'modern' study of the Red Sea to be the ill-fated Danish 'Arabia Felix' expedition of 1761–67 which spent from October 1762 to August 1763 in the Red Sea area. The expedition included the brilliant Swedish naturalist Peter Forsskål, a student of Linnaeus at Uppsala, who made extensive and still important collections, both of plants and animals, particularly fish. The expedition began its study of the Red Sea in Suez and collected there and at several sites on the present Saudi Arabian and North Yemen coast. The terrible privations that the expedition had to endure are recounted by Niebuhr (the sole survivor) in his introduction to Forsskål's posthumous publication (1775), and in his own description of the journey (Wandel 1929). An excellent account of the expedition is given by Hansen (1964) and a valuable English summary prefixes Crossland's (1941) re-analysis of the coral collection. The expedition was faced with the problems of malaria, poor food and bad water, the derision of many local people, and the hostility of petty officials. On one occasion, customs officers at Al Mukha insisted on opening and 'stirring with an iron rod' a bottle of Red Sea fishes preserved inadequately in weak date brandy — 'spilled the spirit and filled the room with the stench of decaying fish and spirit' — the effect on the fastidious muslim customs officers to whom strong spirits were forbidden must have been unfortunate to say the least, and the expedition narrowly avoided being thrown bodily out of the town.

The leader of the expedition died in Al Mukha and Forsskål died soon after. Eventually all the other members of the expedition succumbed except Carsten Niebuhr, who, after extraordinary adventures, managed to return overland to Copenhagen in November 1767. He had earlier sent Forsskål's collections and notes to Copenhagen from Bombay and finally had these published in 1775 (Fig. 1.7).

Misfortune and tragedy seemed to have dogged many Red Sea explorers from Battuta onwards. The French naturalist Savigny accompanied Napoleon I's campaign in Egypt in 1799—1801, and engraved many beautiful plates of marine organisms, but fell victim to blindness before he could produce a text to accompany the plates. The German scientists Christian Gottfried Ehrenberg and Friedrich Hemprich travelled around the Red Sea between 1820 and 1826. Important studies were made in the gulfs of Suez and Aqaba, at Jiddah and at Massawa where Hemprich died in 1825. Ehrenberg's later accounts of the corals and reefs are of great importance; an English translation of his description of the formation of reefs and banks was published in 1844. Charles Darwin made extensive reference to Ehrenberg's descriptions in his own seminal study on coral reefs. Another German traveller in the Red Sea at this period was Eduard Rüppell who later wrote well illustrated monographs including important studies of fish and other vertebrates (Rüppell, 1826—28, 1835).

The most outstanding individual contribution to our knowledge of the Red Sea fauna was that of Carl Benjamin Klunzinger, a German medical doctor who worked as a sanitary or quarantine inspector in the Egyptian Red Sea port of Quseir from 1863—1869 and 1872—1875 (Fig. 1.8). Klunzinger was a true polymath or 'nineteenth century man'. He was a perfect Arabic speaker, an authority on idiom, and an accurate observer of the people and customs of Quseir. His book on *Upper Egypt, its People and Products* was published in German in 1877, and in English translation in 1878. It is superbly written, full of convincing detail, and carries important descriptions of the desert fauna and coral reefs in chapters IV and VI. Klunzinger was a meticulous observer and a talented taxonomist. His publications include an extremely useful three volume tract on the coral fauna, seven publications on the fish and four on the Crustacea, plus a short note on the hemichordates and another on meteorological observations. His descriptions of a zoological excursion on the Quseir reef (Klunzinger, 1870, 1872) are classics of their time, and still worthwhile reading.

Klunzinger's work can be seen as a turning point in the description of the ecology of the Red Sea, and since his time data have accumulated ever more rapidly. Other pioneer naturalists of about the same period include Boutan (1892), Faurot (1888), Haeckel (1876) and Walther (1888), but their total contribution is small compared with that of the great Klunzinger.

In 1895—1896, the Austrian research vessel *Pola*, under the leadership of Franz Steindachner conducted studies in the northern Red Sea, followed in 1897—1898 by a similar cruise in the south (Koske 1895,1898). The work of the *Pola* team was of outstanding importance. They performed the

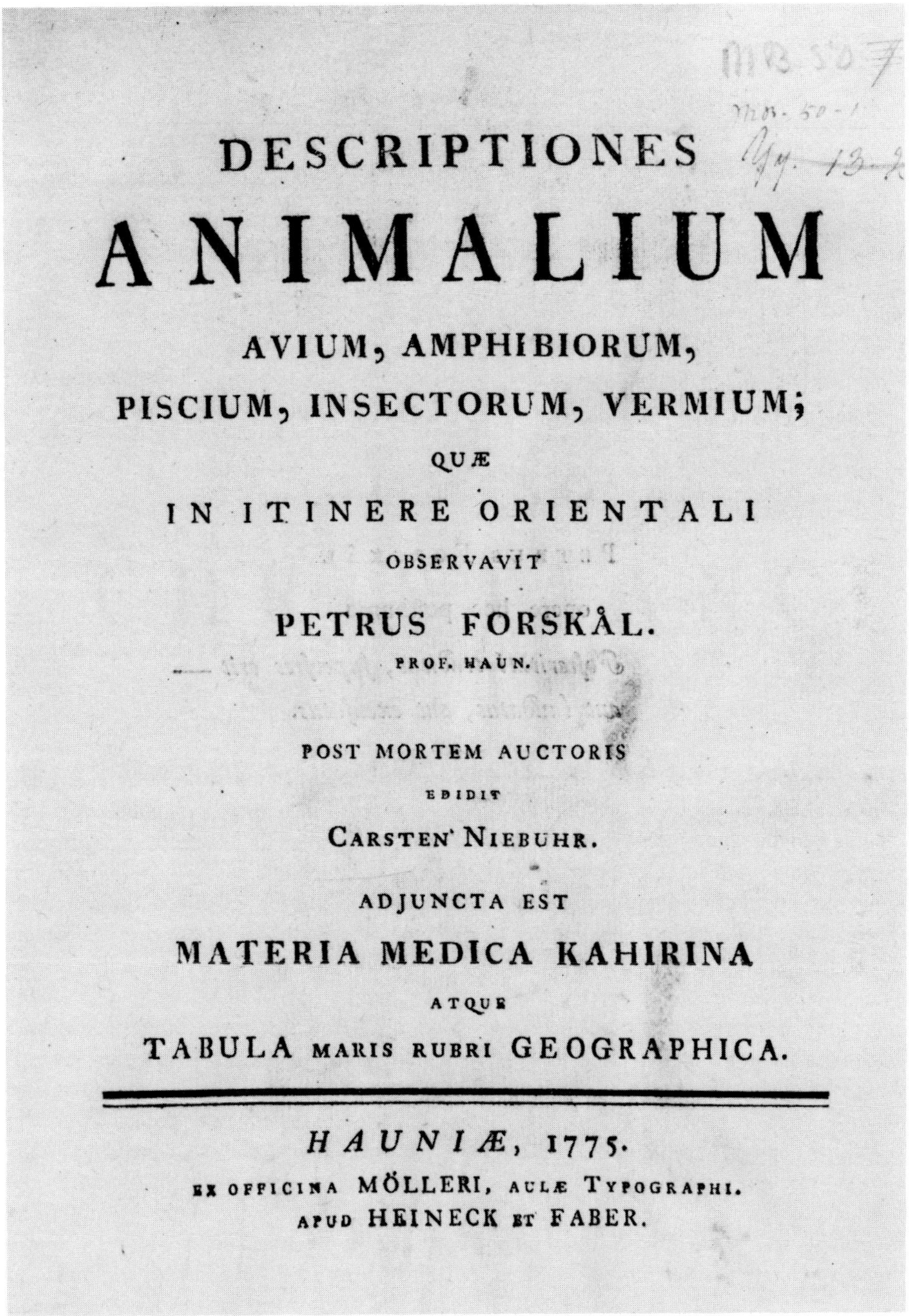

DESCRIPTIONES

ANIMALIUM

AVIUM, AMPHIBIORUM,

PISCIUM, INSECTORUM, VERMIUM;

QUÆ

IN ITINERE ORIENTALI

OBSERVAVIT

PETRUS FORSKÅL.

PROF. HAUN.

POST MORTEM AUCTORIS

EDIDIT

CARSTEN NIEBUHR.

ADJUNCTA EST

MATERIA MEDICA KAHIRINA

ATQUE

TABULA MARIS RUBRI GEOGRAPHICA.

HAUNIÆ, 1775.

EX OFFICINA MÖLLERI, AULÆ TYPOGRAPHI.

APUD HEINECK ET FABER.

Fig. 1.7. Title page of Peter Forsskål's *Descriptiones Animalium* published posthumously in 1775 by Carsten Niebuhr.

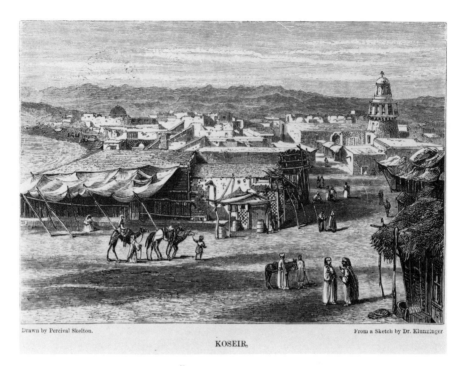

Drawn by Percival Skelton.   From a Sketch by Dr. Klunzinger.

KOSEIR.

Fig. 1.8. View of the market place in Quseir on the Egyptian coast. From Klunzinger (1878) based on a sketch by the good doctor.

first oceanographical studies in the Red Sea at 265 stations, for the first time grabs and dredges were used to sample the deep sea life to 2000 m depth, and a wealth of material was collected from shallow waters. The results were published in a long series of monographs by the Scientific Academy of Vienna; many of these studies are extensively referred to in later chapters of this book.

In 1904, Cyril Crossland, a young British zoologist with collecting experience in Zanzibar and the Cape Verde Islands was sent to the Sudanese Red Sea to investigate the coastal fauna and flora, beginning a long association with Red Sea ecology. Crossland collected briefly at Suez, and in a number of Sudanese sites. His collections were brought to England and worked up by several scientists, the results appearing in a long series published in the Zoological Journal of the Linnean Society, starting with Crossland's narrative of the expedition (Crossland 1907). By this time however, Crossland had returned to Dungunab Bay in northern Sudan, as Director of the Sudan Pearl Fishery, where he remained until 1922. An account of his endeavours during this period was published posthumously much later (Crossland 1957). Crossland's important coral collection was described (in part) by other workers, and his notes and observations were incorporated, somewhat haphazardly, into his posthumous treatise on the Great Barrier Reef corals (Crossland 1952). Crossland went on to found the first biological station in the Red Sea at Hurghada on the Egyptian coast.

The Egyptian research vessel *Mabahiss* carried the John Murray Expedition into the Indian Ocean in 1933–1934, during which important biological collections were made in the Red Sea, and the first investigations completed into the exchange of water over the sill near Bab al Mandab (Thompson, 1939). The *Mabahiss* later carried Crossland and a group of Egyptian scientists on an expedition through the northern Red Sea (Crossland 1936).

Various ship-borne scientific expeditions visited the Red Sea in the years after the Second World War. These included the British *Manihine*, (Anon 1952), the French *Calypso* (Dugan 1955) and the German *Xarifa* (Hass 1961) and *Meteor* (Gerlach, 1967). The International Indian Ocean Expedition of 1959—1964 brought a large number of major research vessels from many nations into the Indian Ocean, and many of these sampled in the Red Sea. The interest in the Red Sea accelerated dramatically with the discovery of anomalous temperatures, salinities and metalliferous brines by the British *Discovery* and American *Atlantis II* expeditions. The later history of studies on the hot brines and sediments is described in Chapter 4 while Morcos (1970) reviewed the oceanographic results of the International Indian Ocean programme.

There has always been considerable interest in the easily accessible northern gulfs of the Red Sea, and in recent years a flood of important studies have originated from the Israeli Marine Biological Station at Elat (Gulf of Aqaba), opened in 1968. Prior to the founding of this laboratory, scientists from Israel conducted two important expeditions to the southern Red Sea in 1962 (Oren, 1962) and 1965 (Lewinsohn and Fishelson, 1967). These expeditions provide most of the data on which our knowledge of the marine biology of the Eritrean area is based. Important studies also commenced by German workers at Aqaba in the Jordanian section of the Gulf of Aqaba (Mergner and Schuhmacher 1974) where there is now another marine station. Research activity at Aqaba has been accelerated by a continuing French—Jordanian Cooperative Research Programme, and over 100 scientific papers have been published from the marine laboratory (Anon, 1985).

In the Sudan, the British Cambridge Coral Starfish Research Group operated from 1968 to 1978 in and around the Port Sudan area. A major focus of their attention was the Towartit reef complex to the south of Port Sudan, where they constructed a unique platform facility which allowed scientists to live for extended periods on top of a coral reef (Head and Ormond 1978). The Cambridge group completed a number of reef-orientated studies, especially on the ecology of the Crown-of-Thorns starfish *Acanthaster planci*, which was at that point a subject of international concern because of accounts of 'plague' numbers destroying large areas of Pacific Ocean reefs. Nowadays, two marine laboratories exist on the Sudanese coast. The Institute of Oceanography is a government laboratory based in Port Sudan with field facilities at Sanganeb Lighthouse; while Khartoum University maintains a small marine laboratory at Suakin. Schroeder (1981) has summarised the problems of marine research in the Sudan.

Until recently, few scientists had worked on the coastal ecology of Saudi Arabia. During the early 1970s, the Saudi Ministry of Agriculture and Water Resources, in collaboration with the University College of North Wales operated a Marine Research Centre. Recently, the impressive new Institute of Oceanography of King Abdulaziz University has opened north of Jiddah. Jiddah is also the headquarters of the Saudi Sudanese Red Sea Joint Commission (SSRSJC). This intergovernmental body is charged with development of jointly owned marine resources, especially the large metal-bearing sediment deposits in the axial trough. The SSRSJC has taken the conservation of the Red Sea environment as being of over-riding importance in deciding the viability of deep sea mining, and has been responsible for conducting one of the most important marine science research programmes ever undertaken in the area. This has involved the German company Preussag, with scientific personnel from Hamburg University and other institutions. Several oceanographic cruises have been conducted with the vessels *Sonne* and *Valdivia*, and a wealth of new data collected. Some of these findings are summarised in the appropriate chapters of this book.

The Red Sea has had a long and impressive record of research, and it is pleasing to record an overall acceleration of studies and improvement of facilities despite economic and political difficulties in some regions. Without doubt, the best documented areas of the Red Sea are the northern gulfs, especially the Gulf of Aqaba, and the northern and central portions of the western shore. Thanks to the International Indian Ocean Expedition and the SSRSJC activities, the oceanography and pelagic ecology of the Red Sea are vastly better documented and understood than was the situation in the mid 1950s. Less coverage exists for the southern and eastern shores of the Red Sea, but at least for the Saudi Arabian coast this is set to improve very rapidly. The least studied coasts are those of North Yemen and Ethiopia, limited descriptions of which suggest some interesting differences from northern Red Sea areas. The coastal shelf is much wider, and marine productivity seems to be greater. More studies in this region would be of great interest.

1.6. THE RED SEA AS A KEY ENVIRONMENT

In what respect is the Red Sea a 'key environment'? Broadly, a key environment is one of international ecological importance. Some areas, such as Galapagos qualify by virtue of their special endemic fauna and flora. Others, such as Antarctica and the South American rain forest, are also very large, and without ecological equivalents elsewhere. The Great Lakes of North America and the Mediterranean are perhaps key environments through their long and chequered relationship with man.

The Red Sea qualifies under none of these heads uniquely, but under all in aggregate. It is not an overwhelmingly large environment, nor yet is it blessed with a wholly unique fauna and flora. It has had a long association with man's activities, but the degree of human impact and exploitation has until recently been negligible. On the other hand, the Red Sea is a very rich and varied environment, and within its confines it encompasses most of the interesting communities of tropical seas. The fauna and flora are not as diverse as in some central Indian Ocean or Indonesian sites, but are highly spectacular. We are still unsure as to the degree of endemism, but for some groups at least it seems to be rather high. Oceanographically, the Red Sea is certainly unique, it is a nascent ocean, it is very warm and salty, and has the warmest deep waters of anywhere in the world.

The Red Sea offers exceptional research potential. It is the best example of a young expanding ocean, and the relationship between its dramatic recent history and its rich fauna has yet to be established. With its strong north-south gradients of productivity (Chapter 5) and physical conditions, it offers great scope for ecophysiological studies of adaptation and distribution.

All environments are key environments to the people who live in and around them. The Red Sea is perhaps unique in being a major navigational channel, with a long record of use, which has not yet deteriorated severely under the influence of man. Although pollution and development stress are beginning to take effect — alarmingly so in recent years — they are still localised and could be contained. Because of its unspoilt nature, clear water and generally very calm seas, the Red Sea offers tremendous amenity and recreational resources to the overcrowded chilly countries of northern Europe, relatively close to hand. Compared with many other tropical seas, the marine life is abundant, undisturbed and largely unafraid of man. But the Red Sea is almost completely enclosed, and thus unfortunately highly susceptible to misuse. This obvious vulnerability also qualifies it for key environment status.

The Red Sea should be valued, not just as a unique environment, but as one of high diversity, great scientific and ecological importance, and of great beauty and value to man. Sadly it is a fragile environment and, since all nations directly or indirectly make commercial use of its waters, then all nations, not just those bordering its shores, should contribute to its protection and preserve it as an environment for future generations to use, study and enjoy.

REFERENCES

Anonymous (1952). The *Manihine* expedition to the Gulf of Aqaba. 1. Foreword, stations lists and collector's notes. *Bull. Br. Mus. nat. Hist. Zool.* 1, 151–8.

Anonymous (1985). *Publications from the Marine Science Station of Aqaba (Jordan)*. Edited by the French–Jordanian Cooperative Research Program in Marine Sciences. Nice University, pp. 8.

Boutan, L. (1892). *Voyage dans La Mer Rouge*, Bigot Frères, Lille.

Briggs, J. C. (1974). *Marine Zoogeography*, McGraw-Hill, New York.

Burton, R. F. (1855). *Personal Narrative of a Pilgrimage to Al Madinah and Meccah*. 1924 edition, G.Bell and Sons, London.

Costello, P (1979). *The Magic Zoo, the Natural History of Fabulous Animals*, Sphere, London.

Crossland, C. (1907). Reports on the marine biology of the Sudanese Red Sea. II. Narrative of the expedition. *Zool. J. Linn. Soc.* 31, 3−10.

Crossland, C. (1936). The Egyptian exploration of the Red Sea. *Nature* 137, 712−3

Crossland, C. (1941). On Forsskål's collection of corals in the Zoological Museum of Copenhagen. *Spolia zool. Mus. Haun.* 1, 1−63.

Crossland, C. (1952). Madreporaria, Hydrocorallinae, *Heliopora* and *Tubipora*. *Sci. Rep. Great Barrier Reef Exped.* 6, 85−257

Crossland, C. (1957). The cultivation of the mother-of-pearl oyster in the Red Sea. *Austr. J. mar. Freshwater Res.* 8, 111−30.

Dugan, J. (1955). *Calypso* oceanographic expeditions. *Trans. Am. Geophys. Un.* 36, 893−6.

Ehrenberg, C. G. (1844). On the nature and formation of the coral islands and coral banks in the Red Sea. *J. Asiat. Soc. Bengal.* 1, 73−83, 129−36, 322−41, 390−402.

Ekman, S. (1953) *Zoogeography of the Sea*. Sidgwick and Jackson, London.

Faurot, L. (1888). Rapport sur une mission dans la mer Rouge (Isle de Kamarane) et dans le golfe d'Aden (Aden et Golfe de Tadjourah). *Arch. zool. Expér. Génér.* (2) 6, 117−33

Forsskål, P (1775). *Descriptiones Animalium, Avium, Amphibiorum, Piscium, Insectorum, Vermium; quae in itineri orientali observavit P. Forsskål, post mortem auctoris edidit Carsten Niebuhr*. Hauniae, Copenhagen.

Foster, W. (1949). *The Red Sea and Adjacent Countries at the Close of the Seventeenth Century*, Hakluyt Society, London.

Fox, H. M. (1926). Cambridge expedition to the Suez Canal 1924. General Part. *Trans. zool. Soc. Lond.* 22, 1−64.

Fox, H. M. (1929). Cambridge expedition to the Suez Canal 1924. Summary of results. *Trans. zool. Soc. Lond.* 22, 843−63.

Gerlach, S. A. (1967). Bericht über den Forschungsaufenthalt der Litoralgruppe auf der Insel Sarso (Rotes Meer). *Meteor Forschungsergebn.* D, 2−6.

Gibb, H. A. R. (1962). *The Travels of Ibn Battuta AD 1325−1355*. Translated by H. A. R. Gibb. Vol. II. Hakluyt Society, London.

Haeckel, E. (1876). *Arabische Korallen*, Berlin.

Hansen, T. (1964). *Arabia Felix. The Danish Expedition of 1751−1767*, Collins, London.

Hass, H. (1961). *Expedition ins Unbekannte*, Berlin.

Head, S. M. and Ormond, R. F. G. (1978). A platform as a base for coral reef studies. In *Coral Reefs : Research Methods*, eds. D. R. Stoddart and R. E. Johannes, pp.109−18, UNESCO, Paris.

Klunzinger, C. B. (1870). Eine zoologische excursion auf ein Korallenriff des rothen Meeres. *Verh. zool. bot. Ges. Wien.* 20, 389−94.

Klunzinger, C. B. (1872). Zoologische excursion auf ein Korallenriff des Rothen Meeres bei Koseir. *Zeitschr. Ges. f. Erdkde, Berlin*, 7, 20−56.

Klunzinger, C. B. (1878). *Upper Egypt : Its People and Products*, Blackie and Son, London.

Koske, P. (1895). Expedition S.M. Schiff *Pola* in das Rothe Meer. *Ber. Commun. Oceanogr. Forsch.* 6, 572.

Koske, P. (1898). Expedition S.M. Schiff *Pola* in das Rothe Meer. *Ber. Commum. Oceanogr. Forsch.* 7, 485.

Lewinsohn, C & Fishelson L. (1967). The second Israel South Red Sea Expedition, 1965 (General Report). *Israel J. Zool.* 16, 59−68.

Mergner, H. and Schuhmacher, H. (1974). Morphologie, Ökologie und Zonierung von Korallenriffen bei Aqaba (Golf von Aqaba, Rotes Meer). *Helgoland. wiss. Meeresunters.* 26, 238−358.

Morcos, S. A. (1970). Physical and chemical oceanography of the Red Sea. *Oceanogr. Mar. Biol. Ann. Rev.* 8, 73−202.

Ogden, J. C. and Gladfelter, E. H. (eds.) (1983). Coral reefs, seagrass beds and mangroves: Their interaction in the coastal zones of the Caribbean. *UNESCO Reports in Marine Science*, 23.

Oren, 0. H. (1962). The Israel South Red Sea Expedition. *Nature* 194, 1134−7.

Por, F. D. (1971). One hundred years of Suez Canal — a century of Lessepsian migration: retrospect and viewpoints. *Syst. Zool.* 20, 138−159.

Por, F. D. (1978). *Lessepsian Migration*, Springer Verlag, Berlin.

Por, F. D., Aron, W., Steinitz, H. & Ferber, I. (1972). The biota of the Red Sea and the Eastern Mediterranean (1967−1972) — a survey of the marine life of Israel and surroundings. *Israel J. Zool.* 21, 459−523.

Rüppell, E. (1826−28). *Atlas zu der Reise im nördlichen Afrika. Zoologie*, Bronner, Frankfurt am Main.

Rüppell, E. (1835). *Neue Wirbelthiere zu der Fauna von Abyssinien gehörig. Fische des Rothen Meeres*, Schmerber, Frankfurt am Main.

Scheer, G. & Pillai, C. S. G. (1983). Report on the stony corals from the Red Sea. *Zoologica* 133, 1−198.

Schroeder, J. H. (1981). Man versus reef in the Sudan: threats, destruction, protection. *Proc. 4th Int. Coral Reef Symp.* 1, 253−7.

Sykes, P. (1949). *A History of Exploration*, Routledge & Kegan Paul, London.

Thomson, J. O. (1948). *History of Ancient Geography*, Cambridge University Press, Cambridge.

Thompson, E. F. (1939). Chemical and physical investigations. The exchange of water between the Red Sea and Gulf of Aden over the "sill". *Sci. Rep. John Murray Exped. 1933−1934*, 2, 105−19.

Walther, J. (1888) Die Korallenriffe der Sinaihalbinsel. *Abh. math. phys. Cl. Konigl. Sachs. Gesellsch. Wissensch.* 14, 439−506.

Wandel, C. F. (1929). *Carsten Niebuhrs Rejse i Jemen 1762−1763*, C. A. Reitzel's Forlag, Kobenhavn.

Year Book (1985). *Collier's Year Book 1985*, Macmillan Educational New York Collier, Macmillan, London.

CHAPTER 2

Geology and Palaeogeography of the Red Sea Region

COLIN J. R. BRAITHWAITE

Department of Geology, Dundee University, U.K.

CONTENTS

2.1. INTRODUCTION: THE ORIGINS OF AN OCEAN

For the geologist the Red Sea is an ocean. This disregard for the realities of scale is rooted in our understanding of the nature of the earth's crust and of what oceans are. For many years the continents, composed of relatively light rocks rich in silica, and with a composition close to that of granite, have been regarded as raft-like bodies 'floating' on a denser and more basic rock layer, which appears on the surface as oceanic basalt.

In 1915 Alfred Wegener (developing ideas which owed something to Alexander von Humbolt and perhaps Francis Bacon (*see* Bishop, 1981)) suggested that the Atlantic Ocean was formed by the break-up and drifting apart of a single continental mass. However, it was not until the 1950s and early 1960s that geophysical evidence compelled a wide acceptance of what is now called 'Plate Tectonics' theory. This suggests that the surface of the Earth is divided into twelve major 'plates' with several smaller

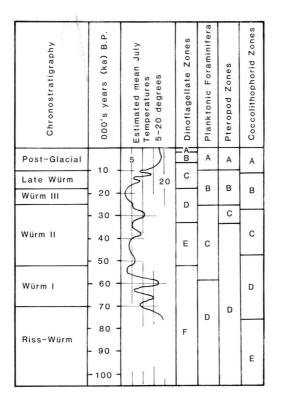

Fig. 2.11. Correlation of Late Pleistocene micropalaeontological zones in deep sea cores in the Red Sea. Based on Berggren and others listed in the text. Note the independence of the different data sets and the sometimes poor correlation of divisions based on separate groups of criteria. The explanations for these divisions are given in the text.

Finally, some 20 species of dinoflagellates have been recorded in two cores (Wall and Warren, 1969). Four cycles of activity have been recognized correlated with climatic and sea-level changes (Fig. 2.11). The first of these began about 70 ka ago with an initial phase of high sea-level in a tropical or sub-tropical climate followed by a gradual retreat of the sea during a period of high evaporation. The second cycle 50 ka ago showed a new marine advance, characterized by *Nematosphaeropsis*, followed by shallowing of the sea and climatic cooling with a relative abundance of *Hemicystodinium*. The marine advance of 20—25 ka ago showed a decline in *Nematosphaeropsis* but no new species. This corresponds with a cold glacial period in Europe. The final advance began 11 ka ago and was characterized by a gradual warming.

The distribution of δO^{18} and δC^{13} ratios, measured independently for carbonates derived from both pteropods and Foraminifera (Deuser and Degens, 1969), show a series of cyclic variations believed to be related to climatic change. The ratio of δO^{18} to δO^{16} increases as a result of the selective loss of O^{16} by evaporation and there is a corresponding increase in the heavier carbon (C^{13}). Together these data suggest the repeated occurrence of periods of evaporation during the last 80—100 ka which are correlated with lowered sea levels, ultimately linked with the glaciations of high latitudes.

D.S.D.P. results show that environmental fluctuations during the late Pleistocene and Holocene have been of considerable importance in modifying sediment character. If evaporation rates exceed precipitation salinity will increase. If sea-level has fallen evaporation will lower it still further. Within the sediments, changes in faunas, in particular of benthic foraminifera, do indeed reflect changes in salinity. However, there is also a direct effect on the sediments themselves. Friedman (1972) describes the results of thirteen deep-sea cores made by the research vessel *Vema*. These show alternating layers of soft and hard aragonite-cemented sediment. Five hard layers have been identified, the oldest extending to perhaps 85 ka and the youngest representing the period 11–13 ka B.P. It is believed that the cemented bands reflect periods of glacially or tectonically (brought about by crustal movements) lowered sea-level. These would restrict inflow of Indian Ocean waters, cutting them off completely if sea-level fell below the sill near the straits of Bab al Mandab, which is at present at about 100 m below sea level. Some authors have suggested that inflow might still have been possible via an inlet through the Danakil depression in the Afar Triangle (Fig. 2.2) but this remains disputed.

Friedman saw conditions during low sea-level periods as being comparable with those of the modern Dead Sea. Apart from the aragonite cements, the dominant carbonate mineral is low magnesium calcite, some of it in the form of coccolith skeletal material. Sulphate minerals (anhydrite and gypsum) are entirely absent which is surprising in view of the evaporative model proposed. It is suggested that they may have been removed by bacterial action. Sulphate decomposition by bacteria would release calcium which could combine with the carbon dioxide released by bacterial oxidation of organic matter to form calcite.

Sediments at the southern end of the Red Sea also show a conspicuous alternation of light and dark layers. The dark layers are from millimetres to 1–2 cm thick. They contain a fauna which is both more abundant and more diverse than that in the light layers, and result from the influx of Indian Ocean waters bringing abundant plankton. In contrast, light coloured layers, poor in fossils, reflect hypersaline conditions. The total number of dark layers is not specified in reports but their confinement to the southern sector of the Red Sea is thought to indicate that organisms carried in from the Indian Ocean were killed off by higher salinities as they moved north and that their remains sank rapidly to the bottom.

World-wide sea-levels have varied considerably in at least the last million years. There is still much discussion of the root causes, but these global changes have now been firmly linked with climatic changes and in particular with the glaciations of higher latitudes. These have been correlated in turn with times of increased rainfall in the tropics, the so-called pluvial intervals. However, evidence presented by Deuser et al. (1976) suggests that during periods of maximum glaciation the Red Sea climate was on average considerably drier than today. In contrast, interglacial periods were apparently wetter. The overall control of climatic changes is thought to lie in astronomical variations. However, the growth of ice sheets leads to an abstraction of water from the world's oceans and thus to a global fall in sea-level. Crustal movements, or mechanisms such as evaporation, can produce only local or regional changes.

2.6.4. Recent (Holocene) reefs

During the last glacial period, sea-level fell by at least 120 m. In the previous interglacial it rose by perhaps as much as 20 m above its present position. Similar positive and negative changes have occurred several times. Pleistocene, Pliocene and indeed Tertiary deposits must have been exposed during low sea-level periods and subjected to weathering and erosion. Braithwaite (1982), suggested that at such times the exposed limestones were sculpted into a complex karst topography. Echo-sounding around reefs on the Sudanese coast suggests that the sea floor within reef areas consists of gently sloping surfaces at 20 to 30 m depth, punctuated by roughly circular closed depressions which now reach 50 to 60 m depth. These depressions are probably solution pits. They are not simply features of past history, but act as receptors of fine grained and poorly sorted Recent sediments. On the Sudanese shore asymmetric distribution suggests that sediments have been transported from an easterly direction, probably by refracted waves driven by prevailing NNW or SE winds. Pits are potentially important to environmental conservation. Once sediment has been deposited within them it is most unlikely that it will come out again. More than this, contained waters, with lower temperatures, also seem to be cut off from surface currents. If pollutants, and especially particulate heavy metals, are introduced into the nearshore zone by offshore workings they may be expected to accumulate within pits to the possible detriment of the biota.

Against this background, contemporary reefs have been built up in only the last 6000—7000 years. The reef forms, which are in part a reflection of the morphology of the underlying eroded limestone surfaces, commonly show no simple topographic distinction between windward and leeward margins. However, on the western shores, the coral assemblages show a clear structural differentiation into a windward *Acropora* community and a leeward *Porites* community (*see* Braithwaite, 1982 and detailed descriptions in Chapter 7). These are characterized respectively by an almost continuous porous growth framework dominated by branching species, and by spaced pinnacles of genera such as *Porites* and *Lobophyllia*. Either assemblage may occur on steep or gently sloping surfaces. Growth on both is more active in shallow water and may thus be expected to steepen existing gradients, theoretically to the point where slopes collapse under their own weight. On leeward margins external growth is limited by mobile sediments spilling over and between pinnacles and preventing colonization. The fact that widespread slope collapse does not occur is taken as additional evidence that these are young systems. In contrast, gravity-induced failure of blocks has been important in maintaining steep margins on submerged cliffs, although it cannot have provided the initial relief.

On fringing reefs sediments merge landwards with the silici-clastic deposits of the coastal plain. Of the surface sediments coarser than 500 μm, 50—80% are coral and algal fragments. Molluscan shell fragments are generally a minor component but may locally form as much as 26% of grains. Benthic foraminifers, Bryozoa, crustacean and echinoderm fragments, rarely rise above a few percent each. On Sudanese reefs distinctive fragments of the benthic foram *Homotrema rubrum*, form important indicators of sediment transport. Sediments are generally coarsest (and contain most *Homotrema*) adjacent to reef edges and show a decrease in grain-size downcurrent. However, most are poorly sorted and size distributions cannot be explained in simple hydrodynamic terms. Statistical analyses of the nature of the grain-size variation within sediments suggests that they result from the mixing of at least two separate size populations. The principal sediment generating zones are the reef edges and sediment characters change downcurrent from these. However, superimposed on this pattern are the additions of varying amounts of material formed on reef flats or elsewhere within the transport system. The grain-size distribution of samples depends on the relative local importance of these two sources.

This pattern of reef formation and sediment accumulation may not apply to all Red Sea reefs. Mergner (1971) reported that on the Saudi Arabian coast near Jiddah reefs bear a richer coral community than those on the Sudanese shores. In addition, he described 'a mighty scree of coral rubble, mud and organogenic sand' which he thought underlay the base of the seaward margin of the living reef. Such features have not been observed by the author in the Sudan and do not accord with the model of limited recent growth on existing eroded limestone blocks suggested here.

2.6.5. Recent oceanic sediments

Offshore, Recent sediments have been extensively studied, not least because of their potential economic value as a source of metals such as silver and copper. The most distinctive features of these regions are hot brine pools, first identified in 1963 by the research vessel *Discovery*. Over fifty reports on the sedimentary environment of these areas have now been published. They are discussed more fully in Chapter 4, but it is worth placing them in their geological context. There are now twenty pools known, principally from the work of the research vessel *Valdivia* (Ross, 1977). Within these areas salinities and temperatures are high. In the Atlantis II deep the temperature was 59.2°C in 1971 and apparently rising, while salinity was 256‰ against the normal Red Sea values of 26–30°C and 36–40‰. Within the sediments, temperatures are even higher and 62.3°C has been recorded from muds in the Atlantis II deep. The heat required to bring about this increase could be transferred from the oceanic crust below either by direct hydrothermal discharge or by convection. The latter (convection) mechanism is thought to be the more likely. Brine pools occur only within the central axial zone and Ross and others (1973) have pointed to a close correspondence between the level of the reflective layer in mid-water marking the top of the brine, and outcrops of reflector S, the seismic marker identified with the top of the Miocene evaporites. The inference of this is that the salts in brines are derived from the Miocene deposits.

Photographs of the sea-floor in the axial zone reveal extensive pillow-lava flows, produced by submarine volcanic eruptions. These are typical of ocean ridges and reflect the continuing creation of new sea floor. The minerals present within the lavas include plagioclase feldspar (labradorite-bytownite), calcite and probable clinopyroxene. Fresh euhedral olivine is recorded in one sample but most are less basic. Chemically, Red Sea basalts resemble oceanic basalts in their low potassium content and lack of normative nepheline but are closer to the alkaline basalts of the African rift in their high total iron and phosphorus. Titanium values are intermediate. The lavas appear fresh and unweathered and this is important since some people assumed initially that these, and associated volcanic emanations, were the sources of heavy metals present within the brines. In fact the rocks show no abnormal alteration or unusual metal content and the source of both the brines and the metals is now thought to lie within the exposed Miocene sediments. Recent sediments in brine pools fall into two groups, one enriched in vanadium and molybdenum, the other chiefly in zinc.

2.7. CONCLUSIONS

In conclusion, the Red Sea Depression cuts across Pre-Cambrian (700 Ma old) basement rocks of the Sudan and Saudi Arabia which were once united as a single mass. It can be seen to have had its beginning in crustal sagging early in the Mesozoic era, perhaps 180 Ma ago but was not established as a linear trough until about the Oligocene Period (only 38 Ma ago). At this time the margins began to move apart at approximately 1.3 cm/year per flank. Deposition during the Miocene Period (from 25–5 Ma ago) was dominated by evaporite sediments (salt deposits) in the central region, with transitions to marginal carbonates and siliciclastic deposits derived by erosion from the rising mountain fronts. Although these have accumulated to a considerable thickness, evidence points to shallow-water deposition and it is thought that the Red Sea formed a vast evaporating basin in which sea level was substantially lowered. Rift movements appear to have ceased or at least been much reduced during the Miocene but they began again about 5 Ma ago with the margins of the depression moving apart at about 0.9 cm/year per flank. This renewed rifting has had important effects in the axial zone. The fault-bounded escarpments formed have exposed Miocene salt deposits and black shales rich in heavy metals. These are believed to be the sources of both the salinity and unusual chemistry of the brine pools which characterize the Red Sea deeps.

Pliocene, Pleistocene and Recent sediments form a relatively thin cover along the margins of the axial zone within which lavas and igneous intrusions are common. Faunal and chemical (isotope) studies of sediments from shallow borings have revealed cyclic variations which are linked with climatic and sea-level changes. These are important since they suggest that present day conditions are in fairly delicate balance, maintained for the moment by the inflow of surface waters through the Straits of Bab al Mandab.

On the coastal margins the same climatic and sea-level variations are reflected in the wide variety of multilayered Pleistocene and Recent sediments. These range from screes and wadi deposits, through alluvial fans and siliciclastic coastal plain deposits, to reefs and associated bioclastic sediments. As sea-level varied, periods of deposition were punctuated by periods of erosion. However, the resulting sediments vary widely in both thickness and distribution. Raised reef deposits extend at least 16 m above present sea level and individual limestone increments have been dated at about 80 ka, 91 ka and 118 ka B.P. in the Sudan, extending to 108–140 ka, 200–250 ka and greater than 250 ka B.P. in the Gulf of Aqaba. Submarine erosion features point to former sea levels at least 100 m below the present datum.

Present day reef deposits have formed within the past 7000 years. Structurally and ecologically two coral communities are recognized, a windward community dominated by species of *Acropora* and a leeward community dominated by *Porites*. Although growth of these associations produces contrasting morphologies and structural frameworks there is no clear correlation between community and reef morphology on the Sudanese shore. The major features of the reefs have been inherited from previous erosion cycles. On the Sudanese coast sediments are generated most consistently within the reef edge zone and are transported onshore. On fringing reefs they merge with the siliciclastic deposits of the coastal plain. Offshore they spill between *Porites* knolls to form an unstable, sand-dominated slope on leeward edges.

Present day environments of the Red Sea region cannot be regarded as static. They are the expression of continuing processes of change open to influence by the activities of man.

REFERENCES

Ahmed, S.S. (1972) Geology and petroleum prospects in eastern Red Sea. *Bull. Amer. Assn. Petrol. Geol.* 56, 707–19.

Battistini, R. (1976) Application des méthodes Th^{230}-Ur^{234} à la datation des dépôts marins anciens de Madagascar et des îles voisines. *Ass. Sénégal Et. Quartern. Afr. Bull. Liason Sénégal* 49, 79–95.

Berggren, W.A. and Boersma, A. (1969) Late Pleistocene and Holocene planktonic foraminifera from the Red Sea. In *Hot Brines and Recent Heavy Metal Deposits in the Red Sea*, Eds. E.T. Degens and D.A. Ross, pp. 282–98. Springer-Verlag, Berlin, 600 pp.

Berry, L., Whiteman, A.J. and Bell, S.V. (1966) Some radiocarbon dates and their geomorphological significance, emerged reef complex of the Sudan. *Zeit. f. Geomorph.* 10, 119–43.

Bishop, A.C. (1981) The development of the concept of continental drift. In *The Evolving Earth*, Ed. L.R.M. Cocks, pp. 155–64. British Museum of Natural History and Cambridge University Press, 264pp.

Braithwaite, C.J.R. (1982) Patterns of accretion of reefs in the Sudanese Red Sea. *Marine Geol.* 46, 297–325.

Braithwaite, C.J.R., Taylor, J.D. and Kennedy, W.J. (1973) The evolution of an atoll: the depositional and erosional history of Aldabra. *Phil. Trans. Roy. Soc. Lond. B* 266, 307–40.

Butzer, K.W. and Hansen, C.L. (1968) *Desert and River in Nubia*. University of Wisconsin Press, Madison and London, 562 pp.

Chin Chen (1969) Pteropods in the Hot Brine Sediments of the Red Sea. In *Hot Brines and Recent Heavy Metal Deposits in the Red Sea*, Eds. E.T. Degens and D.A. Ross, pp. 313–6. Springer-Verlag, Berlin, 600 pp.

Deuser, W.G. and Degens, E.T. (1969) $0^{18}/0^{16}$ and C^{13}/C^{12} ratios of fossils from the Hot Brine deep area of the central Red Sea. In *Hot Brines and Recent Heavy Metal Deposits in the Red Sea*, Eds. E.T. Degens and D.A. Ross, pp. 336–47. Springer-Verlag, Berlin, 600 pp.

Deuser, W.G., Ross, E.H. and Waterman, L.S. (1976) Glacial and Pluvial periods. Their relationship revealed by Pleistocene sediments of the Red Sea and Gulf of Aden. *Science* 191, 1168–70.

Friedman, G.M. (1972) Significance of Red Sea in problem of evaporites and basinal limestones. *Bull. Amer. Assn. Petrol. Geol.* 56, 1072–86.

Furon, R. (1963) *The Geology of Africa*. Translated by A. Hallam and L.A. Stevens, Oliver & Boyd, London, 337 pp.

Girdler, R.W. and Styles, P. (1974) Two stage Red Sea floor spreading. *Nature* 247, 7–11.

Goll, R.M. (1969) Radiolaria: the history of a brief invasion. In *Hot Brines and Recent Heavy Metal Deposits in the Red Sea*, Eds. E.T. Degens and D.A. Ross, pp. 306–12. Springer-Verlag, Berlin, 600 pp.

Graham, A.L. (1981) Plate Tectonics. In *The Evolving Earth*, Ed. L.R.M. Cocks, pp. 165–77. British Museum of Natural History and Cambridge University Press, 264pp.

Guilcher, A. (1955) Géomorphologie de l'extrémité septentrionale de banc corallien Farsan (Mer Rouge). *Ann. Inst. Oceanogr.* 30, 55–100.

Gvirtzman, G. and Buchbinder, B. (1978) Recent and Pleistocene coral reefs and coastal sediments of the Gulf of Eilat. *Tenth Internat. Congr. Sedimentol.* Excursion Y4. Jerusalem, 163–91.

Hsü, K.J. (1972) When the Mediterranean dried up. *Scient. Amer.* 227, 27–36.

Mcintyre, A. (1969) The coccolithophorida in Red Sea sediments. In *Hot Brines and Recent Heavy Metal Deposits in the Red Sea*, Eds. E.T. Degens and D.A. Ross, pp. 299–305. Springer-Verlag, Berlin, 600 pp.

Mergner, H. (1971) Structure, ecology and zonation of Red Sea reefs (in comparison with South Indian and Jamaican reefs). *Symp. Zool. Soc. Lond.* 28, 141–61.

Ross, D.A. (1977) Results of recent expeditions to the Red Sea — *Chain, Glomar Challenger* and *Valdivia* expeditions *Red Sea Research 1970–1975. Min. Resources Bull.* 22, Jiddah, H. 1–14.

Ross, D.A. Whitmarsh, R.B., Ali, S.A., Boudreaux, J.E., Coleman, R., Fleisher, R.L., Girdler, R., Manheim, F., Matter, A., Nigrini, G., Stoffers, P. and Supko, P.R. (1973) Red Sea Drillings. *Science* 179, 277–80.

Ryan, W.B.F. (1976) Quantitative evaluation of the depth of the western Mediterranean before, during and after the late Miocene Salinity crisis. *Sedimentology* 23, 791–813.

Said, R. (1962) *The Geology of Egypt*. Elsevier, Amsterdam, 377 pp.

Stoffers, P. and Ross, D.A. (1977) Sedimentary history of the Red Sea. *Red Sea Research 1970–1975. Min. Resources Bull.* 22, Jiddah, H. 1–19.

Wall, D. and Warren, J.S. (1969) Dinoflagellates in Red Sea Piston cores. In *Hot Brines and Recent Heavy Metal Deposits in the Red Sea*, Eds. E.T. Degens and D.A. Ross, pp. 317–28. Springer-Verlag, Berlin, 600 pp.

Wegener, A. (1915) *Die Entstehung der Kontinente und Ozeane*. Vieweg: Braunschweig 135 pp. An English translation by J.G.A. Skerl was published by Methuen in 1924 as *The Origin of Continents and Oceans*.

Whiteman, A.J. (1968) The formation of the Red Sea Depression. *Geol. Mag.* 105, 231–46.

Whiteman, A.J. (1971) *The Geology of the Sudan Republic*. Clarendon Press, Oxford, 290 pp.

CHAPTER 3

Climate and Oceanography

FREDERICK J. EDWARDS

Thorne's House, Milverton, Somerset, U.K.

CONTENTS

3.1. INTRODUCTION

The great valley in which the Red Sea is situated experiences some of the hottest and most arid conditions which occur in any marine area on Earth. To the west stretches the vast tract of the almost rainless North African desert, over 4,800 km of sand and stone and gravel extending to the Atlantic, relieved only by the thin ribbon of vegetation in the Nile valley which owes its existence to rainfall falling in two areas very far to the south. Eastwards and north-eastwards desert and semi-desert extend even farther — through Arabia, Iran and Afghanistan to central Asia and Mongolia. To the north lies the Mediterranean, almost rainless for six months or more in the year and where, only in winter, do travelling depressions bring periods of unsettled and rainy weather which may briefly effect the northern parts of the Red Sea. To the south, the copious summer rainfall of the Ethiopian Highlands for ever remains distant and the somewhat more benign and more completely 'maritime' regime of the Indian Ocean, with its alternating summer and winter monsoons, barely penetrates to the southern extremities of the Red Sea basin.

The waters of the Red Sea itself, too, are unique. Very high surface temperatures, coupled with extreme salinities make this one of the hottest and saltiest bodies of seawater anywhere; only in the Arabian Gulf are temperatures and salinities sometimes exceeded. But that Gulf is much smaller and, in comparison, is very shallow. Excluding the coastal shallows the average depth of the Red Sea is about 700 m and the main trough everywhere exceeds 1000 m. Several depressions exceeding 2000 m are known. Unlike all of the open oceans of the world where the water at depth is close to freezing point, the waters of the Red Sea are warm throughout. Below the strongly heated surface layers, temperature falls in the top 300 m or so but below about 350 m a remarkably uniform temperature of a little less than 22°C is, with a few notable exceptions, maintained to the bottom. These exceptions, the deep water brine pools, are the subject of Chapter 4.

The Red Sea thus provides a singular environment both meteorologically and oceanographically. The objective of this chapter is to try and describe the salient features of this environment, the climate and weather and the physical characteristics and behaviour of the waters of the sea. It is not intended, neither is it possible in so short a space, to describe in detail the local variations and vagaries which occur over so large a region. Indeed the meagre population and inaccessibility of much of the coastal regions of the area has resulted in a great paucity of such detail. Nor is it proposed to undertake a comprehensive survey of the scientific work in meteorology and particularly in oceanography, which has been attracted to the area, nor will a review of the extensive literature be attempted. For the reader who requires such detail, the review by Morcos (1970), which includes a comprehensive bibliography, is recommended as an excellent overview of the oceanographic scene at that time. No comparable modern account of the meteorology of the Red Sea appears to exist. The British Meteorological Office publication (Anonymous 1951) has long been out of print but may be found in some libraries. This contains much tabulated information some of it derived from records dating back to the middle of the last century accumulated at those places (mainly ports and islands) which were formerly under British influence, and from shipping traversing the area. Limited, but more up to date climatic data are tabulated in the Red Sea and Gulf of Aden Pilot (Anonymous 1980). A useful series of maps showing the distribution, month by month of many meteorological variables (although on a very small scale) is to be found in the Marine Climatic Atlas published by the United States Navy (Anonymous 1976).

The units used for meteorological and oceanographic measurements, and in which climatological data are expressed, have not yet been entirely standardized although in recent years metric units have been generally favoured. The nautical mile and the knot (one nautical mile per hour) remain however in widespread use as does the Beaufort Scale of wind force, of which the limits of each division were originally defined in knots. In this chapter the metric units in common use (not necessarily S.I. units) have been employed and some examples of the equivalent speeds in knots of Beaufort wind forces have been quoted, but for reference, a table of approximate equivalent wind speeds in both kilometres/hour and metres/second for the whole Beaufort Scale up to force 12 is included as an appendix. Atmospheric pressure has been referred to in millibars in conformity with the great majority of data sources although the hectopascal has now been adopted by the World Meteorological Organization. Numerically the millibar and the hectopascal are identical.

3.2. CLIMATE AND WEATHER

3.2.1. Climate

The climate of the Red Sea is largely controlled by the distribution of, and changes in, atmospheric pressure over a very wide area. The pressure centres involved are generally distant from the Red Sea and themselves undergo variations during the course of the year, which result in widespread and sometimes dramatic seasonal changes over extensive areas. It is perhaps surprising that the effects in the Red Sea are so small — with some exceptions the weather characteristics over the whole Red Sea basin show a remarkable uniformity throughout the year, allowing of course for variations due to the quite large range of latitude. Explanation of this may be sought in two main factors. Firstly, although it may come from a number of different directions, almost all the air which enters the Red Sea area is fairly dry. The Red Sea lies broadly within the belt of the North-east Trade Winds which, although distorted by the seasonal pressure changes referred to above, nevertheless form the basis of much of the airflow into the Red Sea. Upwind to the north-east lies little but land with the desert and semi-desert previously mentioned — hence the dryness. Secondly, mountain ranges along the sides of the Red Sea ensure that the main wind systems blow predominantly along the length of the Sea, with only localized air movement at right angles to the shoreline. As a result the prevailing wind directions are remarkably constant and there is virtually no alternation of air masses of different properties which might give rise to changeable conditions or spatial variability in the weather experienced.

3.2.2. General weather characteristics

Before examining the factors which determine weather in some detail, I shall give a short description of the general weather regime along the length of the Sea.

Over the area as a whole the year may be divided into cooler and hotter seasons roughly corresponding to the conventional northern hemisphere winter and summer. 'Winter' may be taken to extend from mid-October to mid-April with 'summer' occupying the rest of the year. The January mean daily maximum temperature ranges from about 20°C in the far north to about 29°C in the far south, the corresponding July figures being 35°C and 40°C respectively. In winter, conditions are not generally too oppressive anywhere; lower night temperatures provide relief from the heat of the day and winds are somewhat stronger and have a cooling effect. In summer, however, it may be most uncomfortable, even in northern areas, and in the south where very high temperatures are combined with high humidity any human activity may prove most difficult, with little, if any, relief being provided at night.

The northern part of the Sea is subject to greater variability of weather than the south, particularly in winter when it may be influenced by disturbances in the Mediterranean. What little rainfall there is, is most likely at this time. This occurs in association with troughs of low pressure which move into the Red Sea from the north and are often accompanied by changes in wind, temperature and humidity and by increased cloud. These fronts may penetrate quite far south — to the latitude of Massawa and even beyond, but south of about 25°N the manifestations referred to above become much less marked and are frequently absent.

In the south, variability is small and is mainly associated with seasonal wind changes. As in the north the occasional outbreaks of rain still tend to be concentrated in the winter months. The salient feature of the weather is monotony; little change from day to day, perhaps the very occasional thunderstorm, and only the slow changes from week to week and month to month as the seasons come and go.

3.3. PRESSURE AND WINDS

3.3.1. Wind direction

In winter, an intense anticyclone centred over Mongolia covers virtually the whole of Asia with ridges of high pressure extending south-westwards to the Arabian peninsula and westwards to eastern Europe and the Balkans. Pressure is also high over North Africa with a centre at about 25°–30°N over western Egypt and Libya. To the south of the equator, over Africa, there is a large area of low pressure from which a shallow trough extends northwards covering the Red Sea. Over the Mediterranean, travelling depressions move eastwards from time to time, from its eastern basin into Asia Minor. These various climatic features all have influence on the Red Sea and produce during this time of the year two quite distinct wind systems which result in the greatest variability in Red Sea weather.

In the south the dominating factor is the Asian anticyclone, on the south side of which, the very extensive North-East Monsoon covers the whole of southern Asia and the north Indian Ocean. Part of this enters the Gulf of Aden from the east as a mainly easterly wind and takes the path of least resistance out at the western end through the Straits of Bab al Mandab and flows towards the low pressure trough over the Red Sea. The southern part of the Red Sea is thus covered by winds from between south and south-east which are, in effect, part of the North-East Monsoon. Over the northern part of the Sea, the main influence is the North African anticyclone which produces, on its eastern flank, a mainly northerly airflow again directed towards the Red Sea low pressure trough, as a result of which the northern part of the Red Sea is covered with predominantly north-west to north winds. These two wind systems meet in an area somewhat to the south of the middle latitude of the Sea, that is at about 18–20°N in January but a little farther south in late autumn and early spring. In this area of convergence there is a belt of light and variable winds with a high frequency of calms.

In the summer months the situation is very different. The great Asian anticyclone is replaced by an extensive low pressure area centred over north-west India, usually referred to as the 'monsoon low'.

From this there extends south-westwards the fairly narrow band of the Inter-Tropical-Convergence Zone (ITCZ) which at this time of the year moves well north of the equator. At its farthest north the ITCZ lies more or less along the southern coast of the Arabian peninsula and just clips the southern end of the Red Sea before turning north-westwards up into the Sudan where pressure is also low. In the Mediterranean pressure is relatively high. Over the whole of the Arabian-Red Sea region the airflow is from a northerly point, circulating anticlockwise around the Indian monsoon low, fed additionally by the Mediterranean anticyclone and directed towards the ITCZ just described. In summer therefore, the winds over the entire Red Sea basin are mainly from between North-west and North. Figure 3.1 shows the salient features of pressure distribution and air flow in January and July.

In summary, the prevailing winds over the Red Sea are along its axis, from north to south, throughout the year, except over the southern part of the sea where the flow is reversed in winter. These two wind systems will be referred to henceforth as the 'North-westerlies' and the 'South-easterlies' respectively although it should be understood that the former include winds from between about north-east and west-north-west and the latter embraces anywhere between east and south. This rather simple picture does however require some embellishment if a complete understanding is to be obtained.

In the south, that is south of latitude 18−20°N where the two wind systems converge in winter, it is not a simple matter of six months of South-easterly winds alternating with six months of North-westerlies. In the Straits of Bab al Mandab the South-easterlies predominate in all months from September to May; only in June, July and August do the North-westerlies take over. As one goes northwards from the Straits, increasing latitude is accompanied by a steadily increasing frequency of the North-westerly winds particularly in the spring and autumn months. The effect of this is not only to decrease the frequency of the South-easterlies but also to erode the period during which they are dominant, so that by about 18°N this is reduced to the six months October to March.

North of the 18−20°N 'convergence area' the South-easterly winds do not penetrate and the North-westerlies are persistent throughout the year. In the winter months, however, disturbances to this rather settled regime do occur. These are associated with depressions travelling eastwards over the Mediterranean or along the north African coast, into Asia Minor. Associated with many of these depressions are cold frontal troughs which extend outwards from the centre in a generally south-westerly direction. These troughs swing from north-west to south-east across the northern part of the Red Sea as the parent depressions move eastwards. Ahead of such fronts the winds back (i.e. change direction in an anticlockwise sense) from north through west, to south-west or south and may become quite strong, even gale force for a time. As the front passes, squalls frequently occur and there is a veer of wind (change in a clockwise sense) back to north-west or north or sometimes northeast. Most departures of the wind direction from the norm are associated with travelling disturbances of this type which generally only affect the Gulf of Suez and the northern part of the Red Sea.

3.3.2. Wind speed

In the main body of the Red Sea, between latitude 16°N and 26°N, i.e. approximately from Massawa to Quseir, more than 75% of winds in all months of the year are of force 4 or less (0−16 kts), and in the summer and autumn months this figure rises generally to over 80% and may exceed 85%. The mean annual wind speed is less than 10 knots and gales are rare. In most months the wind is observed as 'calm' (less than 1 kt), on more than 5% of occasions and south of 20°N this rises to 10−15% particularly in the latter half of the year. In the northern part of this area (north of 20°N) winds of force 5−6 (17−27 kts) account for about 10% of observations in the summer months and about 20% in mid-winter but south of 20°N such winds are much less frequent.

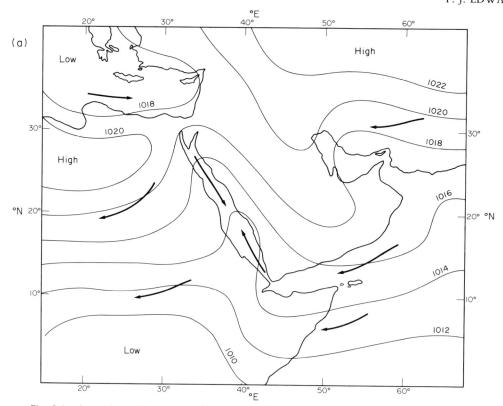

Fig. 3.1a. Approximate Mean Sea Level Pressure (millibars) and Direction of Wind Flow — January.

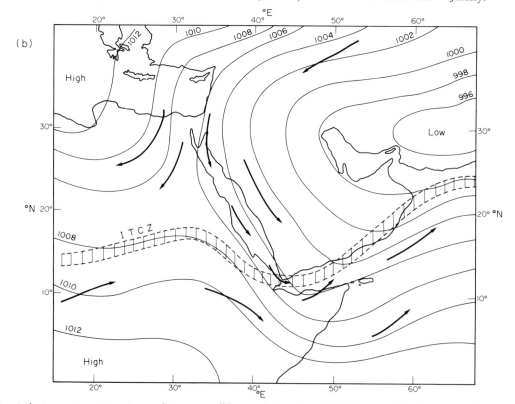

Fig. 3.1b. Approximate Mean Sea Level Pressure (millibars) and Direction of Wind Flow — July. (Intertropical Convergence Zone shaded).

The region to the north of 26°N, including the Gulf of Suez tends to be somewhat windier throughout the year than the area just described. In most months, winds of force 4 or less occur between 60% and 70% of the time. The mean annual speed is a little over 10 knots with September being the windiest month and all winds appear to be rather higher on the Egyptian side. Winds of force 5—6 usually account for some 20-25% of observations; this figure varying from about 15% in mid-winter to about 35% in September. Occasional gales occur throughout the year with isolated instances of force 11 to 12 being on record.

The area between about 16°N and the Straits of Bab al Mandab shows the greatest seasonal variation of wind speed in the Red Sea. In summer, when the North-westerlies are blowing, over 80% of winds are force 4 or less and 10—15% of observations register calm. Gales at this time are virtually unknown. During the winter however, the South-easterlies are generally a good deal stronger particularly just north of the Straits where, from November to January, less than 40% of observations show winds of force 4 or below and over 50% are force 5—6. Gales occur in most years during each of the months November to February.

3.3.3. Land and sea breezes

As in all coastal areas prone to strong solar heating, the shores of the Red Sea experience diurnal wind changes commonly referred to as land and sea breezes. The sea breeze, a flow of air from the sea towards the land, develops during daylight hours, usually reaching maximum strength in early or mid afternoon. The reverse flow, land breeze, occurs at night and is generally most strongly developed around dawn. In the absence of other factors, land and sea breezes tend to flow more or less at right angles to the shoreline. In the Red Sea, where the prevailing winds are very persistent and are parallel to the shoreline, the effect of land and sea breezes is to render the wind direction along the coasts somewhat different from that expected over the adjacent open sea, the difference depending upon the time of day.

On all the coasts on the African side, whenever the North-westerlies blow, the land breeze is usually from between west and north-west. The sea breeze is mainly from between north and north-east, although in the south it may be rather more easterly. The change from one to the other occurs over a period of an hour or two around 0800 and 2000 local time. In the cooler months the morning change — which is seen as a veer of wind, usually occurs after 0800 but in the hotter summer months it frequently commences before this time. The evening change is usually complete by about 2000 but may be somewhat later.

During winter in the southern Red Sea where the South-easterly winds prevail, the land breeze appears to be poorly developed or non-existent. The sea breeze effect is seen by a change from predominantly south to south-east winds in early morning to east or even north-east in the afternoons, but data about this southern coastline are sparse.

The little information that is available for the eastern shores of the Red Sea suggests that during the period of North-westerly winds, these back towards the west by day and become north-east to east by night. Land breezes however, appear to be less well developed than on the African side. This may be due to the greater average elevation of the hinterland to the east compared with that of Egypt and the Sudan, which also is likely to result in greater irregularity and squalliness of the land breezes. In the south, during the South-easterly winds of winter, well developed sea breezes from the south-west occur which may blow from west or north-west in the spring and autumn transition months. Winds by night tend to be in the direction of the seasonal flow.

3.4. AIR TEMPERATURE

3.4.1. Introduction

A generalized indication of the annual range and variation of temperature between the two ends of the Red Sea has already been given in Section 3.2.2 by reference to mean daily maximum temperatures in midsummer and midwinter. The figures given are representative of the somewhat confined waters of the Gulf of Suez in the north and the neighbourhood of Perim Island in the south, but give a far from complete picture. In any area of open sea, well removed from the influence of land, air temperature at two or three metres above the sea surface is usually within a few degrees Celsius of the sea surface temperature which itself rarely changes during the course of a day by more than 2−3°C (usually rather less). In the Red Sea, the amount of reliable temperature information from ships is small and most of the available data are derived from the climatological records of a small number of coastal towns and ports which, though strongly influenced by conditions inland, are reasonably well representative of adjacent inshore water. The only offshore island records in the Red Sea are for Daedalus Reef (Abu el Kizan), which is near the centre of the sea just south of 26°N; Kamaran Island, quite close to the eastern shore at about 15°N and Perim Island in the Straits of Bab al Mandab.

3.4.2. Temperature over the open sea

Later in this chapter some detail of mean sea surface temperature (Ts) is given which provides a good basis for the estimation of likely air temperature (Ta) at any open sea location using the following general rules:

In that part of the Sea south of about 18°N, Ta exceeds Ts throughout the year by a degree or two, the difference being greater in midsummer than in midwinter. North of 18°N the same is true only in the summer months, the greatest difference occurring in the Gulfs of Suez and Aqaba both of which are fairly strongly influenced by conditions over adjacent land. In winter, in this northern area, Ta is less than Ts by about half a degree, again with a rather larger difference over the northern extremities of the Sea and the two landlocked gulfs.

Daedalus reef records indicate a mean diurnal range of air temperature of 4°C from December to April and 5°C in all other months except July when 6°C is attained. These figures are probably representative of the diurnal range over most of the offshore Red Sea waters with a trend towards higher values over the area of warmest water in the south.

From the foregoing, if mean sea temperature is known, an estimate of likely air temperature at any time of the year may be made together with the probable diurnal variation, but it should be noted that even over the open sea occasional extreme deviations from the norm do occur.

3.4.3. Temperature along the shoreline

Coastal climatological records for the Red Sea are confined to about a dozen locations three of which are on the Gulfs of Suez and Aqaba and only one, Jiddah, is on the eastern shore of the Red Sea proper. The remaining eight are spaced at irregular intervals on the western shore.

The hinterland to all the Red Sea coasts consists of vast areas of desert or semi-desert the influence of which is seen as a marked degree of 'continentality' in the temperature regime along the coasts. This is evident, firstly in a greater annual range of temperature than over the open sea in the same latitude, and secondly, by a greater diurnal range of temperature throughout the year, compared to open sea conditions. The extreme values recorded also vary more widely. Nevertheless, examination of the coastal records reveals that the offshore sea temperature can still be used as a useful guide to the mean temperature in the coastal regions. The pattern which emerges is that over the whole length of the Sea there is a 'winter' period during which the mean coastal air temperature is below the offshore sea temperature and a 'summer' period when the reverse prevails. The 'summer' period is May to September from Hurghada to Port Sudan but tends to lengthen somewhat southwards from that latitude, being April to September at Suakin, March to October at Massawa, and February to October at Assab, which is only just north of the Straits of Bab al Mandab. The difference between Ta and Ts is usually only one or two degrees Celsius with the values in the transition months of April and September close to zero. The differences are larger, however, in winter in the north (6−7°C at Hurghada from December to February) and in midsummer in the vicinity of the area of highest sea temperature (about 4°C at Port Sudan and Suakin). Figures for mean air and offshore sea temperatures for a selection of coastal stations are presented in Table 3.1.

It is of interest to compare the figures for Dungunab Bay and Jiddah which are in approximately the same latitude and provide the only comparison between the eastern and western shores. The Jiddah figures show higher values of Ts throughout the year and higher mean air temperatures at all times except from June to September when there is little difference. In every month the mean air temperature at Jiddah is closer to the offshore sea temperature than at Dungunab.

The diurnal range of air temperature on the coasts is usually substantially greater than the 4−5°C to be expected over the open sea. The difference between the mean daily maximum and mean daily minimum is usually least in the winter months at about 7−9°C and greatest in midsummer when ranges of 13−15°C are common, with intermediate values in spring and autumn. There are, of course, exceptions, the magnitude of the diurnal range being influenced by several factors. A sea breeze, for example, when well developed, tempers the afternoon heat by several degrees. On days when no sea breeze occurs therefore, a larger diurnal range of temperature is to be expected, assuming no difference from other causes, e.g. cloud cover. Any effect caused by the land breeze at night is usually less significant. Along the shoreline the extremes of temperature recorded are often very close, or identical to the mean values of highest and lowest temperatures in the hottest and coldest months. For example,

TABLE 3.1. Mean Monthly Air Temperature and Offshore Sea Temperature for Selected Localities.

		J	F	M	A	M	Ju	Jy	A	S	O	N	D
Hurghada	Ta	16.5	15.5	19.0	22.5	26.0	28.5	29.5	30.0	28.0	25.0	21.5	18.0
	Ts	23.5	22.0	22.0	23.5	25.0	27.0	28.0	28.5	27.5	26.5	25.5	24.0
Dungunab	Ta	21.5	21.5	23.0	25.0	28.0	30.5	32.0	32.5	30.5	27.5	25.5	23.0
	Ts	25.5	25.5	25.5	26.5	27.5	29.0	30.0	30.5	30.0	29.5	29.0	27.0
Jiddah	Ta	24.0	23.5	24.0	27.0	29.0	30.0	31.5	32.0	30.5	29.0	27.5	24.5
	Ts	26.0	25.5	26.0	26.5	28.0	29.5	30.5	31.0	30.5	30.0	30.0	27.5
Port Sudan	Ta	23.5	23.0	24.5	26.5	30.0	32.5	34.5	35.0	32.5	29.5	27.5	25.5
	Ts	26.5	26.0	26.5	27.0	28.0	29.5	30.5	31.0	30.5	30.5	30.0	28.0
Suakin	Ta	23.5	23.0	24.0	26.5	29.0	32.5	35.0	35.5	32.5	29.0	27.0	24.0
	Ts	26.5	26.5	26.5	27.5	28.5	29.5	30.5	31.0	30.5	30.5	30.5	28.0
Massawa	Ta	25.5	25.0	26.5	28.0	30.5	33.0	35.0	35.0	33.0	31.0	28.0	26.0
	Ts	26.0	26.0	26.0	27.5	29.0	31.5	32.0	32.0	31.5	30.5	28.5	26.5

Ta; Mean Monthly Air Temperature = ½(Mean Daily Max + Mean Daily Min).
Ts; Mean Monthly Offshore Sea Temperature (Estimated Sea Surface Temperature in Deep Water Offshore).

at Suakin the highest recorded temperature is 48°C and the lowest 12°C. 48°C is also the mean of the highest temperatures recorded in both June and July and 12°C is the mean of the lowest January temperatures, both during a 30 year period. Similar features appear in the figures for most other localities in the Red Sea for which records are available and indicate that, in general, temperatures at or near the extremes recorded must occur in all years. As far as actual values of extreme temperature are concerned, the minimum likely temperature in January to February ranges from about 4–5°C at Hurghada to 17–18°C south of Massawa. Rather lower temperatures occur in the Gulf of Suez, but air frost has not been recorded. In June, July and August temperatures in excess of 40°C occur frequently along all Red Sea Coasts. The hottest area is probably that between about latitudes 16°N and 22°N where temperatures of 45°C (113°F) or more occur in most years and the extreme high temperature recorded is 48°C at Suakin. As might be expected the highest temperatures occur when the winds are from landward.

Before leaving the subject of temperature it is perhaps worth emphasising that the southern part of the Red Sea south of about 16°N is one of the hottest marine areas in the world. There are places where the extreme summer temperatures exceed the figures quoted above, but nowhere else are temperatures so consistently high throughout the year as in the coastal areas of the southern Red Sea.

3.5. HUMIDITY

Although most of the air which enters the Red Sea basin is fairly dry the area has a reputation for hot, moist conditions, which render any prolonged human activity, at best, very difficult. The apparent paradox is explained by the very high rate of evaporation from the warm waters of the Red Sea which raises the humidity of the lower layers of the atmosphere to uncomfortably high values. Humidity may be expressed in several different ways but for simplicity it will be referred to here only as relative humidity. This is an expression of the amount of water vapour present in the air as a percentage of the amount required for saturation at the same temperature. Air containing a constant mass of water vapour will undergo a change of relative humidity as temperature changes, since the amount of water required for saturation increases with increasing temperature; nevertheless relative humidity is a convenient means of expressing the 'dryness' or 'wetness' of air.

Over the open sea, the average annual humidity is about 70%, the annual range about plus or minus 5% from this figure and the daily range a maximum of about 4%. Lowest values of about 65% are recorded in midwinter whilst in summer a fairly constant level of a little over 70% is maintained. In the southern part of the Sea humidity is higher in winter when the South-easterlies prevail than during the summer North-westerlies.

Along the coasts humidity is, usually, rather lower than over the sea and generally increases from north to south. Considerable local variations occur, conditions being much influenced by land and sea breezes and the local temperature range. At night, winds from landward tend to be dry, but lower temperatures raise the relative humidity. By day, higher temperatures lower the relative humidity but sea breezes bring in moister air and have the opposite effect. In some places night cooling is dominant, resulting in maximum humidity at that time or in early morning, in others sea breezes result in a maximum during the afternoon. On the Egyptian coast monthly mean values are around 50% throughout the year for morning observations and some 5–10% higher in the afternoons. In the Port Sudan area, humidity averages 65–70% from October to March but only about 40% from June to August, with only a small daily range at all times. At Jiddah, the humidity is remarkably constant averaging between 50% and 60% in all months with little diurnal variation. The highest average values appear at Perim Island which has a mean of about 80% throughout the year for early morning

The origins of the Atlantis II Deep mining project can be traced to a number of geophysical, geological and geochemical factors resulting in the build-up of hot brines and metal bearing sediments in several deep basins along the central trough of the Red Sea. A series of research studies has revealed the existence and economic potential of these deposits and demonstrated ways of commercially exploiting them. Most of the rest of this chapter will discuss the deposits and their associated environments, a short resumé is also given of the projected mining operation.

The first indications of temperature and salinity anomalies in the Red Sea were obtained in 1948 by Swedish scientists on the research vessel *Albatross*, but their findings went unnoticed by the scientific community. This changed dramatically when scientists from the Woods Hole Oceanographic Institution in their research vessel *Atlantis II* sampled the Red Sea in July 1963. They retrieved a deep water sample with an anomalously high temperature (25.6°C) and salinity (43.3 ‰), initiating large scale scientific interest and the commitment of several research vessel cruises to the Red Sea. The interest increased after the first hot brines and metalliferous sediments were retrieved in September 1964 by scientists on the board the British RV *Discovery*. The most extensive hot-brine research in this phase of study was done by crews of the research vessels *Atlantis II* (1963, 1964), *Discovery* (1963, 1964, 1967), *Meteor* (1964, 1965), *Chain* (1966), *Akademik S. Vavilov* (1966) and *Oceanographer* (1967). Most of the results of the multinational efforts were published in a compendium edited by Degens and Ross (1969).

Detailed mapping of the Red Sea deeps and prospecting for metal rich sediments was conducted by Preussag under license from the coastal states from 1969—1972, using the research vessels *Wando River* and *Valdivia* (Bäcker, 1975). During the same period the area was revisited by RV *Chain*, and stratigraphic data were obtained within the Deep Sea Drilling Project by RV *Glomar Challenger* (Ross et al., 1973). By the end of this phase, a total of about 190 cores had been retrieved by the various research vessels. Twelve deeps containing hot brines and metalliferous sediments, and an additional four with hydrothermal influence had been discovered (Bäcker and Schoell, 1972). The locations and year of discovery of these deeps are shown in Figure 4.1.

The economic phase began in 1974 when the Kingdom of Saudi Arabia and the Democratic Republic of Sudan formally agreed upon joint future exploration and exploitation of the mineral resources of the central Red Sea. The waters beyond the 1000 m depth contour were declared to be a common zone between the two countries, and the Saudi-Sudanese Commission for the Exploitation of the Red Sea resources was created. In 1976 Preussag was awarded the contract for the development programme. At that stage only the Atlantis II Deep was considered to be of economic importance, and the MESEDA programme was initiated. So far, the main activities have been MESEDA I with RV *Sonne* in 1977—78, MESEDA II with RV *Valdivia* accompanying the Pre-Pilot Mining Test using *Sedco 445* in 1979 and MESEDA III with RV *Valdivia* in 1980—81. Some additional studies were conducted from RV *Valdivia* in 1982 and RV *Sonne* in 1984, data from these later studies are included in Figure 4.1. To date, we know of 13 deeps with one or more brine pools and metalliferous sediments (e.g. Atlantis II Deep). Five other deeps contain metal-rich sediments but no brine pools (e.g. Thetis Deep), and there are some areas outside the deeps, such as the Commission Plain, where there are also high sediment metal concentrations.

4.2. THE RED SEA BRINE POOLS

4.2.1. Physical and chemical properties

There are substantial differences in the salt concentrations and temperatures of the various brine pools (Table 4.1), but in each the hottest and saltiest waters are the deepest, lying in contact with the bottom sediments. As can be seen from Figure 4.2, in some deeps the salinity and temperature change in stages,

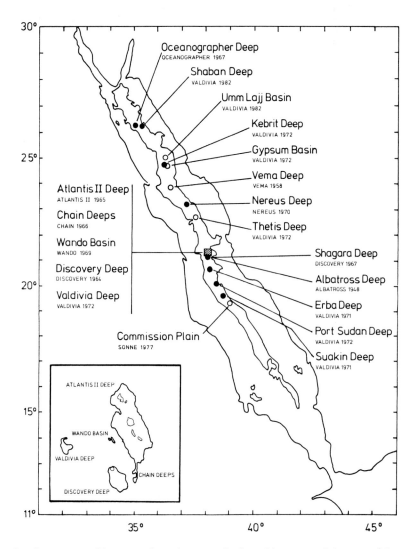

Fig. 4.1. Sites of Red Sea Deeps and basins. Under each name is the date of discovery and the name of the responsible research vessel. Black points are sites with both brine pools and hydrothermally influenced sediments. White circles mark sites with metalliferous sediments only.

forming upper and lower brine layers. These layers with constant salinity and temperature are topped by a transition zone where the hot brines mix with normal Red Sea water and the conditions change continuously.

Studies of the Atlantis II Deep between 1965 and 1979 have shown that in this deep (and possibly in others), conditions are far from stable. During this period, the temperature of the lower brine had increased from 55.9 to 61.7°C, while that of the upper brine had risen from 41.2 to 49.9°C (Hartmann, 1980). Within the same period, the upper boundary of the lower brine had risen from 2050.5 to 2048.0 m depth. Over the whole surface area of the Atlantis II brine pool (43.5 km^2), this corresponds to an increase in the volume of the lower brine of about 9×10^6 m^3 every year, or about one thousand tonnes of water per hour. The temperature changes are shown in Figure 4.3, recent data from RV *Valdivia* and

RV *Sonne* show the trend is continuing. No significant changes in brine salinity have been observed within the observation period.

The various brine pools are also characterised by differences in the concentrations of both major and minor dissolved salts; those of the Atlantis II and neighbouring deeps are generally the most concentrated (Table 4.2). Compared with normal Red Sea water, most of the brines are highly enriched with various heavy metals, such as manganese, iron, zinc, cadmium and copper. The degree of enrichment of these trace constituents of sea water generally exceeds that of the major constituents (sodium, calcium and potassium) by 10 to 1000 times. Most brines are depleted of magnesium, iodine, sulphate and nitrate. The Kebrit Deep brines are also enriched with hydrogen sulphide and carbon dioxide.

4.2.2. Origin of Red Sea hot brines

For many years the origin of the Red Sea brines and deposits has been a source of speculation, but the recent intensive research effort has given us a fairly clear understanding of the processes. In chapter two of this book we learnt of the origins of the Red Sea itself — a nascent ocean developing by crustal

TABLE 4.1. Hydrothermally influenced deeps and basins in the Red Sea, their main physical characteristics. Where no data are given for brines, the deep or basin lacks brines but possesses metalliferous sediments. From Bäcker and Schoell (1972) supplemented with new data.

Name	Coordinates of the centre N	E	Maximal depth m	Thickness of brine layers m	Brine surface area km^2	Chlorinity of lower brine g Cl kg^{-1}	Temperature of lower brine °C
Oceanographer Deep	26° 17.2′	35° 01.0′	>1446	>82	—	—	—
Shaban (Jean Charcot) Deep	26° 14′	35° 20′	1500	163	11.2	154.0	25.3
Um Lajj Basin	25° 02′	36° 17′	—	—	—	—	—
Kebrit Deep	24° 43.35′	36° 16.6′	1573	107	2.5	154.1	23.3
Gypsum Basin	24° 42.1′	36° 24.8′	1196	—	—	—	23.7
Vema Deep	23° 52.2′	36° 30.5′	1611	—	—	—	22.7
Nereus Deep E Basin	23° 11.5′	37° 15′	2458	39	3	—	—
Nereus Deep W Basin	23° 11′	37° 12′	2432	11	1	—	—
Thetis Deep	22° 43′	37° 36′	1970	—	—	—	—
Atlantis II Deep	21° 22.5′	38° 4.5′	2194	200	55	156.7	63.1
Chain Deep A	21° 18.0′	38° 4.9′	2072	80	0.7	153.8	53.2
Chain Deep B	21° 17.17′	38° 4.95′	2130	138	0.8	156.0	46.7
Chain Deep C	21° 16.77′	38° 4.95′	2165	173	0.02	154.6	44.5
Wando Basin W	21° 21.45′	38° 1.80′	2013	28	0.2	—	24.1
Wando Basin E	21° 21.20′	38° 2.40′	2007	22	0.5	73.5	29.3
Discovery Deep	21° 17.0′	38° 3.2′	2224	209	11.5	155.5	44.8
Valdivia Deep	21° 20.5′	37° 57.0′	1550	123	4	144.7	29.8
Albatros Deep	21° 11.9′	38° 7.0′	2133	72	1.5	143.3	24.4
Shagara Deep	21° 07.8′	38° 5.3′	2496	8	1	113	—
Erba Deep	20° 43.8′	38° 11.0′	2395	19	7	86.5	27.9
Volcano Deep	20° 1.3′	38° 27.1′	2350				
Port Sudan Deep	20° 3.8′	38° 30.8′	2836	322	5	125	36.2
Suakin Deep NE Basin	19° 38.0′	38° 46.3′	2776	54	10	86.2	24.3
Suakin Deep SW Basin	19° 36.7′	38° 43.6′	2850	74	2.5	85.8	24.3
Commission Plain	19° 18′	38° 56′	2100	—	—	—	—
Normal deep Red Sea water						22.5	21.8

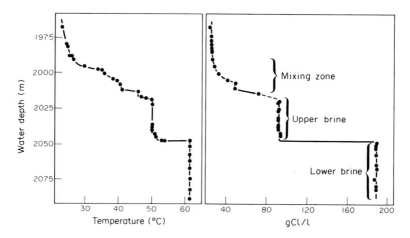

Fig. 4.2. Temperature and chlorosity profiles for the Atlantis II Deep brine pools. Adapted from Hartmann (1980), data for November 1977.

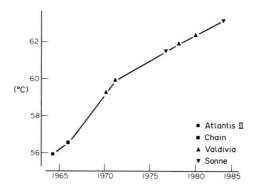

Fig. 4.3. Temperature changes within the lower brines of the Atlantis II Deep, 1964 to 1985. Adapted from Hartmann (1980), supplemented with data from recent cruises.

expansion on either side of a northern arm of the main northwestern Indian Ocean (or Carlsberg) Rift. The deep central trough, along the axis of the Red Sea is typical of all areas of ocean floor spreading, and contains the metalliferous deeps. Within the trough, new ocean crust is slowly but continually being formed by upward and outward flow of hot mantle material injecting lava into the base of the trough. This phenomenon is responsible both for the local high heat input and for the supply of heavy metals to the brines and sediments. We assume that the brines originated from 'fossil waters' which had entered the sea floor at different times and had been trapped there for as long as 15,000 years. Because of submarine volcanism, and through contact with rising lava masses, the trapped water became heated, acidic and corrosive, dissolving material from the lava. The heating would have generated convection currents bringing the water to the sediment surface. The hot metal-bearing waters would have been further enriched with salts leached out from the thick Miocene evaporite deposits (*see* Chapter 2, section

5) at the flanks of the axial trough of the Red Sea. The resulting high density of the hot brines has prevented their general convective mixing with the normal Red Sea water mass above them.

Detailed surveys of the distribution of temperature and heavy metals in the hot brines of the Atlantis II Deep lead to the conclusion (Fig. 4.4) that the site of active brine discharge lies in the south-west basin (Schoell, 1975; Hartmann, 1973). Although vertical mixing is prevented by density stratification, lateral flow and mixing are good. Hartmann (1980) has calculated the temperature of the extruding brines to be as high as 140−280°C. After discharge at the bottom of the deep, chemical reactions take place in the extruded brines as a result of cooling, declining acidity and uptake of oxygen and sulphate, leading to precipitation of metal sulphides, silicates and hydroxides.

The reactions thought to be leading to the formation of sediments in the Atlantis II Deep brines are summarised in Figure 4.5. A sequence of reactions appears, related to progressive loss of temperature and acidity, and the replacement of highly reduced by increasingly oxygenated conditions; this in turn leads to lateral differentiation of the deposits. Part of the sulphides have already been deposited in the channels of the hydrothermal vents through which the rising waters flow (Bäcker and Richter, 1973). Most of the sulphides, mainly containing iron, zinc and copper, have been precipitated from the brines near the vents. Manganese on the other hand has been precipitated within the peripheral areas of the deep and in the surrounding area outside the brine pool. This pattern is shown schematically in Fig. 4.6. which also summarises the geological and geochemical setting.

4.3. METALLIFEROUS SEDIMENTS

As with the overlying hot brines, the sediments of the Red Sea deeps vary in chemical composition, depending on the relative importance of the two main sediment sources. The first source is of course precipitation from hydrothermal fluids, these sediments have high concentrations of iron and manganese

TABLE 4.2. Characteristics of Atlantis II Deep brines. Ranges for average concentrations of dissolved inorganic constituents combined from Brewer and Spencer (1969), Brooks *et al.* (1969), Danielsson (1980), Hartmann (1985) and own measurements.

	Lower brines	Upper brines	Deep Red Sea water	
Na	93−110	47−51	13−17	g/kg
K	1.9−2.6	1.1−1.3	0.4−0.5	
Ca	4.8−5.2	2.3−2.5	0.5	
Mg	0.8	1.2	1.5−1.8	
Cl^-	156−157	80−83	23	g/kg
SO_4^{--}	0.84	2.3	3.1	
HCO_3^-	0.07−0.08	0.08−0.10	0.12−0.16	
Fe	79−95	0.2−7.3	0.0001−0.1	mg/kg
Mn	76−158	72−82	0.0001−0.01	
Zn	2.1−7.9	0.2−8	0.001−0.9	
Cu[1]	0.0002−0.6	0.02−0.3	0.001−0.6	
Pb	0.2−0.6	0.009−0.3	0.00002−0.06	
pH	5.7	6.0	8.1	
Temperature[2]	63.1	53.4	21.8	°C
Salinity	257	135	41	g/kg
Density	1.198	1.100	1.029	kg/dm³

[1] Comparing the copper concentrations determined by Danielsson (1980) and Hartmann (1985) with those of the earlier investigations, a trend of decreasing copper within the brines can be observed.

[2] Data given for temperature and density represent the actual situation in 1984.

ATLANTIS II - DEEP / RED SEA
HEATING AND CONVECTION WITHIN THE 59°C-BRINE

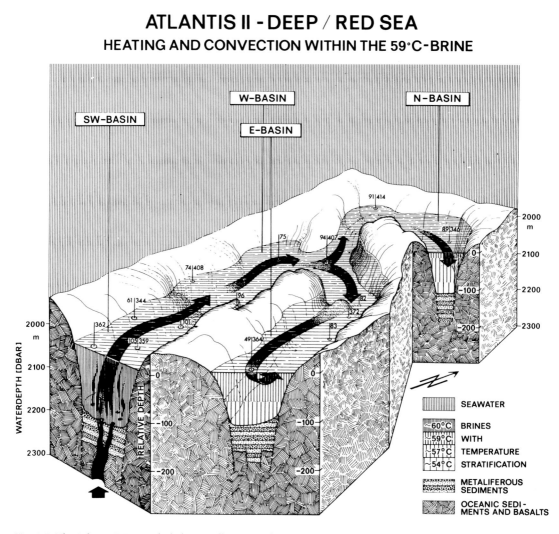

Fig. 4.4. The Atlantis II Deep. Block diagram illustrating the main convective currents within the lower brine layers. Taken from Schoell (1975).

or copper and zinc, depending on the relation of the site to the hydrothermal source. The other sediment source is the normal input of detritus from pelagic organisms. The Atlantis II Deep has particularly dominant hydrothermal input; this is also well marked in the Thetis, Gypsum, Nereus, Erba and Suakin Deeps.

The normal sediments in the central Red Sea trough consist of light coloured marls with a significant microfossil content, especially fossil foraminiferans and pteropod molluscs. The sediments from the deeps most strongly hydrothermally influenced are characterised by a multi-coloured layered structure. The different groups of sediment layers represent repeated changes in the geochemical conditions at the time of deposition. Under normal, well oxygenated conditions, oxidised sediments are formed containing limonite, haematite and manganite. They may contain some sulphides such as pyrite and

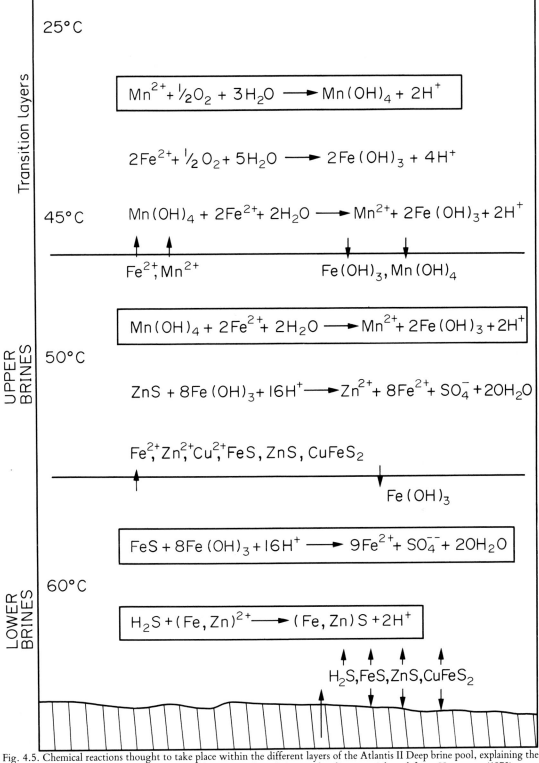

Fig. 4.5. Chemical reactions thought to take place within the different layers of the Atlantis II Deep brine pool, explaining the distribution and precipitation pattern of heavy metals in brines and sediments. Adapted from Hartmann (1973).

chalcopyrite. Under reducing conditions the deposits consist of ferrous sulphide and sphalerite together with iron montmorillonite and manganosiderite.

The sequence of sediment types varies between the different deeps and even in different areas of a single large deep. It can be assumed that major sequence changes come about as a result of changes in the nature of the outflowing brines or because of disturbances in the structure of the brine pools caused by lava flows and tectonic activity. According to Bäcker and Richter (1973), five main strata can be distinguished in the sediments of the Atlantis II Deep. Starting with the oldest (deepest) sediments these are :

a) Detritic-oxidic-pyritic zone (DOP)

Hydrothermal sedimentation began about 25,000 years ago. These lowest layers above the basement basalt consist of biogenic-detritic marls mixed with limonites and pyrites. This layer varies locally between about 2 m and 8 or more metres in thickness.

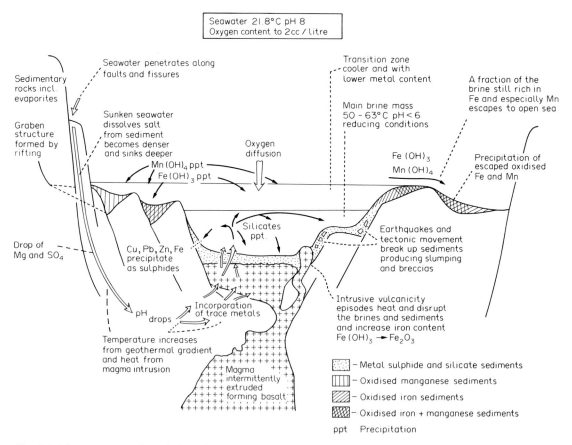

Fig. 4.6. Schematic section through the Atlantis II Deep and environs showing the route taken by groundwaters, and the distribution and formation of the main sediment types. Simplified and adapted from Bäcker (1973).

b) Lower sulphide zone (SU₁)

This is the first economically important metalliferous layer containing zinc, copper and silver. It was formed over a period of about 2000 years starting about 10,000 years ago. During that time, the brine pool must have been without oxygen, and the principal minerals formed were sphalerite, chalcopyrite, pyrite and various silicates.

c) Central oxidised zone (CO)

During the next phase of brine pool instability, oxidising conditions prevailed resulting in the precipitation of iron and manganese oxides. Limonites with an iron content of greater than >50% of the salt-free washed dry material constitute the major part of the sediment.

d) Upper sulphide zone (SU₂)

Anaerobic conditions prevailed once more and sediments were deposited similar to those in SU_1. The main constituents are light-violet coloured sulphides and greenish-brown silicates.

e) Amorphous silicate zone (AM)

This layer, the uppermost and youngest, has a high water content. The sediments are of a dark-brownish colour and are composed mainly of amorphous iron-rich smectites mixed with minor constituents of poorly crystallised oxide and sulphide material.

The metalliferous sediments of the Atlantis II Deep are from 3 to 30 m in thickness, and an average depositional rate of about 1 mm per year has been calculated for them. This is about ten times the rate of normal biogenic-detrital sedimentation in the Red Sea, emphasising both their very fast rate of deposition and the comparative youth of the economic deposit.

4.4. THE BRINE POOLS, A TOXIC ENVIRONMENT?

To what extent is life possible in the various Red Sea brine pools? As we have already seen, the brines differ markedly from normal sea water in their temperature, salinity and metal content. Most lack oxygen and some are enriched with poisonous hydrogen sulphide. Conditions such as these would be unfavourable for most forms of life. Any community inhabiting the brines or underlying sediments would have to be very specialised and probably would contain very few species. Such organisms could either be members of an indigenous community which has adapted itself to the extreme conditions of the brine pools, or could be invaders from the surrounding bathypelagic and deep benthic communities.

Introductions to the plankton and benthos of the deep Red Sea are given in Chapters 5 and 6 of this book. The deep Red Sea benthos is low in biomass and diversity because the surface productivity is low and the food supply to the deep sea bottom is minimal. So far no additional sources of food have been identified in the Red Sea in the areas of hydrothermal activity, although specialist communities based on production by sulphur bacteria have been discovered in association with hydrothermal vents in other ocean areas.

So far little is known of the existence of life in the brine pools, and the few data available from the Atlantis II and nearby deeps are not necessarily representative of other deeps where different geochemical conditions prevail. No living benthic fauna (including macro-, meio- and microfauna) have yet been discovered in Red Sea deeps with strong hydrothermal influence. Likewise no eucaryote species is known to be able to withstand the extremely toxic conditions of the Atlantis II Deep brines. The heavy metals in these brines include some of the most toxic known, and their concentrations exceed those of normal oceanic water by several orders of magnitude. It is not yet known whether any single component parameter of the brines is anathema to life, or if life is excluded by some synergistic effect of the cocktail of toxic metals, high salinity and temperature and lack of oxygen.

Bacteria are highly adaptable to extreme environments and can satisfy their metabolic requirements from a variety of food substrates. Specialised sulphate and nitrate reducing bacteria are known from various marine environments. These bacteria use sulphate or nitrate in anaerobic respiration instead of the more usual oxygen; nitrate bacteria can exist in the presence of oxygen which is toxic to sulphate bacteria. The latter have been isolated from deep sea oil and sulphur wells, and are capable of growing at 65°C under pressures of 200–400 atmospheres, or at 104°C at 1000 atmospheres (ZoBell, 1957).

During the expedition of RV *Chain*, a number of samples were retrieved from the Atlantis II, Chain and Discovery deeps, and were analysed for bacteria by Trüper (1969) and Watson and Waterbury (1969). Bacteria were not detected in samples from the Atlantis II or Discovery Deep brines, but were found in samples from the transition zone covering the Atlantis II brines, and in the ambient deep Red Sea water. Bacteria were also isolated from interstitial water within the sediments of the Discovery Deep. The bacteria proved to be sulphate reducing types of the genus *Desulfovibrio* (Trüper, 1969). Trüper tried to adapt various strains of this bacterium to increased temperature, salinity and mixtures of seawater and Atlantis II Deep brines. None would grow in waters containing 75% brine media or at temperatures greater than 56°C. Only the bacteria isolated from the Atlantis II transition layer could grow well in media containing 100 ‰ salt at 44°C and in half-concentrated brine at 35°C. The temperature tolerance range of the Atlantis II *Desulfovibrio* was 19 to 48°C, with an optimum at about 40°C. Trüper's findings demonstrate that sulphate reducing bacteria may occur in the brine/seawater transition zone. Hydrogen sulphide produced by these bacteria is probably precipitated by the high concentrations of various heavy metals. In this way the bacteria may influence the distribution and precipitation of metals in the different layers of the Atlantis II Deep brine pool and help to reduce their upward diffusion into the surrounding water mass.

Trüper (1969) explained the absence of sulphate reducing bacteria in the main Atlantis II brine pool as the result of several factors. There is a general deficiency in the brines of organic material to meet the rather complex nutritional requirements of the bacteria. Those bacteria which enter the brines do not mutate fast enough to adapt to the hostile conditions they encounter, and there are in any case generally low numbers of bacteria throughout the deep central Red Sea.

During the MESEDA investigations a restricted number of samples was analysed for various organic compounds and bacteria (Karbe and Mycke, 1981; Karbe et al., 1985). Some of the results are summarised in Table 4.3. The data suggest that there is more organic matter in the different brine pools than originally thought. For example there proved to be six types of sugars present in the samples, with glucose and fructose having the highest concentrations. All sugars were more concentrated than in normal Red Sea oceanic water of any depth. It is possible that the sugars accumulate in the brines as a result of metal complexing making them less available for bacterial metabolism, or simply by the pickling effect of the hypersaline waters inhibiting the normal bacterial consumption of these sugars.

Numbers of bacteria proved to be very low, within the extremely oligotrophic deep Red Sea environment. However, investigations on the brine pools of the Suakin, Atlantis II, Kebrit and Shaban Deeps show that all brine pools are inhabited by bacteria enriched in number compared to the ambient waters. This holds specifically for the transition zones between the oxygen containing deep waters and the anaerobic brines. However, no bacteria have been identified, up to now, in samples taken from the lower Atlantis II Deep brines. So far, these are to be considered a toxic environment, even for bacteria adaptable to extreme conditions.

TABLE 4.3. Number of bacteria determined in samples retrieved from deep Red Sea water and different brine pools related to temperature, salinity, oxygen, pH, organic compounds and suspended matter (light attenuation). More detailed data are given in Karbe and Mycke (1981) and Karbe et al. (1985).

	Temperature °C	Salinity g/kg	Oxygen mg/kg	pH	Organic Carbon mg/kg	Sugars mg'kg	Light Attenuation %	Bacteria 10^4/ml
Red Sea deep water	21.8	41.9	3.8 − 4.7	8.1 − 8.2	0.4 − 0.8	0.03 − 0.14	2.6 − 3.7	2.6 − 7.3
Suakin Deep								
Transition zone	22.3 − 23.8	43.1	3.8 − 2.6	8.1	0.6 − 0.8	·	6.0 − 71.6[1]	6.0 − nn
Upper brines	24.1	145	0.3	7.9 − 7.6	0.8	·	6.0	45. − 18.3
Lower brines	24.3	146	0.2	7.6	0.8	·	4.3 − 5.7	6.2 − 17.1
Atlantis II Deep								
Transition zone	21.8 − 38.3	51	1.3 − 0.4	7.4 − 6.5	·	·	3.1 − 46.6[1]	24.4 − 28.1
Upper brines	53,4	119	0.4 − 0.1	6.2 − 6.0	·	·	4.0 − 5.9	14.8 − 3.5
Lower brines	62.8 − 63.1	259	0.1	5.8 − 5.6	2.1 − 3.2	0.24	88.1 − 7.7[1]	none
Kebrit Deep								
Transition zone	21.9 − 22.1	111 − 138	0.0	5.8	·	·	99.3 − 12.5	81.3 − 119.8
Brines	23.3	256	0.0	5.4	·	·	8.8 − 6.9	22.0 − 13.5
Shaban Deep								
Transition zone	22.3 − 23.5	54	3.2	7.6	0.4	·	25.1 − 96.0	42.3
Upper brines	24.1	256	0.0	6.9 − 6.0	0.8	·	5.1 − 5.5	8.7 − 9.9
Lower brines	24.7 − 25.3	258 − 261	0.0	6.2	0.6 − 0.7	·	5.4 − 6.3	2.7

[1]Highest light attenuation at the interface between transition zone and upper brines and between upper and lower brines.

4.5. THE SURROUNDING DEEP SEA ENVIRONMENT

4.5.1. Mechanisms of influence by brine pools

The Red Sea brine pools are at present largely isolated from the normal deep Red Sea water mass by their containment in depressions in the sea floor and by the pronounced vertical stratification in salinity which inhibits mixing. The deep brines could however influence bathypelagic and deep benthic environments by at least three mechanisms.
1. Discharge of hot geochemically anomalous waters from hydrothermal vents could occur at sites where the extruding waters are not conveniently contained in natural basins and depressions.
2. Exchange between the brine pools and the overlaying deep waters can occur by slight but continuous convective mixing, or by short term turbulent intermixing during earthquakes or periods of increased submarine volcanic activity.
3. Exchange processes could occur at the sediment-water interface in areas with geochemically enriched sediment interstitial water but without toxic brine pools. At such sites trace elements could be taken up and accumulated via the food chain throughout the deep benthic and bathypelagic communities.
As it is expected that in some restricted areas the metal content of the Red Sea water may increase as a result of the disposal of mining tailings (section 4.6), it has been essential to establish the natural background metal concentrations in the water and living organisms of the Red Sea. These data serve as an important reference datum for future monitoring of the effects of pilot and commercial mining operations.

4.5.2. Trace elements in Red Sea water

Prior to the MESEDA programme, little was known of the concentrations of metals and other trace elements in Red Sea water except within the various hot brines. Only a few samples from the overlaying

TABLE 4.4. Major and minor inorganic constituents of Red Sea water. Samples taken from different depths between the surface and near-bottom. Analysis by Atomic Absorption Spectrophotometry (AAS). Data from Karbe *et al.* (1981).

Element	Range nmol 1^{-1}	Exceptional	World ocean mean (Bruland, 1983)
Fe	2–57	627	1
Cu	0.9–10	54	4
Zn	5–188	474	6
Ag	0.02–0.4		0.025
Cd	0.04–1.2	2.2	0.7
Hg	0.01–0.11	0.20	0.005

water column had been analysed (Brooks *et al.*, 1969). During the MESEDA II and III cruises water samples were taken from a variety of sites along the Red Sea axial trough and above both eastern and western deep terraces, and the results of trace element determinations are summarised in Table 4.4. All the results refer to unfiltered sea water and so represent the total concentrations of both the dissolved fraction and the fraction leached out from particulate matter during storage. From the data available, it can be concluded that concentrations of most metals and other trace elements are similar to those found in other oceans. Concentrations for some elements are above the world ocean means, but within the range of analytical data published for various marine waters. It must be emphasised that the data given here are summaries of more detailed results discussed elsewhere (Karbe *et al.*, 1981).

The mixing layers above the Atlantis II Deep brines are marked by increased concentrations of iron, zinc, copper, cadmium and other trace elements, the concentrations of which further increase dramatically at the brine surface. It seems obvious that elevated metal concentrations in the deep Red Sea waters overlaying the brines must result from brines and seawater mixing. However, increased concentrations of zinc, copper and cadmium were also found in the deepest samples taken from the water column above the coastal terraces and over the relatively inactive Shagara Deep. It may be that the elevated metal concentrations in waters beyond the direct influence of hydrothermal activity may be due to other causes, principally the sinkage of biogenic detrital material carrying down metal ions from the surface to deep waters in all oceans.

4.5.3. Trace elements in bottom-dwelling organisms

All marine organisms naturally accumulate significant amounts of trace elements within their tissues. These include the most toxic metals and metalloids which were analysed in Red Sea water at concentrations slightly above the limits of detection by chemical analysis. Sampling selected organisms for chemical analysis of contaminants enriched to high concentrations by food-chain accumulation is a common method for monitoring regional and annual trends in environmental pollution.

Of the larger organisms caught during the MESEDA campaigns in the deep central Red Sea, certain species of bottom-dwelling pandalid, penaeid and palaemonid shrimp proved the most suitable for analysis and examination of regional differences in trace element concentration. The shrimp were caught using trawls and baited traps at stations ranging in depth from 350 to 2200 m. Sample stations included areas immediately adjacent to the Atlantis II Deep, while the deepest shrimp were caught in the Shagara Deep.

A detailed account of the results of metal analyses of Red Sea shrimp and other biota is given by Karbe *et al.* (1981), and a summary is given in Table 4.5. Figure 4.7 presents some typical patterns of depthwise distribution of metal concentrations. With some reservations, it is possible to obtain a general trace element composition by amalgamating data for the different species. Generally however, the

TABLE 4.5. Background trace element concentrations in shrimps sampled during the MESEDA I-III cruises in the central Red Sea. Data are overall ranges for pandalid, penaeid and palaemonid shrimp from 353 to 2200 m depth. Concentrations are in mg kg^{-1} dry weight assayed by Atomic Absorption Spectrophotometry (AAS) or Instrumental Neutron Activation Analysis (INAA). Data from Karbe and Schnier (1981).

Element	Method	Concentration range
Ca	INAA	33 − 114 × 10³
Na	"	18 − 40 × 10³
Sr	"	370 −1000
Br	"	240 − 450
As	"	63 − 740
Zn	"	52 − 118
Fe	"	50 −3400
Cu	AAS	29 − 171
Rb	INAA	2.3 − 5.0
Ag	AAS	1.5 − 2.8
Ag	INAA	1.1 − 2.8
Se	"	1.3 − 6.0
Cd	AAS	0.6 − 7.9
Sb	INAA	0.2 − 2.0
Co	"	0.13 − 2.5
Ce	"	0.12 − 3.0
Hg	INAA	0.08 − 5.5
Hg	AAS	0.05 − 7.3
La	INAA	0.05 − 2.0
Cr	"	0.03 −17.9
Lu	"	0.02 − 0.1
Sc	"	7.0 −920 × 10⁻³
Hf	"	4.7 −360 × 10⁻³
Th	"	4.0 −430 × 10⁻³
Yb	"	2.7 −300 × 10⁻³
Ta	"	1.5 − 58 × 10⁻³
Tb	"	0.5 − 61 × 10⁻³
Eu	"	0.4 − 53 × 10⁻³
Au	"	0.2 −23 × 10⁻³

penaeids were more highly contaminated than the other taxa. There are no data for the same species outside the Red Sea, but comparison with data for other small decapods suggest that trace metal concentrations in mid water Red Sea shrimp are approximately the same as in the ocean environments. Little is known of trace metal concentrations in the abyssal fauna of the world's oceans, except that the metal burden of deep sea organisms is generally high. The data collected from sampling deep water Red Sea shrimp fit in with the general scheme of increasing metal concentration with depth. Five categories of trace elements can be recognised from the Red Sea data each with a different depth response.

1. Elements increasing in concentration with depth, reaching highest levels in the palaemonid shrimp near Atlantis II and in Shagara Deep. These include arsenic, cadmium, copper, mercury, silver and zinc.

2. Elements with a less marked tendency to increase with depth, these include selenium and possibly gold.

3. Elements with concentration maxima at intermediate depths (1000−1400 m) and reduced concentrations in the deepest localities. These include chromium, cobalt, europium, hafnium, iron, nickel, thorium and others.

4. Elements which show no relationship between concentration and depth include sodium, bromine and rubidium.

5. Calcium and strontium showed a tendency to decrease in concentration with increasing depth.

Relatively high levels of arsenic, cadmium, copper, mercury, silver and zinc were found in both deep water shrimp and the deep demersal fish of the central trough. It can be assumed that these high levels

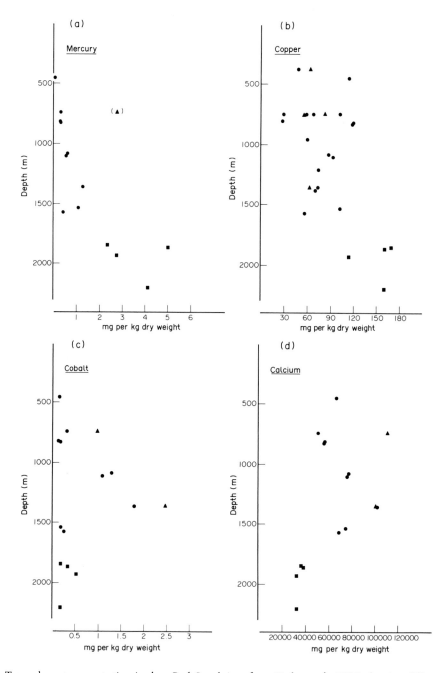

Fig. 4.7. Trace element concentration in deep Red Sea shrimp from Karbe *et al.* (1981) showing different patterns of concentration change with depth. A — mercury and B — copper show an increase in concentration with depth. C — cobalt has highest concentrations at intermediate depths. D — calcium shows a tendency for concentrations to decline with depth. [●] — pandalid shrimps, [▲] — penaeid shrimps, [■] — palaemonid shrimps. Copper was assayed by Atomic Absorption Spectrophotometry, others by Instrumental Neutron Activation Analysis.

result from geothermal activity, which occurs to some extent throughout the length of the central trough. Bioaccumulation from these sources may therefore be characteristic of the whole basin. Trace elements may also accumulate in deep water by the continuous settling of detritus from the overlying euphotic zone.

In order to provide an adequate baseline for future monitoring operations during economic mining, it was necessary to analyse the metal body-burdens of organisms from sites well removed from the projected mining area as well as from the vicinity of the Deeps. An extended survey was conducted during the MESEDA programme, including analyses from the coastal area, and of the concentration in food fish and the rates of fish consumption by local communities. While the results are outside the scope of this publication the interested reader should consult Karbe et al. (1981).

4.6. PLANNED MINING OPERATIONS

4.6.1. Mining methods

Of the various metalliferous deposits located within the axial trough of the Red Sea, only those of the Atlantis II Deep are considered sufficiently concentrated and of large enough volume to be commercially valuable. A successful mining test was conducted in 1979 using a scaled-down version of the commercial system. The techniques to be used are described by Mustafa and Amann (1980) and Nawab and Lück (1979), and a schematic drawing of the system is provided as Figure 4.8. The commercial system will resemble the pre-pilot mining system but scaled up in the order of 100:1. The mining vessel, equipped with computerised position-fixing and manoeuvring equipment supports a long pipe and power string attached to a mid-water pressure pump. Sediments are loosened by high pressure water jets and by active vibration, and converted to a slurry which is sucked into the collecting head and pumped to the surface. The lower pipe sections and the suction head must be equipped with sophisticated instrumentation to allow precise positioning over and within the relatively thin band of deposits.

Figure 4.9 shows in block form the projected quantities of material to be treated during commercial mining. About 100,000 metric tonnes of metalliferous muds would be extracted each day of operation and made fluid by mixing with water from the brine pool and overlaying water mass. About 200,000 tonnes daily would be pumped on board the mining vessel. Transportation of these huge masses of raw material to a processing plant on land would be completely uneconomic, and enormous problems would arise in disposal of tailing (residue after metal extraction) on land. To overcome this problem, a technique has been developed to separate the economically important sulphide fraction from the rest of the sediment on board the mining vessel, using a flotation method after further dilution with surface seawater. With this technique, a concentrate representing about 1% of the mined mud will be transported to the smelter on shore, and the remaining 99% of the material, diluted by seawater and brines and with some processing chemicals added, will be returned to the Red Sea directly from the mining ship. This tailings disposal is considered the major environmental risk of the mining operation, and summaries of the risk evaluation are given by Mustafa and Amann (1978), Karbe, Thiel, Weikert and Mill (1981) and Abu Gideiri (1984).

4.6.2. The problem of tailings disposal

The tailings will be emitted in the form of a slurry with a solid content of 18−30 g per litre, composed of waste metalliferous sediments, brines and deep and surface seawater. Some characteristics of the tailings are given in Table 4.6. The solid constituents are mainly composed of iron hydroxides,

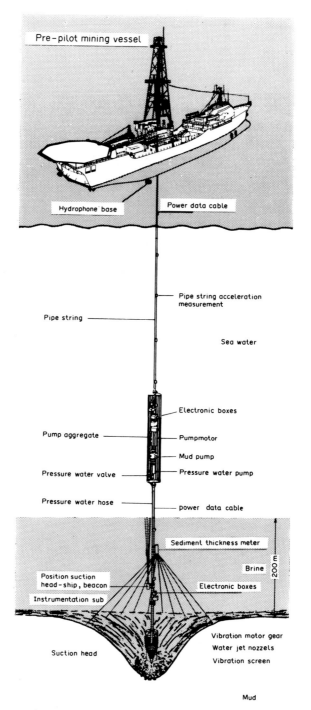

Fig. 4.8. General view of proposed mining system, as used successfully on a reduced scale for pre-pilot mining test in 1979. Figure reproduced from Nawab and Luck (1979) with permission.

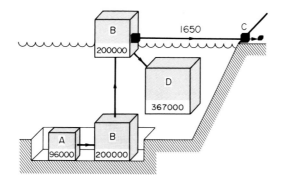

Fig. 4.9. Quantities of muds to be processed daily under commercial operation. A — Muds mined, B — diluted muds pumped aboard ship, C — concentrates carried on shore, and D — tailings disposed of at sea. All quantities are metric tonnes per day. Figures from Mustafa and Amann (1978) and Lange *et al.* (1980).

TABLE 4.6. Constituents of Atlantis II Deep metalliferous sediments, data combined from Bignell (1975) and Bäcker (1976). Also shown are predicted constituents of tailings to be released during future mining, the ranges approximated from analyses of experimentally produced material. Data from Karbe and Nasr (1981). Concentrations refer to washed, salt free, dry components.

		Metalliferous Sediments			
	Sulphidic facies	Silicate facies	Oxidic facies	Tailings	
Fe	157—257	36—287	125—631	250—300	g/kg
Mn	1.6—28.1	0.4—12	0.2—429	4—60	
Zn	53—220	0.04—25.2	0.8—12.9	3—18	
Cu	8.3—36	0.2—5.1	0.02—7.4	2—5	
Pb	1090—1850	70—440	20—280	30—70	mg/kg
Co	125—185	20—25	0—60	40—150	
Ni	60—130	0—7	9—50	10—40	
Ag	75—120	4—68	—	10—50	
Cd	200—1590	8—84	—	5—70	
Hg	1.1—3.1	0.2—2.7	0.05—0.35	0.05—(20) [1]	

[1] Exceptional

oxides and silicates which constitute about 70—80% of the material. From the environmental point of view, some of the minor or trace components of the tailings are of greater significance, these include zinc, copper, lead, arsenic, nickel, antimony, silver, cadmium and mercury. Although the concentrations of these elements in the tailings are small, the huge amount of tailings to be disposed of over the ten to fifteen year life of the project suggests a substantial net addition to the existing metal concentrations present in the Red Sea. Toxicological experiments conducted with various planktonic and benthic animals from the Red Sea have shown that metal ions dissolving in sea water from the tailings are toxic to most test organisms at concentrations corresponding to 1 to 10 mg solids per litre of sea water. This concentration is roughly 0.02% of the average level of solids in the tailings.

Because of the potential toxicity problem, it is generally agreed that a surface disposal of tailings would be unacceptable (Abu Gidieri, 1984). Introduction of the fine-grained tailing particles into the surface water would result in the build-up of a surface plume of suspended material and dissolved contaminants. As well as the toxicity problems mentioned above, the surface plume would cut light penetration and reduce primary production, and would clog the gills and feeding filters of fish and zooplankton. Other problems would arise from the incorporation of nutrient phosphates, nitrates and silicates into the euphotic zone as well as changes of local water temperature, salinity and oxygen content. The prevailing surface currents would tend to carry the plume towards the Saudi Arabian coast, where it would affect the ecologically highly sensitive reef areas. It is obvious that all possible provisions have to be taken to avoid the appalling environmental disaster this would represent.

Two alternative approaches are generally available for waste disposal at sea. One policy is to disperse and dilute the contaminants rapidly to concentrations below any risk of detrimental effects, another approach is to contain the waste within a restricted spoil area. In the case of the Atlantis II Deep project, a simple comparison of the volume of tailings and the volume of the Red Sea itself demonstrates that a dilution policy could result in unacceptable risks to the Red Sea as a whole. It is obvious that confinement strategies are the only feasible option.

It is proposed that tailings should be released well below the surface, and that the axial trough of the Red Sea can be their final resting place. Studies on the plankton stocks and their vertical migration by Weikert (1981) described in Chapter 5 of this book show that at depths below 1000 m plankton densities are very low, suggesting that tailings discharge below 1 km depth may be appropriate. Thiel (1981) has assessed the risks arising for the deep benthos of the axial trough based on hypothetical tailings distribution and sedimentation characteristics. Different numerical models of tailings distribution can be used based on a momentum jet (Mill, 1981) or a gravity flow (Jancke, 1981). Some practical observations on the spreading and sedimentation of tailings were also made during the pre-pilot mining test in 1979.

According to Mill (1981) his modelling studies can be taken as predicting the 'worst case' conditions. Assuming a discharge at 1 km, tailings would be dispersed over an area of about 1500 km^2 of the seabed. Within much of this area, intense sedimentation of tailings (possibly more than 4 cm per year) would blanket the benthic fauna. This will probably eliminate the benthic fauna from the areas of highest sedimentation. Towards the edge of the spoil area, sediment cover may permit some benthic animals to survive. Excluding toxicological effects of metals released from the tailings and carried by deep currents over extended areas, the physical impact of tailings would probably be confined to only 2—4% of the seabed habitat in the central trough below 1500 m. The area to be impacted contains no endemic species, so the remaining 96% of the trough organisms outside the spoil area should represent an adequate stock for later recolonisation of the mining area.

If the discharge depth were reduced to only 800 m, Mill's (1981) model predicts that up to 8—16% of the axial trough would be severely impacted. This clearly demonstrates the importance of an adequately deep discharge point. The deep release point also reduces the passage of potentially toxic tailings through the food webs of the sea, confining them to the impoverished deep benthic and bathypelagic realms.

All the environmental considerations outlined above have been detailed in a report to the Saudi Sudanese Red Sea Joint Commission edited by Karbe, Thiel, Weikert and Mill (1981), which will be separately published in the near future. It can be concluded that the mining of metalliferous muds from the Atlantis II Deep, and their disposal back into a confined area in the axial trough present environmental risks that appear to be acceptable. However, much of our understanding of the potential implications of mining and tailings disposal remains only qualitative and restricted to short term effects, further studies are to be conducted. Monitoring of both pilot-scale and economic mining operations will be an essential precaution to avoid unacceptable long-term effects which could not be predicted from the studies concluded during the pre-pilot phase of the MESEDA programme.

REFERENCES

Abu Gideiri, Y. B. (1984) Implications of sea mining for the Red Sea environment. *Hydrobiologia* 110, 15—19.
Bäcker, H. (1973) Rezente hydrothermal-sedimentäre Lagerstättenbildung. *Erzmetall* 26, 544—55.
Bäcker, H. (1975) Exploration of the Red Sea and Gulf of Aden during the M.S. VALDIVIA Cruises "Erzschlämme A" and "Erzschlämme B". *Geol. Jb. D* 13, 3—78.

Bäcker, H. (1976) Fazies und chemisches Zusammensetzung rezenter Ausfällungen aus Mineralquellen im Roten Meer. *Geol. Jb* D17, 151–72.

Bäcker, H. and Richter, H. (1973) Die rezente hydrothermal-sedimentäre Lagerstätte Atlantis II Tief im Roten Meer. *Geol. Rundschau* 62 3, 697–741.

Bäcker, H. and Schoell, M. (1972) New deeps with brines and metalliferous sediments in the Red Sea. *Nature Physical Science*, Vol. 240, No. 103, 153–8.

Bignell, R. D. (1975) *Geochemistry of Metalliferous Brine Sediments and Other Sediments in the Red Sea.* Ph.D. Thesis, Imperial College, London.

Brewer, P. G. and Spencer, D. W. (1969) A note on the composition of the Red Sea brines. In *Hot Brines and Recent Heavy Metal Deposits in the Red Sea*, ed. by E. T. Degens and D. A. Ross, pp. 174–9. Springer Verlag, Berlin.

Brooks, R. R., Kaplan, I. R. and Paterson, M. N. A. (1969) Trace element composition of Red Sea geothermal brine and interstitial water. In *Hot Brines and Recent Heavy Metal Deposits in the Red Sea*, ed. by E. T. Degens and D. A. Ross, pp. 180–203. Springer Verlag, Berlin.

Bruland, K. W. (1983) Trace elements in seawater. In *Chemical Oceanography* ed. by J. P. Riley and R. Chester, Vol.8, pp. 157–220. Academic Press, London, New York.

Danielsson, L-G., Dryssen, D. and Graneli, A. (1980) Chemical investigations of Atlantis II and Discovery brines in the Red Sea. *Geochem. Cosmoch. Acta* 44, 2041–65.

Degens, E. T. and Ross, D. A. (1969) *Hot Brines and Recent Heavy Metal Deposits in the Red Sea.* Springer Verlag, Berlin.

Hartmann, M. (1973) Untersuchung von suspendiertem Material in den Hydrothermallaugen des Atlantis-II-Tiefs. *Geol. Rundschau* 62, 742–54.

Hartmann, M. (1980) Atlantis II Deep geothermal brine system. Hydrographic situation in 1977 and changes since 1965. *Deep-Sea Research* 27A, 161–71.

Hartmann, M. (1985) Atlantis II deep geochemical brine system. Chemical processes between hydrothermal brines and Red Sea deep water. *Marine Geology* 64, 157–177.

Jancke, K. (1981) Computer simulation of a gravity flow. In *Mining of Metalliferous Sediments from the Atlantis II Deep, Red Sea*, ed. by L. Karbe, H. Thiel, H. Weikert and A. J. B. Mill, pp. 276–9. Hamburg.

Karbe, L., Baufeldt, A. and Meyer-Jenin, M. (1985) Bacterial colonization of Red Sea hot brines. 4th Deep-Sea Biology Symposium. 23–29 June 1985, in Hamburg.

Karbe, L., Moammar, M. O., Nasr, D. and Schnier, Ch. (1981) Heavy meatals in the Red Sea environment. In *Mining of Metalliferous Sediments from the Atlantis II Deep, Red Sea*, ed. by L. Karbe, H. Thiel, H. Weikert and A. J. B. Mill, pp. 195–225. Hamburg.

Karbe, L. and Mycke, B. (1981) Organic compounds and bacteria. In *Mining of Metalliferous Sediments from the Atlantis II Deep, Red Sea*, ed. by L. Karbe, H. Thiel, H. Weikert and A. J. B. Mill, pp. 93–7. Hamburg.

Karbe, L. and Nasr, D. (1981) Chemical and toxicological characteristics of the tailings material. In *Mining of Metalliferous Sediments from the Atlantis II Deep, Red Sea*, ed. by L. Karbe, H. Thiel, H. Weikert and A. J. B. Mill, pp. 245–53. Hamburg.

Karbe, L. and Schnier, Ch. (1981) Trace elements in shrimp. *Preussag Technical Report No. 30 C presented to the Saudi Sudanese Red Sea Joint Commission for the Exploitation of the Red Sea Resources.* Hamburg.

Karbe, L., Thiel, H., Weikert, H. and Mill, A. J. B. (Eds) (1981) *Mining of Metalliferous Sediments from the Atlantis II Deep, Red Sea: Pre-Mining Environmental Conditions and Evaluations of the Risk to the Environment. Environmental Impact Study presented to the Saudi Sudanese Red Sea Joint Commission.* Hamburg.

Mill, A. J. B. (1981) Hydraulics of tailings discharge. Computer simulation of discharge from production scale point. In *Mining of Metalliferous Sediments from the Atlantis II Deep, Red Sea*, ed. by L. Karbe, H. Thiel, H. Weikert and A. J. B. Mill, pp. 264–75. Hamburg.

Mustafa, Z. and Amann, M. (1978) Ocean mining and protection of the marine environment of the Red Sea. *OTC 3188 Offshore Technology Conference*, 1199–1266.

Mustafa, Z. and Amann, M. (1980) Red Sea Pre-Pilot Mining Test 1979. *OTC 3874 Offshore Technology Conference*, 197–210.

Nawab, Z. and Lück, K. (1979) Test mining of metalliferous mud from the Red Sea bottom. *Meerestechnik* No.6, 181–216.

Ross, D. A., Whitmarsh, R. B., Ali, S. A., Boudreaux, J. E., Coleman, R., Fleisher, R. L., Girdler, R., Manheim, F., Matter, A., Nigrini, C., Stoffers, P. and Supko, P. R. (1973) Red Sea drillings. *Science* 179, 377–80.

Scheuch, J. (1974) Zur Goethermik des Roten Meeres und seiner Soletiefs. Marine Wärmestromdichtemessungen und ihre geologisch-geophysikalische Deutung. *Dissertation Freie Universitat Berlin* (Cited according to Schoell, 1975).

Schoell, M. (1975) Heating and convection within the Atlantis II Deep geothermal system of the Red Sea. *Proceedings Second United Nationals Symposium on the Development and Use of Geothermal Energy*, 583–189.

Thiel, H. (1981) Effects on the Benthos. In *Mining of Metalliferous Sediments from the Atlantis II Deep, Red Sea*, ed. by L. Karbe, H. Thiel, H. Weikert and A. J. B. Mill, pp. 313–25. Hamburg.

Trüper, H. G. (1969) Bacterial sulfate reduction in the Red Sea hot brines. In *Hot Brines and Recent Heavy Metal Deposits in the Red Sea*, ed. by E. T. Degens and D. A. Ross, pp. 263–71. Springer Verlag, Berlin.

Watson, S. W. and Waterbury, J. B. (1969) The sterile hot brines of the Red Sea. In *Hot Brines and Recent Heavy Metal Deposits in the Red Sea*, ed. by E. T. Degens and D. A. Ross, pp. 272–81. Springer Verlag, Berlin.

Weikert, H. (1981) Effects on the pelagic organisms. In *Mining of Metalliferous Sediments from the Atlantis II Deep, Red Sea*, ed. by L. Karbe, H. Thiel, H. Weikert and A. J. B. Mill, pp. 298–312. Hamburg.

ZoBell, C. E. (1957) Ecology of sulfate reducing bacteria. In *Sulfate Reducing Bacteria, their Relation to Secondary Recovery of Oil.* Science Symposium, St. Bonaventure University, New York.

CHAPTER 5

Plankton and the Pelagic Environment

HORST WEIKERT

University of Hamburg, Institut für Hydrobiologie und Fischereiwissenschaft, Hydrobiologische Abteilung, Zeiseweg 9, D-2000, Hamburg, Federal Republic of Germany

CONTENTS

5.1. INTRODUCTION

Planktonic organisms live within the water column of the sea, the pelagic zone, and their movements (largely horizontal) are principally dictated by those of the water in which they are suspended. In this way they contrast with the *nekton* — organisms which swim actively such as squids and fish. Planktonic animals may however undergo extensive daily vertical migrations, rising at night to feed near the surface, and moving to deep water during the day.

The term 'plankton' covers a heterogeneous assemblage of organisms, which may usefully be divided into *phytoplankton* (plants) and *zooplankton* (animals). Phytoplankton are the most important primary producers of the sea; they are usually single-celled and very small. Their productivity, like that of terrestrial plants, depends on an adequate supply of light and nutrients. In the tropical setting of the Red

Sea, light is rarely a limiting factor in shallow water. Light does not penetrate water very well, and only the top 100 m or so receive enough sunlight for the photosynthetic energy fixation by plants to exceed their daily energy requirements. The depth at which photosynthesis and energy loss by respiration of a cell balance, is called the compensation depth, and the layer from this depth to the surface, where the net gain of energy allows a cell to synthesize new organic material in surplus, is called the euphotic zone. In the open ocean, photosynthetic primary production is confined to the shallow euphotic zone, and it is this production which is the energy source for zooplankton. Zooplankton are in turn eaten by animals of the nekton. Via a series of overlapping feeding 'ladders' of vertically migrating zooplankton and nekton and by the linkage of falling organic debris and carcasses, energy fixed by photosynthesis is transported from the synthesizing surface layer to the consuming pelagic deep sea and bottom habitats.

In tropical seas, nutrients are often the principal limit to primary production, because, as we will see is the case in much of the Red Sea, tropical waters may be poorly mixed, and the nutrients within material falling out of the euphotic zone cannot be returned to the deficient upper waters. As on land, nitrates and phosphates are among the most important inorganic nutrients, but many other chemicals are needed by some algae, thus silicate is an important requirement for diatoms.

Zooplankton are generally larger than the phytoplankton on which they feed, and some of the largest are carnivores, eating other zooplankton. Some are permanent plankton residents (the *holoplankton*), these include crustacean copepods, predatory chaetognaths or arrow worms and tiny swimming gastropods such as pteropods or sea butterflies, among many other types. Near the coast, the number of species in the zooplankton is greatly enhanced by the larvae of organisms which as adults live intertidally or on the sea bottom. These temporary inhabitants are called *meroplankton*.

A further term has been coined to describe the total particulate content of sea water. The *seston* comprises the plankton including bacteria, organic material (dead organisms and their debris) and inorganic material (silt and dust) within the water. The aggregate term is useful since it is very difficult to separate the three categories of particles when measuring the total particle concentration in natural waters.

Plankton have been sampled in the Red Sea since the last century, by scientists working from oceanographic vessels, and more recently from shore-based laboratories. The chief source of systematic and biogeographic information on Red Sea plankton remains Halim's (1969) detailed review. Since this was published, quantitative planktological research has been enhanced by Russian biologists working along the whole length of the basin, Israelis working in the Gulf of Aqaba, and by a team of Arabian and German scientists who participated in studies of the central Red Sea ecosystem as part of the evaluation of the potential impact of mining metalliferous muds from the deep sea in the Atlantis II Deep (Thiel *et al.*, 1986). Results from these recent studies are described below.

5.2. THE PELAGIC ENVIRONMENT OF THE RED SEA

This section and the next concentrate on the environment and plankton biota of the oceanic parts of the Red Sea. It should be noted at the outset that the coastal or reef environment presents a quite different picture in terms of food webs, plankton diversity and productivity. This special plankton community is described in section 5.4.

5.2.1. Red Sea hydrography and barriers to immigration

From a planktologist's viewpoint, the Red Sea proper, including its two northern branches, the Gulfs of Suez and Aqaba, is an extreme environment. For information on the general physical oceanography of the Red Sea, the reader is referred to Chapter 3.

The unique conditions in the Red Sea become strikingly evident when its deep water habitat, that has some similarities to Cretaceous oceans (Shackleton, 1982), is compared with that in the adjacent Gulf of Aden, in which conditions considered 'normal' for the recent oceans prevail. The Red Sea basin is cut off from the circulation of cold water from the deep layers of the Indian Ocean by the shallow Hanish Sill, north of Bab-el-Mandab. At 150 m depth, only about 10 m below the Sill, the water of the Red Sea is some 7°C warmer and 4‰ more saline than that of the Gulf of Aden. The differences increase with depth, reaching 19°C and 6‰ at about 2000 m. Migrating midwater or deep-sea organisms from the Indian Ocean would be exposed to considerable stress if they entered the 'heated', highly saline environment of the Red Sea across the sill barrier. These organisms are primarily adapted to low temperatures of 10°C or below and can tolerate only narrow amplitudes of variation in temperature and salinity.

For plankton living in the deeper waters of the Indian Ocean, where oxygen levels average about 2.0 ml O_2/l (28% saturation) at 2000 m, the physiological barrier to immigration may be increased by an 'oxygen minimum layer' at 300 to 400 m in the southern Red Sea. Minimum oxygen concentrations of 0.3–0.5 ml O_2/l, corresponding to 6–10% saturation, occur in this layer, the formation of which is described below.

5.2.2. Nutrient and oxygen levels in the Red Sea basin

The formation of the oxygen minimum layer at intermediate depths in the Red Sea results from oxygen consumption by bacteria during the decomposition of particulate organic matter (POM) that sinks from the surface layer. By the remineralization process, nutrient salts are set free in deep water, and the ratio of POM to total inorganic matter in the seston becomes lower as the oxygen concentration decreases. As exemplified in Figure 5.1, the extremes of the concentration curves occur at similar depths except that for silicate, which is released from organic debris less rapidly than nitrate or phosphate, and peaks at greater depths. The concentrations of nutrients remain more or less constant at a relatively high level in deeper water, but the oxygen content increases to 2.0 ml/l (Grasshoff, 1969). The peaks in the curve of POM at about 100 m and below 400 m are explained in section 5.3.3.

The Gulf of Aden water flowing into the Red Sea is enriched with nutrients and POM. The most intense inflow of nutrients is in summer, from July to September, through the Aden subsurface current, by which salts are imported into the lower part of the euphotic zone of the Red Sea (Khimitsa and Bibik, 1979).

In winter, the now prevailing Aden surface current is poor in nutrients (Poisson *et al.*, 1984), but rich in seston. The remineralization of the large quantity of carcasses of imported zooplankton suggests a possible source of plant nutrients in the surface water (Beckmann, 1984) that may add to the supply of nutrients from the subsurface water induced through local upwelling and vertical mixing by strong tidal currents south of 17°N (Khmeleva, 1970).

As a general feature, nutrient concentrations throughout the water column in the southern Red Sea are therefore higher than those in the central and northern regions (Table 5.1). An exception to this are the high nitrate concentrations of some northern waters observed in recent cruises.

On the way through the bottle neck of Bab al Mandab, sediment from the narrow shores and the shallow bottom enters the Gulf water (Shimkus and Trimonis, 1983), thus increasing its high seston load. The introduced particles cause increased turbidity in the southern Red Sea waters and, consequently, the well-illuminated (euphotic) surface layer is only 30 to 60 m, or even less, deep (Khmeleva, 1970; Petzold, 1986).

In the central and northern parts of the Red Sea, the Gulf of Aden water no longer exerts permanent influence. Except in winter, the barrier formed by the thermocline and halocline dominates the water

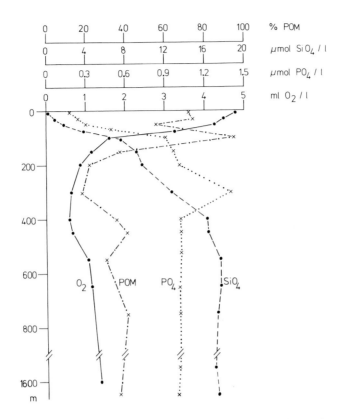

Fig. 5.1. Profiles of dissolved oxygen (O₂), inorganic phosphate (PO₄), silicate (SiO₄) and particulate organic matter (POM) at an oceanic station at the latitude of Gebel Tair, southern Red Sea (Original).

TABLE 5.1. Nutrient salts in oceanic Red Sea water from different regions. Ranges of concentration reflect seasonal fluctuations (modified after Karbe and Lange, 1981). The depth of the surface layer (SL) corresponds to that of the euphotic zone. OM—depth range of minimum oxygen concentrations, DW—deep water.

Nutrients (μmol/l)	Northern Red Sea			Central Red Sea			Southern Red Sea		
	SL ≤100m	OM 400–500m	DW >1000m	SL ≤100m	OM 350–450m	DW >1000m	SL ≤70m	OM 300–400m	DW >1000m
Phosphate	0.01–0.05	0.6– 0.8	0.2– 0.5	0.05–0.1	0.8– 1.1	0.6– 0.7	0.1 –0.3	0.9– 1.2	0.6– 0.8
Nitrate	0.1 –0.4	15.5–18.0	8.0–11.9	0.03–0.2	18.1–20.1	11.9–15.0	0.03–0.2	18.9–22.1	14.0–18.2
Silicate	0.2 –0.4	9.0–11.3	5.2– 8.0	0.3 –0.6	10.3–13.6	8.0–10.2	0.4 –0.7	13.6–16.5	10.2–15.5

body in the Red Sea. This is a layer of abrupt temperature and salinity changes separating warm shallow water of low salinity from cooler deep water of high salinity. Nutrient salts incorporated in organic matter by photosynthesis can be carried below this discontinuity by sinkage and the vertical migration of plankton and nekton. However, the great stability of the discontinuity prevents vertical circulation of water carrying the nutrients back to the euphotic zone, so that nutrient levels and productivity are greatly reduced. This diminishes the seston concentration, so light is scattered less, and the euphotic zone expands to about 100 m in the central and northern Red Sea most of the year.

The overall nutrient decrease in the water column from south to north is inversely related to changes in oxygen content. The surface water is almost oxygen-saturated and the absolute concentration values reflect increasing solubility in the cooler waters to the north (Morcos, 1970). However, since temperature and salinity remain unusually high, oxygen concentrations are low in the Red Sea (4.2–4.8 ml O_2/l) compared with boreal or polar seas (about 8.0 ml O_2/l). In winter, the cooled surface waters in the northern Red Sea sink below 200 m, thereby forming a sub-surface water mass with a relatively large oxygen content. In the north, minimum oxygen concentrations normally do not fall below 1.0 ml O_2/l at medium depths, and increase to nearly 3.0 ml O_2/l in deep water.

The decrease of oxygen concentration in intermediate and deep waters to the south is generally interpreted as reflecting increased biological and chemical consumption rather than diminishing supply. This explanation is also believed to account for an upward shift of about 100 m in the oxygen minimum layer in the southern part of the Red Sea compared with the less biologically rich north (Table 5.1).

5.2.3. Nutrient levels in the northern Gulfs

Compared with the Indian Ocean, nutrient concentrations in the sub-surface waters of the Red Sea proper are low, but they are higher than those in the Gulf of Aqaba (Table 5.2). The Sill of Tiran, 252 m deep, is situated at the entrance of the Gulf and blocks the continuous inflow of deep water from the northern Red Sea (Klinker et al., 1978).

There is, however, an inflow of upper Red Sea water, which distributes nutrients into the approximately 200 m deep surface layer of the Gulf, with decreasing concentrations at the further end. This water from the northernmost Red Sea, which also feeds the surface inflow into the Gulf of Suez, is thus enriched with nutrients from the surrounding shallows which increases its productivity (Table 5.3). This would explain why the nutrient concentrations in the northern Red Sea, quoted by Klinker et al. (1978, Table 5.2), are higher by an order of magnitude as compared with those shown in Table 5.1 which were measured from pure oceanic (and less productive) waters south of 27°N.

In the Gulf of Aqaba in the winter, there is an extensive and large-scale sinking of cooled, nutrient-poor surface water which is saturated with more than 6.0 ml O_2/l. It mixes with the underlying nutrient-rich water to a depth of at least 200 m. Consequently, nutrients are distributed to the surface, and subsurface waters of the Gulf of Aqaba are well oxygenated, with no oxygen-minimum layer, and not less than 3.75 ml O_2/l.

In the shallow Gulf of Suez, there is an almost uniform vertical distribution of oxygen and nutrients due to wind-induced mixing progressing toward the north (Morcos, 1970). Turbidity is increased and the euphotic zone is therefore much shallower than in the clear waters of the Gulf of Aqaba, where the zone extends to below 80 m in summer (Reiss and Hottinger, 1984) and the formation of a thermocline indicates stability in the upper water column (Levanon-Spanier et al., 1979).

Wind seems to be one of the important factors that generates local anomalies in the overall distribution patterns of nutrients and seston. In summer, strong northerly winds are believed to induce

TABLE 5.2. Nutrient salts in the northern Red Sea and different parts of the Gulf of Aqaba (modified after Klinker et al., 1978). Ranges of concentrations reflect seasonal fluctuations. SW—surface water upper 200 m, SSW—subsurface water.

Nutrients (μmol/l)	Northern Red Sea SW	SSW	Southern Gulf SW	SSW	Central Gulf SW	SSW	Northern Gulf SW	SSW
Phosphate	>0.25	<0.5 – >1.0	0.10–0.25	0.15 – 0.30	<0.1 – 0.15	0.10–0.25	<0.10 – 0.15	0.15 – 0.25
Nitrate	<2.0	>5.0 – >10.0	<0.5 – 2.0	2.00 – >5.0	<0.50 – 1.0	>2.0	<0.5 – 1.0	0.5 – 2.0
Silicate	2.0	5.0 – >10.0	<1.0 – 2.0	1.0 – 2.0	<1.0	<1.0 – 2.00	<0.5 – 1.0	0.05 – >2.0

coastal upwelling events in the northernmost parts of the Gulf of Aqaba and the Red Sea off Sinai and the northern Arabian coast (Morcos, 1970; Klinker *et al.*, 1976). At such times, nutrient-rich water from intermediate depths ascends to the surface and penetrates from the northern Red Sea into the southern Gulf of Aqaba. In the extreme northern part of the Gulf of Aqaba, windblown phosphate fertiliser from terrestrial deposits nearby causes permanent enrichment (eutrophication) of the coastal waters (Fishelson, 1973).

5.2.4. Conclusions — environmental gradients in the Red Sea

The water masses of the Red Sea and its two northern branches are impoverished of nutrients as compared with the adjacent Indian Ocean. The concentrations decrease, along the axes of the basins, from their entrances toward the north in oceanic waters, except in the Gulf of Suez, where an irregular distribution has been reported (El-Sabh and Beltagy, 1983). A spectacular drop in the nutrient concentrations in the Red Sea surface water occurs north of 19°N (Khmeleva, 1970). This approximates to the northern extension of the Gulf of Aden inflow. However, in comparison to the overall changes of temperature and salinity in the surface layer, the changes in nutrient concentrations are less pronounced, since the nutrients undergo rapid biological consumption. Oxygen increases throughout the water column towards the north. This is due to the increasing solubility in the cooler waters, and to water circulation (Chapter 3). Cross-sectional gradients in chemical and physical parameters are also present, but they are less pronounced (Morcos, 1970).

The distinctness of gradients in the water column differs from basin to basin. In the shallow, fairly well-mixed Gulf of Suez, gradients are weak, especially in its northern half, so the water body is almost homogeneous. Its division into a surface layer and a sub-surface layer seems justified only on an ecological basis, at the depth of 1% surface irradiation. This depth is usually taken to be the compensation depth and demarks the lower limit of the euphotic zone.

In the deep Gulf of Aqaba and the Red Sea proper, a zone of discontinuities near the sea surface exhibits rapid changes in salinity and temperature, separating a shallow surface layer from a very deep body of water with almost uniform temperature and salinity. This physical zonation is stable in the Red Sea basin, except in its northern part. Here, as in the Gulf of Aqaba, the discontinuity layer is destroyed by the cooling and descent of the surface water in winter. Unlike in the Red Sea, this process initiates overall vertical mixing of water masses in the Gulf by which nutrients and oxygen are almost evenly distributed throughout the water column. The possibly transient import of nutrients through the surface inflow of Red Sea water is of minor importance, consequently the decrease of nutrients in the surface layer to the north is less distinct than in the Red Sea proper. There, the horizontal movement of water from the fertile Gulf of Aden is the key mechanism by which the euphotic zone is replenished with plant nutrients and biogenic particles. However, this supply is not so large as to extinguish the nutrient gradients between the surface and subsurface layers. As in the Gulf of Aqaba, coastal upwelling is of only local importance, except in the southern Red Sea. Mechanical accumulation of seston takes place where surface currents converge (*see* Chapter 3).

5.3 OCEAN PLANKTON

5.3.1 Primary production and plankton abundance

There is strong evidence that the production potential of the Red Sea is low. We have seen that over most of the basin, the development of a thermocline and halocline prevents the recycling of nutrients from deeper waters to the euphotic zone. There is little nutrient input to the pelagic system from land

TABLE 5.3. Daily primary production, phytoplankton and zooplankton biomass and abundance in different parts of the Red Sea at different seasons of the year. Zooplankton biomass as wet weight. In parentheses: Values from eutrophicated northernmost waters. *0–1000 m layer

Season and depth	Northern to 24°N	Central to 18°N(16°N)	Southern	Unit	Size spectrum collected (μm)	Source
Phytoplankton						
Summer						
0–200 m	180	300	3,000	cells/m³	⩾180	Sukhanova, 1969
Euphotic zone	0.08	0.10		gC/m²	Total spectrum	Weikert, 1980a
Autumn						
0–100 m	14, 572	58, 533	21, 217	cells/m³	Total spectrum	Belogorskaya, 1970
Euphotic zone	0.21	0.39	1.60	gC/m²	Total spectrum	modified after Khmeleva, 1970
Winter						
0–160 m	24.3 (73.3)	38.8	59.2	mgchl *a*/m²	Total spectrum	
Euphotic zone	0.08 (0.62)	0.13	0.75	gC/m²	Total spectrum	Petzold, 1986
Zooplankton						
Summer						
0–100 m	42	113		mg/m³	⩾ 65	Gordeyeva, 1970
	43	59			⩾330	
	<1	<3	5	ml/10³	⩾180	Ponomareva, 1968
	43	49		mg/m³	⩾300	Weikert, 1980 a
Winter						
0–100 m	48	81	105	mg/m³	⩾330	Delalo, 1966
0–800 m	11.0*	16.0	23.0	g/m²	⩾330	
0–100 m	1520	1750	4750	No/m³	⩾180	Kornilova and Fedorina, 1970
Spring						
0–100 m		45	91	mg/m³	⩾300	Beckmann, 1984

surface-runoff to compensate for the steady loss of nutrients by sinkage out of the productive zone. On this basis, productivity should be low over most of the Red Sea, increasing somewhat to the north and south (Table 5.3) where some mixing processes are known to occur. Among these, the interoceanic water exchange via the Straits of Bab-el-Mandab is the master process. This exchange is most intense in winter when the plankton-rich Gulf of Aden water flows into the Red Sea at the surface, counterbalanced by an outflow of uppermost Red Sea deep water over Hanish Sill. In summer, when the two layer pattern has turned into a three layer pattern, fertile subsurface water from the Gulf of Aden flows into the basin at intermediate depths compensating the surface and bottom outflow of Red Sea water (*see* section 3.11.1).

The enrichment effect of alien matter in the Red Sea proper is evident from Table 5.3. It documents a marked increase in primary production as well as in phytoplankton and zooplankton abundance from north to south. A similar overall trend may be seen in the Gulf of Aqaba (Klinker *et al.*, 1978; Levanon-Spanier *et al.*, 1979), whereas in the Gulf of Suez, available zooplankton numbers show a reversed pattern in winter (Kornilova and Fedorina, 1970).

In particular in phytoplankton variables, the increase is most pronounced in the southern half of the Red Sea, between about 16°N and 18°N. Differences between the central and northern parts of the Sea are relatively small, and by subtracting the biomass and production values from those recorded in the south, a rough estimation of the amount of plankton contributed directly by the Gulf of Aden and indirectly through the nutrient import can be made. The import rate of medium-sized or *mesozooplankton* in a month may amount to 96×10^{13} organisms and 4.5×10^5 tonnes wet weight, corresponding to 6×10^4 tonnes dry weight. These rates, however, represent maximum estimates, since they are based on zooplankton concentrations above Hanish Sill that were collected in March (Beckmann, 1984) when the inflow of Gulf of Aden water is at its maximum (Khimitsa and Bibik, 1979).

However, most of the Gulf of Aden expatriates are bound to die (Section 5.3.2). Thus, the southern part of the Red Sea is characterized by high rates of plankton mortality, nutrient release from organic particulates in the surface layer due to the high temperatures, and formation of new particulate matter by adapted phytoplankton and zooplankton which maintain a rich stock of fish. About half of the organic carbon fixed by phytoplankton in the southern Red Sea is contributed by symbiotic algae, mainly dinoflagellates, living in the tissue of radiolarian protozoans (Khmeleva, 1967).

From zooplankton data, Ponomareva (1968) assumed summer to be the most productive season in the Red Sea proper. In contrast, in the central Red Sea greatest primary production and phytoplankton standing stocks were observed in winter in oceanic waters (Weikert, 1981), while off the coast of Jiddah there was a secondary peak of phytoplankton production in July in addition to the main peak in winter (Dowidar, 1983a). These findings suggest that there may be two periods of increased plankton production in the course of a year, one in winter and the other in summer. The effects of summer eutrophication on the plankton biota may, however, be confined to the coastal zones and the northern and southern ends of the Red Sea and do not influence the vast oceanic region between 27°N and 18°N. Indeed, the most intense import of phosphates into the Red Sea occurs by subsurface inflow of Gulf of Aden water from July to September (Khimitsa and Bibik, 1979). This has so far only been traced for a few weeks as far as 18°N (Maillard and Soliman, in press). The mass development of blue-green algae (cyanobacteria) in the central Red Sea provides further evidence of a depletion of plant nutrients in oceanic waters during the summer season.

Blooming of *Oscillatoria* (*Trichodesmium*) *erythraeum* and other blue-green algae is a characteristic feature of the phytoplankton of the central Red Sea. Like in the Gulf of Aqaba, this is typical of the summer and autumn seasons. Within the patchy blooms, concentrations of 9×10^5 cells/l are common, accounting for 70% of total phytoplankton cell number (Belogorskaya, 1970), masking the overall trend of a northward decrease in abundance (Table 5.3) that is evident for the other phytoplankton categories. During calm weather, the filaments float to the sea surface and form slick-like rows or a scum called 'sea dust'. The coloured water produced may have given rise to the name 'Red Sea'. *Oscillatoria* species are able to overcome nitrate depletion by absorption of atmospheric nitrogen which is dissolved in the water, thereby providing nitrogen in forms such as ammonia and amino acids which can be assimilated by other phytoplankton organisms. Among the groups, which may succeed *Oscillatoria* blooms, dinoflagellates seem to be more numerous in summer, while diatoms dominate in winter (Belogorskaya, 1970, Dowidar, 1983a).

With the means of nitrogen fixation, patches of *Oscillatoria* blooms achieve a considerable carbon production, although in the surrounding waters production is low and nutrient concentrations are almost nil. Unfortunately, overall primary production estimates of the Red Sea proper are based on data from single widely-spaced measurements, and they may be therefore biased by the patchy algal production in the sea. According to Koblentz-Mishke *et al.* (1970), north of about 17°N, annual production averages in the euphotic zone range within 250 to 500 mg C/m^2/d, but in the southernmost part of the basin, within 500 to more than 1,000 mg C/m^2/d, which would classify very productive waters. Detailed studies that were conducted recently in the central Red Sea at about 21°N, yielded lower annual averages in the euphotic zone that may approximate to 170 mg C/m^2/d in oceanic waters above the axial trough (Weikert,1981), and 258 mg C/m^2/d towards the reef zone off Jiddah (Dowidar, 1983a). Although these figures, including those from Table 5.3, are inconsistent to some extent, they question the common view (e.g. Halim, 1969, 1984) that primary production in the Red Sea is extremely low, averaging 100 mg or less carbon per m^2 a day.

In the Gulf of Aqaba, primary production is between 200 and 900 mg C/m^2/d during the greater part of the year. The maximum values were recorded in the vicinity of coral reefs and in the southernmost basin, near the Straits of Tiran. A significant amount of this relatively high areal carbon production, however, is synthesized beneath the 80 m deep euphotic zone to a depth of about 200 m, where the light level is 0.2% (Reiss and Hottinger, 1984). Chlorophyll *a* and nutrient concentrations are low. All these figures are typical of oligotrophic water bodies, such as the Sargasso Sea. The phytoplankton stock consists mainly of dinoflagellates and blue-green algae of the genus *Oscillatoria* (Levanon-Spanier *et al.*, 1979).

In winter, from December to March, when the euphotic zone is fertilized through the vertical convection of water masses, there is a rapid development of phytoplankton, which benefits particularly from the increased amounts of nitrogenous compounds. Diatoms and dinoflagellates, especially *Ceratium* species, are abundant (Kimor and Golandsky, 1977), but the most important primary producers appear to be nanoplankton organisms, such as coccolithophores. The amount of chlorophyll *a*, which provides a means to estimate the phytoplankton standing stock, increases to 70 mg/m^2, about double the amount recorded under oligotrophic summer conditions. This rich food supply is used by phytoplankton-feeding zooplankton that peaks in March in the northern part (Prado-Por, 1983), i.e. at the end of the phytoplankton bloom. The time lag in the onset of the zooplankton increase is a common feature in prey-predator successions.

During the winter, primary production is significantly increased in the upper 200 m and varies from about 700 to 1,100 mg C/m^2 daily. Over a year, the production would amount to 160 g C/m^2. Roughly half of this is produced from December to February, classifying the Gulf temporarily as moderately productive (Levanon-Spanier *et al.*, 1979).

No information on primary production in the Gulf of Suez is at present available. An incidental production cast made at the entrance of the Gulf in October (Khmeleva, 1970), in the domain of the fertile neritic Red Sea surface water, yielded 220 mg C/m^2 per day which might indicate a moderate production.

5.3.2. Diversity, endemism and immigration patterns

The oceanic plankton in the Red Sea is relatively poor in species compared to other tropical seas. As described in section 5.2, the habitat becomes more hostile to planktonic life as depth increases and with distance from the Gulf of Aden, which partakes of the rich Indian Ocean flora and fauna. Thus, the number of species present decreases with depth and towards the Gulfs of Suez and Aqaba. For example, in calanoid copepods that form the most abundant group in the mesozooplankton, more than 300 species are known from the Arabian Sea, whereas only 60 have been reported as common in the southern Red Sea proper, 46 in its northern part and about 35 in the Gulfs of Suez and Aqaba (Delalo, 1966, Prado-Por, 1983). Of course, such biogeographic comparisons are affected by differences in sampling methods used by different workers, but the increasing poverty of species to the north has been observed for virtually all zooplankton groups investigated (Halim, 1969, Kimor, 1973; Almogi-Labin, 1984). Halim (loc. sit.) has also quoted an overall decrease for some phytoplankton, while Sukhanova (1969) found less species in the northern than in the southern basin, with least diversity in the central part. Recent results on the diversity of the dinoflagellate genus *Ceratium* in the northern half of the Red Sea question the concept of a general paucity of Red Sea phytoplankton species (Dowidar, 1983b).

Only a few species increase in abundance towards the north. This is against the general pattern. Judging from the findings of several investigators (Kornilova and Fedorina, 1970, Prado-Por, 1983, Weikert, 1980a), the copepod, *Rhincalanus nasutus*, is most numerous in the northernmost Red Sea and the Gulf of Aqaba, and huge swarms of the pteropod, *Creseis acicula*, were observed by the author in the northern Red Sea proper in June. Maximum salinity and minimum temperature occur in those northern parts the waters of which display a seasonally increased production, and it is tempting to speculate that these species are "old" ones in the Red Sea, having been there since the Würm glacial period, roughly 40,000 to 13,000 years ago. The biota of that period lived in highly saline, cold water (Por, 1978) of a heightened fertility (Reiss et al., 1980). *R. nasutus* is said to migrate seasonally to cool deeper waters, where it was found in a resting stage in the central Red Sea (Weikert, 1980b), avoiding rising surface temperatures at the end of winter (Halim, 1969). Fossils in the sediments confirm that *Creseis acicula* is one of the few species (*cf* section 2.6.3) that survived the extreme conditions during the late Würm period (Berggren and Boersma, 1969).

The number of endemic planktonic species in the Red Sea, in contrast to the benthic biota appears to be low. The reasons for this are still not clear. If it is not merely a result of incomplete taxonomic evaluation of the Red Sea plankton, it may be related to the geological history of the area (Chapter 2). During the last 100,000 years, the Red Sea has been subjected several times to relatively rapid changes in temperature and salinity. Major climatic changes would be fatal to marine plankton but not necessarily to coastal benthic species, which are less sensitive to environmental stress. After the last glacial period, about 11,000 years ago, the Red Sea has been recolonized by Indian Ocean plankton that has been continually reinforced by organisms carrying new genetic material from the adjoining Gulf of Aden into the basin. Specific Atlantic species are virtually absent, and, if at all, may have immigrated in the recent past from the Mediterranean Sea into the Red Sea via the Suez Canal or its ancient forerunners. In turn, because of eco-physiological reasons, there has been a coincident invasion of Red Sea biota into the eastern Mediterranean Sea, called 'Lessepsian migration' (Por, 1978).

The rate at which Indian Ocean plankton are imported into the Red Sea proper is determined by the prevailing water exchange pattern across the Hanish Sill (Chapter 3). In winter it is most effective, and a large number of oceanic as well as coastal surface plankton species from the Gulf are forced through the bottle-neck at Bab al Mandab into the Red Sea (Beckmann, 1984). They are responsible for the southerly maximum in Red Sea plankton diversity (Halim, 1969) that obviously collapses between 16°N and 18°N. Many of the surface living species attempt to avoid the elevated surface temperatures by invading the somewhat cooler subsurface water. Nevertheless, most of them perish near to the Hanish Sill, as demonstrated by a progressive northward decrease in pteropod shells deposited in the Red Sea sediments (Stubbings, 1939, quoted in Halim, 1969). Imported phytoplankton coccolithophores and copepods of the genus *Eucalanus*, which contribute to the bulk of winter mesozooplankton of the southern Red Sea, often show pathological features and die off (Okada and Honjo, 1975, Beckmann, 1984). In the central Red Sea the presence and abundance of Gulf of Aden species is greatly reduced. The invading zooplankton seem to be unable to reproduce because only late juvenile and adult stages have been observed (Weikert, 1982).

As summer approaches, the number of species in the Red Sea decreases as the surface habitat becomes more hostile, and the recruitment from the Gulf diminishes with the decreasing rate of water exchange. In mid-summer, when the inflow of Gulf water is intermittent, the number of Indian Ocean expatriates in the Red Sea proper is least (Halim, 1969). In addition to the hydrographic causes, the low diversity in Red Sea summer plankton may evidence reduced diversity in the immigrant Gulf plankton itself. When summer upwelling has proceeded in the Gulf, its plankton community appears to be dominated by a few herbivorous species (Gapishko, 1971).

5.3.3. Vertical segregation in Red Sea plankton

In the first parts of this chapter, we saw how physical and chemical conditions differ in discrete depth layers in the Red Sea. Thus light penetration in amounts adequate for photosynthesis is limited to shallow water, and patterns of oxygen and nutrient concentrations show related changes due to decomposition of organic matter at intermediate depths. The water column is also structured physically by the presence of a thermocline and a halocline, which may cause accumulation of seston (Fig. 5.1).

These important environmental differences lead to pronounced vertical partitioning in the plankton, compounded by overlapping vertical migrations from shallow and intermediate depths by the more mobile zooplankton.

Various terms have been coined to describe major habitats or 'life-zones' in the sea, although the boundaries are rather loosely defined. The surface layer is called the *epipelagic zone*. It is above the thermocline in the Red Sea, and includes the euphotic zone, where photosynthesis takes place. The deep sea below the epipelagic zone is divided into the *mesopelagic zone*, and the adjoining *bathypelagic zone*. The deepest *abyssopelagic* and *hadopelagic* zones are absent in the Red Sea. The mesopelagic zone is characterised by strong physical and chemical gradients, while almost no gradients exist in abiotic factors within the bathypelagic zone. The approximate depth ranges of these zones in the central Red Sea are indicated to the right of Figure 5.5.

Light intensities at the sea surface of the Red Sea and Gulf of Aqaba are above the optimum for most of the phytoplankton species during the greater part of the year, so that photoinhibition takes place. Judged by chlorophyll *a* determinations and cell counts (Yentsch and Wood, 1960 quoted in Halim, 1969, Levanon-Spanier *et al.*, 1979, Weikert, 1980a), the bulk of the phytoplankton standing stock occurs between 50 to 80 m, and maximum production was also measured here in the Gulf of Aqaba. In the Red Sea at least, the chlorophyll *a* has been found to accumulate near the bottom of the euphotic zone, within the thermohalocline (Fig. 5.2) where nutrient salts can be clearly detected (Weikert *et al.*, in preparation). On a global scale, the existence of a so-called subsurface chlorophyll maximum (SCM) is typical of nutrient-poor and warm water masses (Lorenzen, 1976). Maximum rates of carbon fixation per unit chlorophyll (assimilation numbers) were observed above the SCM closer to the sea surface. This 'summer' distribution of phytoplankton abundance and assimilation activity in the Red Sea is temporarily changed by the blooming of blue-green algae (*Oscillatoria spp.*). These can resist high light intensities, and thus produce additional chlorophyll and assimilation maxima near to the sea surface, as illustrated in Figure 5.2.

In winter, the zone of photoinhibition is confined to the upper 20 m in the Gulf of Aqaba. Beneath it, high chlorophyll *a* and production values were recorded down to 200 m depth, due to vertical mixing of water masses. Coccolithophore blooms have been found down to 400 m, which is far below the 1% light level (Reiss and Hottinger, 1984). In the relatively stable central Red Sea, phytoplankton peaks in the upper 50 m. A high carbon production per day and unit chlorophyll *a* was measured throughout the upper 40 m (Fig. 5.2).

Unlike the pattern of phytoplankton, the vertical distribution of zooplankton in most of the Red Sea proper is quite different from that in the oligotrophic tropical regions of the great oceans. It exhibits a greater decrease in biomass and individual numbers with depth, and is therefore similar to the distribution patterns known from other partially landlocked marine basins, such as the Mediterranean Sea. However, in contrast to the latter, the Red Sea deep water mass below approximately 1100 m, the lower part of the bathypelagic zone, is virtually empty of zooplankton (Fig. 5.3). Below about 1400 m, zooplankton biomass concentrations of less than 0.04 mg wet weight/m^3 are common. Such low concentrations are not reached until 4000 to 6000 m in other tropical regions of the world's ocean.

By far the largest zooplankton concentration is found in the epipelagic zone above the thermocline. The underlying 100 m layer of the mesopelagic zone is less densely populated and is characterized by conspicuous concentration gradients, termed a *planktocline*. A second peak of zooplankton concentration occurs in the heart of the oxygen minimum layer, between 300–600 m depth in the central Red Sea, where oxygen concentrations are below 1.3 ml O_2/l.

The vertical distribution of species resembles that of total zooplankton abundance. It is illustrated in Figure 5.4 for the important calanoid copepod species, demonstrating also their daily migration patterns. The highest diversity is in the 100 m surface layer, quite unlike the situation in other nutrient-poor tropical seas, where diversity is highest between 500 and 1500 m. This means that the function of

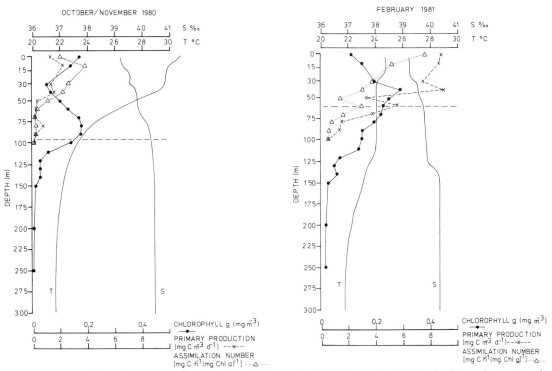

Fig. 5.2. Mean profiles of primary production, chlorophyll *a* and assimilation number in October/November 1980 and February 1981 in the Atlantis II Deep area, central Red Sea. The mean temperature and salinity profiles are also shown. The broken line demarks the 1% light depth. (modified after Weikert, 1981).

the epipelagic zone as the feeding and nursery ground of zooplankton is in particular pronounced in the Red Sea ecosystem. During the night, subsurface species invade the surface zone to reproduce and to crop the organic material that has been synthesized and stored by the epipelagic community during daylight. Over the year, the surface community is strongly affected by seasonal changes in salinity, temperature, plant nutrients and organic food. To a large extent, these changes are coupled with the overall circulation pattern, which also dictates species composition and diversity in the epipelagic zone.

In the mesopelagic zone, below about 100 m, the organic matter produced above is consumed, converted into new substances, and decomposed. Its upper part is characterized by sharp gradients of abiotic factors, the distinctness of which are subject to the seasonal climatic changes in the epipelagic zone. This discontinuity layer, which is about 100 m thick, seems to be unfavourable for zooplankton, as indicated by the coincident strong planktocline and the fact that the population of only a single mesozooplankton species, *Haloptilus longicornis*, congregates there (Fig. 5.4). In the lower part of the mesopelagic zone that is largely governed by the oxygen minimum layer, seasonal changes in the environment are minor.

The oxygen minimum layer contains a secondary diversity maximum, shown in the lower section of Figure 5.4. From this figure it may be noted that the 'band' of least oxygen concentrations at about 400 m (dashed line) seems to separate the distribution patterns of subsurface species into upper (left side of figure) and lower (right side) groups. Within the lower group, the population of *Rhincalanus nasutus*, a large calanoid copepod about 5 mm in size, was found in a resting phase in the oxygen minimum layer of the central Red Sea, and may therefore be used as an indicator for low oxygen concentrations

(Weikert, 1980a, 1980b). Both the upper and lower groups are dominated by the copepod *Pleuromamma indica* which accounts for about 30% of the total mesozooplankton in the oxygen minimum layer and interacts with the epipelagic biota by distinct daily vertical migrations (Fig. 5.4). Both species groups belong to the interzonal or mesopelagic community, which produces the well-known 'deep scattering' effect. During daylight, sonar devices can detect its presence as a distinct stratum of lamellar layers extending from about 300 to 750 or 800 m in the central Red Sea (Weikert, 1980a). At dusk this so-called deep scattering layer moves very fast towards the sea surface. At dawn it quickly descends from the upper 200 m to the day-time residence level, illustrating the impressive phenomenon of daily, large scale vertical migrations in the ocean.

No special plankton fauna is associated with the bathypelagic zone below about 800 m. It is not yet understood why interzonal species have occupied this 'empty' habitat that is extreme to bathypelagic species (section 5.2.1) by only negligible numbers of individuals at depths greater than about 1100 m. Compared with the mesopelagic zone, the bathypelagic zone has similar temperature and salinity, and it is even better supplied with oxygen. A plausible explanation may be that food is too scarce to compensate for the energy loss of an organism by respiration. Respiration is assumed to be increased in zooplankton, as in benthic organisms (Chapter 6), due to the high deep-sea temperature of at least 21.6°C.

Figure 5.5 summarizes the principal features of the vertical congregation of plankton in relation to habitat zones in the central Red Sea; this pattern has been confirmed by studies in the northern and

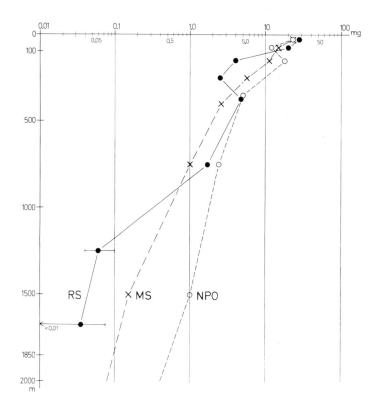

Fig. 5.3. Vertical distribution of zooplankton biomass (in mg wet weight m⁻³) in oligotrophic regions of the (sub)-tropical ocean. RS: Central Red Sea (bars denote ranges of values in deep water); MS: Eastern Mediterranean Sea, single station; NPO: North Pacific Ocean, representative station from the central gyre. Note that the biomass axis is logarithmic, so that concentrations in the shallow Red Sea layers are about 1000 × those in the deepest layers (from Weikert, 1982).

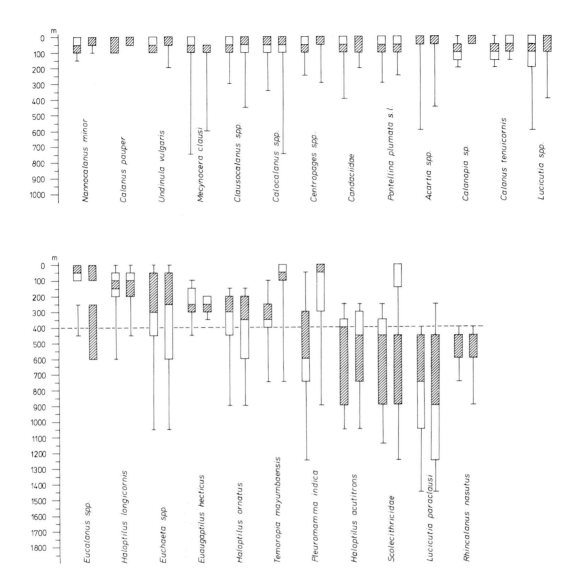

Fig. 5.4. Vertical distribution and preferred range of species of calanoid copepods in the Atlantis II Deep area, central Red Sea. The upper figure shows epipelagic, the lower figure interzonal species. For each species, the left bar indicates midday distribution and the right bar indicates midnight distribution. For each distribution pattern, the preferred depth is indicated as follows: 97.5% of animals are found within the range given by vertical lines, 75% within the whole vertical bar, and 50% within the hatched portion of the vertical bar. The hatched proportions represent the depths of greatest abundance. Broken horizontal line marks the depth of minimum oxygen concentrations (after Weikert, 1982).

southern regions (Weikert, 1980a, Beckmann, 1984). As this Figure and Figure 5.4 show, the bathypelagic zone is virtually unaffected by the daily vertical migration of zooplankton. This means that the water below about 1100 m is not supplied with nutritive particulate matter actively transported from the epipelagic and mesopelagic zones. Therefore, the energy budget of the bathypelagic zone in the Red Sea depends on the 'rain' of faeces, carcasses, moults and their debris that mainly originate from the

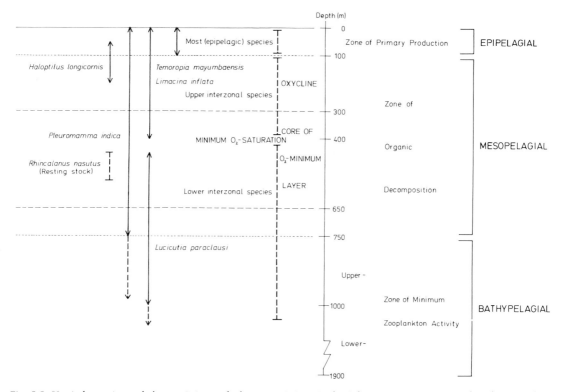

Fig. 5.5. Vertical zonation and characteristic zooplankton associations in the Atlantis II Deep area, central Red Sea. Broken vertical lines: Habitat zones of significant zooplankton assemblages. Associated indicator species are calanoid copepods and the pteropod mollusc *L. inflata*. Vertical arrows: Range of daily vertical migrations. An uncertain distance is indicated by a broken arrow (from Weikert, 1982).

mesopelagic zone. This is concluded from the increase of POM below 400 m (Fig. 5.1) and taxonomic examination of corpses and their larger fragments from net samples (Weikert, 1982).

Small particles that sink from the epipelagic zone are converted into larger ones after consumption by mesopelagic zooplankton. This production of durable, fast-sinking faecal pellets is commonly thought to be a key mechanism for the rapid settling of food to great ocean depths (Eppley and Peterson, 1979). However, experiments indicate that the breakdown of organic particles should be accelerated in the uniquely warm environment of the deep Red Sea. In this case, the available supply of food would amount to less than the 1 to 3% of the surface primary production that are estimated to reach the bathypelagic zones of other warm, oligotrophic seas. The extreme scarcity of plankton and benthic organisms and biomass in the central trough of the Red Sea can be explained by the extremely limited downward transport of energy (Wishner, 1980, Weikert, 1982, *see also* Chapter 6).

Information on the vertical distribution and zonation of zooplankton in the Gulf of Aqaba is scarce and restricted to the upper 600 m of the water column. Unlike in the Red Sea proper, biomass decreases from the surface to deep water, with no indication of a subsurface maximum in the mesopelagic zone (Klinker *et al.*, 1978). Highest diversity seems to exist in the upper 300 m (Prado-Por, 1983). No data on vertical distribution of zooplankton exist for the shallow Gulf of Suez that contains a neritic fauna and flora as well, while in the deep Gulf of Aqaba oceanic species predominate.

5.4 PLANKTON OF CORAL REEFS

5.4.1 Special features of the reef environment

The plankton of coral reefs interacts within an ecosystem radically different from that of the open ocean. The reef environment (*see* Chapter 7) is itself the product of growth of living photosynthetic organisms, principally the corals and their symbiotic zooxanthellae, and various types of calcareous algae. On the reef therefore, the benthos is itself a primary producer, not merely a consumer at second or third hand as in the ocean. Reef phytoplankton plays only a secondary role in primary production. In the open sea, zooplankton holds a key position in the food chain by grazing on phytoplankton, and transferring biomass to the nekton and benthos. On the reef the role of zooplankton is less significant, but more complex. It is heavily preyed upon by planktivorous corals to obtain energy and phosphorus compounds. Conversely, zooplankters feed on suspended mucus released by corals and other benthic animals. This mucus is highly enriched with organic debris, bacteria and small phyto- and zooplankton, which, by a 'mechanical and microbial growth', become available as food for larger animals (Johannes, 1967). Thus, a tight cycling of nutrients is characteristic of the reef food web.

Unlike in the open sea, the direction of energy flow is largely upwards in the reef, from the producers of the benthos to the pelagic community, and while zooplankton is involved in this, it also transfers energy back to the benthos, while many larval zooplankton use it to grow sufficiently to settle and become benthic as adults themselves. A high proportion of reef plankton is made up of these meroplanktonic forms, larvae of the diverse array of organisms living as reef benthos. This increases the diversity of reef plankton samples, reflecting the diversity of reef environment as a whole. In the oceanic plankton permanent or holoplanktonic organisms predominate.

The variety of reef substrates accounts for the abundant cavities and interstices that function as a 'mesh' (Peres and Picard, 1969) entrapping fine organic debris and harbouring many microorganisms. This food depot attracts a large number of plankton animals. In addition, the heterogeneous reef surface gives them shelter from predation by diurnal fish and prevents them from being transported into the open sea by water currents. From their behaviour, most of the reef zooplankters can be termed 'demersal', showing a strong affinity for certain types of substrate during daylight hours (Alldredge and King, 1977). Consequently, the structural heterogeneity of the substrate-water interface should produce a horizontal (or geographical) zonation of assemblages in the reef zooplankton. This is quite unlike the situation in the oceanic zone, where the plankton segregates vertically according to the availability of light and nutrients in the water column, producing conspicuous vertical zonation.

Quantitative information about Red Sea reef plankton is scarce. Much of what is known was learned during two initial studies, conducted along the fringing reef of the Sudanese coast. The reef sites were located at Shaab Baraja (20°50′N), an isolated exposed reef segment of the outer fringing reef (Karbe, 1980), at Dungunab Bay (20°N) inside the reef complex, and in the coastal lagoon of Suakin at 19°N (Nasr, 1980). In the Gulf of Aqaba, Sournia (1977) conducted a cursory phytoplankton assay in the shallow, narrow channel between the fringing reef and the Sinai coast. Off the Jordanian coast, a small pilot study was made for comparison of the zooplankton from the outer ridge of the fringing reef and oceanic waters (Vaissière and Seguin, 1982).

5.4.2 Nutrient levels and phytoplankton production

At the reef sites investigated, nutrient salts were depleted, with the silicate concentration averaging about 2.0 μmol/l, and phosphate and nitrate less than 0.1 μmol/l (compare oceanic data in tables 5.1 and 5.2). Phytoplankton counts were low, averaging about 50 to 70 cells/ml off Sinai. The mean chlorophyll *a* concentration was about 0.25 mg/m^3. These data all suggest that phytoplankton production is very low, indicated also by the daily photosynthetic rates of 30 μgC/l or less, measured in the Sinai coastal waters. These rates correspond to an extremely low mean daily primary production not exceeding 30 mg C/m^2, assuming 1 m depth of water over the reef surface.

The greatest proportion of phytoplankton comprised very small plankters less than 60 μm in diameter (nanoplankton), such as coccolithophores and naked flagellates. The larger microplankton species are reported to be rare. Along the outer Shaab Baraja reef, the diatoms, which were the principal group among the microplankton at all sites, occurred at densities averaging 14 cells/ml in September and October. The main diatom taxa present were *Chaetoceros*, Nitzschiaceae and Naviculaceae, which accounted for 40%, 24% and 22% of the diatoms, respectively. An average of only 0.1 dinoflagellates/ml were counted. The numerical dominance of diatoms to dinoflagellates in neritic waters has been also found in the Bay of Suez (Dowidar, 1976).

The presence of benthic Naviculaceae in the outer reef phytoplankton suggests abrasion of the reef substrate by wave action and currents. This is also indicated by the high proportion (75%) of mainly carbonaceous inorganic particles in the seston. In the open sea at nearly the same latitude and season, the amount of particulate carbon was negligible, and the total inorganic fraction in the seston was only 33% (Weikert, 1982). The amount of seston along the outer reef averaged 2.7 g/m^3, a value 12 times higher than that recorded near the surface in the open sea. There were no significant differences between the concentrations or composition of seston and phytoplankton along the fore-reef and in the lagoon at Shaab Baraja, contrasting with the situation observed on large reefs in other parts of the world's oceans.

5.4.3 Reef zooplankton

At Shaab Baraja, there were conspicuously different catches of zooplankton in emergence traps set over different types of substrate. Traps with 100 μm mesh exposed over living coral substrates caught more zooplankton in terms of both numbers and biomass than those exposed over sand or coral rubble. This finding is similar to those made at sites along the Great Barrier Reef (Alldredge and King, 1977) and in the Philippines (Porter and Porter, 1977).

At all three locations along the Sudanese fringing reef, copepods were predominant in the zooplankton. On the outer reef, they accounted for 95% of all the organisms during both day and night. Like the great majority of other zooplankton groups, they were most abundant in the water column at night. In the traps, the emergence rates of late-larval and adult copepods reached values of 8,600/m^{2}h at night and 550/m^{2}h during the day. Decapod crustaceans were second to copepods in abundance, and the adults were caught only during the night. A pronounced decrease in numbers of adult zooplankters caught during daylight hours is commonly observed in reef as well as pelagic environments. Juveniles are more active than adults and stay closer to the sea surface during the day.

More detailed information on the vertical migration of reef zooplankton at Shaab Baraja was provided by the 100 μm net tows. Sampling during all hours of the day revealed a bimodal pattern of abundance that is well known in other seas. Zooplankton abundance was much reduced during bright daylight, and highest at dawn and dusk. Both numbers and biomass declined at night between the sunset and sunrise peaks. This nocturnal migration, of (mainly) adult zooplankters from the substrate towards the sea surface, resulted in a much greater increase in the number of larger organisms than smaller ones (Table 5.4). Overall, the zooplankton size fraction larger than 300 μm increased from 2% of the samples during the day to about 40% at night.

Preliminary information on the diurnal migration of single species exists for the calanoid copepods, which are generally considered to comprise holoplanktonic species (Table 5.5). The species were more

TABLE 5.4 Zooplankton groups collected in four 100 μm mesh emergence traps at Shaab Baraja during day and night. Numbers are the percentage of individuals larger than 300 μm within each group, a dash indicates that the group was not collected at that time of sampling (Original).

Group	Day	Night
Foraminifera	—	—
Medusae	2.0	0.0
Siphonophora	—	—
Cladocera	—	—
Ostracoda	0.0	60.5
Cirripedia	—	—
Copepoda	2.50	37.5
Amphipoda	0.0	18.5
Isopoda	16.5	35.5
Cumacea	0.0	58.5
Mysidacea	25.0	22.0
Euphausiacea	0.0	16.5
Stomatopoda	—	—
Decapoda	19.0	63.5
Mollusca	0.0	1.5
Polychaeta	0.5	40.5
Phoronida	—	—
Turbellaria	0.0	5.5
Chaetognatha	20.5	53.5
Thaliacea	—	—
Appendicularia	1.5	18.0

abundant in the traps at night than in the traps and net tows operated during the day. This finding is in agreement with visual observations on Pacific reef zooplankton (Hamner and Carleton, 1979) that holoplankton species tend to congregate close to and within corals during daylight. Table 5.5 shows that even individuals of transitional species such as *Undinula vulgaris* and *Calanopia elliptica*, which live in both the oceanic and reef environments, exhibit a different behaviour.

Oceanic zooplankters, such a chaetognaths, siphonophores, medusae and larvae of euphausiids, migrating into the upper layers of the sea at night, were brought to the reef by surface currents, reaching greatest abundance late at night and in the early morning. This external food input to the reef is likely to play a larger role in the food web of exposed reefs, such as Shaab Baraja, than in the inner parts of a reef complex. But even at Shaab Baraja the total amount of drifting zooplankton was low. Copepod species, common about 200 m off the forereef, were absent in the reef samples, with the exception of *Candacia* spp. and related species (Table 5.5). Families and genera found at the open sea and reef sites as well, are known to comprise transitional species, or they were represented by different, but still unidentified species.

Information on the reef plankton in the Red Sea is based on only a few sites, and so no generalizations can be made about other Red Sea reef areas. However, if the data are considered representative for a full year's cycle, then the reef sites investigated in the Gulf of Aqaba and the Sudanese coast are oligotrophic. Phytoplankton biomass and production seem to be maintained to a considerable extent by the nanoplankton rather than the microplankton, and the scarcity of dinoflagellates is noteworthy. A relatively large standing stock and high productivity of benthic flora was observed in the Sinai coastal waters. The role of the phytoplankton in the food web of the reef would appear to be negligible, as it is commonly the case on reefs throughout the world.

Most of the zooplankton investigated near the Sudanese reefs can be termed 'demersal plankton' as the plankters reside in the sediment-water interface during the daylight hours. The relatively great zooplankton abundance recorded at Shaab Baraja compared with reefs in the Pacific Ocean may have resulted from differences in method. Inadequate standardization of procedures and equipment is a general problem in comparative studies of reef zooplankton.

TABLE 5.5 Taxonomic composition and numbers of specimens of calanoid copepods larger than 300 μm collected during midday and midnight at Shaab Baraja reef and near to the reef, central Red Sea. Preliminary reef data from emergence traps and horizontal net tows (Original), off-reef data from the 100 m surface layer by vertical net tows (modified after Weikert, 1982). Numbers of specimens obtained from different devices cannot be compared exactly to each other, because each technique owns different sampling characteristics.

Taxa	Reef			Sea
	Trap Night No/m²h	Trap Day No/m²h	Net tow Day No/30m²	Net tow Day No/30m³
Temora discaudata	4	<0.1	2	
Labidocera spp.	108	1	39	
Calanopia spp. B, C, D	19		1	
Tortanus spp.	33	0.1	2	
Centropages spp.	188	1	31	131
Acartiidae	177	0.5	10	148
Candaciidae	0.4		1	102
Calonopia elliptica	12	<0.1	2	7
Undinula vulgaris	5			85
Calanus tenuicornis				13
Nannocalanus minor				61
Calanus pauper				<0.1
Eucalanus spp.				5
Mecynocera clausi				146
Calocalanus spp.				61
Clausocalanidae				578
Lucicutia spp.				95
Pontellina plumata s. l.				6
Subsurface species				122
Unidentified	48	0.5	36	169

5.5. CONCLUSIONS

Although the investigation of the plankton biota of the Red Sea including its two northern branches has been intensified during the past decade, information on its biology and ecology is still limited. This is particularly true for the reef plankton that has only been investigated during cursory studies. Therefore, general features can be outlined only for the plankton in the open sea by extrapolation of the biological data relative to the somewhat better known hydrographic features of the Red Sea graben system. The continued shortage of quantitative data is most obvious in the field of primary production. The traditional concept that most of the Red Sea is oligotrophic or even ultra-oligotrophic has not yet been verified in a true scientific sense and might turn out to be an exaggeration. No data base exists that would allow an interpretation and evaluation of biological phenomena observed on small and medium scales in terms of overall patterns.

The occurrence of seasonal fluctuations in species composition and stock size of the oceanic plankton is undisputed. Once more, there is no exact answer to the simple question of how large the amplitudes of seasonal oscillations in the diverse plankton parameters are. There is not even any quantitative information on year-to-year fluctuations.

The Red Sea is an extreme environment for plankton. The characteristics of distribution, diversity and abundance all suggest that a substantial proportion of the biota live close to their physiological limits. Knowledge of the function of the pelagic ecosystem in the Red Sea and its stability is urgently needed before it is disturbed by man. The off-shore exploitation of oil and gas fields is proceeding. Rapid urbanization and industrialization in the coastal zones has already produced local effects on the susceptible reef ecosystem. To protect the Red Sea biota, a risk evaluation for the intact ecosystem has to be made before the onset of a disturbance. This pioneering step has been taken for the first time in the history of marine exploitation by the Saudi-Sudanese Red Sea Joint Commission, Jiddah, in preparation for the anticipated mining of metalliferous sediments from the Atlantis II Deep. One of the most fundamental results in the biological field is the detection of the vertical life-zones in the Red Sea proper and their quantitative evaluation.

ACKNOWLEDGEMENTS

The Saudi-Sudanese Red Sea Joint Commission made a significant contribution to the knowledge of the Red Sea by initiating and supporting quantitative ecological research on a broad scale as a basis for setting guidelines for the protection and conservation of the living resources prior to the start of deep sea mining in the Atlantis II Deep area. The Ministry of Research and Technology of the Federal Government of Germany participated in financing the ecological reef and plankton studies during the campaigns 'Reef Survey I' and 'MESEDA I, II and III' (Metalliferous Sediments Development Programme Atlantis II Deep). During the cruises, the author has appreciated the close collaboration and assistance of the chief scientist, Dr. H. Bäcker, Hannover. Thanks are due to Dr. L. Karbe, Hamburg, who provided the reef plankton material to the author. I am also grateful to Prof. Z. Reiss, Jerusalem, whose comments on the production characteristics of the Gulf of Aqaba were of great help to me. Dr. C. Heckman, Hamburg, improved the English of the manuscript.

REFERENCES

Alldredge, A. L., King, J. M. (1977). Distribution, abundance, and substrate preferences of demersal reef zooplankton at Lizard Island Lagoon, Great Barrier Reef. *Mar. Biol.* 41, 317–33.

Almeida Prado-Por, M. S. (1983). Diversity and dynamics of Copepoda Calanoida in the northern Gulf of Elat (Aqaba), Red Sea. *Oceanol. Acta* 6, 139–45.

Almogi-Labin, A. (1984). Population dynamics of planktic Foraminifera and Pteropoda - Gulf of Aqaba, Red Sea. *Palaeontology*, Proceedings B, 87, 481–511.

Beckmann, W. (1984). Mesozooplankton distribution on a transect from the Gulf of Aden to the central Red Sea during the winter monsoon. *Oceanol. Acta* 7, 87–102.

Belogorskaya, E. V. (1970). Qualitative and quantitative distribution of phytoplankton in the Red Sea and Gulf of Aden in October–November 1963. (In Russ.). *Biologiya morja, Kiew* 21, 133–52.

Berggren, W. A., Boersma, A. (1969). Late Pleistocene and Holocene planktonic Foraminifera from the Red Sea. In *Hot brines and recent heavy metal deposits in the Red Sea.* Eds. E. T. Degens and D. A. Ross. Pages 329–35.

Delalo, E. P. (1966). Distribution of the zooplankton biomass in the Red Sea and the Gulf of Aden, winter 1961/62 (In Russ.). *Okeanologicheskiye issled.* 15, 131–39.

Dowidar, N. M. (1976). The phytoplankton of the Suez Canal. *Acta Adr.* 18, 241–56.

Dowidar, N. M. (1983a). Primary production in the central Red Sea off Jiddah. In *Proc. Int. Conf. on Mar. Sci. in the Red Sea.* Eds. A. F. A. Latif, A. R. Bajoumi and M. -F. Thompson. *Bull. Inst. Oceanogr. Fish.* 9, 160–70.

Dowidar, N. M. (1983b). The genus *Ceratium* from the Red Sea. *J. Fac. Mar. Sci. King Abdulaziz Univ.* 3, 5–37.

El-Sabh, M. J. and Beltagy, A. J. (1983). Hydrography and chemistry of the Gulf of Suez during September 1966. In *Proc. Int. Conf. on Mar. Sci. in the Red Sea,* Eds. A. F. A. Latif, A. R., Bajoumi and M. -F. Thompson. *Bull. Inst. Oceanogr. Fish.* 9, 78–82.

Eppley, R. W. and Peterson, B. J. (1979). Particulate organic matter flux and planktonic new production in the deep ocean. *Nature* 282, 677–80.

Fishelson, L. (1973). Ecology of coral reefs in the Gulf of Aqaba (Red Sea) influenced by pollution. *Oecologia* 12, 55—67.

Gapishko, A. J. (1971) Seasonal changes in Gulf of Aden zooplankton in 1965—1966. *Oceanology*, Wash. (Transl. of *Okeanologiya*, Mosk.) 10, 399—403.

Gordeyeva, K. T. (1970). Quantitative distribution of zooplankton in the Red Sea. *Oceanology*, Wash. (Transl. of *Okeanologiya*, Mosk.) 10, 867—71.

Grasshoff, K. (1969). Zur Chemie des Roten Meeres und des inneren Golfs von Aden nach Beobachtungen von FS 'Meteor' während der Indischen Ozean Expedition 1964/65. *'Meteor' Forsch.-Ergebn. A*, 6, 1—76.

Halim, Y. (1969). Plankton of the Red Sea. *Oceanogr. mar. Biol.* 7, 231—75.

Halim, Y. (1984). Plankton of the Red Sea and the Arabian Gulf. *Deep-Sea Res.* 31A, 969—82.

Hamner, W. M. and Carleton, J. H. (1979). Copepod swarms: attributes and role in coral reef ecosystems. *Limnol. Oceanogr.* 24, 1—14.

Johannes, R. E. (1967). Ecology of organic aggregates in the vicinity of a coral reef. *Limnol. Oceanogr.* 12, 189—95.

Jones, E. N. and Browning, D. G. (1971). Cold water layer in the southern Red Sea. *Limnol. Oceanogr.* 16, 503—9.

Karbe, L. (1980). Plankton investigations in an exposed reef of the central Red Sea (Shaab Baraja, Sudan). *Proc. Symp. Coastal and Marine Environment of the Red Sea, Gulf of Aden, and Tropical Western Indian Ocean*, Khartoum 9—15 January 1980, Vol. II, 519—39.

Karbe, L. and Lange, J. (1981). The chemical environment. In *Mining of Metalliferous Sediments from the Atlantis II Deep, Red Sea: Pre-mining Environmental Conditions and Evaluation of the Risk to the Environment*. Environmental impact study presented to Saudi-Sudanese Red Sea Joint Commission, Jiddah. Eds., L. Karbe, H. Thiel, H. Weikert and A.J.B. Mill. pp. 75—99, Hamburg.

Khimitsa, V. A. and Bibik, V. A. (1979). Seasonal exchange in dissolved oxygen and phosphates between the Red Sea and the Gulf of Aden. *Oceanology*, Wash. (Transl. of *Okeanologiya*, Mosk.) 19, 544—6.

Khmeleva, N. N. (1967). Role of radiolarians in the estimation of primary production in the Red Sea and the Gulf of Aden. *Dokl. Akad. Nauk SSSR* 172 (Transl. of *Dokl. Akad. Nauk SSSR* 972, 1430—3).

Khmeleva, N. N. (1970). On the primary production in the Red Sea and the Gulf of Aden (In Russ.). *Biol. morja, Kiew* 21, 107—33.

Kimor, B. (1973). Plankton relations in the Red Sea, Persian Gulf and Arabian Sea. In *The Biology of the Indian Ocean*. Ecological Studies 3. Eds. B. Zeitzschel and S. A. Gerlach, pp. 221—32. Springer, Berlin, Heidelberg, New York.

Kimor, B. and Golandsky, B. (1977). Microplankton of the Gulf of Elat: aspects of seasonal and bathymetric distribution. *Mar. Biol.* 41, 55—67.

Klinker, J., Reiss, Z., Kropach, C., Levanon, J., Harpaz, H. and Halics, E. (1976). Observations on the circulation pattern in the Gulf of Elat (Aqaba), Red Sea. *Israel J. Earth-Sci.* 25, 85—103.

Klinker, J., Reiss, Z., Kropach, C., Levanon, J., Harpaz, H. and Shapiro, Y. (1978). Nutrients and biomass distribution in the Gulf of Aqaba (Elat), Red Sea. *Mar. Biol.* 45, 53—64.

Koblentz-Mishke, O. J., Volkovinsky, V.V. and Kabanova, J.G. (1970). Plankton primary production of the world ocean. In *Scientific Exploration of the South Pacific*. Ed. W.S. Wooster, pp. 183—93. National Academy of Sciences, Washington.

Kornilova, G. N. and Fedorina, A. I. (1970). The zooplankton of the Red Sea (In Russian). *Tr. Azovo-Chernomorskovo Nauchno-Issledovatel'skovo Inta. Rybnova Khozyaistva i Okeanografii* 30, 48—59.

Levanon-Spanier, J., Padan, E. and Reiss, Z. (1979). Primary production in a desert-enclosed sea — the Gulf of Elat (Aqaba), Red Sea. *Deep-Sea Res.* 26, 673—85.

Lorenzen, C.J. (1967). Primary production in the sea. In *The Ecology of the Seas*. Eds. D. H. Cushing and J. J. Walsh, pp. 173—85. Blackwell Scientific Publications, Oxford.

Maillard, C. and Soliman, G. F. (in press). Hydrography of the Red Sea and exchanges with the Indian Ocean in summer. *Oceanol. Acta*, 9.

Morcos, S. A. (1970). Physical and chemical oceanography of the Red Sea. *Oceanogr. mar. Biol.* 8, 73—202.

Nasr, D. H. (1980). Coastal plankton fauna of the Sudanese Red Sea. *Proc. Symp. Coastal and Marine Environment of the Red Sea, Gulf of Aden, and Tropical Western Indian Ocean*, Khartoum 9—14 January 1980, Vol. II, 561—81.

Okada, H. and Honjo, S. (1975). Distribution of coccolithophores in marginal seas along the western Pacific Ocean and in the Red Sea. *Mar. Biol.* 31, 271—85.

Peres, J. M. (1982). Zonations and organismic assemblages. 8. Major benthic assemblages. In *Marine Ecology* 5 (1). Ed. O. Kinne, pp. 373—522. John Wiley and Sons Ltd., Chichester.

Peres, J. M. and Picard, J. (1969). Réflexions sur la structure trophique des édifices récifaux. *Mar. Biol.* 3, 227—32.

Petzold, M. (1986). Untersuchungen zur horizontalen und vertikalen Verteilung des Phytoplanktons im Roten Meer. *Diplomarbeit, Institut für Hydrobiologie und Fischereiwissenschaft, Univ. Hamburg.*

Poisson, A., Morcos, S., Souvermezoglou, E., Papaud, A. and Ivanoff, A. (1984). Some aspects of biochemical cycles in the Red Sea with special reference to new observations made in summer 1982. *Deep-Sea Res.* A31, 707—18.

Ponomareva, L. A. (1968). Quantitative distribution of zooplankton in the Red Sea as observed in the period May to June 1966. *Oceanology*, Wash. (Transl. of *Okeanologiya*, Mosk.) 8, 240—2.

Por, F. D. (1978). Lessepsian migration. Ecological Studies 23, Springer, Berlin, Heidelberg, New York.

Porter, J. W. and Porter, K. G. (1977). Quantitative sampling of demersal plankton migrating from different coral reef substrates. *Limnol. Oceanogr.* 22, 553—6.

Reiss, Z., Luz, B., Almogi-Labin, A., Halicz, E., Winter, A., and Wolf, M. (1980). Late Quaternary paleoceanography of the Gulf of Aqaba (Elat), Red Sea. *Quat. Res.* 14, 294—308.

Reiss, Z. and Hottinger, L. (1984). The Gulf of Aqaba. *Ecological Studies* 50, Springer, Berlin, Heidelberg, New York.

Shackleton, N. J. (1982). The seep-sea sediment record of climatic variability. *Prog. Oceanogr.* 11, 199–218.

Shimkus, K. M. and Trimonis, E. S. (1983). Quantitative distribution of suspended matter in the Red Sea and Gulf of Aden. *Oceanology*, Wash. (Transl. of *Okeanologiya*, Mosk.) 23, 449–52.

Sournia, A. (1977). Notes on primary productivity of coastal waters in the Gulf of Elat (Red Sea). *Int. Revue ges. Hydrobiol.* 62, 813–9.

Sukhanova, J. N. (1969). Some data on the phytoplankton of the Red Sea and the western Gulf of Aden. *Oceanology*, Wash. (Transl. of *Okeanologiya*, Mosk.) 9, 243–7.

Thiel, H. (1979). First quantitative data on Red Sea deep benthos. *Mar. Ecol. Progr. Ser.* 1, 347–50.

Thiel, H. (1980). Community structure and biomass in the Central Red Sea. *Proc. Symp. Coastal and Marine Environment of the Red Sea, Gulf of Aden, and Tropical Western Indian Ocean*, Khartoum 9–14 January 1980, Vol. III, 127–34.

Thiel, H., Weikert, H. and Karbe, L. (1986). Risk assessment for mining metalliferous muds in the deep Red Sea. *Ambio* 15, 24–51.

Vaissière, R. and Seguin, G. (1982). Preliminary study of the zooplankton from the coral reef and the open-sea areas of Jordan in the Gulf of Aqaba (Red Sea). *Vie marine* 4, 1–6.

Weikert, H. (1980a). On the plankton of the Central Red Sea — A first synopsis of results obtained from the cruises MESEDA I and MESEDA II. *Proc. Symp. Coastal and Marine Environment of the Red Sea, Gulf of Aden, and Tropical Western Indian Ocean*, Khartoum 9–14 January 1980, Vol. III, 135–67.

Weikert, H. (1980b). The oxygen minimum layer in the Red Sea: Ecological implications of the zooplankton occurrence in the area of the Atlantis II Deep. *Meeresforsch.* 28, 1–9.

Weikert, H. (1981). The pelagic communites. In *Mining of Metalliferous Sediments from the Atlantis II Deep, Red Sea: Pre-mining Environmental Conditions and Evaluation of the Risk to the Environment*. Environmental impact study presented to Saudi-Sudanese Red Sea Joint Commission, Jiddah. Eds., L. Karbe, H. Thiel, H. Weikert and A.J.B. Mill. pp. 100–54, Hamburg.

Weikert, H. (1982). The vertical distribution of zooplankton in relation to habitat zones in the area of the Atlantis II Deep, central Red Sea. *Mar. Ecol. Prog. Ser.* 8, 129–43.

Wishner, K. F. (1980). The biomass of the deep-sea benthopelagic plankton. *Deep-Sea Res.* 27, 203–16.

CHAPTER 6

Benthos of the Deep Red Sea[1]

HJALMAR THIEL

Institut für Hydrobiologie und Fischereiwissenschaft der Universität Hamburg, Zeiseweg 9, 2000 Hamburg 50, Federal Republic of Germany

CONTENTS

6.1. INTRODUCTION

6.1.1 Benthos of the deep sea

The fauna of the deep parts of the world's oceans comprises a great variety of organisms. Under water pressure as great as two tonnes per square centimetre, in utter darkness, weakly broken only by the phosphorescence of living animals, the deep-sea fauna survives on the fallout of dead organisms and transport of organic matter arriving from the well-lit upper waters. Some food arrives in large parcels, but most of this input is in the form of small particles which some benthic animals filter from the water; others harvest them from the topmost layer of oozes. These animals are eaten by a variety of small predators, which in turn are consumed by larger predators, often present only in very small numbers. Deep-sea communities are limited by food availability. Many deep benthic organisms show bizarre body modifications, while others closely resemble forms extinct since the Palaeozoic era, apparently surviving in the deep sea because of the uniformity and constancy of the environment.

Deep-sea research developed at a slow pace because it is very expensive to study the deep communities, and until recently there was no need to broaden our knowledge except the academic curiosity in this strange world. About 20 years ago commercial interest increased in the deep-sea metal resources,

---

[1] Investigations on the Red Sea Deep Benthos No. 25

manganese nodules and metalliferous muds, and in waste dumping. This simultaneously stimulated deep-sea research as industrial activities in such an environment, undisturbed from time immemorial, introduce great risk to the oceans. The development and use of underwater photography and manned submersibles have helped considerably our understanding of the ecology of the deep benthos, a fauna previously revealed only from laborious and often poorly productive bottom trawling and sediment coring.

Benthic organisms can broadly be divided into two groups, the *epifauna* (sessile, mobile or vagile organism) associated with the surface of the sediment, and the *infauna* which burrows or rests in the sediment. These categories are not exclusive, many animals which forage on the sediment surface for food may return to a burrow for protection. Typical examples of the epifauna include sea anemones or corals, barnacles, bryozoans, and polychaetes, but many species of unicellular organisms as well. The mobile fauna is well represented by fish, larger crustaceans, starfish and sea urchins.

Infaunal organisms vary greatly in size, and for practical purposes are arbitrarily classified into size categories. The larger organisms, of many phyla, but especially the polychaete annelids, crustaceans and molluscs, are termed the *macrofauna* (> 1 mm). They are often present in insufficient numbers for accurate quantitative study. The next smaller group, the *meiofauna*, range in size from 0.04 to 1 mm, and are correspondingly more abundant. They are taxonomically diverse, including larger protozoans such as Foraminifera and representatives of many metazoan phyla as nematodes, crustaceans, kinorhynchs and tardigrades. Finally, the *nanofauna* and the *microorganisms* (not discussed in this report) are less than 40 μm in size, the first mainly comprising ciliates and other protozoans, the latter the fungi and bacteria.

There have been reasons to assume the benthos of the deep Red Sea is relatively low in number and biomass. The tropical-subtropical climate of the region, the perfect underwater visibility and the blue colour of the water throughout the year indicate what may be termed an oceanic desert, exhibiting low primary production in the surface layers. This oceanic poverty stands in constrast to the rich life found in the shallow coral reef areas. The standing stocks of benthos ultimately depend on primary production in the surface waters, as was shown by Sanders and Hessler (1969) and Rowe and Staresinic (1979) for macrofauna, and Thiel (1983a) for meiofauna.

During the oceanographic studies of the central Red Sea for the Saudi Sudanese Red Sea Joint Commission, plankton sampling described in Chapter 5 of this book and by Weikert (1982) confirmed the low standing stock and productivity of Red Sea oceanic plankton. Food input to the bottom water layer and the sediment must then be small. How does the community react to these conditions? Another interesting question concerns the degree of endemism of the fauna. The deep Red Sea is isolated from the rest of the Indo-Pacific deep waters by the 130 m shallow sill near Bab al Mandab. Does this prevent gene flow between the organisms north and south of the sill, and if so have new, endemic species arisen in the Red Sea? Finally, what influence do the unique hydrological conditions (*see* chapters 3 and 4) have on the benthic community?

6.1.2. Investigations in the deep Red Sea

Before the discovery of hot brines and metalliferous sediments in the central Red Sea and their recognition as a source of wealth (Degens and Ross, 1969) prompted major investigations, only two oceanographic cruises had sampled the deep benthos of the Red Sea. The Austrian research vessel *Pola* collected extensively by trawls and dredges between 1895 and 1898, working to depths of more than 2000 metres (Balss, 1915). Among the resulting publications may be noted Fuchs (1901) on general fauna, von Marenzeller (1907) on corals, Balss (1915) on decapod crustaceans, Michaelsen (1918, 1921) on ascidians, and Spandl (1924) on amphipod crustaceans. When Ekman (1953) briefly reviewed the deep Red Sea fauna, the *Pola* results were still the only data available. Since then, Russian workers have described results from conventional sampling with, however, only a few deep water stations, (Murina, 1971), and studies by submersible (Monin *et al.*, 1980, 1982) have been conducted on an east-west transect at 18°N.

The data on which most of this chapter is based were obtained during a programme set up by the Saudi Sudanese Red Sea Joint Commission, to study the oceanography of the Red Sea in order to assess the risk to the ecosystem from the projected mining of the Atlantis II Deep metalliferous muds (Theeg, 1985; Thiel, 1979, 1980, 1983a,b; Thiel and Weikert, 1984; Thiel *et al.*, 1985, 1986). Sampling was conducted between 1977 and 1981, from the research vessels *Valdivia* and *Sonne*. The research concentrated on a quantitative and qualitative description of faunal community components and sediment properties. Sampling the macro- and megafauna was performed using closing trawls and box grabs. The meiofauna and sediment properties were determined from subsamples collected from undisturbed box grab cores. During our last cruise we were able to measure and calculate activity rates of the benthos. For respiration measurements subcores were taken from the box grab samples and oxygen consumption in the water above the sediment was determined in a water bath at ambient temperature using an oxygen electrode. In addition, the respiration potential was calculated from enzymatic determinations of hydrogen transport in the respiratory chain, the so-called electron transport system. Information on the epifauna was obtained by a photosled: a camera and flash rig that automatically photographs the seabed along its path as it is towed slowly over the bottom. Some results from the photosled are shown in Figures 6.1 to 6.5. Finally, a baited phototrap was used in an attempt to catch some of the more agile larger fauna that avoids trawls and the photosled. It operated like a lobster-pot and was equipped with an automatic camera and flash to record the trap occupants at intervals, producing photographs as shown in Figures 6.6 to 6.7. A detailed description of the equipment is presented in Thiel and Rumohr (1979) and Thiel (1980).

A cross section through the central Red Sea characterizes it as a young ocean. A narrow *shelf* carrying the reefs, slopes down to about 100 m depth. This region is followed by a steep gradient termed the *coastal slope* which plunges to 500 or 600 m depth on the *deep terrace*, defined as the upper depth limit of the deep Red Sea for the purposes of this chapter. The reasons for this definition are rather arbitrary and presumably have little biological significance; instead they relate to the difficulty of sampling on a steep slope, and to the danger to damage or loose gear on the rocky bottom. Additionally, the research was directed to the central Atlantis II region and its closer surroundings. Consequently research was confined to the deep terrace, between the base of the coastal slope and the beginning, at 800—1000 m depth, of the *central graben* or *rift*, which slopes steeply to over 2200 m depth.

6.2. SEDIMENTS OF THE DEEP RED SEA

The sediments in the deep Red Sea are characterized by carbonate components, and the most conspicious are the small shells of pelagic pteropod and some heteropod molluscs, which after the animals' death sink to the bottom. Besides the Foraminifera, the Pteropoda and the Heteropoda are the most numerous plankton organisms with durable skeletons or shells (Weikert, 1981). Because of their larger size and mass the pteropod shells are predominant in all the sediment samples collected, which are classified as pteropod oozes (Berger, 1978). Many of the shells are heavily infested from boring fungi or ascomycetes (Thiel, 1983b). Other abundant biogenic particles are the tests of planktonic Protozoa, especially foraminifers and coccolithophorids, less frequent are Radiolaria and diatoms. Shells of the small crustacean ostracods and parts of echinoderms, decapod crustaceans, bivalves and bryozoans are rare. The inorganic component of the deep-sea sediments is composed of quartz, feldspar, mica and other minerals. It is suggested that these particles are wind-born and blown into the sea from the nearby African and Saudi Arabian deserts, or they may be washed in with heavy floods from nearshore land areas. The sediments are poorly size-sorted. Most of the particles are of silt size (4—62 μm), 11—26%

are sand (>62 μm), and 15–31% are clay (<2 μm). Finer grades are more important in the central rift (Bäcker, 1981).

The content of organic matter in the sediments is <1% with higher values generally nearer to the coast and a decrease towards the central area.

The same tendency is exhibited by the values for chloroplastic pigment equivalents (Thiel, 1981) which sum up chlorophyll *a* and phaeopigment components in the sediments. These plant pigments can only be formed in well lit shallow water depths and they indicate relative rates of sedimentation of organic matter originating from primary production. The highest value, 0.214 μg/ml of sediment, was found at the base of the coastal slope where some algae and plant debris may accumulate, washed down from the shallow productive reef zone. Wind induced wave action was observed by Hughes (1977) to destroy reef algae during winter months and part of this organic material will be transported to the open water and settle at slope depth. Between 0.06 and 0.017 μg/ml were measured further offshore, where only phytoplankton may settle, possibly incorporated into faecal pellets of zooplankton. Another source of organic matter, indicated by chloroplastic pigments are cyanobacteria-diatom-mud floats, carrying small organisms. They were observed in April 1979 (Thiel, 1979) abundantly drifting along the surface in the central Red Sea, at least 100 km away from the shore. Sinking of this material would lead to a patchy distribution of pigments in the sediment and to enrichment of the organic matter content.

Photosled transects revealed the surface structure of the unconsolidated sediments. Most of the area is an absolute level and undisturbed bottom with single coarser particles on the surface (Fig. 6.1). Animal tracks in these areas are rare (Fig. 6.2), but bioturbative holes and mounds are abundant (Fig. 6.3). The holes and mounds are not clearly associated with each other and so may have different occupants. Most of the mounds have their tops rounded off, some of them exhibit small depressions. Others (and some of the holes) are capped by a loose fluffy, cottonlike material, which seems to indicate that these at least have not been inhabited for a considerable time. Some 11,200 photographs have failed to reveal the digging species and only a few shrimps were to be seen resting near the holes. Russian investigators likewise could not identify the burrowing organisms (Monin *et al.*, 1982;Vinogradova, pers. comm.). I have the impression that the density of burrowers is low, while the structures may be rather stable in this low current regime, suggesting a falsely high apparent population density.

During our investigations, few hard substrate areas (Fig. 6.4) were trawled or photographed, but the equipment towed along the bottom caught in rocks and anchored the ship several times. Weak links broke and the trawl net was regularly torn. Geologists discovered (Bäcker, 1981) submerged reef platforms, escarpments from tectonic faults, pillow lavas from volcanic activity (Young and Ross, 1974; Monin *et al.*, 1982), salt diapir cap rocks and exposed hard rock aragonite layers. In addition, clinker from coal steamers was regularly brought up with our trawls.

While most of the information on the seabed substrates was collected in the central Red Sea between 20°50′N and 21°40′N, it is believed from a few samples taken in the northern part (at about 24°45′N and 26°45′N) and from the Russian observations noted above at 18°N that the results are also valid for other deep regions of the Red Sea. Por and Lerner-Seggev (1966) described the sediment in the northern Gulf of Aqaba at 500 to 700 m depth as pteropod ooze. Rocky substrates should be more abundant along the steep coastal slope and the steep rift slope, where sampling was avoided for obvious reasons.

6.3. FAUNA OF THE DEEP RED SEA

6.3.1. General considerations

The early sampling by the Austrian RV *Pola* and our research have revealed that nearly all animal groups are present in the deep Red Sea. Major benthic invertebrate taxa missing so far from the collections in the deep Red Sea are the Phoronida, Pogonophora, Hemichordata and Crinoidea.

Fig. 6.1. Flat seabottom with coarse grains on the sediment surface and few bioturbative structures. A cynoglossid flatfish is chased by the photo sled. (Water depth about 1500 m).

However, there seems to be no reason why these taxa should not be adapted to colonize the deep Red Sea. Further sampling may discover other habitats which could constitute the niches for species of these groups. Russian researchers have deployed submersibles (Monin *et al.*, 1980; Monin *et al.*, 1982), and other techniques may help to collect the fauna more thoroughly.

The geological history of the Red Sea (Chapter 2) shows that organisms invaded the Red Sea from the Indian Ocean via the Gulf of Aden and the Straits of Bab al Mandab, presumably during Pleistocene interglacial times, i.e. rather recently, and according to Klausewitz (1980, 1986) modern deep Red Sea species are closely related to or conspecific with those of the Indo-Pacific zoogeographical region. The question arises whether the submarine sill north of Bab al Mandab (130 m depth) effectively closes off the Red Sea from the Indian Ocean for any gene flow in deep-water populations. This problem has to be answered at the species or subspecies level. Distributional records alone are not sufficient to decide for or against endemism in all species. New species discovered in the Red Sea may later be found in less well investigated areas of the Indian Ocean.

In our recent samples the two prawns *Haliporus steindachneri* and *Parapandalus adensameri* are among the most abundant species, and these are clearly endemic (Türkay, pers. comm.). Andres (1981) described three new species of amphipod Crustacea: *Glycerina teretis*, *Socarnes allectus* and *Pseudamaryllis nonconstricta*. *Halcurias sudanensis* (Fig. 6.5) is a new species of sea anemone (Riemann-Zürneck, 1983) and new species of other taxa were discovered, so far not described. These new species could be endemic to the Red Sea, and Janssen (*pers. comm.*) thinks that the benthic molluscs exhibit a high degree of

Fig. 6.2. Animal tracks in level seabottom and *Argonauta* shell (about 1450 m).

Fig. 6.3. Mounds and holes by an unidentified burrower and *Argonauta* shell (about 1500 m).

endemism. New fish species, collected in the deep Red Sea are *Dysomma fuscoventralis* (Karrer and Klausewitz, 1982), *Uroconger erythraeus* (Castle, 1982) and *Harpadon erythraeus* (Klausewitz, 1983a) and *Neocentropogon mesedai* (Klausewitz, 1985). Türkay (1986) estimates 30% of the deep living Red Sea fauna to be endemic.

Another general feature of the Red Sea is the rarity of true benthic deep-sea species. The deep Red Sea is however certainly populated, but most of the species are invaders from shallow water, for several possible reasons. Firstly, the temperature is high throughout all depths (Morcos, 1970). Whereas a species in the Indian Ocean may be limited to a narrow depth range by the decrease in temperature with depth, the warm water in the Red Sea allows deeper penetration. Secondly, cold adapted deep-sea species cannot penetrate into the Red Sea because larvae and adults never appear in shallow waters of the Indian

Ocean to be drifted into the Red Sea basin. Their ecological niches are kept free in the Red Sea, and it may be hypothesized that shallow water species, being able to penetrate deeper down, occupy these niches. Two other explanations may be the lack of specific predators in deep water, allowing a species to penetrate deeper down, or the reduced competition for food through impoverishment of communities in relation to the Indian Ocean. It would be of general interest to study the biology of those Red Sea species with both shallow and deep living populations and to compare their life cycles and behaviour with shallow living Indian Ocean populations.

Examples of Indian Ocean shallow sublittoral species living in the deep Red Sea are compiled in Table 6.1. Further examples are given by Klausewitz (1986) for fishes and by Türkay (1986) for Crustacea Decapoda Reptantia.

Fig. 6.4. Biogenic rock penetrating the sediment surface and gastropod (Fusinus difrons) shells (about 720 m).

Fig. 6.5. An actinian, presumably *Halcurias sudanensis*, possibly fixed to some subsurface hard substrate (about 600 m).

TABLE 6.1. Examples of sublittoral Indian Ocean species being Red Sea sublittoral–deep-water species.

Species	Known depth of distribution in metres		Author
	Red Sea	Indian Ocean	
Crustacea, Stomatopoda (mantis shrimp):			
Kempina zanzibarica	400–(588 to 753) 804	118, 212	Manning 1980
Pisces (fish):			
Lophiodes mutilus	747–804	220–549	Klausewitz 1981
Iago omanensis	1650	92–368 (490?)	Klausewitz 1983b
Muraenesox cinereus	740	shallow coastal and even freshwater to 100 m	Klausewitz & Thiel 1982

6.3.2. The epifauna

According to our experience the epifauna is very sparse. The only sedentary organisms photographed were a few actiniarians (Fig. 6.5) presumably *Halcurias sudanensis*, described from our collections by Riemann-Zürneck (1983) as a new species. The sea anemones in our photographs sit in small sediment depressions and are possibly fastened to hardrock aragonitic layers just below the sediment surface. More frequent but still rare are unidentifiable blue spheres, a few centimetres high and across, possibly ascidians or partly retracted anemones. A single small gorgonian, a few hydroids with rhizoids, and some sponges were also trawled.

Hard substrates photographed were bare of larger epifauna except for the anemones mentioned above. Most of the aragonitic plates recovered in grabs and trawls must have been covered with sediment with no surface exposed. Some organisms (including Foraminifera-like encrusting species) were found on clinker originating from former steam ships. A few specimens of multistalked scleractinian corals were trawled, single Porifera, Hydrozoa and polypoid Scyphozoa of the *Stephanoscyphus*-type, some sedentary polychaetes, barnacles, brachiopods, bryozoans and ascidians were recovered together with small colonies of Bryozoa and Endoprocta growing on the larger sediment particles. Trawl samples have no quantitative value, but with trawling experience from many different habitats, the low density of the fauna becomes evident. The report by Monin *et al.* (1982) conveys the same impression of very low faunal densities.

6.3.3. The vagile (mobile) fauna

The predominant taxa in this ecological group are fish and shrimps, as estimated from trawling and from counts on photographs. Trawl and photosled were towed at only 2 knots and some specimens will have escaped recording. From 5400 m^2 of photographic records a total of 34 specimens were counted: 6 flatfishes (Fig. 6.1) presumably Cynoglossidae, 11 eel-like individuals and 17 other specimens, resulting in a very low density of about 1 specimen per 160 m^2. Larger fish were attracted to the baited phototrap (Klausewitz and Thiel, 1982). A dagger-tooth *Muraenesox cinereus* was only seen twice on the photographic record, on the second occasion it was caught (Fig. 6.6). A small (0.6 m) shark *Iago omanensis* was trapped (Fig. 6.7) and a 3–4 m long shark was seen once outside, when the phototrap was deployed at 456 m depth at the base of the steep coastal slope.

Other vagile fauna belong to the taxa Decapoda, Stomatopoda, Gastropoda, and the Ophiuroidea, Asteroidea, Echinoidea and Holothuroidea, but all are rare and mostly inconspicuous.

Vagile fauna, again mainly fish and shrimps, were also caught in the phototrap (Fig. 6.7) (Thiel and Rumohr, 1979) and in smaller traps (Thiel, 1980). Since the area from which the specimens were attracted to the bait is unknown, no density estimates can be given from these findings. Other taxa found at the bait were lysianassid amphipods (Andres, 1981), cirolanid isopods, Leptostraca and the decapod *Charybdis* sp.. In contrast to the shrimps gathered in the trawl, those caught in the trap were perfectly clean, with no adhering sediment particles, and well suited for the analyses of heavy metal concentrations (*see* Chapter 4, also Karbe *et al.*, 1981).

6.3.4. The infauna

The faunal components in this group do not all live exclusively within the sediment. Small epifauna or vagile species, crawling along the sediment surface may be counted as well. All the species however are relatively small and can be sampled with a grab; furthermore they appear in sufficient numbers for quantitative assessments. This grab-fauna is divided into two size groups, the macrofauna and the meiofauna.

The macrofauna

The main groups belonging to this part of the fauna (> 1 mm) are the Polychaeta, within the Crustacea the Cumacea, Isopoda and Amphipoda, the molluscan classes Aplacophora, Bivalvia and Scaphopoda, the Nemertini, Sipunculida, and Priapulida. None of the taxa are numerous in the Red Sea

TABLE 6.2. Comparison of macrofauna densities (rounded values of numbers of individuals/m^2) for different ocean regions.

Region	Depth range in metres		
	500—750	1000—1250	1750—2000
Red Sea	120—950	225— 750	110—260
off Morocco[1]	1130—1540	930	820
NW African upwelling[2]	1400—3300	1400	500
Peru upwelling[3]	2270—5600	500	—
off Massachusetts[4]	6400	3070—4430	700—2020
Gulf of Mexico[4 5]	420—780	100—400	995

[1] Pfannkuche, Theeg and Thiel 1983
[2] Thiel 1982
[3] Rowe 1971
[4] Rowe, Polloni and Horner 1974
[5] Rowe and Menzel 1971

and most specimens are of small size, summing up to low standing stocks of 120 to 960 specimens/m^2 and to a low biomass of only 0.05 to 0.5 g dry weight/m^2. Standing stock distribution of macrofauna exhibited no clear trend between the regions near the coastal slope and the central part, but decreased down the slope of the central rift. A similar trend was found by Pasternak (1982), who investigated the benthos on an E—W-transect at 18°N. His data are not fully comparable as benthos size groups are differently defined. In depths around 400 m (i.e. the coastal slope) he found slightly more than 5 g/m^2, but on the deep terrace the values decreased to about 1.1 mg/m^2 at around 650 m depth. On hard bottoms, like rock or lava, biomass decreased to 0.02—0.05 g/m^2. All these values seem to be wet weights. The total number of data is still so sparse that an increase to the south is not demonstrable from Pasternak's and my own data.

Macrofauna densities and biomasses from shelf and coastal regions, including only 5 samples between 500 and 850-m depth, were presented by Murina (1971) for the Red Sea. The averaged data indicate an increase to the south which might also occur at deeper stations, as standing stock of plankton is higher in the south, and the inflow of water from the Indian Ocean transports detritus into this region (Weikert, 1982 and Chapter 5).

The low faunal densities in the Red Sea are demonstrated in Table 6.2 by comparison with other regions.

The meiofauna

Like the macrofauna, the meiofauna (< 1 mm), nematodes, harpacticoids, ostracods, small polychaetes and others, including the benthic Foraminifera, exhibits low values in standing stock ranging from 50 to 600 specimens/10 cm^2 and were calculated to 15—180 mg dry weight/10 cm^2 in biomass (Thiel, 1979 and unpublished data). Table 6.3 compares densities from different regions, and the low values in the Red Sea samples are again apparent.

Within the Red Sea, highest values were found near the coastal slope, decreasing towards central regions, reflecting the higher food input to the seabed nearer to the coast. No specific trend was discovered between the central and the northern regions.

A characteristic of regions with low meiofaunal density is the limited depth distribution of these organisms in the sediment (Thiel, 1983a). As the food arrives at the sediment surface, the organisms tend to concentrate there. In the Red Sea deep sediments 85 to 90% of the meiofauna occur in the

TABLE 6.3. A comparison of meiofauna densities (values rounded, omitting Foraminifera, specimens/10 cm^2) from different ocean regions.
(Summarized from various sources, Thiel 1983)

Region	approximate water depth in metres		
	500	1000	2000
Red Sea	400	50	100
Western Mediterannean	—	1200	500
East Atlantic: NW Africa	1500	1200	300
East Atlantic: Bay of Biscay, Portugal	800	700	500
Norwegian Sea Iceland-Faroe Ridge	2600	2000	1000
North Atlantic Iceland-Faroe Ridge	1000	40	300

uppermost 1-cm layer, whereas in areas with higher standing stocks only about 50% of the meiofauna is concentrated in that horizon. Meiofaunal density decreases in deeper sediment layers, with less penetration in the Red Sea than in higher populated areas. This is confirmed by electron transport system activity, a value of the potential respiration level (Theeg, 1985).

6.4. FUNCTIONAL CHARACTERISTICS OF THE DEEP BENTHOS OF THE RED SEA

To understand any marine ecosystem or its benthic component it is essential to consider rates of metabolic activity. No suitable data are available for Red Sea fauna, but a rough estimate can be made from physiological knowledge. Temperature generally increases metabolic rates two to three fold per 10°C rise, so unless animals have specially adapted to the high ambient temperatures, high metabolic rates may be expected in the Red Sea throughout all depths, compared with those of other oceans.

The transport rate of energy from the pelagic realm to the benthic system is low. Weikert (1982 and Chapter 5) describes the low standing stocks of oceanic plankton, especially in deep waters, suggesting a low transfer of organic matter into the deep layers. Wishner (1980) discovered low plankton densities in nearbottom waters. The high temperature should result in high degradation and turnover rates for organic matter in the water column. As waters become deeper the energy input to the benthos should diminish even within the overall low values. Low sedimentation rates of organic matter result in low energy availability for the benthos, and the high temperatures (compared with other deep-sea environments) may result in high respiration rates and a high loss of energy through respiration. It is likely that only a relatively small amount of energy will remain to get fixed as standing stock, i.e. for biomass production in benthic animals.

Comparing the results of shipboard respiration measurements and calculations of respiration potential from the deep Red Sea benthos with respective determinations from samples taken in the Atlantic Ocean of Morocco in equivalent depths, our hypothetical considerations were confirmed. Based on standing stock, respiration is relatively high in the warm Red Sea waters and low in the cold Atlantic waters (Pfannkuche, Theeg and Thiel, 1983; Theeg, 1985, Thiel, Pfannkuche, Theeg and Schriever, in press).

Within this rough scheme, valid in the context of a broad comparison on a world wide scale, the benthic subsystem fits together with the pelagic subsystem (Weikert, 1982) to form a unified oceanic system, which is governed by low standing stocks, high metabolic rates and low production. Although the plankton and the benthos investigations were mainly undertaken in the central Red Sea some data from the literature and from our own researches in other Red Sea regions allow us to conclude that this basic scheme will suit the oceanic system throughout the Red Sea as a whole. However, this is a hypothetic generalization, which has to be confirmed or rejected by future investigations.

Fig. 6.6. *Muraenesox cinereus* (about 1 m long) approaching the bait in a phototrap (depth 456 m). Phototrapping was used to learn about the behaviour of organisms attracted to bait and to aid in the development of smaller free fall traps, which will be essential equipment in later monitoring programmes. (Centre foreground and top: funnel-like entrances; centre middle: battery for camera and flash-operation; left: bait hanging in netting; right: mini trap made from plastic bottle).

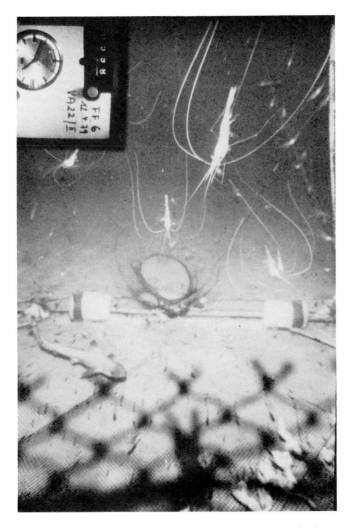

Fig. 6.7. A small shark, *Iago omanensis*, inside the phototrap together with some shrimps (depth 740 m). (Centre: entrance; left and right: mini traps. Netting surrounding the camera fell loose during lowering of the trap; some netting close to the camera lens shows up like a fence at the bottom of the picture).

ACKNOWLEDGEMENTS

I am grateful to Dr. Stephen Head for linguistic and editorial help during the preparation of this article.

REFERENCES

Andres, H. G. (1981) Lysianassidae aus dem Abyssal des Roten Meeres, Bearbeitung der Köderfange von FS "Sonne" — MESEDA I (1977) (Crustacea: Amphipoda: Gammaridea). *Senckenberg. biol.* 61, 429—43.

Bäcker, H. (1981) The geological environment. In: *Mining of Metalliferous Sediments from the Atlantis II Deep, Red Sea: Pre-mining environmental conditions and evaluation of the risk to the environment.* Eds. L. Karbe, H. Thiel, H. Weikert and A. J. B. Mill. Environmental Impact Study presented to the Saudi Sudanese Red Sea Joint Commission, Jeddah. Hamburg, pp. 5—20.

Balss, H. (1915) Expeditionen S.M. Schiff "Pola" in das Rote Meer. Nördliche und südliche Hälfte 1895/96—1897/98. Zoologische Ergebnisse. XXX. Die Decapoden des Roten Meeres. I. Die Macruren. Denkschrift. *Akad. Wiss. math. naturw.* 91., (Ber. Komm. ozeanogr. Forsch.): 1—38; Wien.

Berger, W. H. (1978) Deep-sea carbonate: pteropod distribution and the aragonite compensation depth. *Deep-Sea Res.* 25, 447—452.

Castle, P. H. J. (1982) Tiefenwasser- und Tiefseefische des Roten Meeres. III. A new species of *Uroconger* from Red Sea benthos (Pisces: Teleostei: Congridae). *Senckenberg. biol.* 62, 205—209.

Ekman, S. (1953) *Zoogeography of the Sea.* London, Sidgewick and Jackson. 417 pages.

Fuchs, T. (1901) Über den Charakter der Tiefseefauna des Roten Meeres auf Grund der von den österreichischen Tiefsee-Expeditionen gewonnenen Ausbeute. *S.B. Akad. Wiss. Wien, math-nat. Kl.*, 110, 249—258.

Hughes, R. N. (1977) The biota of reef-flats and limestone cliffs near Jeddah, Saudi Arabia. *J. nat. Hist.* 11, 77—96.

Karbe, L., Moammar, M. O., Nasr, D. and Schnier, C. (1981) Heavy metals in the Red Sea Environment. In *Mining of Metalliferous Sediments from the Atlantis II Deep, Red Sea: Pre-mining Environmental Conditions and Evaluation of the Risk to the Environment.* Eds. L. Karbe, H. Thiel, H. Weikert and A. J. B. Mill. Environmental Impact Study presented to the Saudi Sudanese Red Sea Joint Commission, Jeddah. Hamburg, pp. 195—226.

Karrer, C. and Klausewitz, W. (1982) Tiefenwasser- und Tiefseefische aus dem Roten Meer. II. *Dysomma fuscoventralis* n.sp., ein Tiefsee-Aal aus dem zentralen Roten Meer. (Teleostei: Anguilliformes: Synaphobranchidae: Dysomminae). *Senckenberg. biol.* 62, 199—203.

Klausewitz, W. (1980) Tiefenwasser- und Tiefseefische aus dem Roten Meer. I. Einleitung und Neunachweis für *Bembrops adenensis* Norman 1939 und *Histiopterus spinifer* Gilchrist 1904. (Pisces: Perciformes: Percophidae, Pentacerotidae). *Senckenberg. biol.* 61, 11—24.

Klausewitz, W. (1981) Tiefenwasser- und Tiefseefische aus dem Roten Meer. IV. Neunachweis von *Lophiodes mutilus* (Alcock), mit Bemärkungen über *Lophius (Chirolophius) quinqueradiatus* Brauer und *Chirolophius papillosus* (Weber). (Pisces: Lophiiformes: Lophiidae). *Senckenberg. marit.* 13, 193—203.

Klausewitz, W. (1983a) Tiefenwasser- und Tiefseefische aus dem Roten Meer. VII. *Harpadon erythraeus* n.sp. aus der Tiefsee des zentralen Roten Meeres. *Senckenberg. biol.* 64, 35—45.

Klausewitz, W. (1983b) Tiefenwasser- und Tiefseefische des Roten Meeres. IX. Tiefenrekord des Haifisches *Iago omanensis* (Norman). (Pisces/Chondrichthyes: Elasmobranchii: Carcharhinidae). *Senckenbergiana marit.* 15, 39—42.

Klausewitz, W. (1985) Tiefenwasser- und Tiefseefische aus dem Roten Meer. XI. *Neocentropogon mesedai* n. sp. aus dem Mesobenthos. (Pisces: Teleostei: Scorpaenidae: Tetraroginae). *Senckenberg. marit.* 17, 15—23.

Klausewitz, W. (1986) Zoogeographic analysis of the vertical distribution of the deep Red Sea ichthyofauna, with a new record. Senckenbergiana marit. 17, 279—292.

Klausewitz, W. and Thiel, H. (1982) Tiefenwasser- und Tiefseefische aus dem Roten Meer.VI. Über das Vorkommen des Haifisches *Iago omanensis* (Norman) (Pisces: Chondrichthyes: Elasmobranchii: Carcharhinidae) und des Messerzahnaals *Muraenesox cinereus* (Forsskål) (Teleostei: Apodes: Muraenesocidae) im zentralen Roten Meer, beide mit Hilfe der Fotofalle beobachtet und gefangen. *Senckenberg. marit.* 14, 227—243.

Manning, R.B. (1981) First record of *Kempina zanzibarica* (Chopra 1939) from the Red Sea, with notes on *Lenisquilla gilesi* (Kemp 1911). (Crustacea: Stomatopoda). *Senckenberg. biol.* 61, 297—303.

Marenzeller, E. von (1907) Tierseekorallen des Roten Meeres. *Denkenschrift Akad. Wiss. Wien.* math-nat. Kl. 80, 13—25.

Michaelsen, W. (1918) Ascidia Ptychobranchia und Dictyobranchia des Roten Meeres. *Denkschrift Akad. Wiss., Wien, math-nat.* Kl. 95, 1—128.

Michaelsen, W. (1921) Ascidia Krikobranchiae des Roten Meeres. *Denkschrift Akad. Wiss. Wien. math-nat.* Kl. 97, 1—38.

Monin, A. S., Voitov, V. I. and Yastrobov, V. S. (1980) The Red Sea expedition of the P.P. Shirshov Institute of Oceanology of the USSR Academy of Sciences (Operation Pikar). *Oceanology* 20, 489—93.

Monin, A. S., Litvin, V. M., Podrashansky, A. M., Sagalevich, A. M., Sorokhitin, O. G., Voitov, V.O. Yastrebov, V.S. and Zonenshain, L. P. (1982) Red Sea submersible research expedition. *Deep-Sea Res.* 29, 361—73.

Morcos, S. A. (1970) Physical and chemical oceanography of the Red Sea. *Oceanogr. mar. Biol. Ann. Rev.* 8, 73—202.

Murina, W. W. (1971) Characterization of the Red Sea macrobenthos. In *The Benthos of the Red Sea Shelfs.* Akad. Nauk SSSR. Verlag Naukowa Dunka, Kiew. 3—22.

Pasternak, F. A. (1982) Quantitative distribution of the deep-sea bottom fauna in the southern part of the Red Sea. *Transact. P.P. Shirshov Institute of Oceanology, Akad. Sci. USSR* 117, 42—6.

Pfannkuche, O., Theeg, R. and Thiel, H. (1983) Benthos activity, abundance and biomass under an area of low upwelling off Morocco, Northwest Africa. *"Meteor" Forsch.-Ergebnisse D,* 36, 85—96.

Por, F. D. and Lerner-Seggev, R. (1966) Preliminary data about the benthic fauna of the Gulf of Elat (Aqaba), Red Sea. *Israel J. Zool.* 15, 38—50.

Riemann-Zürneck, K. (1983) *Halcurias sudanensis* n.sp. aus dem Tiefwasser-Benthos des Roten Meeres. (Actiniaria: Halcuriidae). *Senckenberg. biol.* 63, 427—34.

Rowe, G. T. (1971) Benthic biomass in the Pisco, Peru upwelling. *Inv. Pesq.* 35, 127—35.

Rowe, G. T. and Menzel, D. W. (1971) Quantitative benthic samples from the deep Gulf or Mexico with some comments on the measurement of deep-sea biomass. *Bull. Mar. Sci.* 21, 557—66.

Rowe, G. T., Polloni, P. T. and Horner, S. G. (1974) Benthic biomass estimates from the northwestern Atlantic Ocean and the northern Gulf of Mexico. *Deep-Sea Res.* 21, 641–50.

Rowe, G. T. and Staresinic, N. (1979) Sources of organic matter to the deep-sea benthos. *Ambio Spec. Rep.* 6, 19–23.

Sanders, H. L. and Hessler, R. R. (1969) Ecology of the deep-sea benthos. *Science* 163, 1419–24.

Spandl, H. (1924) Die Amphipoden des Roten Meeres. *Denkschr. Akad. Wiss. Wien. math-nat. Kl.* 99, 19–73.

Theeg, R. (1985) Die Aktivität des Elektronen Transport Systems benthischer Lebensgemeinschaften. Diss. Univ. Hamburg, 1–195.

Thiel, H. (1979) First quantitative data on the Red Sea deep benthos. *Mar. Ecol. Progr. Ser.* 1, 347–50.

Thiel, H. (1980) Benthic investigations of the deep Red Sea. Cruise reports: R.V. "Sonne" — MESEDA I (1977), R.V. "Valdivia" — MESEDA II (1979). *Cour. Forsch. -Inst. Senckenberg* 40, 35 pages.

Thiel, H. (1981) The benthic communities. In *Mining of Metalliferous Sediments from the Atlantis II Deep, Red Sea: Pre-mining Environmental Conditions and Evaluation of the Risk to the Environment.* Eds. L. Karbe, H. Thiel, H. Weikert and A. J. B. Mill. Environmental Impact Study presented to the Saudi Sudanese Red Sea Joint Commission, Jeddah. Hamburg, pp.157–71, 174–8.

Thiel, H. (1982) Zoobenthos of the CINECA area and other upwelling regions. *Rapp. P.-v. Réun. Cons. int. Explor. Mer* 180, 323–34.

Thiel, H. (1983a) Meiobenthos and Nanobenthos of the deep sea. In *The Sea.* Ed. G. T. Rowe, Vol. 8, Chapter 5, 167–230, J. Wiley & Sons, Inc. Publ., New York.

Thiel, H. (1983b) Pteropod shells: another food source for deep-sea organisms. *Senckenberg. marit.* 15, 147–55.

Thiel, H., Pfannkuche, O., Theeg, R. and G. Schriever (in press) Benthik metabolism and standing stock in the central and northern Deep Red Sea. P.S.Z.N.I.: Marine Ecology.

Thiel, H. and Rumohr, H. (1979) Photostudio am Meeresboden. *Umschau* 79, 469–72.

Thiel, H. and Weikert, H. (1984) Biological oceanography of the Red Sea oceanic system (Abstract). *Deep-Sea Res.* 31, 829–31.

Thiel, H., Weikert, H. and Karbe, L. (1985) Abschätzung der Umweltrisiken durch Abbau von Erzschlämmen aus dem Atlantis-II-Tief im Roten Meer. *Nature und Museum* 115, 98–110.

Thiel, H., Weikert, H. and Karbe, L. (1986) Risk assessment of mining metalliferous muds in the deep Red Sea. *Ambio.*

Türkay, M. (1986) Crustacea, Decapoda, Reptantia der Tiefsee des Roten Meers. *Senckenberg. marit.* 18, 123–185.

Weikert, H. (1981) The pelagic communities. In *Mining of Metalliferous Sediments from the Atlantis II Deep, Red Sea: Pre-mining Environmental Conditions and Evaluation of the Risk to the Environment.* Eds. L. Karbe, H. Thiel, H. Weikert and A. J. B. Mill. Environmental Impact Study presented to the Saudi Sudanese Red Sea Joint Commission, Jeddah. Hamburg, pp. 100–54.

Weikert, H. (1982) The vertical distribution of zooplankton in relation to habitat zones in the area or the Atlantis II Deep, central Red Sea. *Mar. Ecol. Prog. Ser.* 8, 129–43.

Wishner, K. F. (1980) The biomass of the deep-sea bentho-pelagic plankton. *Deep-Sea Res.* 26, 203–16.

Young, R. A. and Ross, D. A. (1974) Volcanic and sedimentary processes in the Red Sea axial trough. *Deep-Sea Res.* 21, 289–97.

CHAPTER 7

Corals and Coral Reefs of the Red Sea

STEPHEN M. HEAD

Zoology Department, University of the West Indies, Mona, Kingston 7, Jamaica

CONTENTS

7.1. INTRODUCTION

7.1.1. General introduction

The coral reefs of the Red Sea are among the most attractive, most photographed and most studied of any in the world. They represent an immense economic resource for recreational and tourist use, and as fisheries. With the exception of local problems, they are largely unspoilt by pollution or human interference.

Scientific research on Red Sea reefs dates from the ill fated journey of Peter Forsskål, a Swedish scientist who succumbed to malaria in Yemen in 1763. Forsskål's posthumously published account is one of the earliest of all reef studies. Later, the great German biologist Ehrenberg travelled in the area, publishing important accounts of reefs and corals. Ernst Haeckel visited the reefs of Sinai in 1873, and waxed lyrical on the beauty of the marine life 'where every animal is like a flower' (Haeckel, 1876).

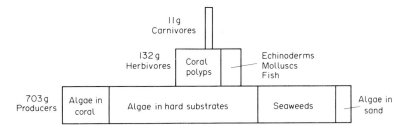

Fig. 7.1. Pyramid of biomass for a coral reef community. Simplified from a figure in Lewis (1977), based on Odum and Odum's data for Eniwetok.

Other early studies include Walther (1888) and Faurot (1888), but the most outstanding was the contribution of C. B. Kluzinger (1879) whose monograph of Red Sea corals is still very useful today. Klunzinger studied corals and many other organisms while working as a doctor in the Egyptian coastal town of Quseir.

Recent scientific studies are summarised elsewhere in this chapter. Popular accounts of Red Sea reefs can be found in Cyril Crossland's (1913) *Desert and water gardens of the Red Sea*, now sadly a rare book, Hass (1952), Roghi and Baschieri (1956) and Cousteau and Diole (1970). An excellent diving guide to the reefs of Israel is available (Cohen, 1975) and a superbly illustrated general account of the reefs and their fauna by Bemert and Ormond (1981).

This chapter aims to provide a brief outline of the nature of reef communities and to review the ecology of Red Sea reefs with particular reference to scleractinian corals. Other reef organisms are extensively discussed in other chapters of this book.

7.1.2. Nature of coral reef communities

Coral reefs are large masses of biogenic rock and sediment, capped with a thin veneer of living organisms, some of which, especially the corals and algae, produce the calcium carbonate of which the reef mass is composed. By their upward growth in well-lit waters, these organisms can compensate for gradually rising sea levels or land subsidence, a simple but far reaching capability first deduced by Darwin (1842).

Reef upward growth is slow, about 0.2—0.7 cm per year (Stoddart, 1969), but reef deposits 1000 m thick are recorded in the Pacific. Some reefs must therefore be over a million years old, while globally reef communities have been in existence since the Triassic period about 200 million years ago. This long period of evolution is one reason why the modern reef community is very diverse and complex, built by comparatively few coral species but inhabited by a bewildering array of fish and invertebrates, many highly specialised in their ecological niche. There are many other possible reasons for the great diversity of reefs, the reader is referred to Rosen (1981) for an excellent analysis from the 'coral's-eye view'.

Modern reefs are highly productive ecosystems, anomalously set in waters of very low nutrient status. Lewis (1977) gives reef production levels as high as 10,000 $gC.m^{-2}.y^{-1}$ in waters typically supporting only 20—50 $gC.m^{-2}.y^{-1}$ production by the pelagic community. High reef productivity seems to be related to the ability of the reef community to retain and recycle all nutrients coming to it. Central to this is the key position of the corals, their ability to absorb nitrogen and phosphorus compounds from seawater, and their role as producers and consumers of detritus (Lewis, 1977).

Corals are only one element of the reef community, although in many respects the most important. The reef ecosystem is structured like all others, with primary producers, herbivores and carnivores. The

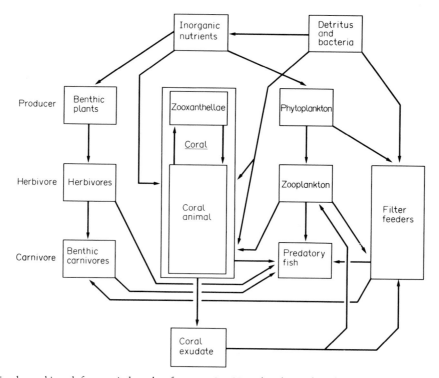

Fig. 7.2. Simple trophic web for a typical coral reef community. Note that the coral symbiosis as a whole occupies all trophic levels from producer to carnivore, and that at least four food chains exist in parallel (Based on data in Lewis, 1977). Pathways of nutrient and detritus regeneration are omitted.

primary producers may be plants or animals, such as corals, containing symbiotic plant cells. Figure 7.1 shows a pyramid of biomass for Eniwetok in the Pacific Ocean. Biomass declines between trophic levels, in the ratio 64 producers : 12 herbivores : 1 carnivore. Comparable data for other communities shows that the 5:1 producer/herbivore ratio is low for a non-planktonic community. One explanation for this may be that as we shall see, corals are carnivorous planktivores as well as functional herbivores, so the large coral polyp biomass shown in Figure 7.1 may be supported partly by other nutrient sources.

Figure 7.2 attempts in a simple way to depict the trophic structure of a reef community. It illustrates the central position of corals as primary and secondary producers, consumers of plankton and detritus, and secreters of energy-rich exudates which are exploited by other benthic and planktonic animals (Richman *et al.*, 1975). It is also striking that on the reef several food chains exist in parallel, with trophic interconnections at many levels; this is probably another contributory factor to the diversity of the community.

7.1.3. Coral reefs of the Red Sea and their form

The Red Sea is peripheral to the main area of the Indo-Pacific, and extends to 30°N, so one might expect cool water temperatures would inhibit reef development. Reefs do not develop in waters with an annual minimum temperature less than about 10°C (Stoddart, 1969), but as we have seen in Chapter 3, the Red Sea is unusually warm, and good reef communities are found as far north as the tip of the Gulf of Aqaba. Sea water temperature there falls to 21.2°C annually (Mergner and Schuhmacher, 1974), while in the central Red Sea the February minimum is 24.5°C (Head, 1980). It is possible that this small difference contributes to the better development of reefs in the central Red Sea, but many other factors such as productivity, wave energy and geological history may also be important.

Figure 7.3 shows very approximately the development of reefs in the Red Sea, based on data in Bemert and Ormond (1981). Reefs occur on most of the length of the Red Sea, on both coasts, but tend to be at their best in the central and northern Red Sea, on the coasts of Sudan, Saudi Arabia and Egypt, and it is there that the greatest variety of reef types can be found.

Since Darwin's early account (1842) three basic reef types have been recognised. Fringing reefs lie close to shore, barrier reefs arise at the edge of an offshore shelf, and atolls are large ovoid reef structures in mid ocean, with a substantial central lagoon. Atolls are not well developed in the Red Sea, perhaps because of its recent and complex geological history, but Sanganeb Reef near Port Sudan is normally recognised as a small true atoll.

The basic reef type in the Red Sea is the fringing reef, and these vary greatly in physical size. Fishelson (1980) has described minimal 'contour' reefs, sometimes only 3—4 m wide, in the Gulf of Aqaba, south of Dahab, they cling precariously to steep Precambrian basement rock cliffs at sea level, water depths below being too great for outward reef growth. In the Sudan by contrast, the seaward edge of the fringing reef may be over 1 km from the shore, with a substantial 10 m deep lagoon in between.

Fringing reefs grow slowly seawards on a slope of biogenic debris, with active coral growth mainly at the seaward crest where water exchange is good. The evolution of a fringing reef is shown in Figure

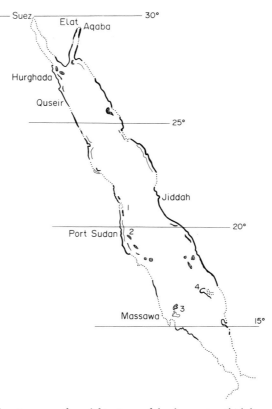

Fig. 7.3. Map of the Red Sea, showing areas of good fringing reef development as thick lines on the coast. Shores without extensive reefs shown as a dotted line, 'barrier' structures and coral islands shown as dark lines offshore. The four areas numbered are 1) Shaab Baraja off Dungunab Bay, 2) is Sanganeb Atoll, 3) is the Dahlak Island group, 4) is the Farasan group (including Sarso Island). Simplified from Bemert and Ormond (1981).

7.4. As the reef flat becomes wider, a shallow lagoon may form if erosion and solution exceeds the input of sediment. Some coral growth may occur in the lagoon, especially if a back-reef is formed where the lagoon meets the fringing reef flat.

Barrier reefs arise on coasts where there is a slow rise of sea level relative to the land, and a previously fringing reef finds itself up to 70 km off the new shore line, where a new fringing reef may develop. True barrier reefs do not occur in the Red Sea, but barrier-like structures, often several kilometres long, arise offshore from deep water off the coasts of Sudan, Egypt and Saudi Arabia. They appear to be formed at the seaward edges of large faulted blocks of Tertiary limestone (Berry *et al.*, 1966; Braithwaite, 1982), and so the distribution and form of the reefs is not the result of steady coral growth. Instead the form depends on tectonic control with extensive sub-aerial erosion some 20,000 years ago during the late Pleistocene when sea level was some 120 m lower than at present and the climate colder (Fishelson, 1980).

Figure 7.5 shows a schematic section through the complex Towartit system off Port Sudan. The shallow offshore platform of which the 'barrier' forms the outer margin carries small pinnacle or patch reefs with flourishing coral communities, interspersed with deep pits. This is a 'fossil landscape', formed by erosion and solution during the Pleistocene low sea-level stand. The major features of all Red Sea reefs must therefore reflect past erosional episodes, and only a few metres of recent coral growth can have occurred in the 5−6000 years during which the Red Sea has stood at its present level (Braithwaite, 1982 and Chapter 2).

Whatever the origins of its structures, reef systems such as Towartit offer the biologist a complex and varied set of habitats, differing in physical scale, depth, water clarity and wave exposure. Consequently

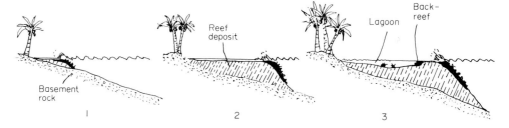

Fig. 7.4. Evolution of a fringing reef. 1) Early stage with little carbonate accumulation. 2) Stage with shallow reef flat, 3) Late stage with lagoon and back-reef developed. Heavy dark areas show good coral growth.

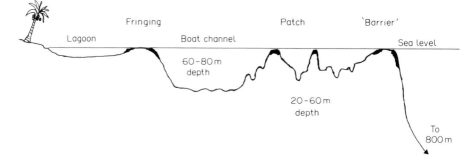

Fig. 7.5. Diagrammatic section from the shoreline through the Towartit reef complex south of Port Sudan, showing the three main reef types. Thick shading shows good coral growth, total width of section about 8 km.

the variety of community types is very great within a small area, an ideal combination for research or recreational diving.

The fringing reefs of the Red Sea are penetrated at intervals by narrow channels, called *marsas* on the western, and *sharms* on the eastern coast. These are interpreted as drowned river valleys cut during the Pleistocene (Berry *et al.*, 1966), and they generally still connect to *wadis* or seasonal rivers. Marsas and sharms make excellent natural harbours but are becoming choked with coral growth, which has contributed to the decline of early ports such as Suakin.

7.2. INTRODUCTION TO CORALS AND THEIR ECOLOGY

The word 'coral' is very imprecise, at least seven cnidarian orders are included in the term at its broadest. In this chapter we are mainly concerned with 'true' corals of the order Scleractinia, class Anthozoa, phylum Cnidaria. Most of these are colonial animals, the settling larva develops first as a single individual, then divides or buds to produce a clonal colony which continues to grow in an integrated fashion to produce one of many characteristic growth forms. The coral tissue secretes a hard exoskeleton of calcium carbonate to support and protect the colony, and this provides 'artificial substrate' for the further growth of the colony, so it is not limited by substrate availability. On the death of the colony, the skeleton is available for settlement by other coral larvae, and so becomes incorporated into the structure of the reef. Useful reviews of coral structure can be found in Barnes (1980) and Schuhmacher (1976).

Two other 'coral groups' are important in the Red Sea. Soft-corals, members of the order Alcyonacea lack a hard skeleton, although their chalky spicules contribute to reef sediments. Alcyonaceans are important space competitors with scleractinians (Benayahu and Loya, 1977, 1981). Fire-corals belong to the genus *Millepora*, they are colonial hydrozoans which build substantial carbonate skeletons and contribute substantially to reef building. *Millepora* is notable in another context — while all cnidarians have stinging cells, those of *Millepora* can sting people, which they do readily and very painfully. Comparative views of the three 'coral' orders are given in Figures 7.6 and 7.7. More is said about *Millepora* and alcyonaceans in chapter 12. Elsewhere in this chapter the term 'coral' can be taken to refer only to scleractinians.

Coral colonies exhibit a great variety of size and shape. Some, such as the important Red Sea genera *Goniastrea*, *Platygyra* and *Favia*, are massive, forming solid, usually hemispherical colonies, with relatively large polyps normally retracted in the daytime. Such corals are rather slow growing (less than 1 cm per year), but are strong and survive heavy wave action. Other corals, including the abundant *Acropora* and *Stylophora* are branching, bush or tree like. Their polyps are small, often partially expanded during the day, and some grow very fast. At Shaab Rumi in the Red Sea, Vine and Head (1977) recorded 39 cm growth in ten months in one *Acropora* species. Such corals are easily broken by wave action, although this can be a major mechanism of asexual reproduction in some species.

Corals enjoy a dual nutritional system of profound importance to reef ecology. They are equipped with the usual anthozoan armoury of stinging cells (nematocytes), and are capable of voracious zooplankton feeding, a method which traps food material imported to the reef ecosystem by water currents. In addition, reef-building corals contain symbiotic algal cells, dinoflagellates called zooxanthellae. These algae contain chlorophyll and perform the function of photosynthesis with great vigour. Although their role in coral nutrition was once doubted, it is now clear that the algae manufacture sufficient food for the symbiosis of alga plus coral animal to be autotrophic to considerable depths on clear-water reefs. Svoboda (1978), working in the Red Sea, found that at 40 m depth, where

Fig. 7.6. Knoll at about 2 m depth in Sudanese Red Sea, showing large colonies of the hydrocoral *Millepora dichotoma*. Note the planar branching and the pure white tips to the buff branches. This coral stings. In the foreground is a large *Stylophora pistillata* colony.

Fig. 7.7. Coral assemblage at about 1.5 m depth on the leeward side of a Sudanese patch reef. In the foreground is a large colony of the alcyonacean *Sarcophyton*, with the polyps partly open. Behind are several scleractinian corals, the fine branching *Seriatopora hystrix*, *Pocillopora verrucosa* and foliate *Turbinaria mesenterina*.

light levels were only 7.75% of surface irradiance, *Acropora* and *Stylophora* were still producing (as plants) about 20% more oxygen than they would consume (as animals), over 24 hours. The coral animal receives a 'balanced diet' of amino acid, glucose and glycerol from the algae, which in turn receive essential nutrients from the coral. The nutrients may be in the form of excretory products from the animal partner's metabolism, but this is supplemented by the ability of the animal to take up nutrient molecules from solution in sea water (Lewis, 1977). In this way the productivity of the reef ecosystem is enhanced, and the wasteful transfer of energy from phytoplankton through zooplankton to animal is neatly by-passed as the symbiosis as a whole becomes autotrophic.

Zooxanthellae have been shown greatly to enhance skeleton formation rates during daylight hours, although how they do so is still obscure and may be complex. Rates of calcification when illuminated may be ten times those in the dark and it is this enhancement of calcification in shallow well-lit waters that allows corals to produce the mass of calcium carbonate necessary to form a solid, wave resistant reef structure.

In deep water, light levels become too low to sustain symbiotic autotrophy and calcification enhancement. Below the compensation depth, where 24 hour production just balances consumption, corals with symbiotic algae (zooxanthellates) disappear, while non-zooxanthellates persist, feeding entirely heterotrophically. Such corals are generally small and solitary, and are not a dominant feature of the benthic community.

Light energy is therefore a major control of coral ecology, although we still have much to learn of its mechanisms and practical range of influence. Another important control is wave energy and water movement in general. Strong wave action can break corals and limit their distribution (Vosburgh, 1977), but more subtle influences are also possible. For example, water movement brings plankton food to corals, but if currents are too high the corals may be unable to feed without special hydromechanical adaptations (Chamberlain and Graus, 1975). Water movement as a control of coral distribution is discussed further in section 7.4, together with the related factor of sedimentation.

Corals interact strongly with each other and with other reef organisms, competing for the limited available substrate. Corals can encroach, kill and overgrow their neighbours using nematocyst rich organs of aggression such as mesenterial filaments. A 'pecking order' exists on some reefs (Lang, 1973) which may materially affect the survival of subordinate species in certain conditions, but as Sheppard (1982) has noted, there is little evidence that this has a great effect on coral assemblage structure. Overtopping growth may be another mode of competition important in some areas, where rapid growth by branching corals shades and stifles slower growing massive species. In the Red Sea, *Acropora hyacinthus* forms huge flat 'table' colonies in medium depths, raised 50 cm or so above the substrate by a stalk. This clearly allows them to escape from substrate competition, but Sheppard (1981a) found no evidence that they inhibited the growth of underlying corals.

Many organisms feed on corals, either by selectively picking off the soft tissues, or by crunching whole chunks of skeleton. Parrotfish are among the most important of the latter, and their faeces contribute substantially to reef sand production. The specialised butterflyfish are important polyp feeders, while among the invertebrates, polychaete worms such as *Hermodice*, and the Crown-of-Thorns starfish (*Acanthaster*) are significant coral predators in the Red Sea. The influence of predation in reef ecology is poorly understood, except perhaps in the case of 'catastrophic' or 'plague' predation by *Acanthaster*. Recently Neudecker (1977) has shown that fish predation exerts a major control on the distribution of individual coral species, while Wellington (1982a) has shown its influence over reef zonation as a whole.

In addition to reproducing asexually by fragmentation, corals reproduce sexually, releasing either large brooded planula larvae or many small eggs which are fertilised and develop in the plankton (Fadlallah, 1983). The larvae always require hard algal-free substrate to settle, so it is important that new bare substrate be continually formed on reefs. In consequence, there is a non-linear relationship between the rate of grazing by such animals as echinoids and the recruitment success of corals. Too little grazing and no bare substrate is available, too much and the newly settled spat are killed before they are properly established (Sammarco, 1980).

7.3. SCLERACTINIAN CORAL FAUNA OF THE RED SEA

The Red Sea coral fauna is exceptionally well known, a result of the long and continuing history of research there. The first permanently valuable monograph was Klunzinger's (1879) study of the corals of Quseir. Von Marenzeller (1907) described corals from many Red Sea areas, subsequent useful lists or illustrated accounts are Rossi (1954) for Dahlak and the Eritrean coast, Loya and Slobodkin (1971) for Elat, Mergner and Schuhmacher (1974) for Aqaba, and Head (1978a, 1980, 1983) and Kühlmann (1983) for Sudanese reefs. Recently, the entire Red Sea coral fauna has been extensively reviewed by Scheer and Pillai (1983), and their well illustrated account should be a standard reference for many years. These authors listed 70 genera and 194 species of scleractinians, including for the first time non-zooxanthellates with the shallow reef building species.

Table 7.1 lists the 53 genera and 177 species of zooxanthellate corals so far known in the Red Sea. This is a rich and impressive fauna, considering how peripheral the Red Sea is to the Indo-Pacific ocean as a whole. The largest genera are *Acropora* and *Montipora*, with 15 species each, but these are taxonomically 'difficult' genera, urgently in need of major revision, and the real number of species could be more or less than the 15 estimated. Also well represented are *Stylophora*, *Pavona*, *Leptoseris*, *Cycloseris*, *Fungia*, *Porites*, *Favia* and *Favites*. *Stylophora* is a genus that may be speciating in the Red Sea, where it is much more diverse than for example on the Great Barrier Reef where only a single species occurs.

It is rather difficult to compare the coral faunas of different regions, because the intensity of sampling is so variable, and workers differ in their taxonomic interpretations. Table 7.2 shows the present documented coral diversity within five Red Sea regions. Diversity is lowest in the Gulf of Suez. While this is relatively undersampled, the very shallow water and cool temperatures prevailing there would not be expected to encourage a diverse fauna. Scheer's (1971) description of Râs Shukheir suggests poor reef development. The low diversity recorded in the southern Red Sea is probably largely a sampling artefact, and further collecting would probably raise the species count considerably.

The central Red Sea, northern Red Sea and Gulf of Aqaba have all been well sampled and have similar high diversities. The central Red Sea is marginally richer, containing for example *Diploastrea*, *Scolymia* and *Physogyra* not recorded further north. Two genera well represented in the Red Sea proper (*Craterastrea* and *Oulophyllia*) do not penetrate into the Gulf of Aqaba, but this is offset by the presence there of *Caulastrea*, *Trachyphyllia* and *Euphyllia*, rare genera not yet recorded in the main Red Sea basin. The slightly higher diversity in the central Red Sea is consistent with its higher ambient water temperatures compared with the north. Rosen (1981) has shown that water temperature is a good predictor of coral generic diversity, although other factors such as proximity to the Indian Ocean proper may also be significant.

The recently described genera *Craterastrea* and *Erythrastrea* are so far known only from Red Sea material; Table 7.3 lists species restricted to the Red Sea. Many of these are rare or very recently described, and may well be eventually encountered elsewhere. Other species on Table 7.3 are well established and rather common in the Red Sea, so that their apparent absence elsewhere may be significant. Thus *Stylophora wellsi*, *Acropora capillaris*, *Montipora stilosa*, *Goniopora klunzingeri*, *Porites nodifera* and *Acanthastrea erythraea* are the most likely candidates for endemic species status. It must be remembered however, that the recent geological history of the Red Sea (Chapter 2) has included periods of isolation from the Indian Ocean and cold, probably hypersaline conditions which would have killed

TABLE 7.1. Zooxanthellate coral genera in the Red Sea, with the total number of species of each genus so far recorded, their depth range and relative abundance. Based on lists in Scheer and Pillai (1983), Head (1980) and other authors. Abundance data from Head (1980), 1 signifies rare, 3 is common, 5 indicates abundant to dominant.

Genus	No. species	Depth range	Abundance
Psammocora	5	1 – 50m	1 – 2
Stylocoeniella	2	1 – 46m	2
Stylophora	6	1 – 80m	1 – 5
Seriatopora	3	2 – 52m	5
Pocillopora	2	1 – 30m	1 – 5
Astreopora	1	12 – 46m	2
Acropora	15	1 – 68m	4 – 5
Montipora	15	1 – 82m	1 – 5
Pavona	7	0 – 52m	1 – 5
Leptoseris	8	4 – 145m	1 – 2
Gardineroseris	1	1 – 40m	2
Pachyseris	2	8 – 67m	3
Siderastrea	1	1 – 40m	1
Coscinaraea	2	1 – 80m	3
Craterastrea	1	36 – 40m	1
Cycloseris	7	0 – 100m	1
Fungia	11	0 – 58m	1 – 3
Ctenactis	1	1 – 24m	3
Herpolitha	1	4 – 30m	2
Podobacia	1	3 – 50m	2
Alveopora	6	1 – 107m	1
Goniopora	6	1 – 65m	1 – 3
Porites	11	1 – 50m	2 – 5
Caulastrea	1	8 – 58m	—
Erythrastrea	1	—	—
Favia	9	0 – 67m	2 – 4
Favites	7	0 – 58m	1 – 3
Goniastrea	4	0 – 84m	1 – 5
Platygyra	3	0 – 40m	2 – 4
Leptoria	1	0 – 6m	2
Oulophyllia	1	3 – 58m	2
Hydnophora	2	0 – 15m	2 – 3
Diploastrea	1	9 – 26m	1
Leptastrea	3	0 – 82m	2 – 4
Cyphastrea	2	0 – 50m	2 – 4
Echinopora	2	1 – 85m	4 – 5
Plesiastrea	1	13m	1
Trachyphyllia	1	—	—
Galaxea	2	0 – 40m	2 – 3
Merulina	1	6 – 76m	3
Cynarina	1	5 – 92m	—
Lobophyllia	2	1 – 52m	2 – 4
Acanthastrea	3	0 – 72m	1 – 2
Blastomussa	3	4 – 98m	1
Scolymia	1	7 – 36m	1
Mycedium	1	2 – 60m	2
Echinophyllia	3	1 – 105m	1 – 3
Oxypora	1	4 – 108m	2
Euphyllia	1	60 – 98m	—
Plerogyra	1	3 – 65m	2
Physogyra	1	7m	1
Gyrosmilia	1	0 – 58m	2
Turbinaria	1	2 – 24m	2

TABLE 7.2. Genera and species of zooxanthellate coral recorded from five regions in the Red Sea. Data largely based on Scheer and Pillai (1983), supplemented by Kühlmann (1983) and Head (1980).

Area	Genera	Species
Gulf of Aqaba	47	130
Gulf of Suez	25	47
Northern Red Sea	45	128
Central Red Sea	49	143
Southern Red Sea	31	74

TABLE 7.3. Coral species so far recorded only from the Red Sea, based on Scheer and Pillai (1983) and Head (1983). Species marked α also represent possible endemic genera. Species marked β are of doubtful status. *P. columnaris* is probably an ecomorph of *P. lutea*, while *F. wisseli* is known from only one specimen.

Species	Year of first description
Stylophora wellsi	1964
Stylophora mammillata	1983
Stylophora kuehlmanni	1983
Acropora capillaris	1879
Montipora spongiosa	1834
Craterastrea levis	1983 α
Cycloseris doederleini	1906
Goniopora klunzingeri	1906
Porites columnaris	1879 β
Porites nodifera	1879
Erythrastrea flabellata	1983 α
Favia wisseli	1983 β
Platygyra crosslandi	1928
Merulina scheeri	1983
Acanthastrea erythraea	1879

all corals. It is possible that the entire Red Sea coral fauna has re-established only within the last 11,000 years, a period not conducive to extensive local speciation and endemism.

Overall, the present zooxanthellate coral diversity is high, but does not approach that of more central Indo-Pacific reef areas. Sheppard (1981b) lists 64 genera and over 200 species from the Chagos area. Nemenzo (1981) recorded an amazing 78 genera and 488 species (and varieties) of corals in the Philippines. The great diversity in these areas may be due to rapid local speciation, warmer waters, or simply to their more central situation, with large reef areas within the dispersion range of coral larvae.

7.4. CORAL DISTRIBUTION ON RED SEA REEFS

This chapter is being written by a coral ecologist who naturally sees reef zonation in terms of coral distributions. Quite different accounts could be given by reef geomorphologists or by specialists in fish or algae. Nevertheless, corals generally form the largest single component of the sessile benthos of Red Sea reef slopes (Benayahu and Loya, 1977, 1981) and impart character and complexity to the substrate, controlling the distribution of other organisms. This partly justifies devoting much space to coral distributions, some alternative viewpoints are available in other chapters.

Of the many studies published on Red Sea reef ecology, over 30 of special importance have been selected and are listed in Table 7.4. Over half the accounts are of work performed in the Gulf of Aqaba, Egypt and Sudan have also had a fair degree of attention, but Saudi Arabia and the southern Red Sea relatively little. The available data suggest that there are no major ecological differences between north, central and southern Red Sea reefs, although the limited accounts of Eritrean reefs (Wainwright, 1965) suggest they may be rather impoverished and poorly developed. The coverage of different reef types is also rather uneven. Most studies, such as those in the Gulf of Aqaba, are confined to fringing reefs. Few accounts are available of patch reef ecology, and even less on the important exposed 'barrier' reefs of the central Red Sea.

7.4.1. Reef top ecology

The Red Sea is a relatively calm water region, with very little tidal range, and corals are able to flourish to within a few centimetres of mean summer sea-level. The physical environmental conditions in very shallow water on the top of reefs are nevertheless extreme compared with the upper reef slope

TABLE 7.4. Important reef and coral assemblage studies in the Red Sea, not including taxonomic studies or accounts of single species.

Benayahu and Loya, 1977 Benayahu and Loya, 1981	Elat	Space partitioning between corals, algae and soft corals.
Bouchon, 1980	Aqaba	Coral assemblages, quantitative study.
Braithwaite, 1982	North and central Sudan	Mainly geomorphological, with outlines of community structure.
Crossland, 1911	North Sudan (Dungunab)	Mainly geomorphological, some reef community data.
Crossland, 1938	Egypt	Outline of Ghardaqa reef ecology.
Fishelson, 1973a	Elat	Reef top community response to pollution.
Fihselson, 1973b	Elat	Major analysis of reef top ecology.
Fishelson, 1980	Gulf of Aqaba	Review of Israeli marine reserves.
Fricke and Schuhmacher, 1983	Elat and Sinai	Distribution of corals in deep water.
Friedman, 1968	Gulf of Aqaba	Mainly geomorphological. Outline of coral assemblages.
Head, 1978b, 1979	North and central Sudan	Coral zonation and distribution.
Head 1980	Central Sudan	Detailed study of coral distributional ecology.
Klausewitz, 1967	Farasan	Shallow reef zonation, mainly fish data.
Klunzinger, 1870	Egypt	Fringing reef zonation at Quseir.
Kühlmann, 1983	Central Sudan	Outline of coral distribution to 70m.
Loya, 1972	Elat	Important study of coral zonation and general ecology.
Loya, 1976b	Elat	Recolonisation of reef flat after stress.
Loya and Slobodkin, 1971	Elat	Coral distribution and zonation from shore to 30m.
Mergner, 1967	Central Sudan and Jiddah	Mainly hydroid ecology, some profiles and hydrography data.
Mergner, 1971	Elat, central Sudan, Jiddah	Outline of reef structure and zonation.
Mergner, 1979	Aqaba	Detailed community analysis of a lagoon reef.
Mergner and Schuhmacher (1974)	Aqaba	Detailed analysis of two traverses over different coastal reefs.
Mergner and Schuhmacher (1981)	Aqaba	Detailed quantitative analysis of a fore-reef site at 10m depth.
Mergner and Svoboda (1977)	Aqaba	Important study of seasonality and productivity in two areas.
Schaefer (1969)	Farasan	Reef area community sequence.
Scheer (1971)	Suez Gulf, Central Sudan, Farasan	Brief outlines of reefs and their communities.
Schuhmacher (1973)	Elat	Zonation of organisms on a shaded harbour pier.
Wainwright (1965)	Eritrea and Dahlak islands	Outline of communities.

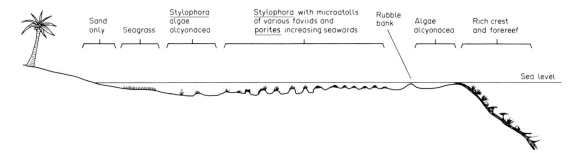

Fig. 7.8. Simplified section through a fringing reef with a shallow reef flat, showing typical zonation described by several authors. Dark shading indicates rich coral growth. Section length 50–200 m.

(Loya, 1972), suffering the full impact of breaking waves in storms, maximum sunlight with potentially damaging levels of ultra-violet irradiation, and irregular emersion stress from catastrophic low tides (Fishelson, 1973b). The ecology of fringing reef flats and shallow reef tops is generally considered to be controlled largely by these physical factors, which may completely exclude sensitive coral species and cause frequent partial or total mortality of colonies of the more resistant types. Lethal conditions are however sufficiently infrequent that fast growing, short lived species can be quite abundant, along with a few highly resistant, but slower growing and reproducing types.

This has been clearly demonstrated on the fringing reef flat at Elat, devastated in September 1970 by a series of catastrophic low tides, some 40 cm below predicted levels, which repeatedly exposed whole coral colonies to desiccation. Most bushy corals such as *Pocillopora*, *Stylophora* and *Acropora* were killed outright, while massive corals with deep calices into which the polyps retreated, such as *Platygyra* and *Favites* survived better (Fishelson, 1973b). The upper parts of such corals may be killed, but growth can continue outwards from the sides, forming characteristic 'micro-atolls' (Scoffin and Stoddart, 1978). Interestingly, the vulnerable branching corals are generally the most abundant reef top species. These may be 'r-strategists', which colonise fast and grow rapidly to reproduce before the next disaster wipes out the local population. Loya (1976a) has identified the abundant Red Sea reef top coral *Stylophora pistillata* as an r-strategist, specialised to exploit unpredictable reef top environments where new substrate surfaces are frequently exposed.

The harsh conditions of the central reef top limit the number of coral species normally found there to some 10% of the total fauna. The substrate cover of living coral is also low (ca. 6%), and the colonies themselves are small, averaging ca. 9 cm diameter (Head, 1980). The assemblages are however not dominated by any individual species, they may show the most even proportions of species over the whole reef. This is thought to result from the extreme physical environmental conditions preventing any species from achieving more than a temporary overabundance, and creating new areas of bare substrate for settlement. In such conditions, corals fare less well than algae (Benayahu and Loya, 1981) and many shallow reef tops are largely covered with lawns of very fast growing and colonising algae, an important food resource for herbivorous fish.

Figure 7.8 shows a representative profile through a fairly exposed coastal fringing reef. In addition to the stress factors already mentioned, water exchange decreases markedly shorewards across the reef flat. This leads to depletion of oxygen and particulate food material, and to extreme diurnal temperature fluctuation, together sufficient to prevent coral growth. Moving seawards from the inshore algal or seagrass zones, the first coral to appear is *Stylophora pistillata*, often accompanied by clumps of the calcareous green alga *Halimeda* and the alcyonarians *Tubipora* and *Xenia*. Coral growth continues to improve seawards, but may be interrupted by belts of storm rubble or by bare areas where heavy waves

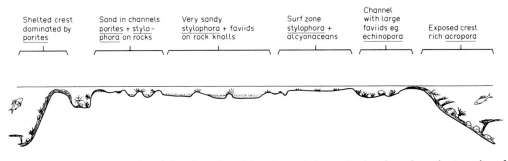

Fig. 7.9. Simplified section through a sheltered patch reef showing typical zonation based on the author's studies of the Towartit system, Sudan. Dark shading shows good coral growth, section length 140 m.

constantly break. Zonation of fringing reef flats is however very patchy, depending perhaps on rather subtle differences of depth and water exchange, and peculiar reversals of zonation may be found in many transects.

On patch or barrier reefs with open water all round, water exchange is much better, and currents up to 0.5 m/sec can be found on exposed reef tops, driven by breaking surf. Figure 7.9 shows schematically the zonation across the top of a fairly sheltered patch reef in the central Red Sea. Coral growth is richest at both leeward and windward reef crests. The central reef top bears typical *Stylophora* and microatoll assemblages similar to those of the fringing reef. Leeward, the proportion of sand covered substrate is high, interspersed with rocky knolls bearing *Porites* dominated assemblages continuous with those of the upper reef slope. The leeward reef crest is formed of massive *Porites solida* colonies forming cliffs, between areas of which sand chutes carry reef top sediment down to deeper water. On the sheltered cliffs, coral cover exceeds 60%, but is almost entirely made up of *Porites solida*, other species are rare. Figure 7.10 shows a typical *Porites* dominated leeward reef top community on a Sudanese patch reef.

The windward reef top community is similar on fringing, patch and barrier reefs, and a typical view is shown in Figure 7.11. The genera *Pocillopora*, *Acropora* and *Millepora* are most important, the balance between them is controlled by exposure as discussed in the next section.

7.4.2. Reef slope ecology

The ecology of the reef top can be largely explained by the physical factors of emersion and wave energy, but on reef slopes the controlling factors become less obvious and more difficult to unravel, not least, because all change progressively with depth. The best recent review of reef slope ecology is that of Sheppard (1982), who discusses the important factors of light energy decline, sedimentation and turbulence outlined below, and also reviews the less well evaluated factors of competition and predation.

The exponential decline in light energy with depth sets an absolute lower depth limit on zooxanthellate coral growth of about 100 m, but it seems to start restricting coral diversity below about 20 m. Porter (1976) has suggested that branching corals with small polyps and a large surface area rely more on light energy than do massive corals with large polyps, which are better adapted to zooplankton feeding. This suggests that branching coral abundance should decline with depth, and this is found to be the case. Head (1980) described the relative abundance of branching corals on Sudanese reef slopes declining from about 40% of the total coral cover at 2 m depth to 10% at 26 m. Simultaneously the abundance of large polyp bearing faviinid corals rose from 20% at 2 m to 40% at 26 m. There may however be other important factors operating, and the real significance of declining light levels is not

Fig. 7.10. Photograph of typical leeward reef top assemblage dominated by colonies of *Porites solida*, growing on grazed algal rock. Some branching colonies of *Pocillopora verrucosa* are also visible.

clear, as Sheppard (1982) points out. As Falkowski and Dubinsky (1981) showed for Red Sea *Stylophora*, corals can photo-adapt to reduced light levels, and some are still autotrophic at 40 m depth (Svoboda, 1978). However, much of the adaptation is based on reduction of metabolic rate (Spencer-Davies, 1977), presumably reducing organic growth rates and perhaps competitive ability. Another adaptation is the

Fig. 7.11. Photograph of windward reeftop assemblage. Principal species are *Pocillopora verrucosa* (lower left, upper centre), *Acropora humilis* (right of centre) and massive *Goniastrea retiformis* (lower right). Rock surfaces are covered with coralline algae.

alteration of colony shape as light levels decline. Hemispherical colonies in shallow water become columnar and finally horizontally expanded in deeper water to trap as much light as possible (Graus and Macintyre, 1976).

Sediment cover is usually high in deep still water at the base of reef slopes, and this restricts the area available for coral growth. Most sediment production however occurs in shallow water, and so the gross (as opposed to net) sedimentation rate is likely to be greater in shallow depths than for example on coral-covered rocks emerging from the sand in deep water. These factors may set an upper limit on the distribution of some very flat plate-like corals such as *Leptoseris* and *Pachyseris* which probably make considerable use of mucus ciliary feeding, and would be severely incapacitated by continuous sedimentation. Branching corals are well adapted to shed sediment precipitated from above, and this may be another factor contributing to their abundance on shallow reef slopes.

Turbulence and wave energy seem to exert a strong over-riding influence on upper reef slopes. Some corals are excluded from these environments by vulnerability to breakage (Vosburgh, 1977); others are particularly well suited to high energy conditions by virtue of their ability to control current velocity within a branching colony (Chamberlain and Graus, 1975), by their reduced dependence on particulate food (Wellington, 1982b), or by alignment to predictable wave surges (Wainwright *et al.*, 1976). Because of this, the assemblages of upper reef slopes vary greatly according to the wave exposure of the site, and specific exposure-related coral associations have been identified in the Indian Ocean (Rosen, 1975). Head (1980), modified Rosen's scheme for Red Sea reefs, based on multivariate analysis of reef slope data. In this model (Fig. 7.12), two dominant axes of exposure and light energy define the environmental continuum in which coral associations can be located.

The Exposed Crest Association (Fig. 7.13) is characterised by dominant *Acropora* species, especially *A. humilis*, *A. haimei* and *A. corymbosa*. *Millepora* spp. and *Pocillopora* are also common, and in the most exposed sites of fringing and barrier reefs may exceed *Acropora* in importance, forming the *Millepora* Sub-association.

Sheltered upper reef slopes (Fig. 7.14) are dominated by *Porites solida*, and this coral dominates the shallowest sites to the exclusion of most other species, as the *Porites* Ridge Sub-association, continuous with the *Porites* dominated leeward reef top assemblages. In slightly deeper water, the Sheltered Shallow

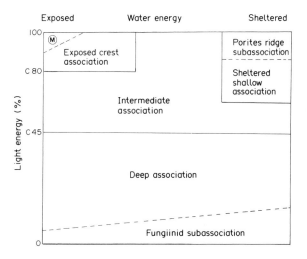

Fig. 7.12. Model of coral association zonation in the Red Sea based on published literature and multivariate analysis of Sudanese data. Simplified from Head (1980). High water/light energy area marked 'M' is the *Millepora* Sub-association.

Fig. 7.13. Photograph of fairly exposed upper reef slope and crest in the Sudan. The assemblage is dominated by *Acropora*, mainly *A. haimei* and *A. humilis*.

Fig. 7.14. Sheltered upper reef slope on a Sudanese reef. In the foreground is a Sheltered Shallow Association, with abundant *Porites solida* and *Acropora hemprichi*. In the middle distance is a *Porites* Ridge Subassociation, dominated by huge *Porites solida* colonies.

Association remains dominated by *Porites*, but includes other corals such as *Acropora hemprichi* and *A. variabilis*, *Seriatopora*, *Pocillopora* and various massive faviids.

The Intermediate Association is found in shallow medium wave energy sites, and below the other shallow associations to ca. 10 m where light energy values are about 45% of surface levels. Here species of *Acropora* co-dominate with *Porites* and other corals. In all sites where light energy levels were less than ca. 45%, including shallow depths in murky waters, the Deep Association was recorded. This is a rich and diverse association, marked by the importance of small colonies of several *Porites* species and large-polyp faviids especially *Goniastrea pectinata*. Branching corals are generally less important, except for *Acropora hyacinthus* and *Acropora variabilis*. In the deeper sites, fungiinids such as *Pavona*, *Pachyseris* and *Leptoseris* become important, forming a sub-association in the deepest sites. The Deep Association (Fig. 7.15) gives a strong impression of diversity, confirmed by sample transects, but in fact there is a marked decline in species richness in the deeper sites. In the Sudan, the greatest species richness was in the upper band of the Deep Association, between about 10 and 20 m depth, away from the extremes of water turbulence in shallow water and low light at greater depths (Head, 1980).

Loya and Slobodkin (1971) and Loya (1972) presented a more detailed zonation for the reefs of Elat, broadly comparable to the general scheme outlined above, but with more divisions in the Deep Association. Such schemes are very valuable on individual reefs, but may not be easily transferable from one area to the next, where for example topographic controls may be quite different. Only a few Red Sea reef slopes have been investigated in any detail, and as more data come in, the scheme of Figure 7.12 may need considerable revision.

7.4.3. Inshore reefs, lagoons and marsas

In the shelter of a fringing reef, shallow still lagoons are often developed with water depths from 1.5 to 10 m. Water exchange is restricted compared with that of open water reefs, but coral growth in the lagoon can be quite strong, especially in the back-reef (Fig. 7.4) where the lagoon meets the rear of the reef flat.

Fig. 7.15. Sudanese reef slope at about 20 m depth, showing typical diverse Deep Association on rock knolls between embayed sand pockets.

Shallow lagoons have been described by Loya and Slobodkin (1971), Mergner and Schuhmacher (1974) and Mergner (1979). Head (1980) investigated a substantial reef developed in a 10 m deep lagoon near Port Sudan. In shallow lagoons, corals occur on rocky outcrops and in assemblages similar to those of the reef flat. *Stylophora pistillata* is much the most important coral, but species of *Favia*, *Platygyra*, *Cyphastrea*, *Porites* and *Millepora* are also present. Alcyonarians may be very abundant, especially *Sinularia*, *Xenia* and *Tubipora*.

The back-reef, where developed, has quite a high coral cover, mainly of the genera already mentioned, but with some additions such as *Pavona*, *Echinopora* and *Psammocora*. The coral growth is very variable, luxuriant in one area, senescent or dead in another, suggesting that environmental parameters may change quickly in both space and time.

In the deepest lagoons, substantial reefs can be built up, benefiting from the wave action developed in the fetch of the lagoon. One such reef in the Sudan (Head, 1980) proved to have an abbreviated zonation pattern condensing the associations normally extending through 30 m or more on an offshore reef within its 6 m depth span. This was attributed to low water clarity and reduced light penetration.

Similar zonation truncation was seen (Head, 1980) in the turbid sharms or marsas common on Red Sea fringing reefs. Branching corals such as *Stylophora*, *Acropora* and *Pocillopora* were practically absent below 1 m depth, while the principal corals were the normally deep water species *Porites (Synaraea) undulata* and *Pavona* spp. At 5 m depth, the coral assemblages had much of the character of open water reefs at 30—40 m, with the added spice of low underwater visibility and the nagging feeling that something large and hungry might at any moment appear around the next coral knoll.

7.5. DEEP WATER REEFS

There is no scientific basis for dividing 'deep water' from 'shallow water' assemblages, since the two intergrade smoothly. The great practical difficulties of SCUBA diving below about 50 m have however provided an artificial cut-off point below which the intensity of studies rapidly declines. Until recently only Yamazato (1972) and Lang (1974) have provided substantial accounts of reef ecology in deeper water. Both used submersibles to penetrate to 300 m or deeper on fringing reef slopes.

Two recent accounts have described deep water assemblages in the Red Sea. Kühlmann (1983) used standard SCUBA to study Wingate reefs off Port Sudan to 70 m depth, a feat not to be emulated by less experienced divers. Below 50 m depth he found an abrupt drop in coral abundance, and only six species were recorded from 50—70 m depth. Kühlmann attributed the drop in diversity to declining light energy, and to a perceptible thermocline, with a temperature drop from 31°C to 28°C in August.

Hans Fricke has recently deployed a small submersible off the coast of Elat and southern Sinai, logging 166 dives to a maximum depth of 215 m. The results of this intensive study have been described by Fricke and Schuhmacher (1983). Bottom topography exerted a strong control on benthic ecology, with the development of several terraces at depths consistent with wave cutting during Pleistocene low sea-level stands. Coral growth was restricted to hard substrates, and was remarkably rich at great depths, with large colonies of the non-zooxanthellate coral *Madracis interjecta*. Below about 50 m however, the contribution of corals to reef building was less than that of Foraminifera, bryozoans and serpulid worms.

Figure 7.16 shows the depthwise diversity changes on the reefs studied by Fricke and Schuhmacher. The diversity of zooxanthellates declined steadily to 100 m, and only *Leptoseris fragilis* persisted to 140 m. The diversity of non-zooxanthellates correspondingly rose from 100 m. The authors did not consider this to be evidence of competitive exclusion by zooxanthellates, since the coral cover was very low.

Instead, they suggested a direct light control, (perhaps influencing settlement of larvae) with the 1% irradiance level (ca. 100–100 m depth) forming a boundary between the distribution of the two types of corals.

7.6. SOFT SUBSTRATES OF CORAL REEFS

As Thomassin (1978) pointed out, descriptions of the fauna of soft substrates (mud, sand and rubble) of coral reefs are few, and often concerned only with the ecology of single groups. Yet soft substrates, including lagoons and sand flats occupy a major proportion of the surface area of coral reefs, Smith (1978) estimates 90–95% of the global reef area. Calcification rates within such areas may be only one quarter to one eighth those of hard substrates with extensive coral cover, but the sand is of immense geological importance as backfill and as the secondary infilling material which consolidates the porous framework laid down by corals. To a biologist, reef sands may appear much less exciting than coral-rich hard substrates, but they are home to a great variety of macrofauna, including sea anemones, molluscs, crustaceans, echinoderms, polychaetes, cephalochordates and fish. There is also a rich and specialised meiofauna feeding on particulate matter and the bacterial populations of sand grains. Many non-burrowing fish and molluscs forage openly over the sediment, collecting food from the surface layers.

The ecology of reef sediments is poorly understood. It is controlled by the 'normal' factors of food supply, wave action, predation and disturbance, but is further complicated by the grain size and sorting characteristics of the sediments themselves, the formation of deep anoxic layers when circulation is reduced, and the stabilisation of the sediment by seagrass or algal mats.

In the Red Sea, soft substrate zones were described by Klausewitz (1967) for Sarso Island, but his description is limited to fish. Hughes' (1977) account of intertidal and reef flat ecology of Jiddah includes observations on the fauna of sand pockets. Mergner (1979) described the sand inhabiting fauna of his lagoon test area at Aqaba, and Wahbeh (1981) gives some data on infaunal densities within seagrass beds.

Fishelson (1971) has provided the most comprehensive review of the ecology of soft substrates in the Red Sea, including extensive species lists and an outline zonation scheme. The first wholly subtidal zone he named the *Hippa picta/Mactra olorina* community after the most characteristic crustacean and mollusc respectively. This zone, found to a depth of ca. 3 m, was typical of steep slopes with a mixture of coarse

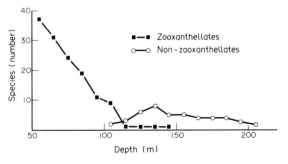

Fig. 7.16. Graph showing changes in species richness of corals with depth on the deep forereefs of the Gulf of Aqaba and northern Red Sea. Plotted from data in Fricke and Schuhmacher (1983).

TABLE 7.5. Coral species found to form coralliths on the Sudanese Red Sea.

Psammocora contigua
haimeana
Montipora monasteriata
Pavona varians
divaricata
Siderastrea savignyana
Favites pentagona
Cyphastrea microphthalma
serailia
Leptastrea purpurea
Millepora exaesa

sand and mud sediments, and exhibited patchy seasonal algal cover. An alternative *Ptychodera flava/Radianthus koseirensis* community is found between 1 and 3 m depth on muddy calcareous bottoms. it is particularly marked by crepuscular digging or burrowing organisms often best observed at night. The next zone extended from about 2 to 40 m depth down gentle slopes and Fishelson called it the *Halophila stipulacea/Asymmetron* community. The seagrass *Halophila* was characteristic, together with other seagrass species, and the fronds of the plants bore a rich colonist fauna. Among larger infaunal animals were the cephalochordate *Asymmetron* (*Epigonichthys*), echinoids, ophiuroids and many other taxa. Fishelson's deepest region was the *Operculina gaimardi/Turritella terebra* community, starting at 40 m and extending to over 200 m depth. The most prominent group was the molluscs, with specialist burrowing echinoids, Crustacea and polychaetes well represented.

Recently Betz and Otte (1980) have quantitatively described soft sediment macrofauna from Shaab Baraja in the Sudan, a reef quite different from the seagrass-rich coastal communities studied by Fishelson. Numerically, polychaetes were overwhelmingly dominant in the samples, followed by crustaceans and molluscs. Nematodes were also abundant. Much greater faunal densities were found in the quiet lagoonal sediments than in exposed areas. The Baraja lagoon showed low biomass compared with other tropical areas with extensive seagrass development. Relatively little production could occur *in situ*, and the sand community was 'fueled' by particulate matter washed from the nearby reef flat. Similar conclusions were reached by Hargraves (1982) in Belize, he found the seagrass beds to be highly autotrophic whereas rubble and sand showed net heterotrophy despite considerable microalgal populations.

Soft substrates are normally considered anathema to coral growth, but under certain conditions, some corals may be found free-living on fairly stable sand substrates. The family Fungiidae are particular specialists at this art, and several Red Sea examples were described by Schuhmacher (1979). Species of *Fungia* and *Cycloseris* attach to small boulders as larvae, but break away and become free living as adults. Many survive well on coarse sand, adopting a cupulate shape to shed sediments passively, while others inflate the polyp or remove sediment by mucus secretion and ciliary action.

Another specialised coral adaptation is the formation of 'coralliths' (Glynn, 1974) in shallow, fairly high energy conditions where sediments are coarse, and unfixed coral chunks often turned over by wave action. An initially fixed colony becomes detached from the substrate but is not smothered as the coarse sands provide sufficient water circulation. It continues to grow, and is periodically turned over by wave surge. Eventually growth over the breakage scar produces a spherical pebble-like object entirely covered with living tissue. Species encountered as coralliths during studies in the Sudan are listed in Table 7.5.

The soft-substrate communities of the Red Sea are only beginning to be investigated. Further studies, especially of their faunal composition, stability, productivity and relation to the rest of the reef community would be extremely interesting.

7.7. CONCLUSIONS

It is readily acknowledged by persons who have dived on reefs in many areas, that those of the Red Sea are among the finest in the world. Certainly, the calm, clear water, spectacular profiles and profusion of life make Red Sea reefs especially attractive to the diver. As we have seen, the coral fauna is rich, but not exceptionally so; the special qualities of Red Sea reefs are therefore probably due to their environmental setting, and above all to their underfished and little-exploited nature. Reefs are however of considerable economic importance. They attract tourists, they can be fished, the coral and rock can be mined for building stone, and their lagoons can be filled in for coastal roads and constructions. Reefs are also susceptible to damage by sedimentation caused by coastal works, by oil or effluent pollution, and to eutrophication from sewage or fertiliser spills. These various problems are discussed in other chapters, so will not be covered further here. The reader is referred to Chapters 17 (fisheries) and 18 (pollution), and especially to Chapter 19, where the whole topic of human impact and conservation is fully aired.

The reefs of the Red Sea must count among the principal assets of the area, together with petroleum and the potential sea-bed riches described in Chapter 4. As these latter non-renewable resources are exploited, and the wealth they create used to improve facilities and increase settlement along the coast, every effort must be taken to preserve intact the reefs and their outstanding faunal riches.

REFERENCES

Barnes, R. D. (1980) *Invertebrate Zoology*. Philadelphia, Saunders College.

Bemert, G. and Ormond, R. (1981) *Red Sea Coral Reefs*. London, Kegan Paul Internation.

Benayahu, Y. and Loya, Y. (1977) Space partitioning by stony corals, soft corals and benthic algae on the coral reefs of the northern Gulf of Eilat (Red Sea). *Helgoländer wiss. Meeresunters* 30, 362–82.

Benayahu, Y. and Loya, Y. (1981) Competition for space among coral-reef sessile organisms. *Bull. Mar. Sci.* 31, 514–22.

Berry, L., Whiteman, A. J. and Bell, S. V. (1966) Some radiocarbon dates and their significance, emerged reef complex in The Sudan. *Zeits. für Geomorphol. N.F.* 10, 119–43.

Betz, K-H. and Otte, G. (1980) Species distribution and biomass of the soft bottom faunal macrobenthos in a coral reef (Shaab Baraja, Central Red Sea, Sudan). *Proc. Symp. Coastal and Marine Environment of the Red Sea, Gulf of Aden and Tropical Western Indian Ocean* 3, 13–37.

Bouchon, C. (1980) Quantitative study of the scleractinian coral communities of the Jordanian coast (Gulf of Aqaba, Red Sea), preliminary results. *Téthys* 9, 243–6.

Braithwaite, C. J. R. (1982) Patterns of accretion of reefs in the Sudanese Red Sea. *Mar. Geol.* 46, 297–325.

Chamberlain, J. A. and Graus, R. R. (1975) Water flow and the hydromechanical adaptations of branched reef corals. *Bull Mar. Sci.* 25, 112–25.

Cohen, S. (1975) *Red Sea Diver's Guide*. Israel, Red Sea Diving (Publications) Ltd.

Cousteau, J.-Y. and Diole, P. (1970) *Life and Death in a Coral Sea*. Cassel, London.

Crossland, C. (1911) A physical description of Khor Dongonab, Red. Sea. *J. Linn. Soc.* 31, 265–86.

Crossland, C. (1913) *Desert and Water Gardens of the Red Sea*. Cambridge.

Crossland, C. (1938) The coral reefs of Ghardaqa, Red Sea. *Proc. zool. Soc. Lond. 1938*, 513–23.

Darwin, C. R. (1842) *The Structure and Distribution of Coral Reefs*. London, Smith, Elder and Co.

Fadlallah, Y. H. (1983) Sexual reproduction, development and larval biology in Scleractinian corals. *Coral Reefs* 2, 129–50.

Falkowski, P.G. and Dubinsky, Z. (1981) Light-shade adaptation of *Stylophora pistillata*, a hermatypic coral from the Gulf of Eilat. *Nature* 289, 172–4.

Faurot, L. (1888) Rapport sur une mission dans la mer Rouge (Isle de Kamarane) et dans le golfe d'Aden (Aden et Golfe de Tadjourah). *Arch. Zool. exp.* (2) 6, 117–33.

Fishelson, L. (1971) Ecology and distribution of the benthic fauna in the shallow waters of the Red Sea. *Mar. Biol.* 10, 113–33.

Fishelson, L. (1973a) Ecology of coral reefs in the Gulf of Aqaba (Red Sea) influenced by pollution. *Oecologia* 12, 55–67.

Fishelson, L. (1973b) Ecological and biological phenomena influencing coral-species composition on the reef tables at Eilat (Gulf of Aqaba, Red Sea). *Mar. Biol.* 19, 183–96.

Fishelson, L. (1980) Marine reserves along the Sinai Peninsula (northern Red Sea). *Helgoländer wiss. Meeresunters* 33, 624–40.

Fricke, H. W. and Schuhmacher, H. (1983) The depth limits of Red Sea Stony corals: an ecophysiological problem (A deep diving survey by submersible). *P.S.Z.N.I. Marine Ecology* 4, 163–94.

Friedman, G. M. (1968) Geology and geochemistry of reefs, carbonate sediments and waters, Gulf of Aqaba, (Elat), Red Sea. *J. Sedim. Petrol.* 38, 895–919.

Glynn, P. W. (1974) Rolling stones among the Scleractinia: mobile coralliths in the Gulf of Panama. *Proc. 2nd Int. Coral Reef Symp.* 2, 183—97.

Graus, R. R. and Macintyre, I. G. (1976) Light control of growth form in colonial reef corals: computer simulation. *Science N.Y.* 193, 895—7.

Haeckel, E. (1876) *Arabische Korallen. Ein Ausflug nach den Korallenbanken des Rothen Meeres und ein Blick in das Leben der Korallenthiere.* Berlin.

Hargraves, P. E. (1982) Production of the benthic communities at Carrie Bow Cay, Belize. In *The Atlantic Barrier Reef Ecosystem at Carrie Bow Cay, Belize.* Eds. K. Rutzler and I. G. Macintyre, pp. 109—14. Washington, Smithsonian Institution Press.

Hass, H. (1952) *Manta, Teufel im Roten Meer.* Berlin.

Head, S. M. (1978a) A cerioid species of *Blastomussa* (Cnidaria, Scleractinia) from the central Red Sea, with a revision of the genus. *J. Nat. Hist.* 12, 633—9.

Head, S. M. (1978b) *Comparative Zonation of Coral Reefs of the Central Red Sea.* Unpublished report to Saudi-Sudanese Joint Red Sea Commission.

Head, S. M. (1979) *A Preliminary Report on the Results of Phase Seven of the CCSRG Reef Study Programme.* Unpublished report to Saudi-Sudanese Joint Red Sea Commission.

Head, S. M. (1980) *The Ecology of Corals in the Sudanese Red sea.* Ph.D. thesis, University of Cambridge.

Head, S. M. (1983) An undescribed species of *Merulina* and a new genus and species of siderastreid coral from the Red Sea. *J. Nat. Hist.* 17, 419—35.

Hughes, R. N. (1977) The biota of reef-flats and limestone cliffs near Jeddah, Saudi Arabia. *J. nat. Hist.* 11, 77—96.

Klausewitz, W. (1967) Die physiographische zonierung der saumriffe von Sarso. 4. Beitrag der arbeitsgruppe litoralforschung. *Meteor Forschungs Ergebn. D 2 Biol,* 44—67.

Klunzinger, C. B. (1870) Eine zoologische excursion auf ein Korallenriff des Rothen Meeres. *Verh. zool. bot. Ges. Wien* 20, 389—94.

Klunzinger, C. B. (1879) *Die Korallthiere des Rothen Meeres.* Zweiter Teil: Die Steinkorallen. Erster Abschnitt: Die Madreporaceen und Oculinaceen. Dritter Theil, Die Steinkorallen. Zweiter Abschnitt. Die Astraceen und Fungiaceen. Berlin, Gutmann'schen Verlag.

Kühlmann, D. H. H. (1983) Composition and ecology of deep-water coral associations. *Helgoländer wiss. Meeresunters* 36, 183—204.

Lang, J. C. (1973) Interspecific aggression by scleractinian corals II, Why the race is not only to the swift. *Bull. Mar. Sci.* 23, 260—79.

Lang, J.C. (1974) Biological zonation at the base of a reef. *Am. Sci.* 62, 272—81.

Lewis, J. B. (1977) Processes of organic production on coral reefs. *Biol. Rev.* 52, 205—47.

Loya, Y. (1972) Community structure and species diversity of hermatypic corals at Eilat, Red Sea. *Mar. Biol.* 13, 100—23.

Loya, Y. (1976a) The Red Sea coral *Stylophora pistillata* is an r strategist. *Nature* 259, 478—80.

Loya, Y. (1976b) Recolonisation of Red Sea corals affected by natural catastrophes and man-made perturbations. *Ecology* 57, 278—89.

Loya, Y. and Slobodkin, L. B. (1971) The coral reefs of Eilat (Gulf of Eilat, Red Sea). *Symp. zool. Soc. Lond.* 28, 117—39.

Marenzeller, E. von (1907) Expeditionen S. M. Schiff 'Pola' in das Rote Meer. Nördliche und südliche Halfte 1895/96 — 1897/97. Zoologische Ergebnisse 26. Riffkorallen. *Denkschr. Akad. Wiss. Wien* 80, 27—97.

Mergner, H. (1967) Über den Hydroidenbewuchs einiger Korallenriffe des Roten Meeres. *Z. Morph. Ökol Tiere* 60, 35—104.

Mergner, H. (1971) Structure, ecology and zonation of Red Sea reefs. (In comparison with South Indian and Jamaican reefs). *Symp. zool. Soc. Lond.* 28, 141—61.

Mergner, H. (1979) Quantitative ökologische analyse eines rifflagunenareals bei Aqaba (Golf von Aqaba, Rotes Meer). *Helgoländer wiss. Meeresunters.* 32, 476—507.

Mergner, H. and Schuhmacher, H. (1974) Morphologie, ökologie und zonierung von korallenriffen bei Aqaba (Golf von Aqaba, Rotes Meer). *Helgoländer wiss. Meeresunters* 26, 238—357.

Mergner, H. and Schuhmacher, H. (1981) Quantitative analyse der korallenbesiedlung eines vorriffareals bei Aqaba (Rotes Meer). *Helgoländer wiss. Meeresunters.* 34, 337—54.

Mergner, H. and Svoboda, A. (1977) Productivity and seasonal changes in selected reef areas in the Gulf of Aqaba (Red Sea). *Helgoländer wiss. Meeresunters.* 30, 383—99.

Nemenzo, F. (1981) Studies on the systematics of scleractinian corals in the Philippines. *Proc. 4th Int. Coral Reef Symp.* 1, 25—32.

Neudecker, S. (1977) Transplant experiments to test the effect of fish grazing on coral distribution. *Proc. 3rd Int. Coral Reef Symp.* 1, 317—23.

Porter, J.W. (1976) Autotrophy, heterotrophy and resource partitioning in Caribbean reef-building corals. *Am. Nat.* 110, 731—42.

Richman, S., Loya, Y. and Slobodkin, L. B. (1975) The rate of mucus production by corals and its assimilation by the coral reef copepod *Acartia negligens. Limnol. Oceanogr.* 20, 918—23.

Roghi, G. and Baschieri, F. (1956) *Dahlak.* Nicholas Kaye. London

Rosen, B. R. (1975) The distribution of reef corals. *Rep. Underwater Ass.* 1 (NS), 1—16.

Rosen, B. R. (1981) The tropical high diversity enigma — the coral's-eye view. In *Chance, Change and Challenge, The evolving Biosphere.* Ed. P. L. Forey, pp.103—30. London, British Museum (Natural History).

Rossi, L. (1954) Madreporari, Stoloniferi e Milleporini. *Riv. Biol. colon.* 14, 23—72.

Sammarco, P. W. (1980) *Diadema* and its relationship to coral spat mortality: Grazing, competition and biological disturbance. *J. exp. mar. Biol. Ecol.* 45, 245—72.

Schaefer, W. (1969) Sarso, Modell der Biofacies Sequenzen im Korallenriff-Bereich des Schelfs. *Senckenberg. marit.* 50, 153—64.

Scheer, G. (1971) Coral reefs and coral genera in the Red Sea and Indian Ocean. *Symp. zool. Soc. Lond.* 28, 329—67.

Scheer, G. and Pillai, C. S. G. (1983) Report on the stony corals from the Red Sea. *Zoologica* 133, 1—197.

Schuhmacher, H. (1973) Die lichtabhängige besiedlung von hafenstützpfeilern durch sessile Tiere und algen aus dem korallenriff bei Eilat (Rotes Meer). *Helgoländer wiss. Meeresunters.* 24, 307—26.

Schuhmacher, H. (1976) Korallenriffe. Ihre Verbreitung, Tierwelt und Ökologie. Munich, BLV Verlagsgesellschaft.

Schuhmacher, H. (1979) Experimentelle untersuchungen zur anpassung von Fungiiden (Scleractinia, Fungiidae) an unterschiedliche sedimentations und bodensverhältnisse. *Int. Revue ges. Hydrobiol.* 64, 207—43.

Scoffin, T. P. and Stoddart, D. R. (1978) Nature and significance of microatolls. *Phil. Trans. R. Soc. Lond.* B 2884, 99—122.

Sheppard, C. R. C. (1981a) Illumination and the coral community beneath tabular *Acropora* species. *Mar. Biol.* 64, 53—7.

Sheppard, C. R. C. (1981b) The reef and soft-substrate coral fauna of Chagos Indian Ocean. *J. Nat. Hist.* 15, 607—21.

Sheppard, C. R. C. (1982) Coral populations on reef slopes and their major controls. *Mar. Ecol. Progr. Ser.* 7, 83—115.

Smith, S. V. (1978) Coral-reef area and the contributions of reefs to processes and resources of the world's oceans. *Nature* 273, 225—6.

Spencer-Davies, P. (1977) Carbon budget and vertical zonation of Atlantic reef corals. *Proc. 3rd Int. Coral Reef Symp.* 1, 391—6.

Stoddart, D. R. (1969) Ecology and morphology of recent coral reefs. *Biol. Rev.* 44, 433—97.

Svoboda, A. (1978) In situ monitoring of oxygen production and respiration in Cnidaria with and without zooxanthellae. In *Physiology and Behaviour of Marine Organisms*. Eds. D. S. McLusky and A. J. Berry, pp.75—92. Oxford, Pergamon Press.

Thomassin, B. (1978) Soft bottom communities. In *Coral Reefs: Research Methods*. Ed. D. R. Stoddart and R.E. Johannes, pp.263—98, Paris, UNESCO.

Vine, P. J. and Head, S. M. (1977) Growth of corals on Commander Cousteau's underwater garage at Shaab Rumi (Sudanese Red Sea). *Jeddah Nature Journal* 1977, 6—17.

Vosburgh, F. (1977) The response to drag of the reef coral *Acropora reticulata*. *Proc. 3rd Int. Coral reef Symp.* 1, 477—82.

Wahbeh, M. I. (1981) Distribution, biomass, biometry and some associated fauna of the seagrass community in the Jordan Gulf of Aqaba. *Proc. 4th Int. Coral Reef Symp.* 2, 453—9.

Wainwright, S. A. (1965) Reef communities visited by the Israel South Red Sea Expedition, 1962. *Bull. Sea Fish Res. Stn. Israel* 38, 40—53.

Wainwright, S. A., Biggs, W. D., Currey, J.D. and Gosline, J. M. (1976) *Mechanical design in Organisms*, London, Edward Arnold.

Walther, J. (1888) Die korallenriffe der Sinaihalbinsel, geologische und biologische beobachtungen. *Abh. sachs. Akad. Wiss.* 14, 437—506.

Wellington, G. M. (1982a) Depth zonation of corals in the Gulf of Panama: control and facilitation by resident reef fishes. *Ecol. Monogr.* 52, 223—41.

Wellington, G. M. (1982b) An experimental analysis of the effects of light and zooplankton on coral zonation. *Oecologia* 52, 311—20.

Yamazato, K. (1972) Bathymetric distribution of corals in the Ryukyu Islands. *Proc. Symp. Corals and Coral Reefs* 1969, 121—34. Mar. biol. Ass. India.

CHAPTER 8

Benthic Algae

DIANA I. WALKER

Department of Botany, The University of Western Australia, Australia

CONTENTS

8.1. INTRODUCTION

The term 'algae' covers a very diverse group of aquatic plants, ranging from Cyanobacteria (blue-green algae), and the single-celled diatoms (Bacillariophyta) through to the more morphologically complex green algae (Chlorophyta), brown algae (Phaeophyta) and red algae (Rhodophyta). These

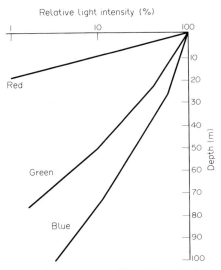

Fig. 8.1. Change in light intensity with depth for red, green and blue light in Red Sea water. Light intensity is expressed as a % of that at the surface (on a logarithmic scale).

organisms share two major features — they absorb nutrients from the water column (they lack true roots) and they convert the energy of sunlight to organic material by photosynthesis. There are three major sources of primary production in the shallow tropical waters of the Red Sea. Firstly, the seagrasses which are flowering plants (angiosperms) possessing true roots, and are discussed in Chapter 9. Secondly, in the water column, the cyanobacteria and diatoms which make up most of the phytoplankton (Chapter 5). And, thirdly, the subject of this chapter, the benthic algae, that is those algae which are attached to a substrate. I intend to discuss both their occurrence and their significance in Red Sea environments.

8.2. FACTORS AFFECTING ALGAL DISTRIBUTION

8.2.1. Light

Algae are dependent on the light penetrating the water column for their energy requirements. Not only does the amount of light reaching the benthos decrease with depth, but the spectral composition of the light also changes. The algae are limited to relatively shallow water, where sufficient light reaches them. The depth of water at which an individual species may survive depends not only on the physiology of the alga, and specifically its ability to photosynthesise effectively at low light levels, but also on the amount and quality (wavelengths) of light present. The low latitudes and comparatively clear skies of the Red Sea ensure high light intensities, and the waters are very clear due to the almost complete absence of freshwater run-off, with associated silt and high nutrients. Light can therefore penetrate to great depths.

Figure 8.1 shows the decrease with depth of red, green and blue light for the clear blue waters typical of the Red Sea. This produces the distribution of wavelengths for each depth shown in Figure 8.2a. The different groups of algae are classified according to the pigments they possess, and it is these pigments which determine those parts of the available spectrum which can best be utilised. Figure 8.2b–d shows

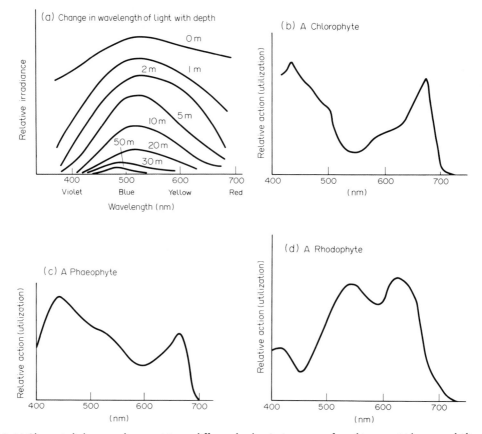

Fig. 8.2. (a) Change in light spectral composition at different depths. Action spectra for a (b) green, (c) brown and (d) red alga.

the wavelengths which are utilised (the action spectra) for a green, a brown and a red alga. As can be seen from these diagrams, the green Chlorophyta absorb most effectively those wavelengths which correspond to the non-green parts of the spectrum, and in particular the shorter red wavelengths, which are rapidly absorbed in the first few metres of the water column; hence the occurrence of most Chlorophyta close to the surface.

The larger brown algae, such as *Turbinaria* and *Sargassum*, which have relatively large amounts of non-photosynthesising tissue to support, generally also occur in shallow water, whereas some of the very delicate red algae, e.g. *Ceramium* and *Polysiphonia*, which consist of very thin filaments and utilize the longer blue and green wavelengths available at greater depths, can be found more than 40 m below the surface of the Red Sea. Below about 50 m, however, benthic algae are relatively rare.

8.2.2. Substrate

In addition to light, algal growth is regulated by other environmental factors. Substrate availability is particularly important. Any stable substrate will be colonized by algae. Stability depends on both the size of particles of which the bottom is composed, and the degree of wave action. For example, in an exposed area of shallow water subject to heavy surf, even large boulders will be unstable, as they may be overturned by a storm; in deeper water near the same site they may well be stable. In contrast, sheltered areas with little water movement may have sandy bottoms which are seldom disturbed and stable enough to permit algal growth. As pointed out in Chapter 3 of this volume, the prevailing winds blow down the length of the Red Sea, and so most of the shallow waters of the shores are relatively sheltered, especially in the more northerly sections. Thus, algae may be found attached to almost any grade of substrate, at least during part of the year.

Although most rocks support a growth of algae,the undersides of rocks and overhangs tend to be very poorly settled, if at all. This is not simply a function of the reduction of light intensity, as sufficient light may occur under such overhangs in shallow water, but of lack of colonization by planktonic spores, which generally settle passively, dropping vertically downwards.

8.2.3. Water movement and grazing

Water movement also influences the distribution of algae. Its effects on substrate stability have already been mentioned, but wave action can also favour algal growth by providing a better supply of dissolved nutrients through increasing the rate of flow of water over the algae. The waters of the Red Sea are low in inorganic nutrients such as phosphorus and nitrogen, and so calm areas rapidly become nutrient depleted, whereas well-mixed areas are not so affected.

One of the most significant factors influencing the standing crop or biomass of algae is the effect of heavy grazing by herbivorous fish, urchins and gastropod molluscs (*see* below). This leads to areas having very low standing crops of algae although primary productivity may be high. In turn, this gives rise to the lack of visual impact shown by the algal populations in most Red Sea habitats. In contrast to the abundance and diversity of most faunal groups, which is dramatically obvious on first seeing a reef environment, algae are thus often inconspicuous and overlooked.

8.3. THE SIGNIFICANCE OF ALGAE IN THE RED SEA

Despite their apparent inconspicuousness, algae play crucial roles in the different habitats of the Red Sea.

8.3.1. Macro-algae

Some macro-algae do occur on reefs, but not on the scale of the kelp forests so characteristic of the rocky coasts of many temperate regions. Those larger individuals that do occur in the Red Sea, e.g. the calcareous green alga, *Halimeda*, and the brown algae, *Padina*, *Sargassum* and *Turbinaria*, are relatively rare, generally occurring as isolated plants in crevices on the reefs. The maximum density that Benayahu and Loya (1977b) observed of *Turbinaria elatensis* in their transects, was 159 in 100 metres. However, they may occasionally become the dominant species in a limited area.

In general, the macro-algae are poorly studied, and little ecological information is available. However, Nasr (1947) provides some data on the distribution of species in terms of substrate preferences, depth, and hydrodynamic conditions, but this is only for those occurring on the Egyptian coastline. Natour *et al.* (1979) also give some ecological data on the macro-algae of the Jordanian coast of the Gulf of Aqaba.

Some seasonality in the occurrence of macro-algae has been recorded. Mergner and Svoboda (1977) showed that in the Gulf of Aqaba the abundance of macro-algae varied considerably at different times of year. At the period of maximum algal abundance (February—May), 25% of their study area had algal cover. However it should be noted that for most of the year, there was no macro-algal presence. Figure 8.3 shows the major trends in species abundance that they observed for the larger species. Such seasonal blooms of macro-algae are not confined to the northern parts of the Red Sea, and have been observed off the Sudanese coast. During the spring, when the algal blooms occur, both day length and light intensity are increasing as the sun moves more directly overhead, making more energy available for photosynthesis. The water temperature, however, is still near its winter minimum — about 20°C in the northern Red Sea. This means that the energy needed for respiration is low, which increases the amount of photosynthate that can be diverted to biomass. There may be some influence of nutrient concentration — generally levels of nitrogen and phosphorus are lower in summer, and they may increase during the winter, and be utilised by the algae.

8.3.2. Reef cementation

Calcareous encrusting algae such as *Porolithon* and *Melobesia*, which form pink coralline patches, are significant as agents of reef cementation. Generally these algae are found only on the reef crest where the effects of wave action allow these slow growing species, which thrive on exposure to water movement, to outcompete faster growing but less robust fleshy and filamentous algae. The hydrodynamic energy of these waves can cause severe erosion. The calcareous algae consolidate and cement coral rubble to the reef surface, forming a rock-hard wave-resistant protective coat. By incorporating loose fragments into the reef structure and lessening erosion these algae contribute significantly to reef building processes.

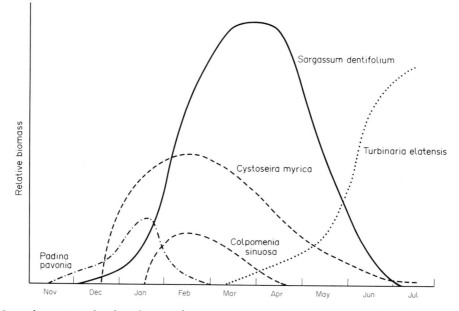

Fig. 8.3. Seasonal variation in the relative biomass of macro-algae in the Gulf of Aqaba. (Redrawn from Mergner and Svoboda, 1977)

However, in the Red Sea, unlike the atoll systems of the oceanic regions of the Indo-Pacific, these calcareous algae do not form 'groove and spur' algal reefs. The extent to which these features develop depends on the degree of exposure to wave action, and in the semi-enclosed and sheltered waters of the Red Sea, insufficient energy is generated to allow their formation.

8.3.3. Contribution to sediments

Although most calcium carbonate in reef systems in the Red Sea is deposited by corals, a variable fraction is attributable to the algae. It is generally derived from the erosion of the chalky skeletons of a group of green algae including *Halimeda* (Fig. 8.4), *Codium*, and *Caulerpa* species. These deposit aragonite needles as a major part of their structure. Calcium carbonate may also be precipitated on and in the cell walls of these species, as well as those of *Padina*, a common brown alga in the Red Sea. Whereas these species deposit calcium carbonate in the form of aragonite, the calcareous red algae e.g. *Porolithon*, *Corallina*, form calcite. Fragments of these algae may be broken off the reef, and can form banks of sediment elsewhere. The coralline red algae may make up 7—51% of the sediments associated with coral reefs, whereas the range for *Halimeda* may be 1—43% (Stoddart, 1969). Braithwaite (1982) found *Halimeda* fragments to be absent from offshore and patch reef sediment samples he examined from the Sudanese Red Sea but to form 14—28% of coarse sediments on fringing reefs; generally their contribution to sediments appears to be less in the Indo-Pacific and Red Sea than in the Caribbean. However, the major contribution of calcareous algal breakdown to the sediments of the Red Sea should not be overlooked.

Fig. 8.4. *Halimeda tuna*, a calcareous green alga composed of rounded segments with flexible joints, commonly found in the crevices on the reef. (Scale Bar = 1 cm)

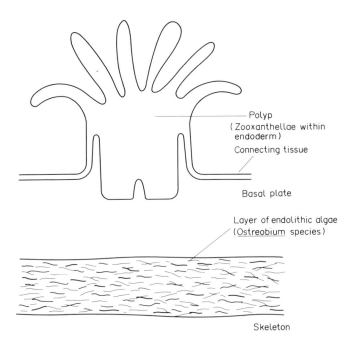

Fig. 8.5. The relative positions of the algae associated with coral — the zooxanthellae within the endoderm of the polyp, and the endolithic algae in the skeleton, underlying the coral tissue.

8.3.4. Association with corals

There are two groups of microscopic algae which are of considerable significance to coral reefs, but which are invisible on an intact reef. Firstly, there are the dinoflagellates belonging to the genus *Symbiodinium*, which occur within the tissues of coral polyps and are referred to as zooxanthellae. They exist as small round vegetative cells in the endoderm of the polyp, but have been shown to have flagellated free-swimming stage for dispersal. Similar algae are also found in the tissues of the giant clam, *Tridacna*. By taking up CO_2 during photosynthesis, these algae assist coral calcification. Also, some of the carbon compounds (e.g. alanine, glucose and glycerol) produced by the algae are transferred to the animal tissue, where they are used as respiratory and metabolic substances.

The second important group of microscopic 'algae' consists of filamentous blue-greens (Cyanobacteria) and greens (Chlorophyta) which are found as a layer below the surface of the coral skeleton — the endolithic algae. Figure 8.5 shows where these algae occur in relation to the coral. These endolithic algae are adapted to the low light intensities of their environment, and have very high concentrations of chlorophyll which can be seen as dense green bands in sectioned coral, and which may even impart a greenish appearance to the surface of intact coral. These algae occur not only under live coral but also under dead coral rock (Fig. 8.6) where red algal species (Rhodophyta) may also be present.

8.3.5. Algal lawns

Various species of small algae can collectively form what are known as algal 'lawns', 'mats', or 'turfs'. These terms describe the thin layers of semi-microscopic species, including juvenile stages of macro-algae, which begin as films of diatoms, and with time, become increasingly diverse assemblages

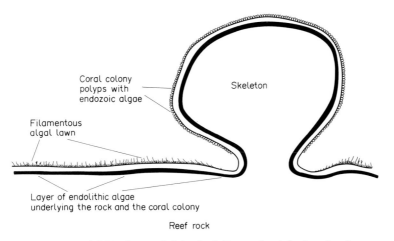

Fig. 8.6. Endolithic algae underlying both live coral and dead coral rock.

of 20—30 larger species. These are opportunistic species, having rapid growth rates which enable them quickly to exploit any substrate when it becomes available. The spores or vegetative fragments of these algae tend to be actively displaced by healthy live coral, but on other substrates, e.g. on coral rock, dead coral heads, on sand where water movement is limited such as within lagoons, on seagrass leaves and even other algae, these spores can grow and develop into an algal lawn. Por *et al.*(1977) record such an algal community on the mangroves of the Sinai Peninsula.

These species do not show any noticeable substrate specificity, and will colonize any surface placed in the water column. Their settlement is a passive process, but is nevertheless very rapid, aided by the broad dispersal abilities and high reproduction/division rates of the algae. After one week, 15—20 algal species may settle on an artificial substrate. As these algal lawns develop, there is a rapid turnover in the species present and progressive increase in the size of the algal filaments. This increase in filament length impedes the flow of water over the algae, reducing the availability of oxygen and nutrients to the inner portions of the mat leading eventually to death and decay in these portions. The reduction in the velocity of the water adjacent to the mat causes sediment carried in the water column to drop out of suspension, and subsequently to be incorporated into the algal mat. This sediment trapping increases both shading and the barrier to diffusion, reducing the availability of oxygen, light and nutrients to the core of the mat. Thus, as the mats become larger their cores inevitably decay and tend to be detached increasingly easily from the substrate (Fishelson, 1973).

For settlement and growth of these mat-forming algae to occur, a suitable substrate must be available, and, in order that a permanent population of spores exists, ready to exploit such a surface, much effort must be put into reproduction. There can be no doubt that a high proportion of the energy devoted to reproduction by the algae, whether it is vegetative, by asexually produced spores, or sexual, is wasted. Passive settlement on sandy areas may allow development of an algal mat (Bathurst, 1967) providing hydrodynamic conditions do not disturb the unstable substrate. Such algal mats may help to bind sediments once established thus aiding their own survival.

8.3.6 Seagrass epiphytes

The seagrass *Halophila stipulacea* stabilizes otherwise unstable sandy substrates in some regions of the Gulf of Aqaba, and in doing so also produces a much greater surface area for algal colonization — up to $12m^2$ in shallow water (1—2m deep) for each m^2 of substrate (Hulings, 1979). There is also a constant 'turnover' of seagrass leaves; on average the life span of a leaf is around 35 days (Wahbeh, 1980). As the new leaves replace those which senesce, a continuously produced substrate is provided. Cyanobacteria are particularly common as epiphytes on seagrasses, with species of *Oscillatoria*, *Phormidium* and *Lyngbya* most abundant.

The relationship between seagrass leaves and their epiphytic cover does not however, seem to be purely one of substrate provision and use. It may be symbiotic with nutrients being exchanged between the seagrass and its epiphytes; this could be of great significance in the nutrient-poor waters of the Red Sea. The possibility of phosphate transfer from the sediment to the epiphytes via the seagrass has been shown to be feasible (Harlin, 1973), and Goering & Parker (1972) have demonstrated that the nitrogen fixed by epiphytic cyanobacteria can be transferred to the host seagrass. Such nutrient transfer may be a major contributory factor in the high productivity of seagrass beds.

8.3.7. Competition for substrate

On hard substrates, there is considerable competition for space between hard corals, soft corals and algae (Benayahu & Loya, 1977a). Hard corals will slough off algal spores and fragments, but coral killed by extreme low tides, or high temperatures, is soon colonized by a rapid growth of algae. Removal of coral tissue by *Acanthaster planci*, the Crown-of-thorns starfish, also leads to algal development on the dead coral skeleton (Belk & Belk, 1975; Biggs & Eminson, 1977). Soft corals occupy substrate, but do not contribute to the reef framework. The nature of their relationship with filamentous algal populations is unclear. Their copious secretion of mucus is assumed to slough off any algal spores which settle but occasional algal growth has been observed by the author on their basal regions. Unlike hard corals, the soft corals leave no trace when they die, disintegrating rapidly. It is unknown whether their planulae are able to settle in the algal mats.

The hard 'corals' — Scleractinia, *Millepora* and *Tubipora* — which are the major reef building organisms, propagate mainly by the release of swimming planktonic planulae. These settle on coral rock, but where algal mats have developed, settlement is reduced and may be almost completely inhibited. This is a result of the planulae being unable to attach to an algal-covered surface. In addition, the sediment associated with the algal mats tends to clog the feeding mechanisms of the newly settled coral.

Although corals grow much more slowly than the shorter-lived filamentous algae with which they compete for space, corals live longer and once established, grow large enough to avoid this competition (Birkeland, 1977), unless environmental conditions become deleterious to coral growth.

8.3.8. Response to pollution

Although parts of the Red Sea are undisturbed some areas have been subjected to the effects of man's activities. The overall significance of this will be discussed in Chapter 19. However, environmental disturbance, and in particular, pollution, tends to exert many of its effects through the algae. Eutrophication — the enrichment of nutrient concentrations — can cause dramatic changes in the balance of reef ecosystems.

Two major sources of excess nutrients occur in the Red Sea. These are sewage outfalls, and in the Gulf of Aqaba and at Quseir the escape of phosphate dust during the loading of crushed phosphate rock on to ships. Eutrophication is particularly significant in the low nutrient concentration waters of the

Red Sea, as the input of nitrogen and phosphorus leads to a dramatic increase in algal biomass. Walker and Ormond (1982) found a doubling of algal biomass in an area where both sewage and phosphorus pollution were occurring, with a 4—5 fold increase in coral death rate. The algae do not generally overgrow live coral, but accelerate the mechanisms which cause coral death. In particular, the increased growth of algae leads to shading of the coral zooxanthellae thus impairing coral growth, and increasing sedimentation which is also deleterious. Corals use their mucus to slough off sediment and this mucus is a continuous drain on their energy reserves. If sedimentation increases they have to divert a considerable and often excessive amount of energy into mucus production (at the expense of growth and reproduction) in order to remove this sediment and as a result survival may be very low (Hubbard and Pocock, 1972; Schuhmacher, 1977). Stimulation of algal growth will also increase the rates of algal spore production, adding to the stress experienced by the corals, as these spores must also be sloughed off. Thus, the enhancement of algal growth by eutrophication is extremely deleterious for coral.

Oil pollution, which occurs at varying degrees throughout the Red Sea but is particularly significant in the Gulf of Suez, influences algal growth as well. Some smothering of algae by direct contact with oil may occur, but this is confined to the immediate area of the spill. In general, there is a greater impact on the fauna, especially where repeated incidents occur. Coral may be killed, providing substrate for algal growth, but of even more significance is the effect on the echinoderms and molluscs which graze the algae. The reduced densities of these organisms can lead to substantial increases in algal standing crop with the concomitant side effects listed above.

8.4. SPECIES OF THE RED SEA

Some 485 species of algae are catalogued by Papenfuss (1968) as having been recorded from the Red Sea. He provides not only the history of the exploration of the marine flora of the area, but also uses all available records to provide information about the general distribution of species in the region. However, some regions of the Red Sea are grossly under-collected, and there are many taxonomic confusions. Table 8.1 lists the sites from which algae have been collected, and Figure 8.7 shows their locations.

8.4.1. Geographic distributions of species

Descriptions of the distributions are based on all available records, but the picture may well be incomplete. A bias to the sites in the Northern Red Sea and the Gulfs of Aqaba and Suez occurs as a result of extensive work by Nasr (1947), Rayss (1959) and Rayss and Dor (1963) in those areas.

A few species show cosmopolitan distributions, being found not only throughout the Red Sea and elsewhere in the Indo-Pacific, but also in the Atlantic and Mediterranean. These include *Ulva lactuca*, *Dictyosphaeria cavernosa*, *Jania rubens* and *Sphacelaria tribuloides* as well as many of the cyanobacteria. In contrast, others are endemic to the Red Sea e.g. *Cystophyllum trinode*, *Sargassum subrepandum* var. *rueppellii* and *Phormidium penicillatum* f. *vaginatum*, or even to particular areas within the region e.g. *Turbinaria elatensis* (Fig. 8.8) and *Dichotrix eylathensis*, which are confined, as their names suggest, to the Gulf of Aqaba (Elat); and *Giffordia ghardaqaensis* which has only been described from Hurghada.

Other species are found in particular sections of the Red Sea, but also commonly in the Indo-Pacific. For example, *Ectocarpus elachistaeformis*, *Enteromorpha flexuosa* and *Centroceras clavulatum* have been found in the Red Sea only in the Gulfs of Suez and Aqaba, although they occur frequently throughout the rest of the Indo-Pacific. Others appear to be restricted to the central and southern Red Sea, although the Saudi Arabian coastline is poorly known. Among such species are several species of *Sargassum*, *Gelidium crinale* and *Corallina tenella*.

Table 8.2 summarises the percentages of species occurring in different sections of the Red Sea for each group of algae, and for all species. The north/south boundary for species distributions seems to be from Jiddah to Suakin, but the dearth of recording sites on the Saudi Arabian coast makes comparison difficult. The patterns of occurrence are probably determined by temperature, although there may be some influence by the higher salinities found in the northern regions. Generally, however, the species in

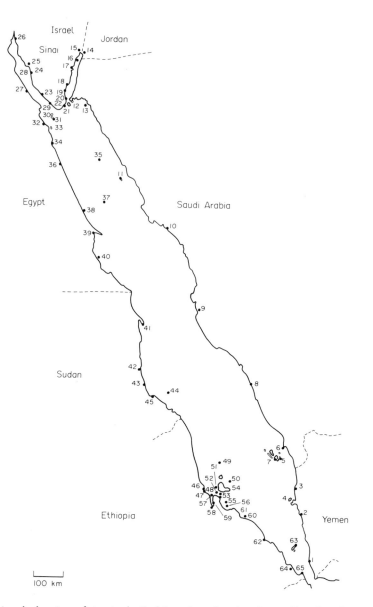

Fig. 8.7. Map showing the locations of sites in the Red Sea where algae have been collected. (After Papenfuss, 1968).

TABLE 8.1. List of collection sites for the algae of the Red Sea

Location	Lat. N	Long. E	Location	Lat. N	Long. E
NORTH YEMEN			**SUDAN**		
1. Al Mukha	13°20'	43°16'	41. Dungunab Bay	21°03'	37°09'
2. Al Hudaydah	14°50'	42°58'	42. Port Sudan	19°38'	37°07'
3. Al Luhayyah	15°44'	42°42'	43. Suakin	19°08'	37°17'
4. KAMARAN ISLAND	15°20'	42°33'	44. Tella Tella Kebir Islet	18°49'	38°09'
			45. Trinkitat	18°41'	37°46'
SAUDI ARABIA					
5. Sulein Islets	16°46'	42°11'	**ETHIOPIA**		
6. Abulad Islets	16°48'	42°10'	46. Massawa	15°37'	39°28'
7. Farasan Archipelago			47. Sheik Said Islet	15°36'	39°29'
8. Al Qunfidhah	19°09'	41°07'	48. Bay of Archico		
9. Jiddah	20°30'	39°10'	49. Romia Islet	16°32'	40°02'
10. Yanbu'al Bahr	24°07'	38°04'	50. Derom Islet	15°54'	40°23'
11. Mashabih Islet	25°38'	36°27'	51. Entedebir Islet	15°43'	39°54'
12. Tiran Island	27°56'	34°33'	52. Assarca Islets	15°32'	39°55'
13. Maqsur Islet	27°56'	35°12'	53. Shumma Islet	15°31'	40°00'
			54. Dahlak Archipelago		
JORDAN			55. Adjuz Islet (S. Massawa		
14. Aqaba	29°31'	35°01'	Channel)	15°14'	40°15'
			56. Onachil Islet (Howakil Islet)	15°10'	40°16'
ISRAEL			57. Dissei Islet	15°27'	39°45'
15. Elat	29°33'	34°57'	58. Arafali	15°03'	39°43'
			59. Gulf of Zula		
EGYPT			60. Mandola Islet	14°44'	40°54'
16. Farun Islet	29°28'	34°52'	61. Bay of Anfila (also as Bay of		
17. Nuweiba	28°59'	34°40'	Amfile)		
18. Shora el Manqata	28°12'	34°25'	62. Edd	13°57'	41°38'
19. Abu Zabad	28°09'	34°27'	63. Hanish Islands	13°43'	42°45'
20. Ras Nusrâni	27°58'	34°25'	64. Assab	13°01'	42°47'
21. Sharm el Sheikh	27°51'	34°16'	65. Bay of Assab		
22. Ras Muhammad	27°44'	34°15'			
23. Tor	28°14'	33°36'			
24. Ras Abu Rudeis	28°54'	33°12'			
25. Ras Abu Zenima	29°03'	33°06'			
26. Suez	29°59'	32°33'			
27. Ras Ghârib	28°21'	33°06'			
28. Gulf of Suez					
29. Strait of Jubal					
30. Ashran Islet	27°47'	33°42'			
31. Shadwan Islet	27°30'	34°00'			
32. Ghardaqa (Hurghada)	27°17'	33°47'			
33. Gifatin Islet	27°14'	33°56'			
34. Safâga	26°43'	33°55'			
35. The Brothers					
36. Quseir	26°04'	34°15'			
37. Daedalus Reef	24°55'	35°51'			
38. Wadi Gimal Islet	24°40'	35°07'			
39. Berenice	23°57'	35°17'			
40. Mirear Islet	23°15'	35°41'			

the southern Red Sea are similar to those recorded in waters with similar warm temperatures throughout the world, and the northern species resemble those from slightly cooler areas.

In general, the species of the Red Sea belong to the circumtropical and subtropical marine flora, occurring in the Caribbean as well as the Mediterranean and Indo-Pacific. Some 64% of the species listed by Rayss(1959) and Rayss and Dor (1963) had this pattern of distribution. A further 9% were endemic, with 14% also occurring in the Indo-Pacific. Examples of these were *Avrainvillea amadelpha, Halimeda opuntia, Udotea argentea, Stoechospermum marqinatum, Hormophysa triquetra* and *Sargassum denticulatum*. Only 1% were restricted to the Mediterranean — Rayss and Dor(1963) suggested that two rhodophytes, *Porphyra umbilicalis* and *Galaxaura cylindrica*, might have passed into the Red Sea via the Suez canal although 9% had distributions in the Atlantic and Mediterranean. These included *Cladophora albida* and *Liagora turneri*. They reported none of the eastern species of algae having passed into the Mediterranean.

8.5. STANDING CROP AND PRODUCTION — THE INFLUENCE OF GRAZING

It is difficult to gather quantitative data on the standing crops of algae on reefs. Distributions are extremely patchy, and the complex irregularities of the reef surface make calculating area very difficult. Removing the algal mats produces more problems, as scraping removes the top layer of rock, which must be separated from the rest of the sample.

Given these problems, and the absence of detailed studies, little quantitative information is available for the standing crops of algae on Red Sea coral reefs. The percentage cover of macro-algae has sometimes been recorded, but this is still extremely variable, both temporally and spatially and no conversions of cover to biomass have been attempted.

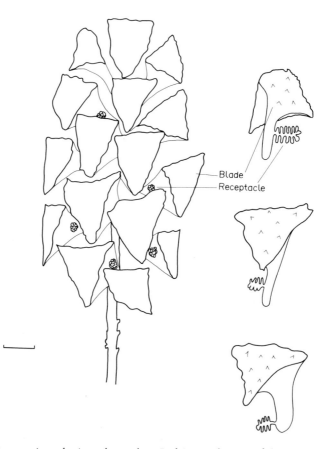

Fig. 8.8. *Turbinaria elatensis*, a species endemic to the northern Red Sea, with views of the roughly triangular blades of the plant. (Scale Bar = 1 cm)

TABLE 8.2. Species Distribution Summary—percentage of species recorded in each region. (Information from Papenfuss, 1969)

Algal Group	Total	Gulfs (G)	North (N)	Central (C)	South (S)	All	G,N	N,C	C,S	G,N,C	N,C,S
Blue-greens	7.2	24.2	33.3	3.0	3.0	9.1	18.2	9.1	—	—	—
Greens	19.7	14.6	20.0	5.6	8.9	21.1	13.3	2.2	1.1	8.9	3.3
Browns	36.1	9.7	7.9	4.8	42.0	20.6	9.0	—	1.8	2.4	1.2
Reds	37.0	17.8	12.4	3.6	18.3	29.6	11.8	—	3.0	2.4	1.2
Total	100.0	14.9	13.8	4.4	26.1	23.2	11.6	1.0	2.0	3.5	1.5

Northern Species (Gulfs of Suez, Aqaba, and Northern Red Sea) = 44.8%
Southern Species (Southern region, south of Jiddah-Suakin)   = 26.1%
All Red Sea Coasts   = 23.2%

8.5.1. Standing crop and productivity

Data obtained from the growth of algae on artificial substrates indicate that algal lawns have standing crops of approximately 25 g-dry/m^2 (grams dry weight per square metre), with similar levels of biomass observable in epiphytic communities on the seagrass *Halophila stipulacea* (Walker, 1982). This is a relatively low value, especially in comparison with temperate region algal standing crops which may be of the order of 1–2 kg-dry/m^2, and seagrasses which tend to range from 0.5–4 kg-dry/m^2.

However, comparison of plant standing crops provides no insight into the productivity of the communities. These algal lawns tend to have very rapid growth rates, with high turnover. Production rates are of the order of 1–3 g-dry/m^2.day (Walker,1982). These are comparable to the values for seagrasses (0.9–4.2 g-dry/m^2.day) and temperate region algae (maximum value 7 g-dry/m^2.day). In comparison, terrestrial production has maximum values of 16 g-dry/m^2.day for tropical rain forests and 11 g-dry/m^2.day for tropical grain. These are, however, the maximum values obtained and the mean production figures would be much lower.

The figures for algal lawns indicate that, given standing crops of 25 g-dry/m^2, the algae should take only 10 days of growth to reach this biomass. This is a very fast turnover time, another indication of the dynamic nature of algal lawns.

8.5.2. Grazing

It is clear that the biomass of algae on reefs does not reflect their production. The explanation for this lies in the observation that reefs in the Red Sea support high densities of herbivores — mainly fish, such as Acanthuridae (Surgeonfish) and Scaridae (Parrotfish), and errant sea-urchins e.g. *Diadema setosum*, and *Tripneustes gratilla*. Each type of herbivore has a different method of grazing and these influence the algal community in different ways. For example, most sea-urchins non-selectively but systematically scrape off not only the algal lawn, but also part of the underlying substrate. Parrotfish behave in a similar way, leaving characteristic broad teethmarks on the coral rock. In contrast, surgeonfish remove individual filaments from the algal mat without clearing the surface, and so may alter the algal community structure.

Grazing is primarily confined to reef areas; elsewhere, in seagrass beds for example, grazing pressure (particularly from fish) is much reduced, as herbivores are reluctant to venture into open areas away from the shelter of the reef. Crevices in the reef structure provide a refuge from predation for the grazers. In areas where coral heads are found within a seagrass meadow, a halo of bare sand can be seen surrounding the coral. A similar band of grazed substrate can be observed along the edges of large patch reefs or fringing reefs. Examination of the gut contents and faeces of such grazers shows the seagrass leaves are ingested primarily as a method of obtaining algal epiphytes, and are not themselves digested.

Several authors have shown that herbivores are important in regulating algal standing crops. Dart (1972) estimated that on Sudanese patch reefs, the Slate-pencil urchin, (*Heterocentrotus mammillatus*) grazed 20 cm^2 of reef rock each night, and that where it and *Diadema setosum* were present, the algal lawn covered only 20% of the available reef surface, as opposed to 100% elsewhere. Benayahu and Loya (1977b) found an interesting relationship between macro-algal cover and the density of *Diadema* on the fringing reefs of Elat.

Intense grazing by sea-urchins, especially in polluted areas where densities of echinoderms are high, may lead to damage to living corals, providing more substrate for algal growth which in turn may damage the corals. Such effects have been recorded from the Gulf of Suez, where repeated oil spills occur and the reefs are in poor condition.

8.5.3. Algal 'gardening' by fish

Fish behaviour can modify the distribution of algae. The establishment of territories by certain of the herbivorous damselfish (Pomacentridae) leads to an increase in algal biomass within the areas they defend. This has a wider significance, in that the dense growths of algae within these territories result in a decrease in settlement, growth and survival of corals (Potts, 1977). Vine (1974) found similar effects for both surgeonfish and damselfish territories, the increased algal standing crop leading to reduced settlement both of coral planulae, and other benthic invertebrates. Removal of the territorial fish rapidly results in removal of the established algal growth by other grazers and subsequent availability of the substrate for settlement by invertebrates.

8.5.4. Significance to community structure

Fish grazing frees substrate for colonization, but invertebrate larvae settling in such grazed areas may later be removed by repeated grazing. Survival thus depends on the frequency of grazing, and the interval between successive 'scrapes' of the same piece of substrate.

In general, grazers prevent algal growth from attaining dominance on the reefs, and allow corals to continue the building of the reef structure. Vine (1974) also showed that the establishment of algal lawns leads to a decrease in the cover of calcareous encrusting red algae. This in turn produces a decrease in the amount of cementation of coral rubble. Thus the environment may become less stable, with greater reef destruction during storms.

The random nature of fish grazing leads to a very patchy distribution of algal mats, and maintains them at early developmental or successional stages with high growth rates. Thus, the grazing of herbivores maintains the rate of algal production at its maximum on the one hand, whilst allowing the settlement of invertebrates in cleared areas on the other. Disruption of this delicate balance by man's activities can have serious effects on the reef ecosystem.

8.6. CONCLUSION

In summary, the algae associated with Red Sea coral reefs are conspicuous by their apparent absence, but play extremely important roles in the habitats in which they occur.

The species found in the Red Sea are mainly cosmopolitan tropical species, the majority occurring in all warm seas. However, there is a fairly high proportion of endemics (nearly 10%) with close relatives in the Indo—Pacific.

Characteristically, the algae have a low standing crop, which is maintained by heavy grazing by herbivores. This is balanced by a high turnover. Cyanobacteria (blue-green algae) contribute significantly to the nutrient budget of both reefs and seagrass beds by nitrogen fixation and their role in nutrient cycling.

Algae can have deleterious effects on reefs. Their ability to colonize all available surfaces makes them highly competitive occupiers of space on reefs. Once established, their sediment trapping capacity leads to a decrease in the settlement of coral planulae, and in areas of eutrophication, they reduce coral survival. Thus, a reef may be regarded as a delicate balance between hard corals which, assisted by calcareous encrusting algae, build up the reef; and the growth of algal lawns, which lead to reef degradation.

The algae represent a poorly studied component of all reef systems, and this is particularly true for the Red Sea. Many taxonomic problems still require attention. Their true contribution to some of the habitats in which they occur remains uncertain, but I have tried, at least qualitatively to outline their influence on the environments of the Red Sea.

ACKNOWLEDGEMENTS

I am grateful to the University of Jordan for permitting me to use the facilities of their Marine Science Station at Aqaba during my studies of the algae of the Gulf of Aqaba and to Dr. Claude and Yolande Bouchon for all their help. I would like to thank Professor A.J. McComb of the University of Western Australia for his very useful criticisms of the manuscript.

REFERENCES

Bathurst, R. G. C. (1967). Subtidal gelatinous mat, sand stabilizer and food, Great Bahama Bank. *J. Geology*, 75, 736—738.

Belk, M. S. and Belk, D. (1975). An observation of algal colonisation on *Acropora aspera* killed by *Acanthaster planci*. *Hydrobiologia*, 46, (1) 29—32.

Benayahu, Y. and Loya, Y. (1977a). Space partitioning by stony corals, soft corals and benthic algae on the coral reefs of the northern Gulf of Eilat (Red Sea), *Helgoländer wiss. Meeresunters*, 30, 362—82.

Benayahu, Y. and Loya, J. (1977b). Seasonal occurrence of benthic algae communities and grazing regulation by sea urchins at the coral reefs of Eilat, Red Sea. *Proc. 3rd International Coral Reef Symposium, Miami*, 383—9.

Biggs, P. and Eminson, D. F. (1977). Studies on algal recolonization of coral predated by the crown of thorns starfish *Acanthaster planci* in the Sudanese Red Sea. *Biological Conservation* 11, 41—7.

Birkeland, C. (1977). The importance of rate of biomass accumulation in the early successional stages of benthic communities to the survival of coral recruits. *Proc. 3rd International Coral Reef Symposium, Miami*, 15—21.

Braithwaite, C. J. R. (1982). Patterns of accretion of reefs in the Sudanese Red Sea. *Mar. Geol.* 46, 297—325.

Dart J. K. G. (1972). Echinoids, algal lawn and coral recolonisation. *Nature* 239, 50—1.

Fishelson, L. (1973). Ecology of coral reefs in the Gulf of Aqaba (Red Sea) influenced by pollution. *Oecologia* 12, 55—67.

Goering, J. J. and Parker, P. L. (1972). Nitrogen fixation by epiphytes on sea grasses. *Limnol. Oceanogr.* 17, 320—3.

Harlin, M. M. (1973). Transfer of products between epiphytic marine algae and host plants. *J. Phycol.* 9, 243—8.

Hubbard, J. A. E. B. and Pocock, Y. P. (1972). Sediment rejection by scleractinian corals: a key to palaeo-environmental reconstruction. *Geologische Rundschau* 61, 598—626.

Hulings, N. C. (1979). The ecology, biometry and biomass of the seagrass *Halophila stipulacea* along the Jordanian coast of the Gulf of Aqaba. *Botanica Marina* 22, 425—30.

Mergner, H. and Svoboda, A. (1977). Productivity and seasonal changes in selected reef areas in the Gulf of Aqaba (Red Sea). *Helgoländer wiss. Meeresunters*, 30, 383—99.

Nasr, A. H. (1947). Synopsis of Marine Algae of the Egyptian Red Sea Coast. Egypt University, *Faculty of Science Bulletin*, 26, 1—155.

Natour, R. M., Gerloff, J. and Nizamuddin, M. (1979). Algae from the Gulf of Aqaba (Jordan). (a) Chlorophyceae and Phaeophyceae, 39—68: (b) Rhodophyceae, 69—93, *Nova Hedwigia* 31 (1 and 2).

Papenfuss, G. F. (1968). A history, catalogue and bibliography of benthic Red Sea algae. *Israel J. Bot.* 17, 1—118.

Por, F. D., Dor, I. and Amir, A. (1977). The mangal of Sinai: limits of an ecosystem. *Helgoländer wiss. Meeresunters.*, 30, 295—314.

Potts, D.C. (1977). Suppression of coral populations by filamentous algae within damselfish territories. *J. Exp. mar. Biol. Ecol.* 28, 207—16.

Rayss, T. (1959). Contribution a la connaissance de la flore marine de la Mer Rouge. *Bull. Sea Fish. Res. Stn.* 23, 1—32.

Rayss, T. and Dor, I. (1963). Nouvelle contribution a la connaissance des algue marines de la Mer Rouge. *Bull Sea Fish. Res. Stn.* 34, 11—42.

Schuhmacher, H. (1977). Ability in fungiid corals to overcome sedimentation. *Proc. 3rd Int. Coral Reef Symp.*, Miami. 503—9.

Stoddart, D. R. (1969). Ecology and morphology of recent coral reefs. *Biol. Rev.* 44, 433—98.

Vine, P. J. (1974). Effects of algal grazing and aggressive behaviour of the fishes *Pomacentrus lividus* and *Acanthurus sohal* on coral reef ecology. *Mar. Biol.* 24, 131—6.

Wahbeh, M. I. (1980). Studies on the ecology and productivity of the seagrass *Halophila stipulacea*, and some associated organisms in the Gulf of Aqaba (Jordan). *Ph.D. Thesis, University of York*.

Walker, D. I. (1982). Algal Lawns in the Gulf of Aqaba, Red Sea. *Ph.D. Thesis, University of York*.

Walker, D. I. and Ormond, R. F. G. (1982). Coral death from sewage and phosphate pollution at Aqaba, Red Sea. *Mar. Pollut. Bull.* 13(1), 21—5.

CHAPTER 9

Littoral and Shallow Subtidal Environments

DAVID A. JONES[*], MOSTAPHA GHAMRAWY[**] AND
MOHAMMAD I. WAHBEH[***]

[*]Department of Marine Biology, Marine Science Laboratories, Menai Bridge, Anglesey LL59 5EH, U.K.
[**]Marine Science Institute, King Abdulaziz University, P.O. Box 1540, Jiddah, Kingdom of Saudi Arabia
[***]Marine Science Station, University of Jordan, P.O. Box 195, Aqaba, Jordan.

CONTENTS

9.1. INTRODUCTION

Oceanographic and biological studies in the Red Sea began well before the opening of the Suez Canal, some of the more important were noted in Chapter 1, section 1.5. Most of the earlier biological studies were taxonomic, but expeditions during the last twenty years have provided some ecological information on shallow water communities, which has been summarised by Fishelson (1971). With the development of marine laboratories along the coasts of most of the countries bordering the Red Sea, there has been an increasing interest in intertidal ecology. Ecological studies have been concentrated in the northern Red Sea, although Hughes (1977) and Jones (1974) have described rocky and sandy shores in the central Red Sea region. However, despite increased interest, large tracts of coast remain unsurveyed and few eco-physiological studies have been made, so that many aspects of the coastal communities remain poorly understood.

This chapter attempts to describe briefly the major Red Sea coastal habitats including rocky shores, sand beaches, mud flats, mangal (D. A. Jones and M. Ghamrawy), and in more detail the seagrass communities (M. I. Wahbeh) together with their physical environmental ranges. The characteristic floral and faunal communities associated with each habitat are described and zonation patterns are compared, both among sites along the Red Sea coast, and with those of other Indian Ocean shores. The shallow sub-tidal soft bottom communities such as seagrasses form important nursery grounds for commercial shrimp and fish species, and their ecological role and production is assessed.

9.2. PHYSICAL ENVIRONMENT

9.2.1. Topography

The shores of the Red Sea are backed by an arid zone and in general are fronted by vigorously growing fringing or patch reefs. In most areas rocky shores are formed from raised Quaternary fossil coral cliffs often undercut by wave action, which has also cut reef platforms extending seawards to terminate in live coral fringing reefs. In other areas such as the coast of Sinai underlying beach rock is exposed and forms plate-like formations which slope at a shallow angle into the sea. Both the fringing and the fossil reefs may be interrupted at intervals by inlets (sharms or marsas) formed by drowned valleys, but in the main provide sheltered lagoons and bays where organogenic marine sediments consisting of broken down corals and shells and terrigenous material including aeolian (wind blown) sands can settle to form sand beaches and mudflats.

The absence of strong tidal currents reduces the occurrence of sublittoral sand bars but, where sand patches are present, seagrass beds grow extensively. Mangroves have their northern hemisphere limit at Ras Muhammad on the tip of Sinai but are never well developed and have a patchy distribution along the Red Sea in muddy lagoons, often based on beach rock.

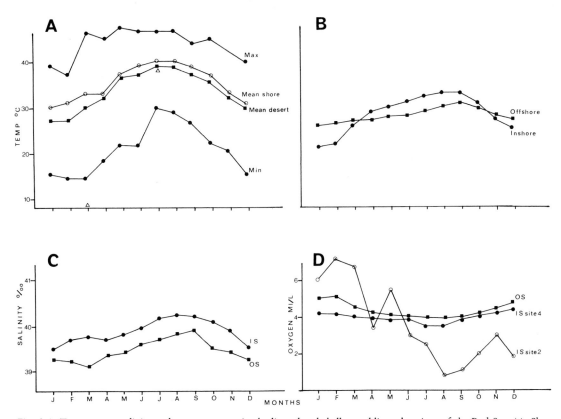

Fig. 9.1. Temperature, salinity and oxygen ranges in the littoral and shallow sublittoral regions of the Red Sea. (a), Shore temperatures, ● max and min ○ mean air temperatures at Jiddah, ■ mean desert air temperatures at Jiddah, △ highest and lowest air temperatures in Northern Red Sea (latter from Por *et al.*, 1977). (b), Sea temperatures at Jiddah. (c), Salinity at Jeddah, ● immediate sublittoral, ■ off shore. (d), water oxygen content at Jiddah, ■ offshore, ● inshore, ○ inshore eutrophic bay.

9.2.2. Temperature

Although sea temperatures fluctuate over a wide range both geographically and seasonally in the Red Sea (Chapter 3), intertidal organisms are exposed to an even wider range of air temperatures (Fig. 9.1a). Shore air temperatures measured in the central Red Sea are somewhat elevated, higher than desert temperatures during the winter months, but nevertheless exhibit a far wider range than inshore sea temperatures, which in turn are more variable than offshore sea temperatures (Fig. 9.1b).

Lowest shore temperatures (8°C) are experienced in the northern Red Sea where diurnal fluctuations may exceed 20°C (Por *et al.*, 1977). At Jiddah the air temperature on the shore rarely falls below 14°C but may reach over 45°C in the summer months. Both daily and seasonal ranges progressively decrease to the south of the Red Sea.

9.2.3. Salinity

Red Sea salinities which range from 39—41‰ in the upper layers of the open sea are the highest found in the world's oceans. Figure 9.1c shows that with little rainfall and high evaporation rates, inshore salinities remain elevated above those of offshore waters during the year. Where there is little interchange between the open sea and coastal lagoons or pools, salinities may rise dramatically, and have given rise to the terms hyperhaline (80–180‰) and hypersaline (80–300‰) (Por, 1972) see Table 9.2. Marine biota occur in the former lagoons but are replaced by continental freshwater organisms under hypersaline conditions.

9.2.4. Oxygen

In general inshore oxygen levels show the same seasonal trend as offshore surface oxygen levels, varying inversely with temperature (Fig. 9.1d). However, in summer strong diurnal fluctuations may occur in lagoons isolated during low tides with almost complete depletion of oxygen occurring during the night. Under natural conditions oxygen is replenished by floodtides and photosynthesis during the day. However, as Figure 9.1d indicates semi-enclosed bays are extremely susceptible to eutrophication arising from pollutants such as sewage, and long-term oxygen depletion may occur.

9.2.5. Tides

Red Sea tides are semi-diurnal and oscillate around a nodal point near 19°N with a spring range in the northern and southern ends of the Red Sea proper of 0.6–0.9 m (Chapter 3). This means that the central Red Sea coasts experience no appreciable semi-diurnal tide and daily changes in sea level are unpredictable as they are controlled by atmospheric pressure and wind direction and strength. These factors often combine to produce tides of up to 75 cm with extreme lows and highs persisting for up to two weeks at a time. Further north and south where tidal amplitudes are larger, climatic factors are less significant and a more normal tidal regime prevails.

In addition the strong variation in the wind system between summer and winter monsoons causes seasonal changes of up to 0.5 m in mean sea levels throughout the Red Sea (Chapter 3). Thus during summer months when other physical factors are combining to produce maximum stress, sea levels are at their minimum, exposing areas of the winter sublittoral to the air and the intense hot sun. Conversely, in winter a large part of the intertidal zone is permanently inundated and wave action may cause erosion of the coastline.

As tides are small even in the north and south of the Red Sea, tidal currents are weak, particularly in the central region. Wave action is also usually moderate in the absence of oceanic fetch, and thus water movements in general are much less severe than are experienced elsewhere in the Indian Ocean.

The seasonal changes in mean sea level, unpredictable daily tides, and lack of water movement in one of the hottest regions of the world, combine to present uniquely severe stresses for intertidal and shallow sublittoral communities. As tidal amplitude becomes greater to the north and south of the Red Sea some amelioration of these stresses is apparent; harshest conditions existing in the central Red Sea region.

9.3. ROCKY SHORES

Shore profiles differ widely along the coasts of the Red Sea, ranging from fossil coral cliffs rising 30–40 m vertically from the sea, to wide horizontal expanses of beach rock producing intertidal reef flats over 100 m in width. Commonly the vertical fossil cliffs are undercut by notches caused by wave and grazing mollusc action and chemical solution by sea water. Reef flats incline gradually to low water, or are eroded to form shallow lagoons behind the fringing reefs (Fig. 9.2a,b). These lagoons, pools and other irregularities trap sediments, and thus patches of sand and mud, with their associated communities, are often found intermingled with the rocky shore organisms.

Intertidal species are relatively well known taxonomically, although there is often confusion over synonyms, and are predominantly of tropical Indian Ocean origin. In addition there is some evidence for colonisation from the Mediterranean since the Suez Canal was opened, although most migrations appear to be in the opposite direction (Safriel and Lipkin, 1975). However, a high level of endemism is present so that the Red Sea has been regarded as a distinct zoogeographical area (Briggs, 1974). A detailed account of the northern and southern Red Sea littoral communities is given by Fishelson (1971), and Hughes (1977) has described central Red Sea intertidal communities, while zonation is discussed by Klausewitz (1967) and Safriel and Lipkin (1964). A summary of the major intertidal zones defined by these authors is presented in Figure 9.2, and Table 9.1 compares these zones with similar zones described elsewhere in the Indian Ocean.

It is clear that northern and southern Red Sea shores, which are subject to regular tides, fit well into the universal scheme for zonation proposed by Stephenson and Stephenson (1949) and Lewis (1964). The terminology used here is that given by the latter author and is based on the fact that certain organisms are characteristically found at certain positions on the shore. Thus the littoral fringe is characterised by periwinkles, the eulittoral by barnacles, limpets and the gastropod *Nerita*, and the sublittoral fringe usually by corals, seagrasses, and the brown algae *Sargassum* and *Turbinaria*. Although these zones may be easily recognised on tidal Red Sea shores, little is known of the physico-chemical factors and biological interactions which determine the limits to each zone.

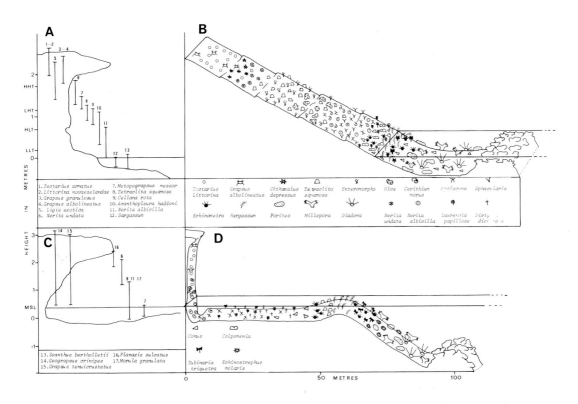

Fig. 9.2. Rocky shore zonation patterns in the Red Sea. (a), vertical cliff exposed to regular tides to north and south of the Red Sea. (b), reef flat exposed to regular tides. (c), vertical cliff exposed to unpredictable sea level changes in the central Red Sea. (d), reef flat in central Red Sea.

TABLE 9.1. Comparison of rocky shore zonation patterns within the Red Sea and Indian Ocean region. Data from: 1. Fishelson (1971), Safriel and Lipkin (1964); 2. Hughes (1977); 3. Jones (1982); 4. Taylor (1971).

[1]Red Sea (Tidal)	[2]Red Sea (Non-tidal)	[3]Kuwait (Arabian Gulf)	[4]Aldabra (Indian Ocean)
Endolithic algae	P. sulcatus	Cyanobacteria	Littorina glabrata
Littorina novaezelandae	Geograpsus crinipes	N. subodosa	Littorina kraussi
Tectarus armatus	Grapsus tenuicrustatus		Geograpsus stormi
Grapsus albolineatus	Ligia pigmenta	Ligia pigmenta	Grapsus tenuicrustatus
Ligia exotica	Nodolittorina subnodosa	Chthamalus sp. nov.	Ligia exotica
Nerita undata			Nerita plicata
			Nerita undata
Chthamalus depressus	Nerita undata	Balanus amphitrite	Tetrachthamalus obliteratus
Planaxis sulcatus		Crassostrea cf. margaritacea	Crassostrea cucullata
Metopograpsus messor		Enteromorpha compressa	Enteromorpha sp.
Enteromorpha sp.		Metopograpsus messor	Acanthopleura brevispina
Tetraclita squamosa			Acmaea sp.
Acanthopleura haddoni	Metopograpsus messor		Morula granulata
Nerita forskali	Cellana rota		Thais aculeata
Cellana rota			Engina mendicaria
Crassostrea cucullata			
Brachidontes variabilis		Pomatoleios kraussi	Nerita albicilla
Sphacelaria tribuloides		Gelidium sp.	Ulva sp.
Laurencia papillosa			Cellana cernica
Ophiocoma scolopendrina	Nerita albicilla	Ectocarpus mitchellae	Metopograpsus messor
Nerita albicilla	Morula granulata	Colpomenia globosa	Brachiodontes variabilis
Thais hippocastaneum		Macrophiothrix elongata	Ophiocoma sp.
Echinometra mathaei		Serpulorbis sulcatus	Laurencia sp.
			E. mathaei
E. mathaei	Engina mendicaria	E. mathaei	E. mathaei
Spyridia Colpomenia	Drupa margariticola	Sargassum	Sargassum
Caulerpa, Sargassum.	Ophiocoma sp.	Zoanthids	Turbinaria
Serpulorbis inopertus	E. mathaei	Balanophyllia sp.	Ascidians
Turbinaria	Sargassum		Corals
Zoanthus bertholleti	Laurencia papillosa		
Corals.	Turbinaria		
	Corals		

9.3.1. Littoral fringe

Cliff tops exposed to the full heat of the sun are devoid of life during the day, but may be visited at night by foraging semi-terrestrial grapsid crabs which emerge from crevices or holes in the shade of the overhang (Fig. 9.2d). Where wave splash occasionally dampens the cliff face the filamentous cyanobacteria (blue-green algae) *Brachytrichia* and *Homaeotrix* may be found together with periwinkles. These become more abundant in the shade of the overhang, and on the most sheltered shores are joined by the small gastropod *Planaxis sulcatus*. The gastropod, *Nerita undata* overlaps with these species but extends further down the rock face, often in dense populations, into the upper eulittoral region reached by the highest tides. Other species characteristic of the littoral fringe include the isopods *Ligia exotica* and *L. pigmenta* and the boring barnacle *Lythotrya valentiana*.

In the central Red Sea, where tides are virtually absent, mobile species such as the grapsid crabs range down to mean sea level throughout the year. A seasonal migration of *Planaxis* and other littoral fringe gastropods also occurs following summer and winter changes in sea levels. At Port Sudan the eulittoral barnacle *Tetrachthamalus oblittoratus* is found in the littoral fringe, possibly as a result of settlement during higher winter sea levels, and survival in summer is due to wave splash created by boat traffic (Hughes, 1977).

9.3.2. Eulittoral zone

To the north and south of the Red Sea the presence of chthamalid barnacles marks the top of the eulittoral, as periwinkles and *Nerita undata* may overlap from the littoral fringe. These barnacles are quickly replaced at a slightly lower level by the large barnacle *Tetraclita squamosa rufotincta*; this level also marks the appearance of macro-algae; namely the green alga *Enteromorpha* and the red alga *Porphyra*, on horizontal reef flats (Fig. 9.2b). *Ulva*, the sea lettuce, occurs at a slightly lower level, and algal cover in general becomes richer in the lower eulittoral with a dense cover formed by red algal species with *Sphacelaria tribuloides*, *Spyridia filamentosa*, and *Laurencia* tufts dominating.

In the upper eulittoral *Nerita forskali* is common along with the limpet *Cellana rota*, the large chiton *Acanthopleura haddoni*, the oyster *Crassostrea*, and the mussels *Brachidontes variabilis* and *Modiolus auriculatus*; these bivalves also extend into the lower eulittoral. *Metopograpsus messor* is the characteristic crab of the eulittoral, foraging widely over a wide range of shores and even extending into the mangal.

At lower levels the brittle-stars *Ophiocoma erinaceus* and *O. scolopendrina* become common, especially on extended eulittoral reef platforms, where they may reach densities of 200 m^{-2}. *Nerita albicilla* replaces *N. forskali* and occurs together with other molluscs including the dog whelk *Thais hippocastaneum*, the cone shell *Conus frigidus*, and in sheltered muddy areas, the small snail *Cerithium morus*. Similar species are found on vertical cliffs at this level (Fig. 9.2a), but bands of macro-algae are absent, and boring and sessile forms such as the bivalves *Lithophaga malaccana* and *Barbatia decussata* predominate. The lower limits of the eulittoral are marked by the occurrence of algae, in particular the bulbous brown *Colpomenia sinuosa*, the green *Caulerpa serrulata* and the brown *Cystoseira myrica* together with the urchin *Echinometra mathaei*, although all of these species are more abundant in the sublittoral fringe.

This rich and intricately zoned eulittoral is absent in the central Red Sea region (Fig. 9.2c,d), where littoral fringe species such as *Nerita undata* overlap with eulittoral species such as *N. albicilla* and *Cellana rota*, which extend down into the submerged reef flat zone. These changes occur abruptly within a metre of the water line, and some characteristic eulittoral species such as the barnacle *Tetrachthamalus*, the chiton *Acanthopleura* and the dog whelk *Thais* are absent. Others such as the barnacle *Tetraclita*, brittle-star *Ophiocoma*, and the algae *Enteromorpha*, *Laurencia* and *Cystoseira* are restricted to the submerged reef flat.

9.3.3. Sublittoral fringe and zone

The transition from the eulittoral to the sublittoral zone in the north and south Red Sea is marked by a distinctive fauna and flora uncovered only on low spring tides, which may be regarded as forming the sublittoral fringe. Characteristic algae are those found in the lower eulittoral together with the red algae *Spyridia*, *Digenea*, *Lophosiphonia*, and *Lithothamnion*, the calcareous green alga *Halimeda*, and the brown algae *Sargassum* and *Turbinaria*. The vermetid *Serpulorbis inopertus* forms a band of cemented tubes in this zone and other characteristic molluscs include *Planaxis lineolatus*, *Conus taeniatus* and *Drupa tuberculata*. Various blennies are common as are crabs, the urchin *Echinometra*, and other invertebrates, all part of a highly diverse community. At the base of vertical cliffs there may be an abrupt transition to *Sargassum* and zoanthids (Fig. 9.2a).

The sublittoral fringe community merges almost imperceptibly with the sublittoral zone which is characterised by the appearance of the corals *Millepora*, *Porites* and *Stylophora*. There is a gradual decline in density of algal species, and seagrasses such as *Cymodocea* and *Enhalis* become common where sand pockets exists. *Echinometra* is replaced by the urchins *Diadema* and *Tripneustes*, and the gastropod groups *Conus* and *Strombus* now occur together with various sea slugs.

At Jiddah (Fig. 9.2d) the submerged inner reef flat becomes almost isolated during summer low water levels, and is thus subject to wide temperature and salinity fluctuations. Organisms found here consist of a unique combination of eulittoral and sublittoral fringe species. Macroalgae include *Dictyosphaera*, *Laurencia* and *Dictyota*, with *Padina* and *Caulerpa* in sand-filled depressions. The brittle-star *Ophiocoma scolopendrina* occurs inshore and is replaced by *O. erinaceus* towards the reef bar. Other inshore molluscs include the Red Sea endemic cowrie *Cypraea pantherina* and the gastropod *Morula granulata*, whilst *Turbo argyrostomus* and the cone shells *Conus catus*, *C. coronatus* and *C. flavidus* occur further out.

The reef bar, which is exposed at periods of low tides and receives the full force of wave action, is characterised by a dense growth of *Sargassum* which is largely destroyed during winter months. The barnacle *Tetraclita* often occurs on boulders in this turbulent zone. On the outside of the bar the reef flat steepens, and a band of *Turbinaria* occurs together with a narrow vermetid zone and *Echinometra* which extends only 1–2 m before it is replaced by the black urchin *Echinostrephus molaris*. At this level the appearance of the corals *Stylophora*, *Millepora*, *Goniastrea* and the organpipe coral *Tubipora* mark the transition to the sublittoral zone.

9.3.4. Comparison with other Indian Ocean rocky shores

It is clear from Table 9.1 that, perhaps with the exception of Jiddah, Red Sea rocky shores show similar zonation patterns to other Indian Ocean shores, although species diversity is often higher in the open Indian Ocean. This lower species diversity may be attributed to several causes. Geographical isolation has prevented the spread of some Indian Ocean species and encouraged a high level of endemism in the Red Sea. The large seasonal temperature changes also act to exclude some species which are also absent from the Arabian Gulf where temperature fluctuations are even higher.

However, the lack of severe wave action and small tidal range are probably most important in restricting species diversity, especially at higher shore levels in the Red Sea. Thus 6 species of *Nerita* occur on the shores of Aldabra Island (Indian Ocean) which have a similar topography, but only 2 in the extremely sheltered lagoon areas (Taylor, 1971). The same 2 species, *N. undata* and *N. albicilla*, are the only representatives of the genus on central Red Sea shores, although 3 or 4 species may occur elsewhere in the Red Sea where tidal amplitude is greater.

Similarly, the range of exposure experienced by eulittoral communities on open Indian Ocean shores increases species richness above that found on tidal Red Sea shores, although at the lowest levels, differences are minimised as community regulation shifts from physical to biological control. The submerged zonation on the reef flats of the central Red Sea presents a uniquely complex situation, for it appears to be determined by physical factors such as turbulence and seasonal sea-level changes, enabling intertidal species to successfully compete with subtidal communities.

9.4 . HIGH SALINITY ENVIRONMENTS

In the north of the Red Sea, around the Sinai peninsula, high salinity conditions have prevailed since the Pleistocene, resulting in the formation of a metahaline (41–44‰) marine fauna with endemic elements. The wealth of lagoons and shore pools of varying salinities in this region has allowed Por (1972) to suggest a classification scheme for these specialised habitats, based on salinity ranges and faunal associations (Table 9.2). These saline environments are of special interest since they may represent relict habitats of extreme glacial communities; when salinities were much higher, and temperatures much lower in the Red Sea.

The Bitter Lake and Sirbonis may be regarded as metahaline lagoons with salinities higher than the Gulf of Suez, which has an impoverished fauna itself, lacking the echinoderms *Tripneustes*, *Echinothrix* and *Ophiocoma*. These metahaline lagoons contain a limited marine flora including the brown alga *Sargassum crispum*, the red *Digenia simplex*, the seagrasses *Halophila* and *Halodule*; the coral *Stylophora*; an impoverished mollusc and echinoderm fauna; and migratory species such as the commercial shrimp *Metapenaeus stebbingi*, swimming crab *Lupa pelagicus*, mullet and the small euryhaline killyfish *Aphanius dispar*. The detailed ecology of a metahaline pool is given by Por and Dor (1975).

Hyperhaline waters are characterised by even higher salinities (Table 9.2), and are found at the landward end of metahaline lagoons or in nearshore pools supplied at some time in the year with water so that they do not dry out. Again their fauna is essentially marine, but extremely impoverished as there can be no population recruitment from the open sea. The flora is represented by *Ruppia maritima* and filamentous algae, and the fauna includes the mudsnail *Pirenella conica*, the mussel *Brachidontes variabilis*, the barnacle *Balanus amphitrite*, specialised copepods and the killyfish *Aphanius dispar*. Some of these species have been recorded from a lake in the Libyan desert some 150 miles inland (Omer-Cooper, 1948).

TABLE 9.2. Classification of high salinity environments in the Red Sea (after Por, 1972).

Classification	Locality	Salinity (‰)	Faunal Type
Metahaline Sea	Red Sea	42–44	Gradual impoverishment with increasing salinity to north. Single mangrove species, absence of mudskippers, coral diversity drastically reduced.
Metahaline lagoons	Bitter Lake Sirbonis Di Zahar pool	40–50 60–70 45–60	Impoverished migrant fauna requiring contact with open sea.
Hyperhaline lagoons	high reaches Sirbonis, El Malaha	80–>180	Marine origin but independent of recruitment from open sea, cannot resist drying out.
Hypersaline waters	Solar Pond nearshore pools north Sinai	80–>300	Continental origin, resistant to drying out.

Hypersaline waters are found in isolated pools which periodically dry up, although Solar Pond near Elat (Table 9.2) is an exception. The flora of these pools consists of cyanobacteria (blue-green algae) and purple bacteria, and the fauna is essentially non-marine and continental in origin, being dominated by the brine shrimp, *Artemia salina*, and salt tolerant beetles.

As a result of their studies Professor Por and others have been able to demonstrate that during the Pleistocene period, when salinities were higher due to increased isolation from the Indian Ocean, the impoverished Red Sea fauna underwent successful adaptation to salinities of around 45‰. Further, by comparison with other similar areas in the world, they found certain regional peculiarities. Thus brine flies (*Epihydra*) are absent, probably due to the ionic composition of Red Sea lagoons, as are nereid polychaetes and amphipods from metahaline and hyperhaline waters, where nematodes are also poorly represented. In addition pond skaters (Hemiptera:Corixidae) and hypersaline diaptomid copepods do not appear to be present in hypersaline Sinai waters. Probably as a result of past history, present Red Sea littoral fauna shows a remarkable resistance to high salinity, indeed upper salinity limits are considerably higher than in any other marine environment in the world, with the exception of the Arabian Gulf (Jones, 1983).

9.5. SAND BEACHES

Almost without exception sand beaches in the Red Sea have been deposited on top of beach rock in bays and back reef areas. Deposits are often thin and interrupted by the beach rock which appears on the surface especially towards low tide. Both low tidal amplitude and the protection received from fringing reefs reduce wave action and tidal currents so that sediments are often poorly sorted, and can vary rapidly from coarse gravels to sands (ranging from 550 to under 200 μm median particle diameter) and to even finer, muddy sand: in sheltered bays at low tide.

To the north and south of the Red Sea, where tides are larger, most beaches have a relatively steep profile, sloping down to horizontal sand flats below midtide level. In central regions the profile remains steeply angled down into the subtidal back-reef lagoons. In all regions the intertidal area is relatively narrow, unlike the extensive sandflats found elsewhere in the Indian Ocean, where tidal range is greater. Trevallion *et al.* (1970) have shown that sand beaches can be divided into broad faunal zones paralleling those of the rocky shore, and their classification is followed in this account. The distribution of characteristic sand beach species is shown diagrammatically in Figure 9.3, and compared with that found in other Indian Ocean beaches in Table 9.3.

9.5.1. Littoral fringe

The level of high water spring tides on Red Sea beaches is clearly marked by the presence of the pyramids produced by the ghost crab *Ocypode saratan*. This large, mainly nocturnal crab lives in burrows during the day, often sealed by a sand plug, and emerges at night to feed on a wide range of organic material. Excavated sand from the male burrow is piled into characteristic mounds during the breeding season as a sexual signal to attract ripe females. This unique example of the use of a 'tool' in decapod social behaviour was first described from the Red Sea (Linsenmair, 1967).

Apart from the rare occurrence of *Ocypode cordimana*, other organisms characteristic of the littoral fringe of all beaches are the land hermit *Coenobita scaevola*, and the amphipod *Talorchestia martensi* found beneath algae washed up on the strand line. *Coenobita* has been recorded foraging close to low water during night time low tides, but may also migrate inland away from the beach, even climbing date palms to feed on ripe dates. The isopod *Tylos exiguus*, like *C. scaevola*, is probably endemic, but is restricted to exposed beaches where it burrows in a zone close to strand line jetsam upon which it feeds at night.

9.5.2. Eulittoral zone

The midtide zone on most exposed clean sand beaches is relatively impoverished (Table 9.3). It is inhabited at higher levels by the endemic cirolanid isopod *Eurydice arabica*, together with a few polychaete worms. More rarely the larger cirolanid *Excirolana orientalis* may occur, although this is far commoner elsewhere in the Indian Ocean. Cirolanid isopods migrate into the water column above the beach at high tide, especially during the night, to feed on plankton, and thus range widely across the midshore occurring at higher levels during spring tides. At lower levels, sand beach amphipods such as *Urothoe*, and the isopod *Exosphaeroma reticulatum* predominate and the mole crabs *Hippa picta* and *H. celaena* mark the edge of low tide. The latter species also migrate tidally, following the rising tide, and may often be trapped in the surf at higher levels on the beach.

Sheltered sand beaches are distinguished not only by their finer and more stable sediments of high organic content, but also by their faunas (Fig. 9.3 and Table 9.3). On the high midshore region the fiddler crab *Uca lactea albimanus* occurs in patches of clean sand, often in dense populations. This species like most *Uca* is a deposit feeder gathering scoops of the substratum and separating out organic material from sand, the latter being rejected by the mouth parts to be deposited in characteristic piles of pseudo-

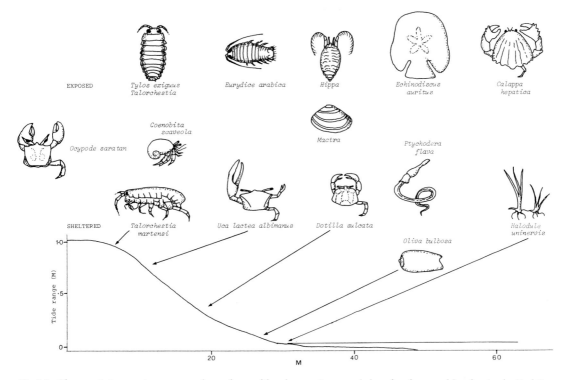

Fig 9.3. Characteristic zonation patterns shown by sand beach organisms on sheltered and exposed beaches in the Red Sea.

faeces. Unlike the ghost crab *Ocypode* and hermit crab *Coenobita*, *Uca* are diurnal and are most active during daytime low tides, when they can be observed feeding and displaying. Males have one enlarged claw of a whitish pink colour, which is waved in a beckoning gesture to attract potential mates. Territories around burrows are defended vigorously and males may often be observed locked in ritualistic combat. Although *Uca* is common in the north and south it has never been recorded from the Jiddah region of the Red Sea, and it is likely that fiddler crabs require regular inundation by tides for survival.

Another deposit feeding crab *Dotilla sulcata* (Fig. 9.3) forms burrows in a zone across the lower eulittoral extending down to low water. Also in this area of wetter sand are various polychaetes (Table 9.3), the gastropods *Nassarius clathratus* and *Oliva bulbosa*, together with bivalves (*Venus* spp). *Dotilla* occurs along the length of the Red Sea, emerging as the tide recedes to deposit-feed in a manner reminiscent of *Uca*, although the details differ considerably. Unlike *Uca*, *Dotilla* is adapted for a life in wet sand, and central Red Sea populations occur very close to mean sea-level, perhaps explaining why this lower shore species tolerates the lack of tides.

9.5.3. Sublittoral fringe

The demarcation of this zone is very indistinct on soft substrate shores, and is best considered as the level at which species found more commonly sublittorally are encountered for the first time. Their intertidal range will depend upon the drainage of the sediment, and the film of standing water left at low tide. In saturated clean sand sediments the sand dollar *Echinodiscus auritus* burrows just below the surface, and the round burrow entrances of the burrowing sea-cucumber *Holothuria arenicola* are common. Patches of the seagrass *Halodule uninervis* may be exposed on the lowest tides, although they flourish only in depressions retaining standing water. Under these saturated conditions the box crab *Calappa hepatica* (Fig. 9.3) is common, half buried in the sand, and the predatory swimming crabs *Thalamita savignyi* and *Portunus longispinus* can be seen swimming at the limit of the low tide.

On sheltered shores the sediment grades into muddy sand, and the fauna changes according to the organic content of the substratum. Thus the crab *Macrophthalmus telescopicus*, burrowing anemone *Cerianthus mana* and acorn worm *Ptychodera flava* are found in muddy sand (200 μm mean particle diameter), and *Macrophthalmus depressus* and *Astropecten polycanthus* are commoner in sandy mud (200 μm particle diameter) with higher organic content. *Halodule uninervis* is again the first seagrass encountered, replaced at lower levels by *Cymodocea serrulata* and *Thalassodendron ciliatum* below lowest tide levels.

9.5.4. Productivity

There is little detailed information on the faunal densities and biomass of tropical beaches in general (MacIntyre, 1968), and for the Red Sea in particular (Jones, 1974). Exposed beaches in the Red Sea may contain some 9–10 macro species with maximum densities of 1200 m^{-2}. Studies on the interstitial meiofauna (which are usually less than 1 mm in length) are even rarer, but Hulings (1975) records mean densities of 460/100 cm^2 for sand beaches along the coast of Jordan, with harpacticoid copepods dominating and free-living flatworms (Turbellaria), roundworms (Nematoda), Archiannelida (a small group of primitive worms), polychaete worms and ostracods also forming a significant contribution.

Sheltered beaches contain a much higher species diversity, and Fishelson (1971) records over 80 macro species for this habitat, although his list includes a subtidal element.

TABLE 9.3. Comparison of exposed and sheltered sand beach zonation patterns in the Indian Ocean region. Data from: 1. Fishelson (1971), Jones (1974); 2. Jones (1982); 3. Pichon (1967), Thomassin (1974); 4. Macnae and Kalk (1962).

	Red Sea[1]		Kuwait[2]		Madagascar[3]	Mozambique[4]
	Exposed	Sheltered	Exposed	Sheltered	Exposed	Sheltered
Littoral fringe	*Coenobita scaevola* *Ocypode saratan* *Ocypode cordimana* *Talorchestia martensi* *Tylos exiguus*		*O. saratan* *T. martensi*	*Orchestia platensis* *Tylos sp.n.*	*Coenobita rugosus* *Ocypode ceratophthalmus* *Ocypode cordimana* *Talorchestia sp.* *Tylos ochri*	*Coenobita rugosus* *Coenobita cavipes* *Ocypode ceratophthalmus* *Talorchestia malayensis*
Upper Eulittoral	*Eurydice arabica* *Excirolana orientalis* *Scholepis lefebvrei* *Goniadides aciculata* *Exosphaeroma reticulatum* *Urothoe sp.*	*Uca lactea albimanus* *Perinereis nuntia* *Nianereis quadraticeps* *Eunice torquata* *Dotilla sulcata* *Lysidice collaris* *Venus spp.*	*Donax scalpellum* *Mitrella blanda* *Platyischnopus herdmani* *U. grimaldi*	*U. lactea annulipes* *Scopimera scabricauda* *Paracleistostoma arabicum* *Dotilla blanfordi* *Ilyoplax frater* *Diplodonta globosa* *Gari rosens*	*Excirolana natalensis* *Excirolana orientalis* *Donax faba* *Nerine cirratulus* *Donax elegans* *Donax madagascariensis*	*Uca marionis* *Phyllochaetopterus elioti* *Mesochaetopterus minutus* *Dotilla fenestrata* *Saccoglossus inhacensis* *Macrophthalmus grandidieri*
Lower Eulittoral	*Hippa picta* *H. celaena*	*Nassarius clathratus* *Oliva bulbosa*	*Perioculodes longimanus* *Emerita holthuisi*	*Solen vagina* *Macrophthalmus depressus*	*Hippa pacificus* *Albunea madagascariensis*	*Halodule uninervis* *Thalassia hemprichii*
Sublittoral fringe	*Echinodiscus auritus* *Holothuria arenicola* *Calappa hepatica* *Thalmita savignyi* *Portunus longispinus*	*Macrophthalmus depressus* *Halodule uninervis* *Astropecten polycanthus* *Thalassodendron ciliatum*	*Matata lunaris* *Branchiostoma sp.* *Portunus pelagicus*	*Halodule univervis* *Pinna bicolor*	*Calappa hepatica* *Portunus emarginatus* *Portunus gladiator* *Portunus granulatus*	*Astropecten granulatus* *Cymodocea serrulata* *Thalassodendron ciliatum*

9.5.5. Comparison with other Indian Ocean beaches

In relation to Indian Ocean exposure scales, Red Sea beaches fall into the moderately exposed category, with some 7−10 macrofaunal species. Although these are mainly endemics, they parallel similar species characterising the major beach zones on shores elsewhere (Table 9.3), and occur in similar densities. The only notable absence is the bivalve *Donax*, characteristic of the midshore region on most Indian Ocean shores. *Donax* species are known to occur in the Red Sea, but do not appear on sand beach faunal lists, so the typical Indian Ocean *OCYPODE−CIROLANID ISOPOD−DONAX-HIPPA/EMERITA* zonation is replaced by *OCYPODE−CIROLANID ISOPOD−HIPPA* on exposed Red Sea sand beaches.

Sheltered beaches again contain a high level of endemics, but follow the *OCYPODE−UCA−DOTILLA−MACROPHTHALMUS−SEAGRASS* zonation pattern found on the other tropical Indian Ocean beaches. Whilst faunal diversity appears to be higher than that found on Indian shores (MacIntyre, 1968), it falls far short of the rich sand flats of African shores where almost 400 macro species have been found (Table 9.3). It appears that this is partly due to the virtual absence of extensive low-shore, soft-substrate, tidal flats in the Red Sea, but also to the restriction of more of the typical low-shore fauna to the sublittoral. Thus whilst *Halodule* commonly extends up to low water neap-tide level in Mozambique and Kuwait, it is restricted to the edge of the sublittoral in the Red Sea. Meiofaunal densities appear low in the Red Sea, but this may be due to predation by the high numbers of deposit feeding crabs, as these replace the lamellibranch populations which dominate temperate sheltered sand beaches.

9.6. MANGROVES AND MUDFLATS

In tropical regions mangrove trees replace the temperate salt marsh community to a large extent and dominate over reed swamp and salt marsh vegetation types in the Red Sea (Zahran, 1977). These salt tolerant trees may grow in forests along the upper part of the intertidal zone in sheltered areas, especially estuaries, where soft muddy sediments occur, and form a self-contained community known as the mangal ecosystem (Chapman, 1975).

This mangal ecosystem is poorly developed in the Red Sea where beach rock is usually only thinly overlaid with sediment even in the most sheltered areas. These reef mangroves or 'Shura' based on rock extend along the length of the Red Sea, terminating in Sinai at 28°N to form the northernmost mangal in the Indo-Pacific. At this northern limit the mangal is composed solely of bushy growths of *Avicennia marina*, but further south in the Dahlak Archipelago in areas of deep soft mud the trees *Rhizophora mucronata* and *Bruguiera gymnorrhiza* also occur, forming a more typical mangal ecosystem (Zahran, 1977).

In the Red Sea where rivers and estuaries are absent, mangroves have developed in high rather than low salinity areas, and salinity levels together with high temperatures and low oxygen levels combine to limit the faunal diversity. However, as Fishelson (1971) and Por *et al.* (1977) have shown, reef dwelling species may live in close proximity to Shura in some areas, and changes in environmental conditions may move sand into established mangroves, bringing sand and mud faunas together.

The intertidal community (Fig. 9.4) on the landward edge of the mangal at Sinai is dominated by the fiddler crabs *Uca tetragonon* and *U. inversa inversa* where soft sediments occur. The former prefers more exposed sediments containing coral rubble, the latter softer, muddier areas. Primary productivity is high in this zone due to the presence of blue-green algal mats, and dense populations of the mud snail *Pirenella cailliaudi* occur in patches. This zone is absent in the virtually tideless central Red Sea, where bushes of *Avicennia* extend down to mean sea-level.

The well developed mangroves to the south of the Red Sea contain a characteristic fauna amongst the trees. This includes several polychaetes and crustaceans, and the mud skipper *Periophthalmus koelreuteri*. A full faunal list is given by Fishelson (1971) and Por *et al.* (1977). A progressive impoverishment of this mangrove fauna occurs towards the north of the Red Sea.

The seaward edge of the mangal, where mud is present, is characterised by the burrows of the ocypodid crab *Macrophthalmus boscii* and at lower levels where the water table reaches the surface, by another similar crab *M. depressus* (Fig. 9.4). Lowest tide levels are indicated by the appearance of the seagrasses *Halodule* and *Halophila*, the presence of the large bivalve *Pinna*, and the mangrove crab *Scylla serrata*.

In addition to this horizontal zonation pattern extending across the mangal, the aerial roots and trunks of the mangroves provide hard surfaces on which a vertical zonation of typical rocky shore species can be seen. This takes the form of bands of periwinkles, barnacles, the oyster *Crassostrea*, and at lower levels sponges, tunicates and the coral *Stylophora pistillata*.

In the absence of nutrient input from rivers and the oligotrophic waters of the Red Sea, it is evident that Red Sea mangals must constitute a nutrient conserving and accumulating ecosystem. Although they are impoverished faunistically, they accumulate and retain sediments, preventing coastal erosion, and more importantly, they form an oasis of high primary productivity in an otherwise barren zone. As yet there do not appear to be any productivity measurements for Red Sea mangals, despite the fact that they form the nursery grounds for important food fish such as mullets (Mugilidae) and porgies (Sparidae).

9.7. SHALLOW SUBLITTORAL COMMUNITIES

There are, apart from coral reefs (Chapter 7) several other shallow sublittoral communities. These Red Sea sublittoral communities are extremely rich in species, due in the main to the removal of the harsh intertidal environmental conditions, but also to the wide range of habitats, which include lagoons,

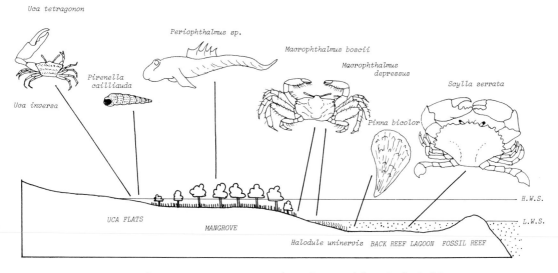

Fig. 9.4. Characteristic zonation pattern shown by mangal fauna in the Red Sea.

TABLE 9.4. Red Sea seagrass species.

Family:	Potamogetonaceae	Hydrocharitaceae
	Cymodocea rotundata	*Enhalus acoroides*
	Cymodocea serrulata	*Halophila ovalis*
	Halodule uninervis	*Halophila ovata*
	Syringodium isoetifolium	*Halophila stipulacea*
	Thalassodendron ciliatum	*Thalassia hemprichii*

channels, fringing and patch reefs. Wave surge and weak tidal currents combine to produce a wide variety of substrates within these habitats, ranging from bare rock, dead coral blocks and rubble, to muds. In addition, in these shallow, well-lit areas where seagrasses reach their maximum development, primary productivity is extremely high.

Fishelson (1971) has described several faunal communities from the shallow sublittoral in the Red Sea. A *Hippa/Mactra* community occurs on coarse sands and gravels mixed with mud, and contains the algae *Caulerpa*, *Padina* and *Cystoseira*. It is dominated by mole crabs (*Hippa*), portunid crabs and stomatopods together with the molluscs *Mesoderma*, *Mactra* and *Strombus*. A *Ptychodera* (acornworm)/*Radianthus* (anemone) community extends sublittorally from sheltered sand beaches on muddy calcareous sediments. This is mainly a burrowing fauna consisting of worms and bivalves penetrating the substratum to depths of up to 40 cm in between beds of the seagrass *Thalassodendron ciliatum*. Subtidal areas of fine muddy sediment extending down to depths of 40 m are dominated by seagrass meadows, with *Halophila* as the most common representative, together with dense populations of a lancelet, *Asymmetron*, which lives half buried in the sand. These seagrass communities, the most productive of the shallow sublittoral soft sediments, are discussed in detail in the next section.

Where the substratum is predominately of dead coral blocks or rubble interspersed with pockets of sand Fishelson recognises a community dominated by the trochid gastropod *Gena* and the sea urchin *Echinometra*. It is a rich community with many species of molluscs either sheltering under boulders or in the sand. The urchins *Echinometra*, *Diadema* and *Tripneustes* appear from crevices at night moving into the open to feed. Where bedrock predominates the spiny lobster *Panulirus penicillatus* is common, together with xanthid crabs and various eleotridid fish species.

Measurements of faunal biomass at 4 shallow sublittoral sites in the Jiddah region have produced figures ranging from 5–50 g dry organic weight m^{-2}. These faunal biomasses are comparable with those found in similar habitats elsewhere in the world.

9.8. SEAGRASSES

9.8.1. Introduction

Seagrasses are the only group of higher plants that have adapted to life submerged under the sea. As their name implies they are grass-like, and belong to a single order of the Monocotyledons. They inhabit soft-bottomed, shallow water areas of temperate, subtropical and tropical seas where they may form large meadows. Altogether 49 species of seagrass are known from around the world; these are grouped into 12 genera, nine of which are placed in one family, the Potamogetonaceae, and three in another, the Hydrocharitaceae (den Hartog, 1977). Five genera (all in the Potamogetonaceae) are more or less confined to temperate and subtropical seas and do not concern us further. The remaining seven tropical genera all have representatives in the Red Sea (den Hartog, 1970). In all ten species of seagrass have been recorded from the Red Sea (Table 9.4) — (Lipkin, 1975a, 1977). One species, *Halophila ovata*, has been recorded in the Red Sea only recently (Aleem, 1979).

At the turn of the century seagrasses were considered to be of fundamental ecological importance as primary producers in the sea. However, when the mass destruction of eelgrass (*Zostera*) beds in the

North Atlantic by a slime mould (*Labyrinthula*) failed to produce the predicted catastrophic decline in fisheries, interest in seagrasses faded; this decrease in interest was accelerated by the growing realisation of the prime importance of phytoplankton in marine productivity. Only in recent years has the importance of seagrass communities again been recognised and many features of the community are still poorly understood. In this section I shall discuss firstly, how seagrasses have adapted to life in the sea, secondly, the pattern of distribution of seagrass species in the Red Sea, and thirdly, the major features of the seagrass community.

9.8.2. The seagrass plant

Seagrasses are found from mid-tidal level, on shores receiving regular tides, to about 70 m depth in the Red Sea, although any one species tends to be restricted to a narrower depth range (Lipkin, 1977). Most seagrasses have linear or strap-like leaves, similar to those of terrestial grasses, with species in the genus *Halophila* being the exception. Their rather limited range of leaf morphology, as exemplified by Red Sea species, is shown in Figure 9.5.

The precise form of the leaf may vary according to the environment of the seagrass plant. For instance, plants of the common Red Sea species, *Halophila stipulacea*, have larger (120 mm by 10 mm), greener leaves in dimly-lit, deeper water than in shallow water, where leaves may be only 30 mm by 5 mm (Lipkin, 1979; Hulings, 1979). Similarly, the leaves of shallow water *Halodule uninervis* plants are generally 0.5–1 mm in width but those of deeper water plants are up to 3 mm wide and tend also to be longer. The exception is shallow water *H. uninervis* shaded by mangroves which also has broader longer leaves. This indicates that light, and not wave action, is probably the prime influence on leaf form in this species (Lipkin, 1977). Plants of the large seagrass, *Thalassodendron ciliatum*, grow to only about 20 cm height in shallow water but may reach 60–100 cm in deeper areas. The decreased wave action in deeper water may be the prime influence in this case. The leaves generally contain internal, gas-filled channels which buoy them up in the water, again some *Halophila* species are an exception.

The plants are anchored to the soft sediments on which they grow by a well-developed system of underground rhizomes and roots. These form an excellent anchorage which is especially important to plants subjected to wave action in shallow water. The narrow, thin, flexible leaves are probably also significant in allowing the plants' survival in such habitats. The rhizomes and roots of seagrass plants stabilise the seabed in a similar manner to marram grass, used by man to bind shifting sand dunes. The leaf : rhizome and root biomass ratio may vary considerably with depth. In *Halophila stipulacea* plants in the intertidal zone there is on average twice as much rhizome and root as leaf, while at 10–30 m depth there is about twice as much leaf as rhizome and root (Lipkin, 1979). This reflects the need for secure anchorage in the shallows and maximum light gathering in the deeps.

Seagrass plants appear to reproduce asexually (or vegetatively) more readily than sexually. Thus, quite often all the plants of a sizeable stand may be the vegetatively-produced progeny of a single seedling. The relatively common occurrence of single species stands, and the less frequent occurrence of truly mixed populations of seagrasses in the Red Sea probably results from this phenomenon (Lipkin, 1977). Most seagrass species flower underwater and release their pollen into the water. Following pollination by water-borne pollen grains, seeds are set, and later, these too are distributed by water currents. The whole life-cycle is thus carried out underwater. In the northern Red Sea the majority of seagrasses flower in the summer, between June and September.

9.8.3. Distribution of Red Sea seagrasses

Seagrasses generally have extensive distributions. Of the seven tropical genera, four (*Halodule, Halophila, Syringodium* and *Thalassia*) are found in both the Atlantic and Indo-Pacific regions, whilst three (*Enhalus, Cymodocea* and *Thalassodendron*) are found only in the Indo-Pacific (den Hartog, 1970).

All the species recorded from the Red Sea (Table 9.4), except *Halophila stipulacea* which is confined to the western Indian Ocean, are widely distributed in the tropical Indian and West Pacific Oceans.

The commonest species in the Red Sea are *Halophila stipulacea*, *Halodule uninervis*, *Thalassodendron ciliatum*, *Syringodium isoetifolium* and *Halophila ovalis*. Photographs of three of these are shown in Figure 9.6. *Halophila stipulacea* and *Thalassodendron ciliatum* have the greatest vertical distribution, the former

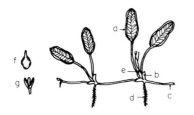

1. Halophila ovalis (R. Brown) Hooker fil

2. Halophila stipulacea (Forsskål) Ashers

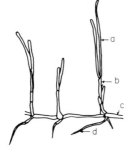

3. Halodule uninervis (Forsskål) Ashers

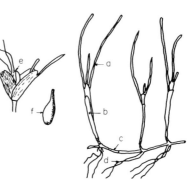

4. Syringodium isoetifolium (Aschers) Dandy

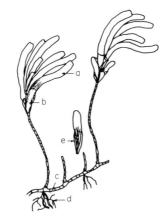

5. Thalassodendron ciliatum (Forsskål) den Hartog

Fig. 9.5. Line drawings of five species of Red Sea seagrasses, showing: a- foliage leaf, b- scale leaf, c- rhizome, droot, e- female flower, f- fruit, and g- male flower.

Fig. 9.6. (a). Pure stand of *Halophila stipulacea*; (b). Mixed bed of *H. stipulacea* and *Halodule uninervis*; (c). Pure stand of *Halodule uninervis*; (d) The sea-cucumber *Holothuria martensi*; covered in detached *Halophila stipulacea* leaves. Photos: M. Wahbeh.

extending from the lower shore to at least 70 m depth, and the latter from extreme low-water level to at least 40 m depth (Lipkin, 1979). The remaining species are restricted to seabed under less than 10 m of water.

Red Sea seagrasses have not been adequately mapped, but some general points have emerged from the few studies of their distribution. First, conditions in the Gulfs of Aqaba and Suez (particularly the latter) appear to be at the limits of temperature tolerance for the majority of the seagrass species. Seven species have been recorded at the mouths of the Gulfs. Half way up the Gulf of Aqaba five or six species may be found but at the extreme northern end only *Halophila stipulacea*, *H. ovalis* and *Halodule uninervis* occur (Wahbeh, 1980). In the Gulf of Suez only these latter three species are found half way up the Gulf, and only *Halophila stipulacea* and *Halodule uninervis* occur at the northern extremity (Lipkin, 1977). Both these species have also invaded the Suez Canal and *H. stipulacea*, having reached Port Said, has successfully colonised several areas in the eastern Mediterranean (Fox, 1926; Aleem, 1962; Lipkin, 1972).

Halophila stipulacea appears to be most abundant towards the northern and southern ends of its range in the western Indian Ocean and is perhaps best regarded as a subtropical rather than a truly tropical seagrass. This would account for its predominance in the cooler waters of the northern Red Sea and for why it has been the only Red Sea seagrass able to colonise the Mediterranean (Lipkin, 1977). The other seagrasses appear to have more restricted ecological ranges in the Gulfs of Aqaba and Suez than they do elsewhere in the Indo-Pacific, a common feature of plant species near the limits of their distributions.

Neither *Cymodocea serrulata* nor *Enhalus acoroides* seem to have been reliably recorded from the Sinai coast (Lipkin, 1977), though both species are known from the Saudi Arabian coast in the vicinity of Jiddah (Aleem, 1979). The only records for *Halophila ovata* in the Red Sea are those of Aleem (1979), who reported it from three sites near Jiddah. The occurrences of the seagrass species are summarised in Table 9.5.

TABLE 9.5. The occurrence of seagrass species in the northern (north of 29°N), central (18−25°N) and southern (south of 18°N) Red Sea. For each area the percentage of sites at which each species occurs is given.

	Northern	Central	Southern
No. of sites studied	36	23	15
Species			
Cymodocea rotundata	19	65	20
Cymodocea serrulata	6	30	20
Enhalus acoroides	0	9	27
Halodule uninervis	81	44	33
Halophila ovalis	67	35	33
Halophila ovata	0	17	0
Halophila stipulacea	78	48	20
Syringodium isoetifolium	19	13	27
Thalassia hemprichii	17	22	53
Thalassodendron ciliatum	44	22	40

9.8.4. Features of the seagrass ecosystem

Seagrasses are able to colonise large areas of unconsolidated sediments, ranging from muds to coarse coralligenous sands, which may be to a large extent unuseable by other attached macrophytes. In so doing, vast areas of soft bottom are stabilised and a highly productive, plant dominated ecosystem is created in otherwise relatively barren areas. The fauna living buried in the sediments (infauna) does not seem to be adversely affected by the establishment of a seagrass bed though its species composition may be altered (Kikuchi and Peres, 1977).

Wahbeh (1982) found more than 49 species of invertebrates in *Halophila stipulacea* beds in the northern Gulf of Aqaba, nearly 70% of which were molluscs. These lived either on the seagrass (e.g. small gastropods, such as *Phaseanella nivosa*, *Smaragdia rangiana* and *Cerithium rostratum*), or buried in the sediment (e.g. predatory gastropods and various filter- or deposit-feeding bivalves). Echinoderms also constituted a significant component of the seagrass associated fauna, notably the urchin *Tripneustes gratilla* and various sea-cucumbers, some species of which habitually covered themselves with detached *Halophila* leaves (Fig. 9.6d).

The dense stands of vegetation act as baffles to water flow, reducing the velocity of water movements and promoting sedimentation. Seagrass beds thus act as sediment traps, the trapped particles being subsequently bound in place by the rhizome and root networks (Burrell and Schubel, 1977). Only the top layer of sediment remains oxidised, the remainder is anoxic and has a characteristic grey or black colour. The rhizome and root network by reducing sediment mobility (and consuming oxygen) promotes and maintains these anoxic conditions, often in areas where the sediment would otherwise be continually stirred up by wave action and oxidised to a much greater depth.

The degree of damping of water movements, and hence the amount of sedimentation, depends on both the density of plants and the species. Wahbeh (1980) reported the mean density of *Halophila stipulacea* in the northern Gulf of Aqaba to be just over 400 shoots/m^2. This density is comparatively low (up to 4000 shoots/m^2 are recorded for *Zostera* in Alaska) and higher densities may be found elsewhere in the Red Sea. The larger, relatively broad-leaved species, such as *Thalassodendron ciliatum* and *Enhalus acoroides*, provide more effective baffles than the smaller, narrow leaved species, such as *Halodule uninervis*. Sediment trapping is also selective; fine grained sediments being relatively abundant in seagrass beds. As the sediments accumulate the seagrass beds may become elevated a few centimetres above the surrounding seabed, or in certain conditions even form substantial mounds (Burrell and Schubel, 1977).

In areas which are otherwise unsuitable, seagrass plants provide a surface on which many species of micro-algae and macroalgae (including benthic diatoms and algae-like blue-green bacteria), discoid foraminifers (particularly in shallow water), and a variety of sessile invertebrate animals such as bryozoans, hydroids (e.g. *Dynamena cornicina* on *H. stipulacea*) and barnacles can become established (McRoy and McMillan, 1977; Wahbeh, 1982). These species attract others, notably gastropod molluscs, polychaete worms and crustaceans, which feed on the attached plants and animals. The leaf area available for colonisation by epiphytes and epifauna may be up to 20 times the area of the sea bottom on which the seagrass plants are growing, the ratio of leaf area to substrate area being greatest in dense stands of the broad-leaved species such as *Thalassodendron ciliatum* but only reaching 4.1—12 in *Halophila stipulacea* communities in the northern Gulf of Aqaba (Wahbeh, 1980; Hulings, 1979).

The organisms using the seagrass leaves as a surface for attachment are an extremely important component of the seagrass community, photosynthetic epiphytes accounting for up to a fifth of the primary production of a seagrass bed.

The readily available food supply and calming effect on water movement makes seagrass beds an ideal nursery ground for fish, shrimps and other invertebrates, some of which may be of commercial importance (den Hartog, 1977). For instance, the juvenile stages of the commercially fished shrimp, *Penaeus semisulcatus*, use seagrass beds as nursery grounds both in the Red Sea and in the Arabian Gulf. Sometimes it is only when seagrass beds are destroyed that the reliance of commercial fisheries on them is belatedly recognised; as happened in Florida with a shrimp fishery, destroyed after seagrass beds were dredged away.

9.8.5. Standing crop and primary production

Figures for the standing crops (biomasses) of tropical seagrass beds given by McRoy and McMillan (1977) range from 20 g to 8.1 kg dry weight per square metre (kg-dry/m^2). In general the standing crops of the larger, broad-leaved species tend to lie at the higher end of the range, while those of the smaller, narrow-leaved species (even though they can form very dense stands) lie at the lower end. Several figures are available for the standing crops of seagrasses in the Red Sea. On the Sinai coast, Lipkin (1979) reports pure stands of *Thalassodendron ciliatum* with average standing crops of 70 kg-dry/m^2 and occasionally biomasses of over 100 kg-dry/m^2, figures similar to those for dense tropical rain forest. Lipkin also gives average standing crop values for seagrass communities (often single species stands) dominated by *Thalassia hemprichii* (2.5 kg-dry/m^2), *Syringodium isoetifolium* (1.2 kg-dry/m^2), *Cymodocea rotundata* (0.5 kg-dry/m^2), *Halophila stipulacea* (0.33 kg-dry/m^2), *Halodule uninervis* (0.23 kg-dry/m^2) and *Halophila ovalis* (35 g-dry/m^2). Aleem (1979) records a value of 3 kg-fresh/m^2 (equivalent to around 0.5 kg-dry/m^2) for *Thalassia hemprichii* on the Saudi Arabian coast near Jiddah. Wahbeh (1980), working on smaller species in the northern Gulf of Aqaba, reported maximum standing crops of 0.26 kg-dry/m^2 for *Halophila stipulacea*, 0.40 kg-dry/m^2 for *Halodule uninervis* and only 0.01 kg-dry/m^2 for the sparse growths of *Halophila ovalis*. For comparison, the standing crops of temperate grassland generally range between 0.2 and 5 kg-dry/m^2.

Perhaps more informative are estimates of productivity, which tell us how fast new plant material is produced rather than just how much is there. Conventionally, rates of production are expressed as number of grams of carbon fixed (by photosynthesis) per unit area of substrate per unit time (for example, g-C/m^2.year). The most highly productive areas of the ocean such as areas of upwelling and coral reefs have productivities averaging 500—2500 g-C/m^2.year. For dense *Thalassia* beds in the Caribbean net productivities of 1000 g-C/m^2.year are reported (McRoy and McMillan, 1977). The only figures for seagrass productivity in the Red Sea are those of Wahbeh (1980) from the Gulf of Aqaba. Using measurements of oxygen release he estimated annual productivities at 617 g-C/m^2 for *Halophila stipulacea*, 1326 g-C/m^2 for *Halodule uninervis* and only 11 g-C/m^2 for the sparse growths of *Halophila ovalis.* Thus primary production is high where beds are moderately dense, ranking with that of the most productive ocean areas. This is without even taking into account the production of associated phytoplankton and benthic algae which may be substantial. Wahbeh (1980) also measured the growth of *Halophila stipulacea* leaves and estimated production from these measurements. He found a value of 370 g-C/m^2.year, about half of the estimate obtained from oxygen release experiments. Thus a substantial amount of the carbon fixed during photosynthesis may not be incorporated as new leaf tissue. This production may be used in the growth of the rhizome and root network, stored in the rhizomes or excreted into the surrounding water as dissolved organic material.

9.8.6. Grazing of seagrasses

The vast primary production of the seagrass beds, broadly, forms the basis of two food webs: in one, live plant material is grazed directly by herbivorous animals; in the other, dead plant material is first broken down by bacteria and fungi and these micro-organisms then used as the basis of decomposer food chains.

Somewhat surprisingly, considering the amount of food available, few herbivorous species appear to feed on seagrasses directly. This may relate to the presence in seagrasses of distasteful phenolic compounds, such as are used by certain terrestrial plants to deter herbivores (McMillan *et al.*, 1980). In general, only a minority of seagrass production seems to be consumed by herbivores, estimates ranging from 3% (Kirkman and Reid, 1979) to 62% (Greenway, 1976). Many sea urchins and gastropods which have been implicated as seagrass grazers, have been found on closer study to feed primarily on the epiphytes and epifauna on the seagrass leaves and/or on benthic algae growing with the seagrasses (Kikuchi and Peres, 1977).

There is little information on seagrass grazing in the Red Sea. Wahbeh and Ormond (1980) report significant grazing by a common sea urchin, *Tripneustes gratilla*, and a little grazing by two species of surgeonfish (*Zebrasoma xanthurum* and *Ctenochaetus striatus*) and a rabbitfish (*Siganus rivulatus*) in the Gulf of Aqaba. In a *Halophila stipulacea* bed, where the density of the urchin was about 1/m^2, Wahbeh (1980) estimated seagrass consumption by *T. gratilla* at 124 g-C/m^2.year, equivalent to a third of the total seagrass leaf growth. By contrast, consumption by grazing fishes was estimated at only 18 g-C/m^2.year or less than 5% of total plant growth. The remainder (about 60%) of the seagrass growth must presumably eventually enter the decomposer food web.

Other significant seagrass grazers in the Red Sea are dugongs (*Dugong dugong*) and sea-turtles (*see* Chapter 14). *Halodule uninervis*, and to a lesser extent *Syringodium isoetifolium*, appear to be the favourite seagrass food of the dugong in the Red Sea (Gohar, 1957; Lipkin, 1975b). In the southern Red Sea the Green turtle, *Chelonia mydas*, is a major seagrass grazer (Hirth *et al.*, 1973).

9.8.7. Decomposition of seagrasses

As already indicated, at least half, and often maybe much more, of the plant material produced in a seagrass bed each year is not eaten by herbivores. Thus in a square kilometre of a *Halophila stipulacea* bed, something in the order of 5000 metric tonnes (fresh weight) of seagrass may be produced each year, over and above that accounted for by grazers. During the course of the year wave action will remove substantial amounts of material (broken off leaves, uprooted plants, naturally senescent leaves etc.) from the seagrass beds. Some of these fragments may end up on the strand line of the shore where insects, sand-hoppers such as *Talorchestia martensi*, (Amphipoda:Talitridae), and other invertebrates may eat them; others may be transported far out to sea and end up on the deep ocean floor, and yet others may accumulate in hollows in the seabed in and around seagrass beds. All will eventually be decomposed.

Soluble constituents quickly leak out of the senescing seagrass which is soon broken down mechanically by wave action. Senescent leaves are rapidly colonised by bacteria and fungi which chemically break down the seagrass detritus, converting it from dead, particulate plant material poor in both nitrogen and phosphorus to living particulate microbial material rich in both these nutrients. In order to grow efficiently the micro-organisms have to absorb the additional mineral nutrients from the surrounding seawater. The bacteria and fungi attract a whole microfauna ranging from protozoans such as zooflagellates, which feed exclusively on the bacteria, to flatworms which might prey on the organisms eating the zooflagellates (Fenchel, 1977). But this is only the base of the decomposer food web. Larger animals such as small gastropods, amphipods and polychaete worms feed on the microfauna, often indiscriminately ingesting the rotting seagrass, which they cannot digest, as well. Their activity speeds up the mechanical breakdown of the seagrass fragments, increasing their surface area and hence the rate of decomposition. However, not all seagrass fragments remain in the oxygenated water or surface sediment layer where this sort of decomposition (aerobic) can occur; much is buried in anoxic sediments before it is fully decomposed. Decomposition is continued by anaerobic bacteria, but this is a slow process so that rotting leaves may accumulate to depths of 30 cm. In the northern Gulf of Aqaba, Wahbeh (1980) found that the rate of addition of dead *Halophila stipulacea* leaves was almost twice the rate of decomposition. At his study site it appeared that a high proportion of the dead seagrass not accounted for by decomposition was transported away from the seagrass bed by water movements. A lot more work needs to be done before it will be clear what proportions of the sun's energy trapped by seagrasses are recycled by grazers, aerobic decomposers or anaerobic decomposers.

9.9. CONCLUSION

Whilst Red Sea coastal regions include examples of most of the typical tropical habitats found elsewhere in the Indian Ocean, they are often impoverished, both faunistically and in the range of microhabitats available. This is almost certainly due to the severe environmental regime imposed by the arid zone climate, the sheltered nature of the Red Sea, and small tidal range. As a consequence of these factors and recent periods of isolation of the Red Sea from the Indian Ocean a high degree of endemism has developed, and many species show far higher tolerances to temperature and salinity than their Indian Ocean counterparts. Few ecophysiological studies have been made as yet, but it would appear likely that many intertidal organisms must be close to their physiological limits. For this reason alone it is important that the additional stresses imposed by man's activities and pollution in the coastal zone are restricted. There is increasing evidence (Fishelson, 1977) that industrial and domestic pollution can bring about the rapid collapse of Red Sea intertidal and shallow sublittoral communities, which act as nursery grounds for many important food fish and shrimp.

REFERENCES

Aleem, A. A. (1962). The occurrence of the sea-grass: *Halophila stipulacea* (Forssk.) Asch. on the west coast of Egypt. *Bull. Fac. Sci. Univ. Alexandria* 4, 79–84.
Aleem, A. A. (1979). A contribution to the study of seagrasses along the Red Sea coast of Saudi Arabia. *Aquat. Bot.* 7, 71–8.

Briggs, J. C. (1974). *Marine Zoogeography*. New York. McGraw Hill Book Co.

Burrell, D.C. and Schubel, J. R. (1977). Seagrass ecosystem oceanography. In *Seagrass Ecosystems: A Scientific Perspective*. Ed. C. P. McRoy and C. Helfferich, pp. 195−232, Dekker, New York.

Chapman, V. J. (1975). Introduction. In *Ecosystems of the World 1: Wet Coastal Ecosystems*. Ed. V.J. Chapman, pp.1−29. Elsevier, Amsterdam.

Fenchel, T. (1977). Aspects of the decomposition of seagrasses. In *Seagrass Ecosystems: A Scientific Perspective*. Ed. C. P. McRoy and C. Helfferich, pp. 123−45, Dekker, New York.

Fishelson, L. (1971). Ecology and distribution of the benthic fauna in the shallow waters of the Red Sea. *Mar. Biol.* 10, 113−33.

Fishelson, L. (1977). Stability and instability of marine ecosystems, illustrated by examples from the Red Sea. *Helgoländer wiss. Meeresunters.* 30, 18−29.

Fox, H. M. (1926). Cambridge Expedition to the Suez Canal, I. General part. *Trans. zool. Soc. Lond.* 22, 1−64.

Gohar, H. A. F. (1957). The Red Sea dugong, *Dugong dugong* (Erxlb.), subspecies *tabernaculi* (Rupelli). *Publs mar. biol. Stn, Ghardaqa* 9, 3−49.

Greenway, M. (1976). The grazing of *Thalassia testudinum* (Konig) in Kingston Harbour, Jamaica. *Aquat. Bot.* 2, 117−39.

Hartog, C. den (1970). *The Sea-grasses of the World*, North-Holland, Amsterdam.

Hartog, C. den (1977). Structure, function, and classification in seagrass communities. In *Seagrass Ecosystems: A Scientific Perspective*. Ed. C. P. McRoy and C. Helfferich, pp. 89−121, Dekker, New York.

Hirth, H. F., Klikoff, L. G. and Harper, K. T. (1973). Seagrasses of Khor Umaira, People's Democratic Republic of Yemen with reference to their role in the diet of the green turtle, *Chelonia mydas*. *Fish. Bull.* 71, 1093−7.

Hughes, R. N. (1977). The biota of reef-flats and limestone cliffs near Jeddah, Saudi Arabia. *J. nat. Hist.* 11, 77−96.

Hulings, N. C. (1975). Spatial and quantitative distribution of sand beach meiofauna in the Northern Gulf of Aqaba. *Rapp. Comm. int. Mer. Medit.* 23, 163.

Hulings, N. C. (1979). The ecology, biometry and biomass of the seagrass *Halophila stipulacea* along the Jordanian coast of the Gulf of Aqaba. *Botanica mar.* 22, 425−30.

Jones, D. A. (1974). The systematics and ecology of some sand beach isopods (Family Cirolanidae) from the coasts of Saudi Arabia. *Crustaceana* 26, 2-1-11.

Jones, D. A. (1986). Ecology of rocky and sandy shores of Kuwait. *Proc. First Gulf Conference on Environment and Pollution*. Feb. 7−9, 1982. In press.

Kikuchi, T. and Peres, J. M. (1977). Consumer ecology of seagrass beds. In *Seagrass Ecosystems: A Scientific Perspective*. Ed. C. P. McRoy and C. Helfferich, pp.147−93, Dekker, New York.

Kirkman, H. and Reid, D. D. (1979). A study of the role of the seagrass *Posidonia australis* in the carbon budget of an estuary. *Aquat. Bot.* 7, 173−83.

Klausewitz, W. (1967). Die physiographische Zonierung der Saumriffe von Sarso. 4. Beitrag der Arbeitsgruppe Littoralforschung. *'Meteor' Forshungsergebnisse D* 2, 44−68.

Lewis, J. R. (1964). *The Ecology of Rocky Shores*. English Universities Press, London.

Linsenmair, K. E. (1967). Konstruktion und Signalfunktion der Sandpyramide der Reiterkrabbe *Ocypode saratan* Forsk. *Z. Tierpsychol.* 24, 403−56.

Lipkin, Y. (1972). Marine algal and sea-grass flora of the Suez Canal. *Israel J. Zool.* 21, 405−46.

Lipkin, Y. (1975a). A history, catalogue and bibliography of Red Sea seagrasses. *Israel J. Bot.* 21, 89−105.

Lipkin, Y. (1975b). Food of the Red Sea *Dugong* (Mammalia: Sirenia) from Sinai. *Israel J. Zool.* 24, 81−98.

Lipkin, Y. (1977). Seagrass vegetation of Sinai and Israel. In *Seagrass Ecosystems: A Scientific Perspective*. Ed. C.P. McRoy and C. Helfferich, pp. 263−93, Dekker, New York.

Lipkin, Y. (1979). Quantitative aspects of seagrass communities, particularly of those dominated by *Halophila stipulacea*, in Sinai (northern Red Sea). *Aquat. Bot.* 7, 119−28.

MacIntyre, A. D. (1968). The meiofauna and macrofauna of some tropical beaches. *J. zool., Lond.* 156, 377−92.

Macnae, W. and Kalk, M. (1962). The fauna and flora of sand flats in Inhaca Island, Mozambique. *J. Anim. Ecol.* 31, 93−128.

McMillan, K., Zapata, O. and Escobara, L. (1980). Sulphated phenolic compounds in seagrasses. *Aquat. Bot.* 8, 267−78.

McRoy, C. P. and McMillan, G. (1977). Production ecology and physiology of seagrasses. In *Seagrass Ecosystems: A Scientific Perspective*. Ed. C.P. McRoy and C. Helfferich, pp. 53−88, Dekker, New York.

Omer-Cooper, J. (1948). The Armstrong College Expedition to Siva Oasis (Lybian Desert) 1935, General Report. *Proc. Egypt Acad. Sci.* 3, 1−51.

Pichon, M. (1967). Contribution á l'ètude des peuplements de la zone intertidale sur sables fins et sables vaseux non fixes dans la region de Tulèar. *Annls. Fac. Sci. Univ. Madagascar* 5, 171−214.

Por, F. D. (1972). Hydrobiological notes on the high-salinity waters of the Sinai Peninsula. *Mar. Biol.* 14, 111−9.

Por, F. D., Dor, I and Amir, A. (1977). The mangal of Sinai: Limits of an ecosystem. *Helgoländer wiss. Meeresunters.* 30, 395-414.

Safriel, U. and Lipkin, Y. (1964). Note on the intertidal zonation of the rocky shores at Eilat (Red Sea, Israel). *Israel J. Zool.* 13, 187−90.

Safriel, U. and Lipkin, Y. (1975). Patterns of colonisation of the eastern Mediterranean intertidal zone by Red Sea immigrants. *J. Ecol.* 63, 61−3.

Stephenson, T. A. and Stephenson, A. (1949). The universal features of zonation between tide marks on rocky coasts. *J. Ecol.* 37, 289−305.

Taylor, J. D. (1971). Intertidal zonation at Aldabra Atoll. *Phil. Trans. Roy. Soc. Lond. B.* 260, 173−213.

Thomassin, B. A. (1974). Soft bottom carcinological fauna on Tulear coral reef complexes (S.W. Madagascar): distribution, importance, roles played in trophic food-chains and in bottom deposits. *Proc. Second Int. Coral Reef Symp.* 1, 297–320.

Trevallion, A., Ansell, A. D, Sivadas, P. and Narayanan, B. (1970). A preliminary account of two sandy beaches in South West India. *Mar. Biol.* 6, 268–79.

Wahbeh, M. I. (1980). Studies on the ecology and productivity of the seagrass *Halophila stipulacea*, and some associated organisms in the Gulf of Aqaba (Jordan). Ph.D. thesis, University of York, U.K.

Wahbeh, M. I. (1982). Distribution, biomass, biometry and some associated fauna of the seagrass community in the Jordan Gulf of Aqaba. *Proc. 4th Int. Coral Reef Symp.* (In press).

Wahbeh, M. I. and Ormond, R. F. G. (1980). Distribution and productivity of a Red Sea seagrass community. In *Proc. coast. mar. Environ., Khartoum, 1980* (In press).

Zahran, M. A. (1977). Africa A. Wet Formations of the African Red Sea Coast. In *Ecosystems of the World*, 1. Ed. V.J. Chapman, pp.215–32. Elsevier, Amsterdam.

CHAPTER 10

Molluscs of the Red Sea

MICHAEL MASTALLER

Staatliches Museum für Naturkunde, Arsenalplatz 3, 7140 Ludwigsburg, Federal Republic of Germany

CONTENTS

10.1. INTRODUCTION

The coral reefs and the shallow water zones of the Red Sea represent extremely heterogeneous ecosystems. Accordingly, a great variety of species and number of individuals live there within a relatively narrow spatial range. In fact, the diversity of Red Sea molluscs corresponds with the multiplicity of their habitats. Although the molluscs lag far behind the insects in the number of species, the 80,000 species so far described represent perhaps the most widespread group of animals on the Earth. In the Red Sea this animal phylum shows how perfectly its members have adjusted themselves to a wide variety of environmental conditions. Molluscs have occupied almost every ecological niche provided by the complex structures of different reef zones and substrates.

Unlike many organisms, the beauty of shells — particularly those from tropical seas — remains well-preserved in dried specimens. As the hobby of shell collecting has become more and more popular, once common shells are now reported to be scarce. Unfortunately, due to excessive trade and demand, nowadays many shallow reef areas around the world have become deprived of some of their most attractive and biologically important constituents. Strict regulations concerning the collection of shells have therefore become necessary. Thus conservation measures have had to be established, for example, along the coasts of the Sinai peninsula. In keeping with this background of marine conservation the following presentation of the Red Sea molluscs is intended to stimulate more naturalists to build up collections of underwater photographs and records of live molluscs rather than to pile up sets of dead shells, albeit colourful marine souvenirs.

10.2. THE INVESTIGATION OF THE RED SEA MOLLUSCS. A HISTORICAL REVIEW

At least 5000 years ago the Red Sea was incorporated into the trading activities of the classical world and during the last few centuries this sea has drawn the interest of generations of shell collectors and conchologists in Europe. Thus molluscs originating from the Red Sea were among the first in the famous shell cabinets of the old museums.

The literature on the Erythraean malacofauna (i.e. the molluscs of the entire Red Sea region) indicates that a large number of scientists has studied this group.* I shall briefly categorize first, those who investigated that fauna only as part of a general regional survey, and secondly, those who worked specifically on the systematic, zoogeographic, ecological and behavioural aspects. One of the first investigators of the Red Sea coastal environment was the Swede Forsskål whose expedition came to a tragic end in 1763. In 1775 the only survivor, Niebuhr, published part of the results which included the description of 133 marine molluscan species. Although it is necessary to subject this number to a critical examination the description of 29 new species represents a remarkable achievement of the expedition.

At the beginning of the 19th century studies concentrated on the fauna of the Gulf of Suez, particularly the northern part. The most outstanding result of these activities were the unique engravings by the French naturalist Savigny. Unfortunately blindness later prevented Savigny from adding the legends to his fine illustrations of some 250 marine molluscs. Therefore the taxonomic interpretation was left to a number of subsequent scientists, namely Audouin, Jonas, Pfeiffer, Fischer, Issel, Pallary, Lamy and Moazzo. As yet approximately 70 species designed by Savigny have not been reconfirmed in the Red Sea.

From 1820 to 1826, the Germans Ehrenberg and Hemprich made extensive collections on marine invertebrates near Jiddah and Massawa. They reported finding a total of 375 species of molluscs. However, their species list should be examined with some scepticism since it shows great conformity with the shells of Sicily, enumerated by Philippi (1836). This publication serves as example of the confusion which often can be noted in the malacological literature of the past century; comparing his 'collections' Philippi reports that '74 species are to be found in both the Red Sea and the Tyrrhenian Sea'. Ehrenberg and Hemprich seemed to be strongly influenced by Philippi's work, and Cooke (1886) scoffed in his analysis that 'their Red Sea collections had in some way become impregnated with very strong Mediterranean leaven.' In this connection we may mention the most remarkable and accurate malacologist who worked last century on that fauna, the Italian Issel. He wrote in 1869 (i.e. the year of the opening of the Suez Canal) that 'there could not be found any common species living in both seas'. Although we know today that quite a number of marine organisms have been able to migrate from the Red Sea to the Eastern Mediterranean we can assume the correctness of Issel's statements. He compiled a comprehensive synopsis of all malacological data of his period and presented faunistic details for 528 Erythraean species (comprising 356 gastropods and 172 bivalves). About half of that number he confirmed for the Gulf of Suez. The great value of his regional monograph is that Issel collected all systematic and zoogeographical data which were at his disposal, referring mainly to the plates of Savigny (1817) and the original descriptions of the great encyclopaedists such as Linné, Chemnitz, Gmelin, Lamarck, Sowerby, Reeve and Kiener. Although the number of species which Issel listed has to be reduced according to our present knowledge of taxonomy, his 'Malacologia del Mar Rosso' remains a unique source of information for everyone who is interested in the mollusc fauna of the Red Sea.

A major source of knowledge about the shells of the entire Red Sea originates from the dredging and collecting operations which were performed from 1895 to 1898 by the Austrian research vessel *Pola*. The description of the molluscan material obtained was published by Sturany (1899, 1903). He

---

*A comprehensive review on the malacological literature of the Red Sea has been published by Mastaller (1979).

summarized the records of 472 species found in 88 localities in both coastal and deep sea areas. Apart from the general taxonomic accuracy of these articles the great scientific value of the *Pola* Expedition is that for the first time (and almost the only time) abyssal dredgings were carried out in different areas of the Red Sea basin. The 54 mollusc species dredged from great depths by the *Pola* are still the main source of information regarding this faunal aspect.

In the 20th century faunistic studies on molluscs became increasingly the result of the great expeditions to the Red Sea. Although most of them were primarily oceanographic explorations they also brought forth a number of detailed regional and zoogeographical studies. Table 10.1 summarizes the main studies.

The prevailing literature on Red Sea molluscs describes the distribution of specific families or genera in certain areas, while in many cases the faunistic aspect of a research programme is restricted to mentioning only lists of species names without further reference. It seems, however, worth to mention briefly some studies which are of local importance.

For approximately two centuries the Gulf of Suez including the Canal has been one of the most intensively investigated areas in tropical seas. The reasons were both the relative closeness to Europe and political interests. Apart from German, Austrian and Scandinavian researchers it was mainly the French, British and the Italians who explored the northern Red Sea. Prominent names in this field of science are Fischer (1865, 1870, 1871), Vaillant (1865), MacAndrew (1870), Cooke (1886), Martens (1887), Vassel (1891), Bavay (1898), Lamy (1938), Moazzo (1939), Barletta (1969. 1974), and Barash and Danin (1972) in recent times. The cephalopods of both the Gulf of Suez and the Gulf of Aqaba have been described by Adam (1959, 1960). Many of these publications discuss also the so-called 'Lessepsian Migration', i.e. the intermingling of Erythraean and Mediterranean faunas caused by the migration of marine organisms through the Suez Canal.

In comparison with other regions in the Red Sea the Gulf of Aqaba was only sporadically visited by malacologists. Apart from some general surveys published by Fischer (1870) only the British 'Manihine'-Expedition focussed also on molluscs. Recently the Gulf of Aqaba experienced a number of comprehensive studies by Israeli ecologists and shell collectors (c.f. Fishelson, Fridman, Dafni, Barash),

TABLE 10.1.

Year	Name of the Expedition or the Research Vessel	Area investigated	Published by
1761—63	Danish Expedition	Egypt, S. Arabia, Yemen	Forsskål (Niebuhr), 1775
1826—29	Astrolabe	Indo-Pac., incl. Red Sea	Quoy & Gaimard, 1834
1865—69	MacAndrew Expedition	Gulf of Suez, Canal	MacAndrew, 1870. Cook, 1886
1895—98	Pola Expedition	entire Red Sea	Sturany, 1899; 1903. Weindl, 1912
1923—24	Ammiraglio Managhi	Gulf of Suez	Bisacchi, 1931
1924	Cambridge Expedition	Gulf of Suez, Canal	Tomlin, 1927. O'Donoghue, 1929
1933—34	John Murray Expedition	Red Sea + Gulf of Aden	Eales, 1938
1948—49	Manihine Expedition	Gulf of Aqaba	Rees & Stuckey, 1952
1951—52	Calypso	central and southern Red Sea	Adam, 1955. Franc, 1956
1957—58	Xarifa	southern Red Sea	Gerlach, 1959
1962, 1965	Israel South Red Sea Expedition	Dahlak Archipelago	Kohn, 1965. Schilder, 1965 Mienis, 1971
1964	Meteor	Red Sea, Gulf of Aden	Klausewitz, 1967
1977, 1979	Sonne, Valdivia	central deep Red Sea	Mastaller, 1979

while the reefs along the Jordanian coast were investigated mostly from the biophysiographic viewpoint (Mergner and Schuhmacher, 1974; Mergner and Mastaller,1982). Mastaller (1979) gathered comprehensive data on the ecology and faunistics of littoral molluscs on the reefs at Aqaba, reporting a total of 531 species of which 248 were first records for the Gulf of Aqaba.

Faunistic studies in the Central Red Sea are rather scarce. Apart from the results of the French *Calypso* Expedition to the Farasan Islands there are only a few record stations from the *Pola* survey. A fairly good idea of the subfossil shell fauna of raised Pleistocene beaches at Suakin, Sudan has been given by Hall and Standen (1907). Eliot (1908) described nudibranches (sea-slugs) from localities along the Sudanese coast and recently this faunal group was studied by Engel and Van Eeken (1962) who focussed on the Opisthobranchia from the Sinai coasts. Mastaller (1978) carried out studies on molluscan assemblages on reefs at 19 localities in the Port Sudan area. 282 littoral species were studied with special reference to their microhabitats. Similar investigations were continued in the vicinity of Suakin, Sudan. Sudanese mollusc assemblages were also described by Taylor and Reid (1984).

Jousseaume (1888) and Anthony (1905) collected large numbers of littoral shells in the southernmost parts of the Red Sea, namely in the regions of Massawa, Dahlak and the adjacent Gulf of Tadjourah near Djibouti, and at Aden. A most comprehensive systematic elaboration of the bivalve material was published by Lamy from 1905 until 1930 (*see* Lamy, 1938). Recently the Dahlak archipelago has been investigated by the 'Israel South Red Sea Expedition' in 1962/65, yielding several papers on the regional distribution of well-known gastropods.

Regarding the general zoogeography of the molluscs in the Red Sea it is necessary also to take the adjacent faunal distribution into consideration. Therefore the species lists of shells from the southern part of the Arabian peninsula are sources of valuable information (c.f. Smith, 1891; Shopland, 1902; Melvill, 1928; Biggs, 1965).

10.3. THE USE OF RED SEA MOLLUSCS BY MAN

Some 4000 years ago Egyptian vessels regularly crossed the Red Sea. A fine relief in the Temple of Queen Hatshepsut (1470 B.C.) at Luxor documents the arrival of the Pharaoh's fleet at the Land of Punt (probably nowadays the Islands of Shadwan and Tiran, but *see* Chapter 16): The marine fauna is represented by fish and squids, while the riches which are brought back contain pearls and shells. The mother-of-pearl and the pearls themselves always played an important role in the jewelleries of the ancient dynasties. Last, but not least it is reported that the famous Queen of Sheba used pearls from the Red Sea to underline the preciousness of her gifts when, in about 950 B.C., she visited King Solomon of Judaea. The pearls originate from the pearl oyster *Pinctada margaritifera*, a bivalve that is not naturally very common in the Red Sea. In 1904, a British scientist, Crossland, started successfully to cultivate pearl oysters in Dungunab Bay north of Port Sudan. The maricultural enterprise was yielding good harvests for many decades, but a mass mortality in 1969 and the increasing influx of pearls from the Far East on the world market caused the Sudanese pearl fishery to collapse. In Port Sudan there are still some small-scale factories which produce shirt buttons out of mother-of-pearl. While in former times they used the pearl oyster and related wing oysters as raw material, they now cut most of the buttons out of the nacreous layers of the common top shell *Trochus dentatus* (Figure 10.1).

Red Sea shells have probably been used since ancient times for ornamental purposes. Shell decorations, necklaces, bangles etc. were excavated in the archaeological sites of the ancient Nabatean city called Petra. These items probably originated from the Nabatean trade with seafarers and merchants coming from the south, from the land hitherto known as Arabia Felix (modern Yemen).

Today many of the Sudanese, Sinai and Saudi-Arabian nomadic tribes living along the coastal lands around the Red Sea, like their ancestors, still appreciate the ornamental value of shells from the adjacent

Fig. 10.1. The top shell *Trochus dentatus* whose nacreous layers are cut into mother-of-pearl buttons.

reefs. A more modern version of shell decoration results in local mass collection of giant clams *Tridacna squamosa* and spider conches *Lambis truncata* which are widely used as ashtrays and souvenirs.

Traditionally there is only little use of Red Sea molluscs for nutritional purposes. Hardly any mussel or snail is taken as sea food, and only sporadically Red Sea fishermen can be seen to enrich their daily diet with some sun-dried gastropod flesh. The shells which serve for this procedure are the common shallow water molluscs which reach the largest size, such as Ramose dye shells (*Chicoreus* (*Murex*) *ramosus*) or conches (e.g. *Lambis truncata*). Locally there is also intensive fishing for the stromb *Strombus tricornis* to obtain the horny plate called the operculum which is attached to the muscular foot of the animal and used to close the entrance to the shell. This clawlike item is processed into a perfume which is highly esteemed by Arab women.

10.4. ZOOGEOGRAPHY OF THE RED SEA MOLLUSCS

In total, some 2500 species names are encountered while dealing with the malacological literature on the Red Sea. Naturally the major part of the records are subject to discussion on the basis of their taxonomic validity. Modern revisions indicate that approximately 950 to 1000 species of marine shells are living in the Red Sea. This number will certainly be considerably increased when specialists work out the identifications of micromolluscs (size less than 3 mm), especially those originating from great depths, such as those dredged from the Atlantis II basin, Central Red Sea.

The molluscan fauna of the Red Sea is entirely of Indo-Pacific origin. Tentative estimates (Mastaller, 1979) rate the endemic proportion as less than 5%. The Red Sea molluscs are not homogeneously distributed within the whole basin. The reasons for these peculiarities are presumably due to the very unusual geological history and the environmental stresses of the Red Sea (*see* Chapters 1, 2 and 3). The repeated isolation of the Red Sea basin has created an environment in which the living organisms have had long and undisturbed periods for the process of their specific evolution. Moreover the isolated populations were favoured in their tendency to develop a number of endemic species by two ecological factors which are unique to the Red Sea. First, the temperature patterns are quite different from those of the Indian Ocean; for instance, Red Sea surface water temperatures show significant fluctuations depending on the seasonal climate (e.g. Gulf of Aqaba: 16—20°C in winter / 25—33°C in summer). On the other hand the temperature of the deep waters is almost constant and extraordinarily warm (about 20°C). Accordingly, deep sea organisms which are usually restricted to water bodies of low temperature face completely different living conditions in the deeps of the Red Sea. Secondly, while the salinity of the Indian Ocean is mostly constant at 35‰ the Red Sea is characterized by considerable higher values, ranging up to 41‰ in the northern parts.

Both environmental factors, the pattern of the water temperature and the high salinity influence the survival and distribution of molluscs, especially of the juvenile forms. Therefore it is not surprising that the mollusc fauna of the Red Sea is a small fraction of the number of species living in the entire Indo—Pacific region, only a part of this fauna could develop adaptive strategies to live under the special conditions of the Red Sea. The high salinity values seem to be a major barrier for both the adults and the free-swimming mollusc larvae. Accordingly there is a gradient in the number of species from south to north — as the salinity increases and the shallow water temperature shows greater fluctuations. Out of a total of 950 species of molluscs occurring in the whole Red Sea basin about 850 live in the southern and central part, while only 637 species are recorded from the Gulf of Aqaba (Mastaller, 1979). Similar zoogeographic results are obtained when we analyse the early records of littoral gastropods: Sturany (1903) mentioned altogether 294 species; 253 species were collected in the southern and central Red Sea, while the Gulf of Suez (199) and the Gulf of Aqaba (150) yielded far fewer species.

A lot of research has been done on 'Lessepsian Migration' (i.e. the migration of Red Sea species to the eastern Mediterranean via the Suez Canal). At least 27 Erythraean molluscs have succeeded in the transition between the Red Sea and the Mediterranean (Por, 1978). The increasing number of records of live shells (mostly cowries) from the Indo-Pacific fauna along the coasts of the Eastern Mediterranean seems to confirm that the successful settlement of marine molluscs in new environments may not require a long larval existence, but rather the ability to adjust to changed temperatures and salinities. On the other hand it appears that the extreme hydrological conditions in the northern part of the Gulf of Suez, particularly the high salinities of the Bitter Lakes, prevent Mediterranean species from colonizing the Red Sea.

The small vertical temperature gradient in the Red Sea explains another characteristic feature in the zoogeographical distribution of its molluscs. In contrast to other tropical oceans no distinct and diverse deep-sea mollusc fauna has developed in the Red Sea. Although the corresponding data from the *Pola* operations are not so numerous we may conclude the following: Some species such as *Murex tribulus*, *Nassarius albescens* and *Cantharus fumosus*, which are widely distributed in shallow waters in the entire Indo-Pacific, live also in abyssal depths in the Red Sea. Moreover, there are only a few species known to be restricted to the deep sea habitat.

A third striking phenomenon is that there is little conformity between the typical deep-sea forms of the Red Sea and those of the adjacent Arabian Sea (Melvill, 1928).

I have indicated already that there are several molluscs restricted to the Red Sea. This brings us to the discussion of the future development of the Red Sea fauna and the continuance of the endemic forms. In the Pliocene a connection with the Indian Ocean was established which allowed the initial invasion of

Indo-Pacific forms into the Red Sea. During the Pleistocene glaciations lowering of the mean sea level narrowed the Straits of Bab al Mandab, which may have periodically become a land barrier. This would have produced the isolation necessary for the evolution of various sibling and endemic species in the Erythraean province. At the end of the last ice age the Red Sea became reconnected to the Indian Ocean, thus enabling a new wave of Indo-Pacific invasion and reverse migration by Red Sea species. Accordingly it is suggested by Schilder and Schilder (1939) that hybridisation with sibling species migrating to the Red Sea via the Straits of Bab al Mandab will occur leading to the introgression of previously allopatric species. For example, it is argued that the invasion of the Red Sea by the Tiger cowrie *Cypraea tigris* from the Gulf of Aden should lead to the extinction of the sibling *C. pantherina* as a distinct species.

The evidence of higher species diversities in the southern parts of the Red Sea favours the view that the mollusc fauna of the Erythraean province is gradually increasing towards an equilibrium with the Indian Ocean proper, depending mostly on the competitive strategies of sibling species and species of similar synecological functions once their distribution becomes overlapping. If this is the case it may be predicted that the endemic species will persist for longest in the remotest parts of the Red Sea, such as the Gulf of Aqaba. Therefore the Gulf of Aqaba will be of particular interest for further evolutionary studies on marine life, as it repeats morphogenetically the history of the whole Red Sea basin. During the major tectonic events this Gulf has been separated several times from the main basin, leaving each time a small isolated marine branch which might have evolved its own faunal adaptations.

A problem which still remains to be discussed is the composition of the Red Sea palaeo-fauna. The raised fossil reef benches along many coasts of the Red Sea are excellent documents of the presence and abundance of former reef communities. The age of most fossil reefs located close to the actual shoreline is rather young — dating back to the late Pleistocene. All shelled or calcareous fossilized organisms within these reef-benches are generally well-preserved and easy to excavate from the corallogenic sediments. The predominant members of these fossil communities are corals and molluscs. They belong mostly to species which still inhabit neighbouring living reefs. Striking, however, are the great differences in species composition when we compare them quantitatively with recent communities. This is especially true for the raised coral reefs along the Jordanian coast where, in the Pleistocene soft-bottom communities, shells like *Strombus fasciatus*, *Oliva* cf. *inflata* and *Cypraea annulus*, and in hard substrate biotopes, *Diodora rueppelli*, *Colubraria tortuosa*, *Cardita muricata* and *Tridacna crocea* were quantitatively among the most abundant forms (Mastaller, 1979). Nowadays, by contrast, some of these species rank as great rarities in the Gulf of Aqaba.

Somewhat different is the situation in the central Red Sea: Analysis of both the fossil and the recent coral reefs in the area of Suakin, Sudan, indicated no significant difference regarding the composition and abundance between the two faunal elements. We may therefore argue that during the last 100,000 years major zoogeographical changes took place in the Red Sea, but it remains to clarify why the most drastic alterations of mollusc distributions happened in the Gulf of Aqaba. Very little is known about whether molluscs have had, and still have, problems in adapting to some of the peculiarities in the northernmost areas of the Red Sea, or whether there exist physico-chemical and/or biological factors which limit the distribution of mollusc species within the Red Sea.

10.5. THE CHARACTERISTIC SEA SHELLS OF THE RED SEA

The phylum Mollusca includes five major classes: the familiar snails and slugs (Gastropoda), the chitons (Polyplacophora), the tusk shells (Scaphopoda) the bivalve shells (Bivalvia or Lamellibranchia) and the head-footed molluscs (Cephalopoda).

In the following section a general morphological and physiological outline is given for each of these subdivisions. In addition, a few representatives of specific families are mentioned which are either peculiar to the Red Sea, or simply are the species most frequently detected by anyone who is interested in this animal group.

The chitons (or coat-of-mail shells) are slug-like organisms protected by eight shell plates. The head which lacks eyes and tentacles bears a central mouth with a narrow chitinous tongue called a radula. The radula is inserted with rows of teeth which are used for scraping the algal cover from rocks. Most of the 20 chitons known from the Red Sea are commonly found clinging to rocks in shallow water zones, such as *Acanthochiton penicillatus*, *Chiton platei* and *Tonicia suezensis*. They usually remain concealed under boulders or in rock crevices during the day and emerge to feed at night. The largest and also the most conspicuous chiton is the olive-green *Acanthopleura haddoni* which grows up to 5 cm long (Figure 10.2). It lives exclusively on the upper intertidal rocky shore, hiding between large barnacles (*Tetraclita*) or in shady crevices during daytime. Like many chitons these animals always return to the same position on 'their' rock after feeding at high tide.

Similar patterns of behaviour can be observed in two archaic gastropod families; the limpets (Patellidae) and the abalones (Haliotidae) which are classified as Archaeogastropoda (snails of a rather primitive anatomical organisation regarding their mantle cavity, radular apparatus and gill system). The two other major groups of gastropod are the Mesogastropoda and the Neogastropoda which will be discussed later.

Many archaeogastropods, which also include the top shells (Trochidae) the turban shells (Turbinidae) and the slipper-winkles (Neritidae) represent the most abundant molluscs in shallow water zones. Typical for the rocky intertidal all over the Red Sea are the limpets *Cellana rota* (Figure 10.3). These animals feed during the night in the close vicinity of their homing place, to the contours of which they even adjust the edge of their roof-like shell. Sculpture and colour patterns of this shell vary according to the background. This is why these limpets can hardly be distinguished from the rocky surface. Certainly this camouflage is a good protection against predatory shore birds.

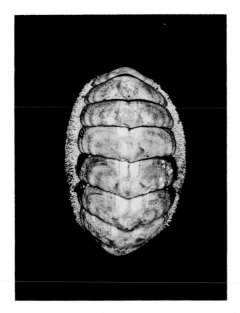

Fig. 10.2. The largest of the Red Sea Chitons, *Acanthopleura haddoni*.

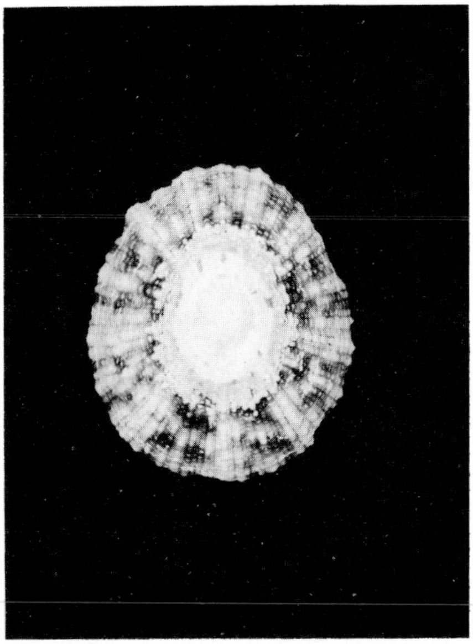

Fig. 10.3. The common limpet of Red Sea rocky shores, *Cellana rota*.

The top shells are widely distributed on any hard substrate in the shallow waters. We already have mentioned that some of the *Trochus* shells are of commercial interest because of the mother-of-pearl in the inner nacreous shell layers. The species that are fished for this purpose are the heavy-shelled *Trochus dentatus* and *T. niloticus*. A very attractive member of the top shell family is the small Strawberry top, *Clanculus pharaonis*. It is bright red, bearing black and white pearl-like patterns running in bands around the whorls of the shell. Rarely, in winter time, these small marine jewels can be found in aggregations under rocky slabs where they congregate for spawning. In similar habitats live the turban shells *Turbo argyrostomus* and *T. radiatus*. Snorkellers are likely to notice their presence on a reef because of occasional findings of the polished operculum of turbinids in the sand. These items are calcareous discs which are fixed to the live animal on the foot. The disc's diameter fits tightly with the aperture when the animal withdraws into its shell. In *Turbo petholatus* the operculum is called 'catseye' because of its attractive and distinctive coloration. Like all archaeogastropods this turban shell has exclusively herbivorous diet, feeding on filamentous algae which cover hard substrates.

A related snail is *Gena varia* which superficially can be mistaken for a small abalone. It ranks among the most abundant inhabitants of the littoral zones in the Red Sea. These cryptic animals mostly live aggregated under big rocks and dead coral slabs. During daytime they avoid any motion that might bring them to the attention of predators. However, any turning over of a boulder in the shallow water demonstrates the rapid escape reaction of the animals once they are exposed and unprotected; they nimbly glide over the rocky substrate in order to reach the underside of the boulder. If *Gena varia* is disturbed it presses the shell closely to the substratum leaving only the hind part of the muscular foot uncovered. It reacts to strong touch by autotomizing and discarding this part of the foot which can be regenerated. Specimens without complete feet can be regularly observed in their microhabitat. This indicates that active loss of a body part may help in escape when the animal is attacked by a predator. The coloration of the shell is cryptic and forms patterns of dark and light spots. A similar perfect matching with the colours of the rocky environment is found in the slipper winkles *Nerita polita* and *N. albicilla*. *N. polita* especially shows a striking colour variation which corresponds to the red, black and grey conglomerate pebbles which are typical for many rocky shores in the northern Red Sea. Also it appears that the variability of shell colouration has some protective advantages for animals inhabiting a biotope that is frequently visited by sea birds. Both these neritids graze on microorganisms splashed on the shore by waves. In areas with a marked tidal range the two species show some partitioning of the habitat: While *Nerita albicilla* rarely crawls higher than mid-tidal level, the sibling *Nerita polita* prefers the upper half of the shore. During night the latter species migrates towards the shore at a speed corresponding to the incoming tide. In daytime these snails remain inactive, partly or totally burrowed in the sand.

The next order of gastropods are the Mesogastropoda, a large group occupying a great variety of niches on the coral reefs. The periwinkles, slipper and worm shells, horn and moon shells, wentletraps, conchs, tuns and highly prized cowries all belong to this group of molluscs. Common features are the spirally coiled shells which are closed usually by a horny operculum. In contrast to the previous group they all lack mother-of-pearl in the inner surface. The radula has only seven teeth in each row and is modified according to the mode of nutrition. Some species have an elongated snout to capture cryptic prey, while others use their radula to drill holes into the shells of their victims. A most peculiar way of feeding has been developed by the worm shells (Vermetidae) which filter feed by periodically spreading out a sticky mucous net which is then swallowed along with all the organic particles stuck to it. Typical and abundant Red Sea worms of this family are *Dendropoma maximum* (Figure 10.4) and *Serpulorbis inopertus*; both live embedded in rocky substrates or associated with hermatypic corals on the reef flat.

The periwinkles of the Red Sea *Littorina scabra*, *Nodolittorina millegrana* and related forms like *Planaxis lineolata* and *P. sulcatus* are shore inhabitants like all members of the family in other oceans of the world. Common on all supratidal rocks or in tide pools these animals usually form aggregations of sometimes

 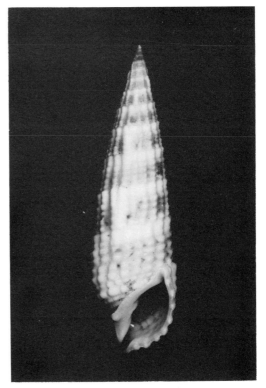

Fig. 10.4. The curious worm shell *Dendropoma maximum* which lives embedded in the reef and filter feeds using a sticky mucous net.

Fig. 10.5. A typical horn shell (Cerithiidae). This species *Rhinoclavis aspersa*, is generally found on coarse sand in fore-reef areas.

hundreds of individuals during inactive periods in daytime. It has been shown that this behaviour is advantageous for those marine animals which regularly have to endure several hours of extreme insolation. By clumping together the winkles create a slightly humid microclimate which protects them against desiccation. Marking of *Planaxis sulcatus* has shown that individuals do not always return to the same resting place after their nocturnal feeding excursions nor do they always remain with the same aggregation.

The horn shell family (Cerithiidae) is represented by 24 species in the Red Sea. All members are algal grazers. They dominate hard bottoms of shallow water zones in the whole region. The most abundant and prominent species are the Columnar horn shell *Cerithium columna* the Knobby horn shell *C. nodulosum erythraeonense* and the infratidal grazers *Clypeomorus morus* and *Cerithium caeruleum*. The ceriths, however, have also successfully entered soft-bottomed biotopes where they often occur as the predominant species. Thus we find *Cerithium scabridum* and *C. rueppelli* in seagrass beds and between macroalgae, while *Rhinoclavis aspersa* (Figure 10.5), *R. fasciata* and *Cerithium kochi* prefer the coarse sand in fore-reef areas.

Another group of typical sand-dwellers are the strombs (Strombidae). For locomotion and digging for food they use their claw-like operculum. This claw is also used for defence, and the inexperienced collector of larger specimens can be hurt by the animal. Bringing together an octopus and a large strombid in an aquarium can demonstrate the efficiency of the operculum as a defence weapon, but also as an instrument enabling the animal to right itself when it is turned onto its back. Noticeable Red Sea species are the endemic *Strombus fasciatus*, the purple-lipped white stromb *S. gibberulus albus* and the small and irregularly coloured *S. mutabilis*. As for other strombids the typical habitat of the thick-shelled *S. tricornis* (Figure 10.6) is the vast seagrass beds formed by the spermatophytes *Halodule* and *Halophila*. A common member of the conch family is the spider-conch *Lambis truncata* which occurs in the Red Sea as the subspecies *L. t. sebae*. In contrast to most other molluscs where it is not possible to distinguish externally between the sexes, in spider conchs the female is characterised by the much longer fingerlike spines and the body whorl is more humped than that of the male.

The shark's eyes (Naticidae) are easily recognized by the smooth and polished surface of the shell, which is why they also are called moon shells. The animals usually are large compared with their shells. Naticids are able to crawl rapidly over long distances on soft bottoms; mostly they hunt for prey under the surface of sandy areas. While *Natica marochiensis* has specialized more on sandy depressions on the reef flat, *Polinices mamilla* and *P. melanostoma* prefer seagrass beds. Shell collectors commonly find that many dead specimens of molluscs, either washed up on the shore or scattered on the sea floor, show a characteristic bevelled drillhole about 1 mm in diameter. These circular holes mostly originate from the predatory activities of moon shells. The snails hold their prey tightly with their muscular foot while the radula gently rasps a hole into the victim's shell.

Fig. 10.6. One of several impressive strombs in the Red Sea, *Strombus tricornis* is typically found on seagrass beds.

The cowries (Cypraeidae) have the reputation of being the most attractive among tropical sea shells, so they need little introduction because of their popularity. The distinct feature of this group is the bilobed mantle which can completely envelop the shell when fully extended. This is the reason why the outer shell remains shiny and lustrous in living specimens. The mantle may be thin and translucent or, as in many Red Sea cowries, it may have thick warts and fleshy appendages. The colouration of the mantle is often completely different from that of the shell, as in the mole cowrie *Cypraea talpa* where it is a deep black. In some species the mantle is able to secrete sulphuric acid which acts presumably as a good defence mechanism for a sluggish animal which cannot protect its soft inner parts by closing the shell with an operculum. Little is known about the diet of cowries but most seem to be omnivorous. During daytime the animals rest under rocks or in crevices, while at night time they crawl around on the reef flat. In this context it seems worthwhile to mention a few words about shell collecting which unfortunately is not always performed in an appropriate, i.e. ecological manner.

Occasionally cowrie hunters recklessly use crow-bars to break up branched corals and turn over large boulders in order to obtain their souvenirs. It is especially due to the latter 'method' that a few unscrupulous shell collectors can destroy a reef flat to such an extent that recovery may take many years or even decades. By turning over big rocks and coral slabs in shallow water the complex community structure of the 'shade-loving' sessile fauna (including also the egg masses of most marine molluscs) will be killed both by direct illumination and by creating easy access for predators. With the extinction of these important members of any reef community a number of other inhabitants will be deprived of both their food supply and suitable hiding places. Because of this, any rock which must be examined on its underside should immediately be replaced in its initial position.

The Red Sea harbours a number of endemic cowries such as *Cypraea erythraeensis*, *C. camelopardalis*, *C. macandrewi*, *C. pantherina* and *C. pulchra*. The Red Sea Tiger cowrie *C. pantherina* is one of the commonest and largest cypraeids found in shallow waters in rocky areas. This species exhibits considerable variation in the coloration of the shell; from nearly pure white with few yellowish spots to almost completely black. Typical forms in the southern Red Sea are golden brown with black spots.

Similar and most puzzling colour variations are shown by another typical Red Sea cowrie, *C. arabica grayana*. Conchologists still argue whether it is taxonomically justified to distinguish two separate species or if the variations in the coloration of the Arabian cowrie should be designated as ecomorphs, i.e. local varieties.

Other attractive Erythraean cowries, especially when observed with their mantle extended, are the Carnelion cowrie *Cypraea carneola*, the Eroded cowrie *C. erosa nebrites*, the Tapering cowrie *C. teres*, the Fringed cowrie *C. fimbriata*, the Lynx cowrie *C. lynx* and the Thrush cowrie *C. turdus*.

The tritons (Cymatiidae), frog shells (Bursidae) and tuns (Tonnidae) represent a group of mesogastropods which are all predators. *Cymatium pileare* (Figure 10.7) is a triton shell with a wide distribution all over the Indian Ocean. In the Red Sea it has occupied a range of quite different habitats such as rocky substrates, beds of macro-algae and mangrove swamps. Sometimes it even can be found between scleractinian corals in the upper fore-reef. Its common name 'Hairy triton' indicates a characteristic feature of this animal: The spiral ribs of the shell are separated by longitudinal rows of tubercles which are covered by a thin bristly epidermis. A related and very prized species is the Giant triton *Charonia tritonis*. This shell is rather uncommon in the Red Sea, occurring on reef slopes and on the fore-reef. Formerly an appreciated souvenir item for collectors this shell is now included on the 'International List of Endangered Species'. The Giant triton as known to be an effective natural enemy of the coral-eating Crown-of-thorns starfish *Acanthaster planci* which already has devasted many coral reefs in tropical seas.

Members of the frog shell family Bursidae are not very conspicuous,but they are abundant in rocky shallow waters where they play an important role in the ecological balance of those biotopes. Bursids are scavengers which remove dead or decaying organic material, *Bursa affinis granularis* can occasionally be

Fig. 10.7. The Hairy triton, *Cymatium pileare*, occurs in a variety of shallow water habitats.

observed feeding on fish carrion. Tun shells are infrequently seen in the Red Sea because most of them are crepuscular. *Malea pomum*, the Apple tun shell, hunts for small burrowing organisms in deep sandy areas.

The third subdivision of the marine snails is the Neogastropoda, including the familiar dye shells (Muricidae), the coral loving shells (Coralliophilidae), the whelks (Buccinidae), the basket shells (Nassariidae), the spindle shells (Fasciolariidae), the mitres (Mitridae), the cones (Conidae), the augers (Terebridae) and the turrets (Turridae). Neogastropods are highly advanced predators, feeding on other gastropods, bivalves, worms, and even fish. The spirally coiled shell usually has a tube-like groove (= siphonal canal) near the shell aperture. This siphon has become a sensitive chemoreceptory organ for detecting prey or possible predators at a distance.

The murex shells kill their prey, mostly other gastropods and bivalves, by drilling holes into the calcareous shell through which they secrete digestive enzymes to dissolve their victim. While boring, the narrow radula is assisted by acid secreted by the surrounding tissue of the head which dissolves the limestone compounds of the prey's shell. A great variety of small and cryptic muricids have specialized in this mode of feeding, preying on small organisms of the reef flat. In tidal and lagoonal zones the most common species are *Morula granulata*, *M. anaxeres*, and *Chicoreus corrugatus*. Under large boulders a great number of very small forms which resemble ceriths are often found. These are *Maculotriton serriale*. The same habitat is shared by the drupe shells *Drupa morum* and *D. ricinula*, and the Francolin rock shell *Nassa serta.* The most conspicuous dye shell of the Red Sea is the heavy Ramose dye shell *Chicoreus (Murex) ramosus* which is intensively collected in some areas for drying the flesh or for tourist markets. *Murex tribulus,* a shell with a long siphonal shaft inserted with spines, is often entangled in the gill-nets of fishermen. In most cases, however, this is due to hermit crabs who adopt the dead shells as their portable refuge.

The coral shells (Coralliophilidae) are a highly specialised mollusc family. They are restricted to a more or less sessile way of life, and all of them are harmful to corals. *Magilus antiguus* lives within scleractinian corals and is overgrown by the coral skeleton. Between the branches of *Acropora* and *Pocillopora* coral colonies, *Quoyula madreporarum* can often be observed resting tightly on a calcareous pedestal which is probably built up by the host coral under the influence of the snail. Similarly *Coralliophilis violacea* sits on the surfaces of spherical corals like *Porites lutea.* Although these coralliophilids are able to creep they seem never to abandon their home place. They possess a long and extendable snout with which they suck out all coral polyps which they can reach. This sort of parasitic nutrition has resulted in the reduction of the radular organ in this family.

Only few species of whelks have successfully entered the Red Sea. The commonest of them is a small littoral species, *Engina mendicaria.* This species lives in sandy depressions of the rocky intertidal and is easily recognized by the black and yellow bands on the shell whorls. The most active among the crawling neogastropods are the basket shells *Nassarius albescens gemmuliferus* and *N. arcularia plicatus.* They live in fine sand in shallow water, roaming through the upper layers of the seafloor. The highly mobile siphon always projects above the surface of the sand, continually testing the surrounding water for evidence of carrion. Having detected a possible source of food the animals move immediately in the direction of the taste gradient. To test the very sensitive chemoreaction of the basket shells one can put a piece of carrion on the reef flat: after a short while various nassariids will congregate at the carrion and feed.

Because of their specialised feeding and locomotory behaviour the predatory gastropods show the greatest species diversity among the Red Sea molluscs, however, none are as numerous as many of the herbivorous animals we have described previously. A few neogastropods which may frequently be met with during night dives over sandy areas are the spindle shells *Fasciolaria trapezium, Fusinus polygonoides* and *Latirus turritus* as well as the olive shell *Ancilla cinnamomea.* Here also occurs the largest of the Red Sea mitres, *Mitra fissurata.* Most of the 47 recorded mitres are small and cryptic, living under boulders, in rock crevices or among macro-algae, like *Mitra litterata* and *Vexillum (Pusia) amabile.*

In the large lagoons behind the fringing reefs along the central Red Sea coasts, the vase shell *Vasum turbinellum* is common. The heavy and compact shell is thickened with knots and short spines (Figure 10.8), and its aperture has yellow lips. Usually the shell is heavily encrusted with algae, so it is difficult to recognize the animal in the field.

A very peculiar way of feeding has been developed by the cone shells. In this family the radula has become reduced to a series of dart-like teeth which are kept in a pouch near the snout. One of these teeth, barbed like a harpoon, is held in readiness in the tip of the trunklike proboscis. In addition, poison is secreted from a separate gland and pumped to the proboscis when the animal shoots the prepared dart into the body of its prey. The poison of the fish-hunting cones is neurotoxic, and two such species which are reasonably common on the Red Sea, *Conus geographus* and *C. textile* (Figure 10.9), have been reported to be responsible for a number of human fatalities. The latter species is regularly found in shallow lagoons and in seagrass beds. Most cones prey either on polychaete worms or on other molluscs, including their own family members. A total of 38 cone species have been recorded from the Red Sea; the most common are *C. coronatus, C. flavidus, C. lividus, C. taeniatus, C. musicus, C. arenatus* and *C. vexillum.*

Many of the auger shells have a poisoned feeding apparatus similar to that of the cones. These animals live entirely burrowed into soft bottoms where they prey on worms. Auger shells are very rarely seen on the surface of the sea floor. *Terebra crenulata* is one of the few exceptions. It lives in sandy depressions on the reef flat where it sometimes becomes visible from the surface when strong waves disturb the sandy sediments. The salient members of the Red Sea terebrid family are *Terebra babylonia*, *T. dimidiata* and *T. maculata*, while most of the rare and endemic auger shells are small and predominantly restricted to deep sandy bottoms.

While the molluscs we have discussed up to now are united in the subclass Prosobranchia (meaning that by body torsion the gills became situated in front of the internal organs), the second subclass of the gastropods are designated as Opisthobranchia. In this taxon the gills are located at the posterior end of the body. They are commonly known as sea-slugs, which indicates a typical feature of their morphology. In this group the characteristic molluscan shell is often reduced or absent. The muscular foot has a flat creeping sole, and some forms possess large lobes which can even be used in swimming. Dorsal processes called cerata are often used for defence. As the animals no longer have a protective shell into which they can withdraw when threatened they have found other ways of becoming less vulnerable. Certainly the lack of a hard shell may not always be disadvantageous as it enables the animal to creep into the smallest crevices (which is also necessary in many species when they search for their prey such as small sea-squirts or encrusting sponges). To make up for the loss of one means of passive protection many opisthobranchs have developed a number of other defence mechanisms. Most of them deposit poisonous chemical compounds under their skin. This usually goes together with a particular bright 'warning' coloration and patterning which makes this group a favourite subject for underwater photography. These gaudy patterns indicate to potential predators to ignore this distasteful but pretty-looking prey. In some sea-slugs like the eolid family the stinging cells of certain cnidarians are deposited

Fig. 10.8. The vase shell *Vasum turbinellum*, a common inhabitant of the large lagoons behind fringing reefs in the central Red Sea.

Fig. 10.9. The venomous cone shell *Conus textile* which has been reported as responsible for a number of human deaths.

in the dorsal processes. It is not understood yet how this delicate transport can be managed through the entire intestinal system of the slugs. Normally the stinging cells of a coral will discharge their poisonous harpoon-like threads immediately they are touched. By incorporating the intact stinging cells the sea-slugs probably have a rather 'fiery' taste to any predator. In addition, it has been observed in several Red Sea slugs that the animals, when they are disturbed, first autotomize some of their cerata. Any predator who is mistaken enough to taste these wormlike structures as an appetizer will certainly leave the rest of the intended meal untouched.

A few interesting opisthobranchss will now be mentioned. I start with those representatives who have partly retained their calcareous shell, i.e. the Tectibranchia. Commonly *Bulla ampulla* (the bubble shell) is found in all kinds of soft bottoms. The sea butterflies (Cavoliniidae) are a highly specialised group; they are one of the few examples of molluscs that live their whole life cycle in the oceanic plankton and periodically *Cavolinia uncinata* and *Diacra quadridentata* are superabundant in Red Sea plankton samples. These transparent animals move slowly through the water like small butterflies by means of two 1 cm long body lobes. These wings are coated with mucus which serves to trap the organic particles and phytoplankton on which these species feed.

The members of the family of sea hares (Aplysiidae) rank among the largest opisthobranchs of the Red Sea. The elongated head bears two pairs of tentacles and a well-developed radula is present. The Red Sea sea-hares *Aplysia oculifera*, *Dolabrifera dolabrifera* and *Notarchus indicus* are olive-brown sluggish molluscs which can hardly be detected in their natural environment. They live and feed on large brown and green algae. The aplysiids are characterized by glands which can produce slimy and bad-smelling secretions when the animals are disturbed. This may explain why, for example, Sudanese fishermen believe that the capture of sea hares in their nets is a bad omen. Some even believe that personal fatalities will happen to those persons who come into contact with these animals.

A closely related species from the family Pleurobranchidae is an orange sea-slug which is often found in groups under coral slabs. This is *Berthellina citrina*, an animal which feeds exclusively on small sea-squirts. This slug is able to produce sulphuric acid if it is disturbed.

The largest and most attractive group of sea-slugs are the nudibranchs (meaning naked gills). They occupy a great variety of ecological niches in the reef community. All of them are predators, feeding on a wide range of marine invertebrates. Many species are highly selective, feeding exclusively on one sort of prey. A very familiar example is the Pyjama nudibranch *Glossodoris quadricolor* which predominantly occurs on one species of red sponge. Nudibranchs are mostly shade-loving animals and often occur on the reef slope and in deeper coral zones where they frequently can be observed feeding on corals, sponges or sea anemones. A rather common species here is *Phyllidia varicosa*, a black animal bearing warty white and blueish protuberances with orange tips. *Casella obsoleta* is one of the few nudibranchs which seems to live permanently on sandy bottoms in the deeper zones.

Hexabranchus sanguineus is the most spectacular Erythraean sea-slug. This is the largest nudibranch in the world and reaches 30 cm in length. It feeds on the soft coral *Sarcophyton*. In swimming the animal propels itself with synchronized undulations of the mantle. These elegant movements plus the scarlet red colour are presumably the reason why the vernacular name 'Spanish Dancer' is attributed to this animal. Usually a couple of commensal shrimps (*Periclimenes imperator*) lives permanently in the external gill area of *Hexabranchus*. They utilize the host both as transport and as a relative safe protection against predators since the flesh of the slug is reported to be rather poisonous or at least distasteful for reef fish.

To complete the systematic review of the gastropods I will briefly mention the largely terrestrial lung snails (Pulmonata), a few of which are also encountered in the Red Sea. *Siphonaria kurracheensis* can easily be mistaken for a true limpet as it has adopted a limpet-like structured shell. These small snails are generally abundant on any rocky shore where they cling to the underside of big boulders. A common pulmonate snail in a similar habitat is *Melampus castaneus*, while a closely related species, *M. flavus*, is a member of the sandy infauna in reef lagoons. The general appearance of both species resembles more a freshwater than a marine mollusc. In this context it should be mentioned that the Red Sea lacks a brackish water fauna due to the absence of any permanent freshwater influx.

One small group of shelled marine molluscs has developed such distinct morphological features that they have been united in a separate class. These are the elephant tusk shells or Scaphopoda. They occur in the soft bottoms of all reef zones from shallow to abyssal depths. Their calcareous shell resembles an elephant tusk. It is a curved tapered tube open at both ends. The head and muscular foot project from the wider end of the shell. The head which has no eyes (unlike most of the gastropods) carries a number of sensory and food-collecting tentacles. Scaphopods burrow actively into soft sediments where they feed on microscopic organisms, mainly foraminiferans. A Red Sea visitor will always find a number of dead tusk shells, belonging to the common species *Dentalium bisexangulatum* and *D. longirostrum*, in the flotsam of the shore.

The Bivalvia or pelecypods are the fourth major class of the mollusc phylum. Bivalves are represented by a large number of species (about 270) which inhabit all reef zones in the Red Sea and occur from upper tidal down to abyssal depths of almost 2000 m. Most of these molluscs are not so obvious to anyone who swims over a reef as, for instance, the gastropods. Apart from a few exceptions, such as the giant clams, the wing and the true oysters, the majority live burrowed into sandy and muddy substrates. In general, their physiological reactions and mode of feeding are quite uniform. Nevertheless the heterogeneity of a tropical marine environment like the Red Sea allows a fairly large number of species to coexist, in spite of the fact that they may all compete for a similar source of food.

The common features of bivalves include a laterally compressed body enclosed within two calcareous shell plates or valves which are dorsally hinged by an elastic ligament. Hinge teeth are usually significant characteristics which are used in the classification of the group. Each tooth fits into a corresponding socket of the opposing valve, fitting tightly when both valves are closed by the aid of the strong adductor muscles. The inner mantle cavity encloses a pair of gills which act as respiratory organs and usually also as food gathering organs. Most bivalves feed by filtering small food particles which are passed to the gills via a respiratory siphon. According to the specific mode of life the muscular foot is either held permanently within the shell, or can be extended outside the valves for digging and locomotion. In the majority of bivalves the sexes are separate, but there are a few hermaphrodite forms.

As already indicated almost all types of Red Sea bivalves exhibit a rather uniform way of life as sand dwellers. The most common species of these soft bottom communities are *Glycymeris pectunculus* (Figure 10.10), *Lucina dentifera*, *L. semperiana*, *Codakia divergens*, *Loripes erythraeus*, *Trachycardium pectiniforme*, *Laevicardium orbita*, *Mactra olorina* and *Asaphis violascens*. Two superfamilies are particularly diverse in these biotypes: The wedge shells (Tellinacea) and the carpet or Venus shells (Veneracea). Regionally there are considerable variations regarding the abundance of these families. Out of a total of 35 recorded species of tellinids. *Tellina edentula*, *T. isseli*, *T. virgata*, *Cyclotellina scobinata* and *Quidnipagus palatum* are the most common species which can be obtained by examining the upper layers of sandy substrates. The same sample may yield the venerids *Gafrarium pectinatum*, *Callista florida*, *Dosinia erythraea*, *Tapes literatus* and *Timoclea marica*. Two prized Red Sea Venus shells are *Lioconcha picta* and *L. castrensis*, both characterized by fine brown zigzag patterns on an ivory coloured valve.

Among the sand dwellers the fan shells *Pinna muricata* (Figure 10.11) and *P. bicolor* often attract the interest of shell collectors. These bivalves grow up to 30 cm or more, with the shape of a partially opened ladies fan. They are buried upright in the mud with the top (posterior) end protruding. The anterior end is attached to some piece of hard substrate by means of horny thread-like secretions, called byssus threads. The examination of the interior of these pinnids often reveals a couple of crab commensals (*Anchistus*).

In general, bivalves are fairly immobile organisms, but some species like the scallops (Pectinidae) are able to swim by rapidly flapping their valves together. *Gloripallium pallium*, *Chlamys sanguinolentus* and *C. senatorius* are the common Red Sea species of this family. A similar mode of locomotion can be

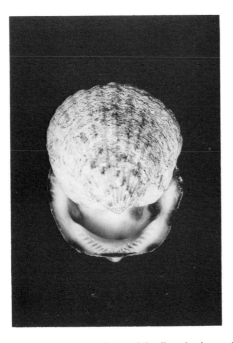

Fig. 10.10. A common bivalve sand-dweller, the dog-cockle *Glycymeris pectunculus*.

Fig. 10.11. The fan shell *Pinna muricata* which may grow to 30 cm or more in length and lies with its pointed anterior end buried in the sediment.

observed in file shells (Limidae). In addition to this typical escape reaction *Lima fragilis* uses a bundle of detachable sticky tentacles to ward off predators when it is disturbed.

I will leave now the soft bottom dwellers and describe some bivalves which are typically found on hard substrates in the Red Sea. An intensive search in crevices of rocks and boulders, under coral slabs, on artificial ramparts or even in the interstices of branched corals will reveal a surprising variety of bivalves. The tightest attachment to the hard substrate is achieved by the true oysters (Ostreidae), the thorny oysters (Spondylidae) and the jewel box shells (Chamidae). Very familiar to the Red Sea visitor are the infratidal species *Ostrea forskali*, *Saccostrea cucullata*, *Spondylus gaederopus* and *Chama imbricata*. All of them have one valve firmly cemented to shallow water rocks or to the abraded dead coral rock on the reef flat. The typical tidal species open their shell slightly during high tides.

Attachment to the rock by means of byssal threads is common in many shallow water bivalves. Thus, the ark shells (Arcidae) like *Arca plicata* and *Barbatia setigera* have successfully populated many hard substrates. This also holds true for the mussels *Modiolus auriculatus*, *M. cinnamomeus* and *Brachydontes variabilis*. None of the Red Sea mussels are consumed by Man.

The mussel family (Mytilidae) includes a highly specialized group which is able to bore actively into limestone. These so-called date mussels (e.g. *Lithophaga lessepsiana*, *L. hanleyana* and *Botula cinnamomea*) are found in any large coral block or in the basement and the ramifications of branched stone corals like *Stylophora*, *Pocillopora* and *Acropora*. The gaping clams (or flask shells) of the family Gastrochaenidae live in a similar habitat although they are anatomically quite different from the date mussels. They burrow into spherical corals such as *Montipora*, *Cyphastrea* and *Favia*. Their presence is easily recognizeable on the surface of the corals by the typical 8-shaped tubes of the inhalant and exhalant siphons.

A number of bivalves live in a less destructive association with scleractinian corals. Hammer oysters (Malleidae) and their relatives (Isognomonidae), and the fan shells *Atrina vexillum* and *Streptopinna saccata*

Fig. 10.12. The giant clam *Tridacna squamosa*. Giant clams are a common feature of reef flats in the Red Sea and commonly reach 30 cm diameter.

often settle between the branches of shallow water corals. The small mussel *Septifer bilocularis* is one of the most abundant Red Sea bivalves and lives not only in a loose association with corals but also in rock crevices at a wide range of depths.

Hydrocorals (*Millepora*), whip corals (*Cirripathes*) and black corals (*Antipatharia*) are the preferred site of attachment for a number of bivalves belonging to the family of wing oysters (Pteriidae). They are regarded as indicator species for strong longshore currents; for example, *Pteria aegyptiaca* settles predominantly on the branches of the fire coral *Millepora dichotoma*. The Rooster-comb oyster *Lopha cristagalli* has similar ecological demands. This distinctive species with a zigzag-edged shell is quite familiar to Red Sea divers who explore the dimly lit deeper reef zones. The bivalve is fixed on exposed whip corals and gorgonarians. Usually the valves of this species are heavily encrusted by a red sponge or by colonial ascidians.

The final bivalve to be discussed is the giant clam *Tridacna squamosa* (Figure 10.12) which is found embedded on reef flats all over the Red Sea. A most conspicuous feature of this animal is the bright turquoise coloration of the fleshy mantle. On the margin of the mantle a row of sensitive light organs are embedded, which enable the valves to close rapidly when the shadow of a potential predator suddenly appears above the animal. *Tridacna* reaches 30 cm or more in diameter in the Red Sea. Its mode of nutrition is of great interest. In the mantle tissue innumerable single-celled algae are embedded and these algae play a similar symbiotic role to the zooxanthellae in the tissues of hermatypic scleractinian corals (*see* Chapter 7). The unicellular plants produce nutrients both from their photosynthetic activities and the metabolic products of their host. The mantle of the clam is therefore highly modified for this purpose. Special lens-like structures in the upper mantle layers allow light to penetrate to deeper tissues, providing the algae there with sufficient light energy. In addition the clam seems to protect the algae from harmful radiation by a number of colourful pigments ranging from blue, green and brown to black. As the clams are dependent on the symbiotic algae the distribution of *Tridacna* is restricted to clear and shallow waters. It is reported that the clams actively digest a certain surplus of their algae. We may therefore say that these algae are truly farmed by the bivalve. It is thought that the zooxanthellae by using up carbon dioxide in photosynthesis assist in the calcification of the shell which accounts for the size of this, the largest of all Red Sea molluscs.

The last and most specialized class of the molluscs are the Cephalopoda, an entirely marine group which includes the nautilus, squids, cuttlefish and octopus. Cephalopods are characterized by a well-developed head bearing prehensile tentacles which often have one or more rows of suckers. The mouth which lies in the centre of the tentacle ring is armed with a strong parrot-like beak, used to crack the shelled organisms on which these molluscs mainly prey. Many cephalopods have an ink sac opening into the pouch-like mantle cavity and can produce a 'smoke screen' of ink when disturbed. The cephalopods rival or surpass most vertebrates in the quality of their sense organs, the complexity of their behaviour (especially in the speed and precision of movements) as well as their ability to change colour to match almost any background. They possess a well developed brain which accounts for the extraordinary intelligence achievements of this group when compared with all other invertebrates.

In the Red Sea some cephalopods such as the common octopuses *Octopus aegina* and *O. macropus* are benthic, living mainly in grottoes and spacious crevices on reef flats. A related species is *Octopus horridus* which carries out nocturnal feeding excursions into the very shallow infratidal zones. All three species seem to be rather territorial, swimming only when they are forced to escape danger. Other cephalopods, like the Red Sea squid *Sepia pharaonis* and the cuttlefish *Sepioteuthis lessoniana* are capable of rapid and steady swimming, but they also remain in close vicinity to particular benthic communities. The latter species is often gregarious.

In general, Red Sea fishermen rarely capture cephalopods, so it is not clear whether this group of molluscs is rare in the whole region, or lives in greater depths where they are seldom detected.

REFERENCES

Adam, W. (1955). Céphalopodes. Résultats Scientifiques des Campagnes de la 'Calypso'. *Ann. Inst. Océan, Monaco.* 30, 185–94.

Adam, W. (1959). Les Céphalopodes de la mer Rouge. Mission R. Ph. Dollfus en Égypte (1927–1929). *Rés. Sci.* (3) 28, 125–93.

Adam, W. (1960). Cephalopoda from the Gulf of Aqaba. *Bull. Sea Fish. res. Stn. Haifa* 26, 1–26.

Adams, H. (1870). On some genera and species of Gastropoda (Mollusca) collected by R. MacAndrew in the Gulf of Suez. *Ann. Mag. nat. Hist.* (4) 6, 121–9.

Anthony, R. (1905). Liste des mollusques acéphales du golfe de Tadjourah. *Bull. Mus. natn. Hist. nat. Paris* 11, 490–500.

Audouin, V. (1826–1827). Explication sommaire des planches des Mollusques (Echinodermes) de l'Égypte et de la Syrie dont les dessins ont été fournis par J. C. Ssvigny. *Déscr. Égypte*, 22.

Barash, Al. and Danin, Z. (1972). The Indopacific species of Mollusca in the Mediterranean and notes on a collection from the Suez Canal. *Israel J. Zool.* 21, 301-74.

Barletta, G. (1969). Malacofauna del Mar Rosso I. Hurghada. *Conchiglie* 5, (9–10), 145–54.

Barletta, G. (1974). Considerazioni critiche intorno alla ipotetica presenza di forme indo-pacifiche di Cypraeidae nelle aque Mediterraneo (Moll.-Gastropoda). *Quaderni della Civica Stat. Idrobiol. Milano* 5, 79–84.

Bavay, M. (1898). Notes sur les mollusques du Canal de Suez. *Bull. Soc. Zool. Fr.* 23, 161–4.

Biggs, H. E. J. (1965). Mollusca from the Dahlak Archipelago, Red Sea. *J. Conch.* 25(8), 337–41.

Bisacchi, J. (1931). Le Nassariidae del Mar Rosso e del Golfo di Aden. *Ann. Mus. Civ. Storia nat. Genova* 55, 43–70.

Cooke, A. H. (1885–1886). Report on the Testaceous Mollusca obtained during a dredging excursion in the Gulf of Suez in 1869 by R. MacAndrew. *Ann. Mag. nat. Hist.* (5) 15, 322–39; (5) 16, 32–50, 262–276; (5) 17, 128–142; (5) 18, 92–109.

Eales, N. B. (1938). A systematic and anatomical account of the Opisthobranchia. *J. Murray Exp. sci. Rep.* 5(4), 77–122.

Eliot, C. N. (1908). Notes on a collection of Nudibranchs from the Red Sea; Reports on the marine biology of the Sudanese Red Sea XI. *J. Linn. Soc. Lond. (Zool.)* 31(204), 86–122.

Engel, H. and Van Eeken, C. J. (1962). Red Sea Opisthobranchia from the coasts of Israel and Sinai. *Bull. Sea Fish. res. Stn. Haifa* 30, 15–34.

Fischer, P. (1865). Notes sur les faunes conchyliologiques des deux rivages de l'Isthme de Suez. *J. Conch.* 13, 241–8.

Fischer, P. (1870). Sur la faune conchyliologique marine des baies de Suez et de l'Akabah. *J. Conch.* 18, 161–79.

Fischer, P. (1871). Sur la faune conchyliologique marine de la baie de Suez. *J. Conch.* 19, 209–26.

Forsskål, P. (1775). *Descriptiones animalium quae in itinere orientali observavit P. Forskål*, Salvius, Copenhagen, 164 pp.

Franc, A. (1956). Résultats scientifiques des campagnes de la 'Calypso' en Mer Rouge. IX. Mollusques marines. *Ann. Inst. Océan Monaco.* 32, 19–60.

Gerlach, S. (1959). Über das tropische Korallenriff als Lebensraum. *Verh. Dt. Zool. Ges. Münster* 39, 356–63.

Hall, W. J. and Standen, R. (1907). On the Mollusca of a raised coral reef on the Red Sea coast. *J. Conch.* 12(3), 65–8.

Issel, A. (1869). *Malacologia del Mar Rosso*, Pisa, 387pp.

Jousseaume, F. (1888). Déscription des mollusques recueillis par le Dr. Faurot claus la mer Rouge et le Golfe d'Aden. *Mëm. Soc. zool. France* 1, 165–223.

Klausewitz, W. (1967). Die physiographische Zonierung der Saumriffe von Sarso. *Meteor Forschungsergebnisse* D(2), 44–68.

Kohn, A.J. (1965). *Conus* (Mollusca, Gastropoda) collected by the Israel South Red Sea Expedition 1962, with notes on collections from the Gulf of Aqaba and Sinai Peninsula. *Bull. Sea Fish. res. Stn. Haifa* 38, 54–9.

Lamy, E. (1938). Mollusca Testacea (Molluscs collected by Dollfus) from Gulfs of Suez and Aqaba). *Mém. Inst. Égypte* 37, 1–89.

MacAndrew, R. (1870). Report on the testaceous Mollusca obtained during a dredging excursion in the Gulf of Suez in the months of February and March 1869. *Ann. Mag. nat. Hist.* (4) 6, 429–50.

Martens, E. V. (1887). Conchylien aus dem Suezkanal. *Sitzungsber. Ges. Naturf. Berlin* 6, 96.

Mastaller, M. (1978). The marine molluscan assemblages of Port Sudan, Red Sea. *Zool. Meded.* 53(13), 117–44.

Mastaller, M. (1979). *Beiträge zur Faunistik und Ökologie der Mollusken und Echinodermen in den Korallenriffen bei Aqaba, Rotes Meer*. Ph.D. Dissertation, Univ. Bochum, 344 pp.

Melvill, J. C. (1928). The marine mollusca of the Persian Gulf, Gulf of Oman and North Arabian Sea. *Proc. Malac. Soc. Lond.* 18, 93–117.

Mergner, H. and Mastaller, M. (1980). Ecology of a Reef lagoon area near Aqaba (Red Sea). *Symp. Coast. Mar. Env. Red Sea, Gulf of Aden and Trop. Western Ind. Oc., Khartoum,* Proc. Vol.I, 39–76.

Mergner, H. and Schuhmacher, H. (1974). Morphologie, Ökologie und Zonierung von Korallenriffen bei Aqaba (Golf von Aqaba, Rotes Meer). *Helg. wiss. Meeresunters.* 26, 238–58.

Mienis, H. K. (1971). Strombidae (Mollusca, Gastropoda) collected by the Israel South Red Sea Expedition. *Argamon* 2(3–4), 87–94.

Moazzo, P. G. (1939). Mollusques testacés marines du Canal de Suez. *Mém. Inst. Égypte* 38, 1–283.

O'Donoghue, C. H. (1929). Report on the Opisthobranchiata, Cambridge Expedition to the Suez Canal, 1924. *Trans. zool. Soc. Lond.* 22, 713–841.

Pallary, P. (1926). Explication des planches de J.C. Savigny. *Mém. Inst. Égypte* 11, 21–130.

Pfeiffer, L. (1846). Bemerkungen über Audouin's Bearbeitung der Savigny'-schen Tafeln. *Zietschr. Malakozool.* 3, 67–9.

Philippi, F. (1836). *Enumeratio Molluscorum Siciliae*. Berlin, 1936, 248–9.

Por, F. D. (1978). *Lessepian Migration*. Berlin, Springer Verlag.

Quoy, J. R. and Gaimard, P. (1834). *Voyage de découvertes de "L'Astrolabe", executé par ordre du Roi, pendant les années 1826–29.* Paris.

Rees, W. J. and Stuckey, A. (1952). The 'Manihine' Expedition to the Gulf of Aqaba. VI. Mollusca. *Bull. Brit. Mus. (nat. Hist.), Zool.* 1(8), 183–201.

Savigny, J. C. (1817). *Déscription de l'Égypte, Histoire naturelle, II. Atlas des Mollusques, Echinodermes.* Paris.

Schilder, F. A. (1965). Cypraeidea. Israel South Red Sea Exped. 1962, Rep. 16. *Bull. Sea Fish. Res. Sta. Haifa* 40, 75–8.

Schilder, F. A. and Schilder, M. (1939). Prodrome of a monograph of living Cypraeidae. *Proc. malac. Soc. Lond.* 23(3), 119–80; 23(4), 181–231.

Shopland, E.R. (1902). List of marine shells collected in the neighbourhood of Aden between 1892 and 1901. *Proc. Malac. Soc. Lond.* 5(2), 171–9.

Smith, E. A. (1891). On a collection of marine shells from Aden, with some remarks upon the relationship of the molluscan fauna of the Red Sea and the Mediterranean. *Proc. zool. Soc. Lond..* 28, 390–435.

Sturany, R. (1899). Lamellibranchiaten des Rothen Meeres. *Denkschr. Math. Naturw. Cl. Kais. Akad. Wiss. Wien, Zool. Ergebn.* 14, 1–41.

Sturany, R. (1903). Gastropoden des Rothen Meeres. *Denkschr. Math. Naturw. Cl. Kais. Akad. Wiss. Wien, Zool. Ergebn.* 23, 1–75.

Taylor, J.D. and Reid, D.G. (1984). The abundance and trophic classification of molluscs upon coral reefs in the Sudanese Red Sea. *J. nat. Hist.* 18, 175–209.

Tomlin, J. R. (1927). Report on the Mollusca (Amphineura, Gastropoda, Scaphopoda, Pelecypoda). *Trans. zool. Soc. Lond.* 22, 291–320.

Vaillant, R. (1865). Recherches sur la faune malacologique de la Baie de Suez. *J. Conch.* 13, 97–127.

Vassel, E. (1891). Les Faunes de l'Isthme de Suez. *Bull. Soc. Hist. nat. Autun* 3, 21.

Weindl, T. (1912). Vorläufige Mitteilung über die von SM Schiff *Pola* im Rothen Meere gefundenen Cephalopoden. *Proc. Malac. Soc. Lond.* 22, 16–23.

Wülkner, G. (1920). Über Cephalopoden des Roten Meers. *Senckenbergiana* 2, 48–58.

CHAPTER 11

Echinoderms of the Red Sea

ANDREW C. CAMPBELL

Queen Mary College, University of London, Mile End Road, London, U.K.

CONTENTS

11.1. INTRODUCTION

Echinoderms are distinctive animals showing unique qualities of structure and function. About 6000 species are recognised from today's seas and oceans where they live as bottom dwellers at all depths and all latitudes. They thrive particularly well in warm shallow waters, and nearly 200 species have been recorded from the Red Sea, which provides many suitable habitats for them. The ecological niche diversity of the phylum is perhaps somewhat restricted; echinoderms have never really escaped a benthic existence (although there are a handful of aberrant pelagic sea cucumbers), and their whole radiation is confined to life in the sea because echinoderms are unable to regulate their body salt concentration. The most characteristic features of the phylum include their unusual pentamerous symmetry, the endoskeleton of reticulate calcium carbonate ossicles, and their unique water vascular system. The latter is a hydraulic system powering small tentacles or tube feet, originally developed for filter feeding, respiration and in many modern forms it is used for locomotion.

Echinoderms are better documented than most Red Sea invertebrates, selected taxonomic or distribution studies include Chadwick (1908), Campbell (1971), Cherbonnier (1967), Clark (1952, 1967), James and Pearse (1969) and Mortensen (1926, 1938). Clark and Rowe (1971) contains an extensive bibliography and should be consulted for species identification.

11.2. OUTLINE OF ECHINODERM BIOLOGY

Space limitations allow only the briefest outline of the biology of echinoderms here, although details of the ecology of certain taxa appear in sections 11.4 to 11.6. For an elementary description the reader is referred to Barnes' (1980) textbook, more complete accounts are Hyman (1955) or Nichols (1969). Many aspects of echinoderm physiology are excellently reviewed by Boolootian (1966) or by Binyon (1972).

11.2.1. The five classes of echinoderms

The surviving echinoderms fall naturally into five distinctive classes. These are the Crinoidea (sea lilies and feather stars), Asteroidea (starfish), Ophiuroidea (brittlestars), Echinoidea (sea urchins, heart urchins and sand dollars) and the Holothuroidea (sea cucumbers). The appearance of typical members of each class is shown in Figure 11.1, and details of the symmetry, orientation, anatomy and feeding of each class are summarised in Table 11.1. It will be noted that while most echinoderms preserve the curious five-fold radial symmetry from filter-feeding ancestors, only the crinoids remain orientated mouth-up, using the water vascular system in its original form, for filter feeding. The other classes, no longer sessile, have inverted their bodies, bringing the mouth to the substrate, and use the water vascular system largely for locomotion, or specialist types of feeding. The skeleton is variously modified. In most groups, it takes the form of ossicles, embedded in the dermis and linked by muscles that allow considerable flexibility when relaxed, but afford surprising rigidity when contracted. In ophiuroids and crinoids the arms contain large internal ossicles which function like vertebrae, forming an articulated 'backbone' within the arms. The skeleton is best developed in echinoids, forming a continuous rigid test, and the formidable battery of defensive spines and pedicellariae which give the common name of 'sea urchin' to the group.

11.2.2. Feeding

Echinoderms exploit a great variety of food resources, and some are discussed in detail in the sections dealing with the ecology of echinoderms in the Red Sea. Generally, the sea lilies and feather stars are suspension or filter feeders. Most starfish are carnivorous, often feeding by eversion of their lower stomach on to suitable prey. Many are notorious predators of bivalves, others take various food items such as coral or sponges, not favoured by most other predators. Brittle stars are flexible in their feeding and can be carnivores, scavengers and feeders on small particles, (deposited or suspended in the water column). Sea cucumbers are also deposit or suspension feeders, some are important reworkers of reef sediments for their small organic component. Sea urchins have two distinct feeding strategies. The radially symmetrical or 'regular' urchins use the elaborate array of ossicles, muscles and teeth called the 'Aristotle's lantern' to graze algae from rock surfaces, and some habitually feed on marine angiosperms (e.g. *Thalassia*), many opportunistically scavenge dead animal matter. By contrast, the sand inhabiting sand dollars and heart urchins (the bilaterally symmetrical 'irregular' echinoids) have evolved various strategies for extracting small particles of organic material from sand. Useful reviews of echinoderm food relations are provided by Lawrence (1975), and Jangoux and Lawrence (1982). No echinoderms live parasitically, a few brittlestars live almost as commensals on other organisms such as sponges and sea fans.

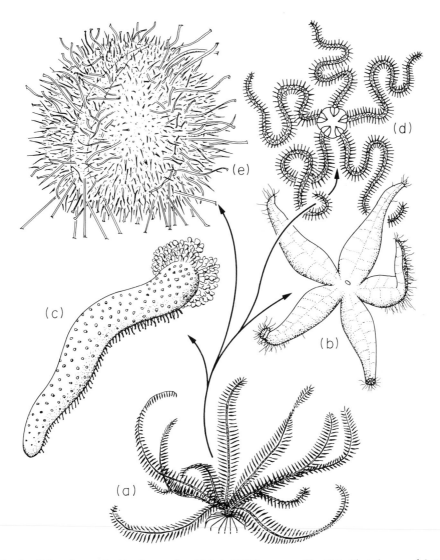

Fig 11.1. The five living classes of echinoderms, after Nichols (1969). a — sea lily (Crinoidea), b — starfish (Asteroidea), c — sea cucumber (Holothuroidea), d — brittle-star (Ophiuroidea), e — sea urchin (Echinoidea).

TABLE 11.1. Characteristic features of living echinoderm classes. This table summarises the major features of living echinoderm classes, further details are given in the text and in the general references quoted.

ANATOMICAL & PHYSIOLOGICAL CHARACTERS	CRINOIDS—Sea lillies and feather stars	ASTEROIDS—Starfish	OPHIUROIDS—basket stars and brittle stars	ECHINOIDS—sea urchins, heart urchins, sand dollars	HOLOTHUROIDS—sea cucumbers
SYMMETRY	Penta-radial central cup-like body either borne on stalk or free living: bears pairs of jointed arms, with side branches, arranged in fives or multiples: some attached to substrate by stalk, other free.	Penta-radial central body not always distinct from arms which are arranged in fives, multiples or irregularly: body sometimes pentagonal.	Penta-radial central discoidal body; 5 repeatedly branching arms in basket stars; 5 or rarely more unbranched arms in brittlestars, often spiny.	Penta-radial or bilateral, body globular, discoidal or heart shaped, radial in sea urchins, bilateral in sand dollars and heart urchins.	Bilateral cucumber shaped bodies: upper and lower sides may be differentiated.
ORIENTATION, MOUTH AND ANUS	Central mouth and anus away from substrate.	Central mouth faces substrate, anus faces away.	Central mouth faces substrate; anus lacking.	Mouth faces substrate, anus away; in heart urchins mouth may be forward pointing; in sand dollars and heart urchins anus is posterior.	Mouth anterior, anus, in cloaca, faces posterior.
SKELETON	Flexible, jointed arms	Flexible, skeletal elements embedded in leathery test	Flexible disc, jointed arms used for locomotion	Rigid test of fused ossicles, rarely flexible. Special jaws in sea urchins and sand dollars.	Flexible to varying degrees according to extent of embedded ossicles.
WATER VASCULAR SYSTEM	Tube feet without suckers arranged along arms and side branches, used for food gathering and respiration, not locomotion.	Tube feet with suckers arranged in rows on underside of rays, used for locomotion and sometimes for food gathering.	Tube feet suckerless, used for food gathering, not locomotion.	Tube feet arranged in paired rows, used for locomotion and sometimes for food gathering.	Tube feet arranged in rows or reduced, used for locomotion in many species; special ones around mouth used for food gathering.
APPENDAGES	Pinnules; cirri and hooks may be present.	Spines, paxillae, pedicellariae may be present.	Spines and hooks.	Well developed spines and pedicellariae.	Hooked spicules may protrude through skin.
FOOD Sources	Suspension feeders.	Carnivores, scavengers and detrital feeders.	Basket stars are suspension feeders. Brittle stars are scavengers and eat detritus.	Grazers, drift feeders, scrapers and deposit feeders.	Deposit and suspension feeders.
HABITATS	Rock and reef.	Epifaunal on rock and reef. Some burrow in sand and mud.	All benthic sites.	Regular (round) echinoids on rock and reef. Irregular (bilateral) types burrow in sand.	All benthic habitats.
HABITS	Nocturnal in Red Sea, hidden by day.	Mainly nocturnal feeders, some are cryptic.	Brittle stars all cryptic, basket stars climb on to corals at night.	Many are active at night and hide by day. Some use debris to cover themselves.	Large types are obvious by day, others may burrow or live under rocks.

11.2.3. Reproduction, development and recruitment

Our knowledge of echinoderm reproduction and recruitment is poor, Mortensen (1938) for example noted that while adults of the long spined sea urchin *Diadema setosum* were common on reefs, their juveniles were rare. This paradoxical situation is common in reef echinoderms and reflects our poor understanding of their reproduction and recruitment.

Most echinoderms species have separate sexes, although this can be unstable and sex change can occur. Only a few species (except ophiuroids) exhibit hermaphroditism, the Red Sea asteroid *Fromia ghardaqana* is an example, starting its sexual career as a male and changing to female in later life.

Echinoderm reproductive periodicity was reviewed by Boolootian (1966). Red Sea species seem rather variable in their spawning seasons (Mortensen, 1937, 1938). Most holothurians spawn from June until August, some echinoids such as *Eucidaris metularia* are less restricted, spawning from April through to September. Crinoids show similar variability. The asteroid *Asterina burtoni* spawns in the Red Sea from December to March, and takes two years to reach maturity (Achituv, 1973). *Acanthaster planci* probably spawns from July to August (Crump, 1971).

Several authors have described reproductive periodicity in Red Sea echinoids. Pearse (1968) found that periodicity and seasonality depended on the species, latitude and temperature regime. In the head of the Gulf of Suez Pearse (1969b) found that *Echinometra mathaei* showed well defined reproductive periodicity, spawning in summer and fall. At the mouth of the Gulf of Suez the synchrony was poorly defined, and disappeared further south in the Red Sea. *Prioncidaris baculosa* (Fig. 11.2) in the Gulf of Suez showed synchronous gametogenesis in April, with spawning occurring synchronously in July and August. *Lovenia elongata* was not synchronous in this Gulf, with gametogenesis from mid to late winter and spawning from April to September (Pearse, 1969a). Pearse has suggested that accumulations of nutrients in the sea, and specific minimal temperatures may control these patterns, and that the nutrient levels are probably also temperature and photoperiod controlled.

Fertilisation in echinoderms is usually external, and most species have a lengthy planktonic larval

Fig 11.2. Primitive echinoid *Prioncidaris baculosa* from Port Sudan. This urchin has very large primary spines which bear lateral thorns, in addition to smaller secondary spines visible surrounding the primaries on the test. The primaries help to wedge the urchin in crevices and may deter predators. The secondaries are protective.

stage before settlement and metamorphosis. The larvae are often complex in shape and elegantly ciliated. Full accounts of the larval forms and larval biology of many tropical echinoderms, including Red Sea species, were given by Mortensen (1931, 1937 and 1938). Little is known of the conditions required by larvae for optimal survival and recruitment to the adult population except for the Crown-of-Thorns starfish, *Acanthaster*. Lucas (1973) found under experimental conditions that low temperature and salinity (28°C, 30‰) led most larvae through the metamorphosis; both these levels are below ambient for the Red Sea during the breeding season for the species. Yamaguchi (1973) has described settlement and metamorphosis in species of *Acanthaster*, *Culcita* and *Linckia* known from the Red Sea. These animals favour dead coral encrusted with coralline and other algae for settlement and metamorphosis. Kenchington (1976) concluded that recruitment to adult populations of *Acanthaster planci* on the Great Barrier Reef was due mainly to larval immigration, not to adult migration from other areas.

Little is known of longevity or age of maturation in tropical echinoderms. According to Hyman (1955) temperate echinoids become sexually mature after one year, and live up to eight years; some holothurians may attain an age of ten years.

11.2.4. Predators, defence and toxicity

Despite their spiny arsenal and chalk-bound flesh, echinoderms are eaten by a variety of animals. Little is known of crinoid predators and those of the Red Sea probably escape most predators from their nocturnal habits. Asteroids are eaten by puffer- and trigger-fish (Ormond and Campbell, 1974), and in the aquarium can be attacked by cone-shells. Brown (1972) records that in Australia the helmet shell *Cassis cornuta* attacks *Acanthaster planci*, as does the Trumpet triton *Charonia tritonis* (Endean, 1969). Another gastropod, *Stilifer linckiae* was found several times near Port Sudan with its shell buried in the body wall of *Linckia multifora* or the peristomial membrane of *Heterocentrotus mammillatus*, while its mouth consumed the inner tissues (Campbell, *pers. obs.*). Starfish also fall victim to decapod crustaceans and the remarkable account by Wickler and Seibt (1970) showed that the painted shrimp *Hymenocera picta* will destroy *Acanthaster* and *Echinaster*. *Hymenocera* is not a native of the Red Sea, but other starfish-eating species may exist there.

Sand dwelling ophiuroids are almost certainly taken by foraging fish in much the same way as they are in temperate waters, while cryptic crevice dwellers will fall victim to foraging fish such as wrasse.

Echinoids are attacked by starfish in many parts of the world, but this behaviour has not yet been recorded in the Red Sea, neither have attacks by helmet or cone shells. The formidable armament of *Diadema* does not protect it from attacks by triggerfish, and pieces of echinoid test and spines were commonly encountered in gut samples taken from the Humphead wrasse *Cheilinus undulatus* caught off Port Sudan (S.M. Head, *pers.comm.*).

Sea cucumbers appear to have few natural enemies, although in the past Red Sea stocks have been extensively fished and exported as *bêche-de-mer*.

Echinoderms have a variety of defence mechanisms against predators. Many are retiring in habits, burrowing in sand or seeking refuge in crevices. In the Red Sea most are active mainly at night. Some species such as *Linckia multifora* are cryptically coloured, others, especially holothurians, have very thick and tough skins. Echinoids and some asteroids are well protected by sharp spines, those of the Crown-of-Thorns starfish appear to secrete toxic mucus. The spines of the flexible echinoid *Asthenosoma* bear terminal venom sacs, while the fragile hollow spines of *Chaetodiadema*, *Diadema* and *Echinothrix* are barbed with whorls of fine teeth making them impossible to extract from wounds. These spines are also venomous, although the precise form and distribution of the venom is not known.

Some asteroids and all echinoids bear remarkable minute pincer-like organs called pedicellariae, some of which are equipped with venom glands. They are involved in grooming the body, keeping it free of

Fig 11.3. Unidentified crab associated with crinoid in the central Red Sea. Note the camouflaged shape and striped pattern.

sediments, trapping minute animals like amphipods and isopods and preventing settlement of larvae of larger animals such as barnacles on the outside of the body. In some cases the pedicellariae also deter larger predators. A full review has recently been given by Campbell (1983).

The tissues of echinoids and holothuroids are palatable and form part of the human diet in many cultures, but asteroids have a repellent taste (S.M. Head, *pers. comm.*). This may serve as a protection from predators even in the cases of brilliantly coloured asteroids like *Fromia ghardaqana*, whose conspicuousness may serve to warn visually cued predators.

11.2.5. Symbiosis and parasitism

While no species of echinoderm is known to exist as a true parasite, many are hosts to other organisms living either as parasites or symbionts. Polunin (1971) listed symbiotic and commensal decapod crustaceans found on echinoderms in the Red Sea. The small shrimp *Periclimenes soror* was identified in association with the asteroids *Culcita* and *Acanthaster*. A strange association of a comatulid crinoid and trapeziid crab (Fig. 11.3) was recorded from Port Sudan (A.P. Sanders, *pers. comm.*), and ophiuroids are often found living associated with sponges and gorgonians. The pearlfish *Encheliophis* lives within the cloacae of sea cucumbers, and has been found in the Port Sudan area. It is likely that careful searches of deeper reef slope organisms may reveal many further cases of echinoderms living in association with other animals.

11.3. ORIGINS AND CHARACTER OF THE ECHINODERM FAUNA OF THE RED SEA

The Indo-West Pacific region supports the richest marine fauna in the world, much more diverse than the Atlantic at the same latitudes. Clark and Rowe (1971) list approximately 1090 echinoderms from the shallow water (i.e. down to a depth of 20 m) of the Indo-West Pacific region, and it is certain that the number of species recorded from the tropical Atlantic will be less than this. Clark (1976) refers to a

TABLE 11.2. Red Sea echinoderm fauna in relation to the rest of the Indo-West Pacific. Data for this table are from Clark and Rowe (1971) and refer only to shallow water (<20m depth) species. Percentages are rounded to the nearest whole number. Ekman (1967), taking deep water crinoid data into account found the endemism in crinoids to be 70%.

	Total IWP Shallow spp	East Indies & Malaysian	Red Sea	Endemics	% Red Sea which is endemic
Crinoids	139	81 (58%)	12 (9%)	4	33
Asteroids	229	88 (38%)	30 (13%)	4	13
Ophiuroids	290	126 (43%)	40 (14%)	3	8
Echinoids	146	80 (54%)	43 (30%)	1	2
Holothuroids	287	121 (42%)	64 (22%)	15	23
Totals	1091	496 (45%)	189 (17%)	27	14

West Indian echinoderm fauna of about 150 species. The figure of 1090 represents approximately one sixth of the total number of known living echinoderm species.

Ekman (1967) believes that the rich fauna of the Indo-West Pacific has survived from Tertiary times without catastrophe and he suggests that this inherited richness has continued into the present age along with more recently developed forms. He attributes the lower diversity of the tropical Atlantic to various factors, including distance from the East Indies and Malaysian region, the effect of barriers and catastrophes such as ocean cooling which have overtaken earlier tropical faunas. The East Indies and Malaysian region appear to be the nucleus of the Indo-West Pacific fauna, and as one moves away from this area, so the species diversity declines both to the east and the west.

The distribution of shallow water echinoderm species in some of the sub-regions of the Indo-West Pacific is given in Table 11.2, where the Red Sea, lying on the north western extremity, is compared with the East Indies and Malaysian region. The proportion of endemism in the Red Sea is small, probably reflecting its disturbed recent history (Chapter 2). Taking deep water species into account may raise the proportion of endemics, but few data are available apart from the deep water crinoids, which Ekman (1967) notes as comprising 70% of the Red Sea crinoid fauna.

Without doubt the high water temperatures and salinities occurring in the Red Sea are responsible for some specialisation of the fauna. Ekman (1967) gives data showing that below 300 m the Red Sea is the warmest marine region in the world. At 2000 m temperatures of 21.5°C have been recorded, whereas for comparable depths of the adjacent Indian Ocean 2.5°C is typical.

The Red Sea is cut off from the deep basins of the Indian Ocean by a shallow sill of about 100 m depth at its southern end. Unfortunately data on the deep water echinoderms of the Indo-West Pacific are scattered through the literature and have not been well reviewed unlike the shallow water species. Information on other deep water groups is more available, e.g. decapod crustaceans and gastropod molluscs. Ekman (1967) suggests that the deep sea decapods of the Indian Ocean are excluded from the Red Sea because they either cannot tolerate the high temperatures of the deep water there or they cannot cross the shallow sill separating the two regions, or both (see also Chapter 6). In the Red Sea itself many decapods and gastropods which are elsewhere restricted to shallow water are able to penetrate into the depths because of the favourable temperatures there, the same may be true for the echinoderms.

Red Sea echinoderms are predominantly shallow water species. Many of them are widely distributed throughout the Indo-West Pacific and the ubiquitous long spined sea urchin Diadema setosum epitomises this general distribution, but this is not to say that varieties and appearances do not change through the region. Tripneustes gratilla, a sea urchin common on lagoon floors amongst seagrasses is pale reddish brown in the central Red Sea, whilst in East Africa it is darker brown with occasional white spines. The Crown-of-Thorns starfish, Acanthaster planci, generally has more, shorter arms on the Great Barrier Reef and fewer, longer arms in the Red Sea (pers. obs.).

The various habitats of the Red Sea provide a wealth of niches for echinoderms. Table 11.3 summarises the genera recorded from the Red Sea by Clark and Rowe (1971), and notes the numbers of species known from each genus. Table 11.4, from the data given by the same authors, shows the geographical distribution within five key areas in the Red Sea. The holothurian fauna is particularly diverse, followed by the echinoids. The ophiuroid diversity at present known from the Red Sea is rather low compared with that of other regions, and the same applies to the asteroids and crinoids. These figures may reflect differences in sampling intensity rather than real faunal patterns; this is certainly demonstrated in Table 11.4, in which the heavily studied Gulf of Aqaba is seen to have the highest recorded diversity. In reality, the central and southern Red Sea probably have the richest fauna, but for these regions data are lacking for the holothurians. The Gulf of Suez has a reduced echinoderm fauna, and tends to contain species not found in the nearby Gulf of Aqaba. Of the 68 species recorded from one or both of the northern gulfs, only 20 (30%) are recorded from both. There are major hydrographic differences between the two gulfs (*see* Chapter 3) associated with the shallower depths of Suez. Fishelson (1971) suggests that prevailing southerly winds may blow inhibiting silts from the main sea into the Gulf of Suez. The reefs of Suez are very poorly developed compared with those of Aqaba (*see*

TABLE 11.3. Red Sea echinoderm genera. The figures give the number of species recorded in the Red Sea, and an asterisk indicates at least one Red Sea endemic in the genus. (Data from Clark and Rowe, 1971).

Genus	#	Genus	#	Genus	#
Crinoidea		*Amphipholis*	1	*Clypeaster*	5*
Capillaster	1	*Amphiura*	1*	*Echinocyamus*	1
Comissia	1*	*Paracrocnida*	1	*Fibularia*	2
Heterometra	2*	*Ophiactis*	3	*Laganum*	2
Lamprometra	1	*Macrophiothrix*	2	*Echinodiscus*	2
Stephanometra	2	*Ophiomaza*	1	*Echinolampas*	1
Decametra	2*	*Ophiopsammium*	1	*Maretia*	1
Oligometra	1	*Ophiothela*	1	*Palaeostoma*	1
Tropiometra	1	*Ophiothrix*	5*	*Lovenia*	1
Dorometra	1*	*Ophiocoma*	5	*Diploraster*	1
		Ophiocomella	1	*Moira*	1
Asteroidea		*Ophiomastix*	1	*Paraster*	1
Luidia	3	*Ophiopsila*	1	*Brissopsis*	1
Astropecten	6*	*Ophionereis*	2	*Brissus*	1
Ogmaster	1	*Ophioconis*	1	*Metalia*	2
Stellaster	1	*Ophiopeza*	1		
Stellasteropsis	1*	*Ophiolepis*	2	Holothuroidea	
Choriaster	1	*Ophiura*	1	*Actinopyga*	5
Culcita	1			*Bohadschia*	6*
Pentaceraster	2	Echinoidea		*Holothuria*	34*
Fromia	2*	*Eucidaris*	1	*Stichopus*	2
Gomophia	1	*Phyllacanthus*	1	*Pentacta*	1*
Leiaster	1	*Prioncidaris*	1	*Pseudocnus*	1
Linckia	2	*Asthenosoma*	1	*Trachythyone*	2*
Ophidiaster	1	*Chaetodiadema*	1	*Ohshimella*	1
Asteropsis	1	*Diadema*	1	*Phyllophorus*	1*
Asterina	1	*Echinothrix*	2	*Semperiella*	1
Acanthaster	1	*Microcyphus*	1	*Euapta*	1
Euretaster	1	*Salmaciella*	1	*Leptosynapta*	1*
Mithrodia	1	*Salmacis*	1	*Opheodesoma*	2*
Echinaster	2	*Temnopleurus*	1	*Patinapta*	2*
		Nudechinus	2	*Polyplectana*	1
Ophiuroidea		*Tripneustes*	1	*Potankyra*	1
Astroboa	2	*Parasalenia*	1	*Synapta*	1
Amphilycus	1	*Echinometra*	1	*Synaptula*	2
Amphiodia	1	*Echinostrephus*	1	*Athyone*	1
Amphioplus	5*	*Heterocentrotus*	2	*Thyone*	1*

TABLE 11.4. Geographical distribution of echinoderm species in the Red Sea. The data are from Clark and Rowe (1971), based on published accounts and specimens in collections. Coverage is very uneven, no data on holothurians are recorded for central and southern sites where they certainly occur.

	AQABA	SUEZ	NORTHERN	CENTRAL	SOUTH
Crinoidea	2	2	1	3	1
Asteroidea	10	7	5	11	6
Ophiuroidea	10	5	8	14	6
Echinoidea	12	8	12	13	12
Holothuroidea	22	10	10	?	?
Total	56	32	36	41	25

TABLE 11.5. Red Sea echinoderm habitats. Data from Clark and Rowe (1971): again by no means a complete listing, and asterisks indicate habitats in which groups are definitely present but not formally recorded.

	Crinoidea	Asteroidea	Ophiuroidea	Echinoidea	Holothuroidea	TOTAL
Seagrass beds	1	*	*	1	1	3
Lagoons		1	10	8	*	19
Mangrove swamp			2	3		5
Reefs	*	5	7	9	*	21
Live coral	*	3	3	4		10
Rock	2	11	11	9	7	40
Rubble		3	*	4	1	40
Gravel					2	2
Sand		4	3	10	4	21
Mud	2	4	1	4	3	14

Chapter 7).

A resumé of available habitat data for Red Sea echinoderm species is given in Table 11.5. Again, data are very incomplete, but it can be seen that the reefal habitats of coral reef and coastal lagoon support more species than seagrass or mangrove. In terms of substrate, rock surfaces support the largest fauna, followed by the sand and mud substrates inhabited by burrowers.

11.4. ECOLOGY OF ECHINODERMS ON CORAL REEFS

The echinoderms are a conspicuous element of the reef fauna in many parts of the tropics and this is particularly true in the Red Sea. The existence of a reef is due to two interacting processes, reef genesis and reef breakdown. Echinoderms contribute to the former only to a very modest extent, but the aggregation of their skeletal remains along with those of other organisms augments reef formation. The echinoderms contribute far more to the process of reef breakdown where their activities fall into the following categories: direct coral polyp predation, grazing and browsing on algae and calcareous invertebrates which are secondary reef contributors, drilling and boring into coral limestone and removal of organic matter from coral sand. The role of echinoderms in Red Sea reef breakdown was described by Bertram (1936).

11.4.1. Coral polyp predation

The most notorious example of this is the Crown-of-Thorns starfish *Acanthaster planci* (Fig. 11.4). This species, considered a rarity before its appearance in plague proportions in northern Australia in the early 1960s, is present in most parts of the Red Sea and was the subject of extensive research there between 1968 and 1975 by workers based at Port Sudan (*see* Ormond and Campbell, 1974). The first accurate description of its feeding habits was given by Goreau (1964), also working in the Red Sea. *Acanthaster* feeds nocturnally by everting the cardiac portion of the stomach through the mouth and spreading this over the coral colony under attack, digesting the soft tissue of the polyps *in situ*. After feeding the stomach is withdrawn leaving a characteristic white predation scar. When present in great numbers (e.g. 1000 countable in 20 minutes (Pearson, 1970)) the starfish can kill a high percentage of the corals present. Because their dead skeletons quickly become colonised by algae the nature of the reef can be changed from a coral dominated one to an algal dominated one. This change caused great concern in the Great Barrier Reef region because of its implications for the tourist industry and because of suggestions that the so called plagues of starfish might have arisen because of the effects of pollution or other human interference (Endean, 1969). Vine (1970), who worked in the Pacific and the Red Sea favoured natural causes. Much of the Red Sea work was reviewed by Ormond *et al.* (1973) and Ormond and Campbell (1974). *A. planci* seems never to have been a serious threat to corals in the Red Sea. Population levels in 1970 were generally between 5 and 44 individuals per km of reef face, but occasionally this was greatly exceeded, as on the Towartit reef complex, where densities approached 1000 per km. There is growing support for the view that *A. planci* outbreaks are periodical phenomena like many other natural population explosions and immigrations. It is interesting to note that in the Red Sea the Trumpet triton shell *Charonia tritonis* is nowhere common, and feeds on a variety of other animals. Endean (1969) has suggested that *Charonia* is an important predator of *Acanthaster* on the Great Barrier Reef, and its collection for the shell trade may have caused the celebrated *Acanthaster* plagues. Triton shells were never found attacking *Acanthaster* in natural conditions in the Red Sea although they would do so rapidly and efficiently in the aquarium. Puffer- and trigger-fish seem to be more important predators in the Red Sea (Ormond and Campbell, 1974).

There are probably other asteroids which habitually prey on coral tissue, one such is *Culcita novaeguinae* (Endean, 1976). There are several records of echinoids attacking coral polyps. *Prioncidaris baculosa* feeds on live and dead coral (James and Pearse, 1969), and Herring (1972) has found coral in the gut of *Diadema setosum*. Echinoids can seriously limit coral growth and cause extensive erosion of colonies. *Diadema setosum* is common in the Red Sea but whether it is an important coral predator there is open to question. Its behaviour at Suakin on the Sudanese coast was described by Thornton (1956). *Echinothrix diadema* and *Echinothrix calamaris* which are closely related to *Diadema* may be found in similar coral habitats and they might also be occasional coral predators, but there are no accounts of this. These sea urchins are active by night and hide if possible by day. In certain habitats such as lagoons and wharf walls where there is no shelter *D. setosum* remains in the open, often in great flocks.

11.4.2. Effects of grazing echinoids

Virtually all non sand-inhabiting reef echinoids must fall into this category although there are notable exceptions such as *Echinostrephus molaris* which is restricted to a burrowing existence in coral rock.

Yonge (1954) provided information on food recorded from the guts of certain reef dwelling species of echinoids and Lawrence (1975) has dealt fully with echinoid-algal relations.

In the central Red Sea several small echinoids populate the algal covered coral rubble of the reefs. These include *Eucidaris metularia*, *Echinometra mathaei* and *Parasalenia poehli*. The larger *Diadema setosum*, *Echinothrix diadema* and *Heterocentrotus mammillatus* are also active nocturnal browsers on algae and encrusting invertebrates. Dart (1972) used exclusion cages to demonstrate the relationship between echinoids, algal growth and invertebrate settlement in the Red Sea, studies taken further by Sammarco (1980) in the Caribbean. Sammarco found that *Diadema antillarum* exerted a strong influence on coral development as a result of its influence on coral larva settlement success. The richest coral communities and best recruitment occurred in areas where echinoid grazing pressure was at intermediate levels. Parallel studies on Red Sea *Diadema setosum* and other grazing echinoids would be of great value.

11.4.3. Drilling and boring into coral limestone

In the Red Sea two echinoids, *Echinometra mathaei* and *Echinostrephus molaris* penetrate reef limestone. *E. mathaei* may be found in several habitats but in more exposed areas it excavates shallow groove-like runs in the rock which appear to provide it with some protection against wave action. It may emerge from these to browse elsewhere but Campbell *et al.* (1973) noted that it could capture drifting particles

Fig 11.4. Crown-of-Thorns starfish *Acanthaster planci* on a Sudanese reef. The animal is emerging from its daytime hiding place in the crevice to the bottom right. The large rounded coral to the left is *Goniastrea pectinata*, a less favoured food than the *Acropora variabilis* colony to the upper right.

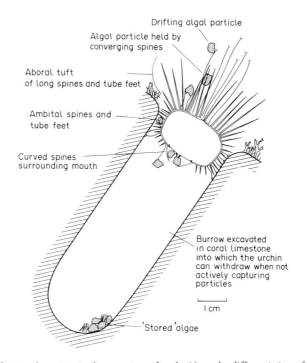

Drifting algal particle

Algal particle held by converging spines

Aboral tuft of long spines and tube feet

Ambital spines and tube feet

Curved spines surrounding mouth

Burrow excavated in coral limestone into which the urchin can withdraw when not actively capturing particles

1 cm

'Stored' algae

Fig 11.5. *Echinostrephus molaris* 'at home' in its burrow in reef rock. Note the differentiation of spines and tube feet, long aboral spines catch particles by convergence, these are passed around the test by short ambital ('equatorial') tube feet and held over the mouth by curved oral spines. Redrawn from Campbell *et al.* (1973).

of algae by converging its spines upon them and then passing them to the mouth. Near Port Sudan *Echinostrephus molaris* occurs only in burrows about 70 mm by 25 mm. It appears well adapted to burrowing life and can survive the most exposed surf washed reef crest where it collects drifting algal fragments by spine convergence (Campbell *et al.*, 1973), see Fig. 11.5. We know of nothing of the life of juvenile *E. molaris* and a study of burrow construction would be most interesting.

11.4.4. Removal of organic matter from sand

Particles of organic matter accumulated within coral sand provide a rich food source for those animals able to exploit them. Chief among these are the holothuroids which are conspicuous members of the reef associated fauna. Bakus (1973) points out their importance in moving sediments and in reworking them to eliminate stratification. Few papers discuss tropical holothurian ecology, but the work of Rowe and Doty (1977) must be mentioned as it is of value for identification. It is still not clear what niche specialisations separate the many closely related co-existing holothuroid species present in the Red Sea. Particle feeding species may collect material from suspension or from deposits. Suspension feeders need a current to bring food to them, but they may inhabit sand or rock surfaces. Roberts and Bryce (1982) noted that deposit feeders collect their food in a variety of ways using peltate, pinnate or digitate tentacles, so niche partitioning in these species may be due to distinctive feeding techniques or particle selectivity. Tropical holothuroids are probably of great importance because of their abundance and role in recycling detritus locked up in sediments, and a detailed investigation into their ecology is long overdue.

Few studies have been published on the ecology of burrowing echinoids, the clypeastroids and spatangoids. Clypeastroids use their small spines to sort edible detrital particles from the sediments; these are passed to the mouth and consumed. Spatangoids ingest the substrate and consume the contained organic matter. Again, the ecological importance of these animals is still unclear (De Ridder and Lawrence, 1982; Ghiold, 1979; Nichols, 1959).

11.4.5. Coral reefs as platforms for suspension feeders

Suspension feeders utilise currents to bring them a supply of food particles. Crinoids, some ophiuroids and some holothuroids are suspension feeders. Most of the suspension feeding echinoderms in the Red Sea are nocturnal unlike many species on the Great Barrier Reef which may be seen feeding in the day time. Comatulid crinoids and basket stars need a firm platform to hold on to whilst feeding. Live branching coral colonies and rocky out-crops serve this purpose well. These animals seek out sites where a favourable water movements occurs. Meyer and Lane (1976) note that comatulid crinoids form a planar filtration fan with the arms and pinnules held normal to the currents or wave surge. The sides of the arms bearing the particle collecting tubefeet are held facing the current. The basket stars (order Euryalea of the ophiuroids) are larger but may be mistaken by the uninitiated for enormous crinoids. *Astroboa nuda* occurs in the Red Sea and was investigated at Elat by Tsurnamal and Marder (1966). A few arms are used to adhere to the substrate while the rest form a parabolic filtration fan held vertically with the concave side facing the current (Fricke, 1968). This author noted that *Astroboa nuda* can feed in currents which are too strong for comatulid crinoids. Fricke found particle velocities of 4 to 8 cm/sec in basket star habitats, while Magnus (1963), also working in the Red Sea, found speeds of 2 to 5 cm/sec for the comatulid crinoid *Heterometra savignii*.

Meyer and Lane (1976) report that the coarse mesh feeding net of basket stars has quite large gaps in it. The flexible arm tips wave from side to side and wrap around larger objects in the 10 to 30 mm size range. Fricke (1968) found their diet to include fish larvae, chaetognaths and polychaetes. Rutman and Fishelson (1969) working at Elat on *Lamprometra klunzingeri* found 90% of a sample of food organisms from this comatulid measured less than 0.4 mm maximum dimension. Such crinoids feed on a mixture of phyto- and zooplankton, much of which must be in the size ranges utilised by some coral polyps.

The suspension feeding activity of tropical holothuroids from Aldabra has been described by Sloan (1979) who found differing sized food particles in the guts of *Afrocucumis africana* and *Holothuria cinerascens*. Sloan found *A. africana* on the rough undersides of boulders regardless of the substrate underlying the boulder. *H. cinerascens* occurred under boulders on slabrock indicating a partitioning habitat as well as food selection. Little is known of suspension feeding holothuroids in the Red Sea.

11.4.6. Non-gorgonocephalid ophiuroids on coral reefs

Any study of ophiuroids is beset with identification difficulties and this reason probably explains why the ophiuroids of the tropics have been accorded far less attention than they deserve. Their abundance and diversity means that they are likely to be important in the ecosystem. In reef habitats they abound in live coral, coral rubble, under rocks and in sediments as well as in associations with other organisms. Sloan *et al.* (1979) give useful accounts of ophiuroids from Aldabra including a number of species which occur in the Red Sea. Fishelson (1971), Hughes (1977) and Magnus (1967) should be consulted for

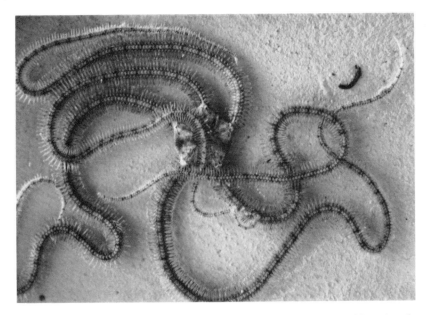

Fig 11.6. *Macrophiothrix demessa* from Port Sudan. Ophiuroids like this with long arms and large lateral spines may make considerable use of mucus suspension feeding.

details of Red Sea ophiuroids. Sloan (1982) investigated the size and structure of echinoderm populations associated with corals in Aldabra where he found ophiuroids dominating the coral associated echinoderm fauna.

11.5. ECHINODERM ECOLOGY IN LAGOONS AND SEAGRASS BEDS

Along much of its coastline the fringing reefs of the Red Sea are separated from the shore by lagoons of varying width. Frequently the fringing lagoon is shallow (less than 3 m deep in the central region). The lagoon floor may be covered by coral sand rich in detritus and by outcrops of rubble, rock and live coral (*see* Chapter 7). The lagoon is an extreme environment where temperatures and salinities are frequently much higher than in the open sea, and for most part they are well sheltered with little current. Such calm conditions encourage predators such as stingrays and eagle rays, and a variety of wrasses. The lagoon fauna therefore needs to be well protected in order to survive, or to be cryptic or burrow. The lagoon floor is frequently marked by burrow openings and spoil heaps indicating that this last strategy is a popular one. Many burrowing ophiuroids, echinoids and holothuroids are found there pursuing lifestyles of suspension feeding, deposit feeding, carrion feeding or carnivory. Prominent lagoon echinoderms include *Linckia multifora* (cryptically marked), *Diadema setosum* (spiny armament), *Clypeaster* species and synaptid holothuroids (burrowers).

Seagrasses occur in lagoons (*see* Chapter 9) and provide a ready source of food for herbivorous echinoids such as *Tripneustes gratilla* as well as a massive detrital input arising from their rotting leaves. *T. gratilla* is not restricted to seagrass beds and has been recorded in the open sea reefs of the Port Sudan region (*pers. obs.*) as well as from the reef flats of the northern Red Sea (James and Pearse, 1969). This species has been used by Dafni (1980) in his work on abnormal growth patterns under the effects of pollution.

11.6. MUDDY SUBSTRATES AND MANGROVE

Fishelson (1971) investigated a range of habitats in the Red Sea. Muddy situations harboured either surface forms able to utilise detritus like some holothuroids, or burrowing ophiuroids, echinoids and holothuroids. The clypeastroids such as *Clypeaster humilis* and *C. fervens* occur in the silty lagoon floors of the central Red Sea. Fishelson also noted the occurrence of large regular echinoids in the mangrove swamps where a suitable substrate occurred for them to move over and presumably graze on.

11.7. CONCLUSIONS

The foregoing account shows that the echinoderm fauna of the Red Sea is comparatively well known within the invertebrates. This is due to the abundance and accessibility of these animals in this region. However, echinoderms as a phylum are difficult to work on, and many aspects of their biology are poorly understood, particularly their physiology and behaviour. The Red Sea provides an ideal tropical environment for the research worker, with a wide range of habitats in relatively easy reach of land. This could be exploited by scientists a great deal more than it is and the advantages to echinoderm biologists and reef biologists would be great. Particularly studies on sociobiology of echinoderms, reproductive physiology and erosive influences would bear fruit. Little is known about echinoderm venoms, sensory physiology and specialised activities such as burrowing and boring. The wide range of Red Sea echinoderm life styles would allow many of these topics to be investigated.

REFERENCES

Achituv, Y. (1973). The genital cycle of *Asterina burtoni* Gray (Asteroidea) from the Gulf of Eilat, Red Sea. *Cah. Biol. mar.* 14, 547–53.

Bakus, G. J. (1973). The biology and ecology of tropical holothurians. In *Biology and Geology of Coral Reefs 2, Biology*. Eds. O. A. Jones and E. Endean, Vol.1, pp. 325–67. Academic Press, New York.

Barnes, R. D. (1980). *Invertebrate Zoology*, Saunders College, Philadelphia.

Bertram, G. C. L. (1936). Some aspects of the breakdown of coral at Ghardaqa, Red Sea. *Proc. zool. Soc. Lond.* 106, 1011–26.

Binyon, J. (1972). *Physiology of Echinoderms*, Pergamon Press, Oxford.

Boolootian, R. A. (ed.) (1966). *The Physiology of Echinodermata*. Wiley Interscience, New York.

Brown, T. (1972). *Crown of Thorns, The Death of the Great Barrier Reef?* Angus & Robertson, Sydney.

Campbell, A. C. (1971). Echinoderms of the Red Sea. In *New Studies on the Coral Predating Crown of Thorns Starfish*. Eds. C. H. Roads & R. F. G. Ormond, pp. 72–84. Cambridge Coral Starfish Research Group.

Campbell, A. C. (1983). Form and function of pedicellariae. *Echinoderm Studies* 1, 139–67.

Campbell, A. C., Dart, J. K. G., Head, S. M. and Ormond, R. F. G. (1973). The feeding activity of *Echinostrephus molaris* (de Blainville) in the central Red Sea. *Mar. Behav. Physiol.* 2, 155–69.

Chadwick, H. C. (1908). Reports on the marine biology of the Sudanese Red Sea VII. The Crinoidea. *J. Linn. Soc. Lond.* 31, 44–7.

Cherbonnier, G. (1967). Deuxième contribution à l'étude des Holothuries de la Mer Rouge collectées par des Israelians. *Bull. Sea Fish. res. Stn. Israel* 43, 55–68.

Clark, A. M. (1952). The Manihine Expedition to the Gulf of Aqaba VII. Echinodermata. *Bull. Brit. Mus. (nat. Hist.) Zool.* 1, 203–14.

Clark, A. M. (1967). Echinoderms from the Red Sea. Part 2. Crinoids, Ophiuroids, Echinoids and more Asteroids. *Bull. Sea Fish. res. Stn. Israel* 41, 26–58.

Clark, A. M. (1976). Echinoderms of Coral Reefs. In *Biology and Geology of Coral Reefs 3, Biology*. Eds. O. A. Jones and R. Endean, vol.2, pp. 95–123. Academic Press, New York.

Clark, A. M. and Rowe, F. W. E. (1971). *Monograph of the Shallow Water Indo West Pacific Echinoderms*, 1–238 pp. Brit. Mus. (Nat. Hist.), London.

Crump, R. (1971). Reproductive condition of *Acanthaster planci* in the Port Sudan area of the Red Sea. In *New Studies on the Coral Predating Crown of Thorns Starfish*. Eds. C. H. Roads and R. F. G. Ormond, pp. 50–2. Cambridge Coral Starfish Research Group.

Dafni, J. (1980). Abnormal growth patterns in the sea urchin *Tripneustes gratilla* (L) under pollution (Echinodermata, Echinoidea). *J. exp. mar. Biol. Ecol.* 47, 259–79.

Dart, J. K. G. (1972). Echinoids, algal lawn and coral recolonisation. *Nature, Lond.* 239, 50–1.

De Ridder, C. and Lawrence, J. M. (1982). Food and feeding mechanisms Echinoidea. In *Echinoderm Nutrition*. Eds. M. Jamgoth and J. M. Lawrence, pp. 57–115. A. A. Balkema, Rotterdam.

Ekman, S. (1967). *Zoogeography of the Sea*. Sidgwick and Jackson, London.

Endean, R. (1969). Report on investigations made into aspects of the current *Acanthaster planci* (Crown of Thorns) infestation of certain reefs of the Great Barrier Reef. *Fisheries Branch Queensland Department of Primary Industries, Brisbane*, 35pp.

Endean, R. (1976). Destruction and recovery of coral reef communities. In *Biology and Geology of Coral Reefs 3, Biology*. Eds. O. A. Jones and R. Endean, vol.2, pp. 215–54. Academic Press, New York.

Fishelson, L. (1971). Ecology and distribution of benthic fauna in shallow waters of the Red Sea. *Mar. Biol.* 10(2), 113–33.

Fricke, H. W. (1968). *Beiträge zur Biologie der Gorgonenhaupter*. Astrophyton muricatum (*Lamarck*) und Astroboa nuda (*Lyman*) (*Ophiuroidea, Gorgonocephalidae*). Ernst Reuter Gesellschaft, Berlin.

Ghiold, J. (1979). Spine morphology and its significance in feeding and burrowing in the Sand Dollar *Mellita quinquiesperforata* (Echinodermata Echinoidea). *Bull. mar. Sci.* 29, 481–90.

Goreau, T. F. (1964). On the predation of coral by the Spiny Starfish *Acanthaster planci* (L.) in the Southern Red Sea. *Sea Fish res. Stn. Haifa Bull.* 35, 23–6.

Herring, P. J. (1972). Observations on the distribution and feeding habits of some littoral echinoids from Zanzibar. *J. nat. Hist.* 6, 169–75.

Hughes, R.N. (1977). The biota of reef-flats and limestone cliffs near Jeddah, Saudi Arabia. *J. nat. Hist.* 11, 77–96.

Hyman, L. H. (1955). *The Invertebrates Vol. IV Echinoderms*. McGraw Hill, New York.

James, D. B. and Pearse, J. S. (1969). Echinoderms from the Gulf of Suez and Northern Red Sea. *J. mar. biol. Ass. India* 11, 78–125.

Jangoux, M. and Lawrence, J. M. (eds.) (1982). *Echinoderm Nutrition*. A. A. Balkema, Rotterdam.

Kenchington, R. A. (1976). *Acanthaster planci* on the Great Barrier Reef detailed surveys for four transects between 19° and 20°S. *Biol. Conserv.* 9, 165–79.

Lawrence, J. M. (1975). On the relationships between marine plants and sea-urchins. *Oceanogr. mar. Biol. A. Rev.* 13, 213–86.

Lucas, J. S. (1973). Reproductive and larval biology of *Acanthaster planci* (L.) in Great Barrier Reef waters. *Micronesica* 9(2), 197–203.

Magnus, D. B. E. (1963). Der Federstern *Heterometra savignyi* in Roten Meer. *Nat. Mus. Frankfurt* 93(9), 355–68.

Magnus, D. B. E. (1967). Ecological and ethological studies and experiments on the echinoderms of the Red. Sea. *Stud. trop. Oceanogr.* 5, 635–64.

Meyer, D. L. and Lane, N. G. (1976). The feeding behaviour of some paleozoic crinoids and recent basket stars. *J. Paleontol.* 50(3), 472–80.

Mortensen, T. (1926). Cambridge expedition to the Suez Canal, 1924. VI. Report on the Echinoderms. *Trans. zool. Soc. Lond.* 22, 117–31.

Mortensen, T. (1931). Contributions to the study of the development and larval forms of echinoderms. I and II. *Kgl. Dan. Vidensk. Selsk., Skr. Naturvid. Math. Afd.* (9), 4(1), 1–39.

Mortensen, T. (1937). Contributions to the study of the development and larval forms of echinoderms. III. *Kgl. Dan. Vidensk. Selsk., Skr. Naturvid. Math. Afd.* (9), 7(1), 1–65.

Mortensen, T. (1938). Contributions to the study of the development and larval forms of echinoderms. IV. *Kgl. Dan. Vidensk. Selsk., Skr. Naturvid. Math. Afd.* (9), 7(3), 1–59.

Nichols, D. (1959). Mode of life and taxonomy in irregular sea-urchins. *Systematics Association Publ. 3. Function and Taxonomic Importance*, pp. 61–80.

Nichols, D. (1969). *Echinoderms*. Hutchison University Library, London.

Ormond, R. F. G. and Campbell, A. C. (1974). Formation and breakdown of *Acanthaster planci* aggregations in the Red Sea. *Proc. 2nd Int. Coral Reef Symp.* 1, 596–619.

Ormond, R. F. G., Campbell, A. C., Head, S. M., Moore, R. J., Rainbow, P. S. and Sanders, A. P. (1973). Formation and breakdown of aggregations of the crown-of-thorns starfish *Acanthaster planci* (L.). *Nature, Lond.* 246, 167–9.

Pearse, J. S. (1968). Patterns of reproduction in four species of Indo-Pacific echinoderms. *Proc. Indian Acad. Sci.* Ser. B 67, 247–9.

Pearse, J. S. (1969a). Reproductive periodicities of Indo-Pacific invertebrates in the Gulf of Suez. I. The echinoids *Prioncidaris baculosa* (Lamarck) and *Lovenia elongata* (Gray). *Bull. mar. Sci.* 19(2), 323–50.

Pearse, J. A. (1969b). Reproductive periodicities of Indo-Pacific invertebrates in the Gulf of Suez. II. The echinoid *Echinometra mathaei* (de Blainville). *Bull. mar. Sci.* 19(3), 580–613.

Pearson, R. (1970). Studies on the Crown-of-Thorns Starfish on the Great Barrier Reef. *Queensland Littoral Society Newsletter* 36, 1–10.

Polunin, N. V. C. (1971). Symbolic and commensal decapods collected in the Sudanese Red Sea. In *New Studies on the Coral Predating Crown-of-Thorns Starfish*. Eds. C. H. Roads and R. F. G. Ormond, pp. 87–9. Cambridge Coral Starfish Research Group.

Roberts, D. and Bryce, C. (1982). Further observations on tentacular feeding mechanisms in Holothurians. *J. exp. mar. Biol. Ecol.* 59, 151–63.

Rowe, F. W. E. and Doty, J. E. (1977). The shallow-water Holothurians of Guam. *Micronesica* 13(2), 217–50.

Rutman, J. and Fishelson, L. (1969). Food composition and feeding behaviour of shallow-water crinoids at Eilat (Red Sea). *Mar. Biol.* 3, 45–57.

Sammarco, P. W. (1980). *Diadema* and its relationship to coral spat mortality grazing, competition and biological disturbance. *J. exp. mar. Biol. Ecol.* 45, 245–72.

Sloan, N. A. (1979). Microhabitat and resource utilisation in cryptic rocky intertidal echinoderms at Aldabra Atoll, Seychelles. *Mar. Biol.* 54, 269–79.

Sloan, N. A. (1982). Size and structure of Echinoderm populations associated with different co-existing coral species at Aldabra Atoll, Seychelles. *Mar. Biol.* 66, 67–75.

Sloan, N. A., Clark, A. M. and Taylor, J. D. (1979). The echinoderms of Aldabra and their habitats. *Bull. Brit. Mus. (nat. Hist.) Zool.* 37(2), 81–128.

Thornton, I. W. B. (1956). Diurnal migrations of the echinoid *Diadema setosum* (Leske). *Brit. J. anim. Behav.* 4(4), 143–6.

Tsurnamal, M. and Marder, J. (1966). Observations on the basket star *Astroboa nuda* (Lyman) on coral reefs at Eilat (Gulf of Aquaba). *Israel J. Zool.* 15, 9–17.

Vine, P. F. (1970). Densities of *Acanthaster planci* in the Pacific Ocean. *Nature, Lond.* 228, 341–2.

Wickler, W. Von and Seibt, V. (1970). Das verhalten von *Hymenocera picta* Dana, einer seesterne fressenden gamele (Decapoda, Natantia, Gnathophyllidae). *Z. Tierpsychol.* 27, 352–68.

Yamaguchi, M. (1973). Early life histories of coral reef asteroids with special reference to *Acanthaster planci* (L.). In *Biology and Geology of Coral Reefs 2, Biology*. Eds. O. A. Jones and R. Endean, vol.1, pp. 369–87. Academic Press, New York.

Yonge, C. M. (1954). Feeding mechanisms in the invertebrates. *Tabulae Biologicae* 21, 3 and 4. Nos. 21 and 22, 25–68.

CHAPTER 12

Minor Invertebrate Groups

STEPHEN M. HEAD

Zoology Department, University of the West Indies, Kingston 7, Jamaica

CONTENTS

12.1. INTRODUCTION AND REVIEW OF PHYLA

The Red Sea occupies a surface area of over 400,000 square kilometres, is over 2000 m deep in places, and has a long shoreline by virtue of its linear shape. Within its tropical waters are a great variety of habitats, deep and shallow, sandy and rocky, reef and shelf, clear and murky, exposed and sheltered. Onshore are a variety of intertidal and shallow subtidal environments, including seagrass and mangrove communities. These factors combine to suggest that the Red Sea faunal list should be large and diverse. Mitigating against this might be the relative isolation of the Red Sea from the Indo-Pacific in general, its rather low productivity and chequered recent geological history, topics discussed elsewhere in this book. No complete faunal list is yet available for the Red Sea, or is likely to be so for many years. Amirthalingam's (1970) list is a useful source of references, but very incomplete in most taxa recorded.

Table 12.1 lists the phyla and some of the classes in the animal kingdom, and indicates those which are known to inhabit the Red Sea, those which probably do so, and some introductory references dealing with the Red Sea fauna. Many references are very old, and some may have been supplemented by recent studies inaccessible to the present author. In many cases the reference given may not be the most substantial study on the subject, but a more recent work with bibliography giving access to the available literature. For the larger and more important taxa, other chapters within this volume give substantial accounts and bibliographies. Useful general sources of information on Red Sea organisms include Jones' (1971) valuable Indian Ocean Bibliography, Amirthalingam (1970), Ferber (1977) and Paldi (1969a,b, 1971). The geographical indices in Zoological Record represent a valuable and rapid source of information.

The remaining sections of this chapter illustrate in a little more detail the ecology of some of the more important 'minor' invertebrate groups living within the Red Sea. For a general background to these and the taxa listed on Table 12.1, the reader is referred to Barnes' (1980) excellent textbook. There is as yet no general account of Indian Ocean benthic invertebrates available, but readers may find Colin's (1978) Caribbean field guide a useful introduction to tropical reefal organisms, although naturally the fauna is quite different at the species level. For the Red Sea, some invertebrates are discussed and figured within Bemert and Ormond (1981), the best general reviews remain Fishelson (1971) on the benthic communities, and Halim (1969) for plankton.

12.2 PROTOZOA

Protozoa of many groups exist in all marine habitats around the Red Sea including plankton and benthos. A group much in need of attention is the ciliates which are abundant as epiphytic organisms on algal turf and as interstitial fauna within the voids between sand grains. Ciliates are important detritivores in marine environments, but have received little study in reefal environments. Other ciliates are associated with other organisms as ectocommensals; this habit is particularly common on mucus-secreting cnidarians and a study of their distribution on Red Sea corals and soft corals might be interesting.

The most widely studied benthic protozoans are members of the Sarcomastigophora, Order Foraminiferida. These animals are related to the familiar amoeba, and feed on small particles using long reticulopodia, extensions of the cell surface. Forams live enclosed within a calcium carbonate shell, often bearing successive sculptured growth chambers of great beauty. The shells often contribute substantially to marine sediment deposits, and they can always be encountered when sorting through coral reef sand. Descriptions of Red Sea Foraminiferida include Said (1949, 1950a,b), Bahafzallah (1979) and Reiss (1977).

Although most forams are free living, some are sessile, and the Red Sea *Homotrema rubrum* is a common and conspicuous example. *Homotrema* forms irregular bright red skeletal masses up to 5 mm diameter on the underside of coral colonies. Its skeletal debris is very obvious in coral sediments, and has been used as a sediment transport indicator, since the animal lives only in relatively shallow water (Mackenzie *et al.*, 1965).

Many protozoans of the Sarcomastigophora are abundant in Red Sea plankton. For a general account the reader is referred to chapter 6 in this book, and to the review by Halim (1969), from which references to the literature can be obtained.

TABLE 12.1. Animal phyla and selected classes of the animal kingdom, with references to the Red Sea fauna and brief notes on their ecology. Groups definitely or probably not present are shown in brackets.

Phylum	Subphylum/Class	Reference	Notes
Protozoa	Sarcomastigophora	*See below*	Important benthic and planktonic types.
	Ciliophora	—	Present, in sand, on rocks, algae.
	Sporozoa +) Cnidospora)	—	Probably present, parasitic.
(Placozoa)		—	Not recorded.
Porifera		*See below*	Abundant and important.
Cnidaria	Hydrozoa	*See below*	Important, calcified or soft spp.
	Scyphozoa	*See below*	Common in plankton, 1 benthic.
	Anthozoa	*See below*	Vastly important, see Chapter 7.
Ctenophora		—	Present, common in plankton.
Platyhelminthes		Palombi (1928)	Benthic carnivores, some can swim.
Mesozoa		—	Probably present, parasitic.
Nemertini		Gibson (1974)	Understudied. Burrow in sand, porous rock. Predatores, detritivores.
Gnathostomulida		—	Probably present, sand meiofauna.
Rotifera		—	Mostly fresh water.
Gastrotricha		—	Almost certainly present, very small meiofaunal animals.
Kinoryncha		—	Present, sand meiofauna.
Nematoda		Betz and Otte (1980)	Important meiofauna and parasites.
(Nematomorpha)		—	Probably absent, only 1 marine genus (*Nectonema*)
Acanthocephala		Schmidt and Paperna (1978)	Present as fish parasites.
(Priapulida)		—	Cold water animals.
Sipunculida		Murina (1971) + Stephen, (1965)	Borers in coral rock and in sand.
Mollusca		Chapter 10 and Taylor and Reid (1984)	See chapter 10 for details.
Echiura		Stephen, (1965)	Live in porous rock or in stable sand.
Annelida	Polychaeta	*See below*	Very abundant in all habitats.
	Oligochaeta	—	Probably present, small benthic.
	Hirudinea	—	May be present, fish ectoparasites.
Pogonophora		Ivanov and Gureeva (1980)	Present only in deep water.
Tardigrada		—	Probably present as meiofauna.
(Onychophora)		—	Purely terrestrial.
Anthropoda	Crustacea	*See below*	Very abundant in all habitats.
Phoronida		Fishelson (1970)	Present, underinvestigated.
Bryozoa		*See below*	Important benthic animals on rock and under live corals.
Entoprocta		Waters, (1910)	Benthic, on rocks and commensal.
Brachiopoda		Jackson et al (1971)	Present, significant in shaded caves, probably in deep water.
Chaetognatha		Ducret (1973)	Important predators in plankton.
Echinodermata		*See* Chapter 11	Very important benthic animals.
Hemichordata		Klunzinger (1902)	Present, sand macrofauna.
Chordata	Urochordata	Pérès (1960) + Godeaux (1973) + Halim (1969)	Quite important cryptic benthic and planktonic filter feeders.
	Cephalochordata	Fishelson (1971)	Important in sand.
	Vertebrata	*See* chapters 13, 14, 15 and 17.	Abundant and important.

12.3. PORIFERA — SPONGES

The sponges are very simple organisms compared with other multicellular animals, and until recently have been rather neglected by reef ecologists. They are however extremely important ecologically, and many species occur in the Red Sea. Unfortunately sponge taxonomy is difficult even for the expert, and it may be some while before the sponges are routinely surveyed as part of community descriptions. Studies such as Mergner and Schuhmacher (1981) in which the sponges are included within a general ecological description remain the exception. Row (1909, 1911) published extensive monographs on the Red Sea sponges of The Sudan. Among recent general studies may be mentioned Levi (1965) and Wilkinson (1980a). Wilkinson listed 21 common Sudanese sponges from coastal and offshore reefs, and gave notes on their distribution. Figure 12.1 illustrates in a very simple fashion the variety of habitats and roles of sponges on coral reefs.

Sponges are important in tropical environments for several reasons. They are primarily filter-feeders, and specialise in the smallest sizes of particles, often below the limit of resolution of the optical microscope (Reiswig, 1971). A very high proportion of the total available seston in tropical waters is within this size class, and the sponges probably play a vital role in trapping this material and preventing its loss from the reef community. Many sponges, like corals, contain symbiotic algal cells and are at least partly autotrophic. Wilkinson (1980b) lists seven Red Sea sponges found to contain significant numbers of blue-green algae (Cyanobacteria), and demonstrated their ability to fix carbon and transfer it to the sponge host. Not all reef sponges have this ability. While those with cyanobacteria are found in shallow, well-lit conditions, many other species are cryptic or hide from light in dark caves, possibly avoiding competition from light-utilising organisms. Vasseur (1974) listed no less than 65 cryptic sponge species from Madagascan reefs.

The symbiotic cyanobacteria within sponges have recently been shown to be able to fix molecular nitrogen. The studies were conducted in the Red Sea, and two symbiotic species tested (*Siphonochalina tabernacula* and *Theonella swinhoei*) had the ability (Wilkinson and Fay, 1979). *S. tabernacula* is a

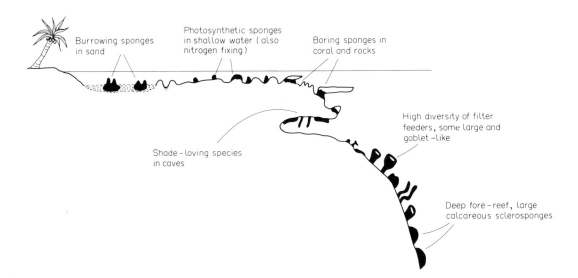

Fig. 12.1. Simple diagram to suggest some of the habitats typical of Red Sea sponges, not drawn to scale. Encrusting and small tubular sponges are found on all the hard substrates of the fringing reef illustrated.

particularly common and prominent shallow water species. Tropical waters are notoriously low in nutrients, and the ability to fix nitrogen as well as carbon dioxide could be of great adaptive value to a reef sponge. Nitrogen fixation by reef organisms is becoming more widely recognised. Capone and Taylor (1977) for example recorded significant fixation by cyanobacteria on seagrass leaves, providing nearly 25% of the total seagrass nitrogen requirement.

The extent to which the trapped nutrients, fixed carbon and nitrogen within sponges are available to the rest of the community is not clear. Few organisms eat sponges, which are gritty and often evil tasting, containing specific antibiotics and toxic chemicals (Bergquist, 1978). Hiatt and Strasburg (1960) found evidence of sponge feeding in only two fish families, the filefishes (Monacanthidae) and the pufferfishes, (Tetraodontidae). Surprisingly, the Hawksbill turtle eats sponges (Witzell, 1983), and the author has observed Red Sea asteroids (*Fromia* and *Gomorphia*) with stomachs everted on encrusting sponges. Perhaps because of their unpalatability, sponges form an important habitat and retreat for many small animal species, especially fish, ophiuroids and crustaceans, so they contribute materially to the spatial heterogeneity and hence the diversity of reef systems.

Sponges compete for space with corals and other reef organisms. They are usually fairly insignificant space occupiers in shallow water but can become much more important in deeper water and shaded caves, penetrating well below the depth range of scleractinian corals. In deep water the recently discovered sclerosponges (Hartman and Goreau, 1970) are important reef frame builders, secreting extremely dense calcium carbonate skeletons. It may be assumed that sclerosponges are present in deep water in the Red Sea; Bemert and Ormond (1981) mention their presence, but they were not noted by Fricke and Schuhmacher (1983) following deep submersible dives in the northern Red Sea. At such depths in Jamaica, sclerosponges are very prominent benthic organisms.

Most sponges are found fixed on the surface of rocky substrates, but a few types inhabit the substrate itself, or live within sand. As an example of the latter the common Red Sea sponge *Biemna* is found in sand on fringing and patch reefs, with most of the 'body' buried and with tubular siphons extending to the sediment surface.

Sponges of the genus *Cliona* live within rock or coral substrates, boring by chemical action and creating significant quantities of fine sediment (Futterer, 1974). *Cliona* is a major factor in the breakdown of reef rock (Otter, 1931), and by burrowing into the dead areas of coral skeletons can seriously weaken the structure and cause the death of the coral by fracture (Chamberlain, 1978). While this obviously causes coral mortality, the new substrate created is available for colonisation and this may increase the local diversity of the community.

12.4. CNIDARIA, HYDROZOA AND SCYPHOZOA

Although Red Sea reefs are extremely well populated by cnidarians of the class Anthozoa (*see* Chapter 7 and below), representatives of the Hydrozoa are also important. Pre-eminent among these are the hydrocorals of the genus *Millepora*. These are similar ecologically to the scleractinian corals, and like them contain abundant symbiotic zooxanthellae, and secrete substantial calcium carbonate skeletons. The polyps of *Millepora* are however minute in comparison, and of two main types. Feeding polyps with mouths and tentacles (gastrozooids) are surrounded by a ring of five or six even more slender defensive polyps called dactylozooids, which can deliver an unpleasant sting rather worse than severe nettle rash.

There are three important *Millepora* species in the Red Sea, fan-like *M. dichotoma* (*see* Chapter 7, Fig. 6), *M. platyphylla* (Figure 12.2) which is heavily ridged, and the encrusting *M. exaesa*. Mergner (1977) has shown that the first two species are indicators of bright illumination and strong water currents. Head (1980) has designated a *Millepora* Subassociation on Red Sea reefs, characteristic of very shallow

Fig. 12.2. Colonies of *Millepora platyphylla* growing on an exposed reef crest at about 2 m depth in The Sudan. The sharp ridges fusing to give a somewhat honey-combed effect are characteristic of this species which is only found in exposed areas.

high energy reef environments. Boschma (1968) has reviewed the Red Sea hydrocorals, he mentions a fourth species (*M. tenera*) but this has not been recorded by recent field workers.

In addition to *Millepora* (sub-order Milleporina), another hydrocoral sub-order is found in the Red Sea, in the form of the stylasterine species *Distichophora violacea*. This non-symbiotic colonial animal is generally cryptic, occurring in caves or under overhangs. The skeleton, generally fan-branched in one plane recalling *M. dichotoma*, is a beautiful violet-blue or occasionally pink colour.

The Red Sea contains a rich fauna of soft hydroids, reviewed by Mergner and Wedler (1977). They list 46 species from 28 genera. Some, such as *Gymnangium eximium*, are extremely photophobic in contrast to the milleporines, but most share the preference for good water exchange, which brings in a supply of planktonic food.

The planktonic hydromedusae of the Red Sea were briefly reviewed by Halim (1969), and some data on scyphozoans can be found in Stiasny (1938). The most interesting scyphozoan is however the unusual *Cassiopea* or 'upside-down jellyfish', common in fringing reef lagoons on sandy bottoms (Gohar and Eisawy, 1961). This attractive animal reclines upside-down on the substrate, exposing its oral lobes to full sunlight. The lobes contain large numbers of symbiotic zooxanthellae, and the symbiosis as a whole is a net autotroph (Drew, 1972). The animal is not fixed to the sand, and can swim in a leisurely but conventional fashion if provoked.

12.5. CNIDARIA, ANTHOZOA

Most scientific attention has been focussed on the abundant Scleractinia of the Red Sea (Chapter 7), but several other orders are also important. The class Anthozoa is divided into two subclasses, the octoradiate Alcyonaria, and the hexaradiate Zoantharia.

Fig. 12.3. Sheltered back-reef environment in The Sudan, showing the substrate at about 1 m depth completely dominated by fleshy colonies of the alcyonacean *Lithophyton* sp.

12.5.1. Order Alcyonacea, the soft corals

Alcyonaceans or soft corals are colonial organisms rather similar in many respects to scleractinians, but lacking the hard skeletal support of the true corals. Instead, the colony is very soft or rubbery to the touch. Alcyonaceans secrete calcium carbonate in the form of spicules embedded within their tissue; in some areas these contribute substantially to reef sediment production. Most spicules are very small but in some genera such as the common Red Sea *Dendronephthya*, they are up to one centimetre long and easily visible through the tissue mass.

Soft corals are very important and prominent organisms on the reefs of the Red Sea. They occupy a high proportion of the substrate in some reef zones (Figure 12.3), but their cover is usually much greater in shallow water than deeper down (15 m) where they are outranked by the scleractinians (Benayahu and Loya, 1981). Many soft corals contain symbiotic zooxanthellae, and in some cases seem to rely on them completely for nutrition. Gohar (1940) was unable to elicit any reactions to animal food from the xeniid soft corals he studied, and they rapidly deteriorated if kept in darkened aquaria. Svoboda (1978) recently confirmed this finding by measuring the oxygen production of Red Sea alcyonaceans, which proved net autotrophs to 10 or 20 m depth.

Several authors have noticed the importance of soft corals, particularly species of *Xenia* (Figure 12.4), in shallow, low predictability reef flat areas where wave action can be intense. Benayahu and Loya (1981) have shown that colonies of *Xenia macrospiculata* can migrate over the substrate, making the animal well adapted to colonise newly cleared surfaces. Other evidence suggests that alcyonaceans can overgrow and smother scleractinian colonies; clearly, the ecology of these interesting animals needs more study.

Many papers deal with the taxonomy of Red Sea alcyonaceans, a difficult task even for the specialist since the colonies are hard to preserve in a life-like state, and much of the taxonomy is based on details of spicule form. Indeed one biologist is rumoured to have sent two halves of the same alcyonacean colony on separate occasions to a prominent taxonomist, to be informed that they belonged to different genera!

Let us hope this story is apocryphal. Useful papers include Klunzinger (1877), and more recently Verseveldt (1965, 1974) and Verseveldt and Benayahu (1978). A total of about 60 species of alcyonaceans are known from the Red Sea, but it is possible that more collections from deeper water could increase this number substantially. The best general accounts of their ecology are Fishelson (1970) and Benayahu and Loya (1981).

12.5.2. Other alcyonarians

In addition to the Alcyonacea, two other orders of alcyonarians are important in Red Sea reefs. *Tubipora musica* is the only representative of the order Stolonifera. This organism has the common name of 'organ-pipe coral', because of the characteristic long parallel tubes of hard endoskeleton which it secretes for support. The skeleton is an attractive ruby red colour, against which the white tentacles of the polyps stand out in sharp contrast. *Tubipora* is common on Red Sea reefs, from the reef top down to at least 50 m depth (Head, 1980). It can be particularly abundant and form large colonies in shallow lagoons behind offshore fringing reefs.

Gorgonians or fan corals belong to the order Gorgonacea. Their colonies are usually tree-like or take the form of flat fans, and are supported by an extremely tough axial skeleton of tanned protein and calcareous spicules. This gives them support against strong water currents allowing them to feed on zooplankton. Gorgonians require hard substrates, and are usually most common on steep reef slopes or submarine cliffs on exposed fringing reefs where water energy is high. Schuhmacher (1973) has given an interesting account of the hydromechanical adaptations of two Red Sea gorgonians of the genus *Acabaria* on the reefs of Elat. General accounts of the Red Sea gorgonian fauna include Hickson (1940) and Stiasny (1959).

Fig. 12.4. Colony of a Red Sea *Xenia* species, showing the fleshy columns and many tentaculate polyp 'heads'. In these soft corals the tentacles are continuously opening and closing over the oral region, the reasons for this behaviour are unknown.

It is interesting to note that in the Red Sea and the Indo-Pacific Ocean in general, gorgonians are much less conspicuous than alcyonaceans on reefs. In the Caribbean on the other hand, the alcyonaceans are practically absent, and gorgonians are a very important part of the reef community (Kinzie, 1973).

12.5.3. Order Actiniaria, the sea anemones

The indefatigable Klunzinger (1877) published an important monograph on the soft corals and anemones of the Red Sea, and this is still a valuable reference. More recently Fishelson (1970) has described the sea anemones of Elat and their ecology, in a well illustrated paper. Fishelson recorded 22 species from a variety of habitats. On rocks, the large anemone *Gyrostoma helianthus* is conspicuous, and is almost always host to symbiotic anemone fish, especially the clown-fish *Amphiprion bicinctus* (Fishelson, 1964). Another common shallow water species is the large *Cryptodermum adhesivum*, which has very short sticky tentacles, and which withdraws rapidly into holes in the reef rock when disturbed.

Many anemones burrow within sand, and some, such as *Megalactis* are markedly nocturnal. *Radianthus koseirensis* is a burrowing species with long tentacles, often host to juvenile anemone fish, the adults of which inhabit *Gyrostoma*. Like so many other cnidarians, many of the reef anemones are found to contain symbiotic algal cells and to supplement their zooplankton feeding with photo-autotrophy.

12.5.4. Other zoantharians

In addition to the sea anemones and scleractinian corals, the subclass Zoantharia includes four orders, Zoanthidea, Corallimorpharia, Ceriantharia and Antipatharia, with Red Sea members. The ecology of corallimorpharians and cerianthids is in most respects similar to that of the sea anemones. Corallimorpharians such as *Corynactis* and *Rhodactis* live on rocks, while the beautiful *Cerianthus* species burrow deep within muddy sediments inside slime tubes.

The zoanthid genera *Zoanthus* and *Palythoa* are pan-tropical in distribution, forming dense mat-like colonies covering rocks in shallow high water energy areas. Zoanthids contain zooxanthellae, but do not rely on autotrophy alone, being active plankton feeders (Sebens, 1977). Zoanthids have been shown to be extremely fast growing and aggressive colonists, with the ability to overgrow and kill nearly all other reef organisms (Suchanek and Green, 1981).

Antipatharians are the black corals which form large arborescent colonies in deep water. Like gorgonians, they have an extremely tough protein skeleton, which can be polished and used in jewelry. Little is known of the biology of antipatharians, the best general ecological account of the group is probably Warner (1981). Black corals grow very slowly, and are only of commercial value when the colony is large, so urgent conservation measures have to be taken when a local population comes under commercial pressure.

12.6. ANNELIDA, POLYCHAETE WORMS

Many types of 'worm' occur in the Red Sea, including members of the Annelida, Nemertini, Platyhelminthes, Sipunculida and Echiura. With the exception of a few polychaete annelids they are not conspicuous in shallow water habitats, because they tend to inhabit sand or burrow within porous rock. They may however be present in considerable numbers.

Polychaetes are a highly diverse and ecologically important group abundant on reefs and in shallow environments. They can be loosely divided into three categories, errant predators and scavengers, infaunal burrowers and tube-dwelling filter feeders. The polychaete fauna of the Red Sea is well known, important recent works include Hartmann-Schroder (1960), Amoureux *et al.*, (1978) and Amoureux (1983). Amoureux (1983) lists 81 species from rock, coral and algal substrates in the Gulf of Aqaba. The algal turf habitats contained by far the richest fauna, and the relatively large errant species *Nereis ghardaqae* and *Platynereis dumerilii* were the most abundant species. Errant polychaetes are important scavengers and benthic predators, some such as the large bristle-worms *Hermodice* and *Eurythoe* are significant coral predators. Errant worms are not usually active during the day time when predators are actively foraging, but can be discovered by turning over stones or by night snorkelling.

Infaunal polychaetes are extremely common and very diverse. Betz and Otte (1980) recorded 31 species in their Sudanese reef sand samples, in densities up to 35,000 per square metre. Further details of polychaete distribution in sand can be found in Fishelson (1971).

Another very diverse worm community inhabits porous coral rock. Polychaetes, nemertines, sipunculids and echiurans are all readily found when breaking up reef rock. Few studies have yet examined this specialised but important community, but Brock and Smith (1983) have shown it to be a useful indicator of sewage pollution, the rock fauna biomass rising with sewage enrichment.

Many Red Sea polychaetes are tube-dwellers, using an attractive crown of tentacles to catch food material. Important species include the large fan worm *Sabellastarte indica* which is fairly common among coral rubble and on rocks. The abundant *Spirobranchus giganteus* is found on living corals, especially *Porites solida* and *Millepora* species. *Spirobranchus* feeds using two elegant spiral fans, vividly coloured in yellow, red or blue. They are highly sensitive to shadows that may indicate a hovering predator, and it is instructive to shade them with a hand and observe their extraordinarily fast retraction within the tube. Recently Vine and Bailey-Brock (1984) have reviewed the taxonomy and distribution of two important tubeworm families in the Red Sea.

Terebellid polychaetes live in sand or in tubes constructed of debris under large stones. They feed on particulate matter, not by use of a tentaculate crown to filter water, but by spreading long grooved tentacles over the substrate. The Red Sea species *Reteterebella queenslandia* is common on shallow reef slopes in relatively still water. Its tentacle span may exceed one metre, but the tentacles are rapidly withdrawn if touched by a diver.

Most polychaetes have planktonic larval stages, but one family (Tomopteridae) is permanently planktonic, and the worms are flattened to improve their swimming. The genus *Tomopteris* is particularly well represented in the Red Sea, which contains over one third of the whole world fauna (Halim, 1969).

12.7. CRUSTACEA AND OTHER ARTHROPODS

12.7.1. Introduction and minor groups

Among the arthropods, while the insects dominate the land, the Crustacea are supreme in the sea, although the ecological roles of the two groups are not precisely equivalent. Very few insects make a living in or are associated with the sea. There are a few wingless species found on the strand or intertidally, some flies have marine larvae, and in the warmer seas there is the unusual water strider bug *Halobates halobates* which lives above the surface of the sea, feeding as a predator on animals just below the surface, and laying its eggs on floating feathers. The genus is recorded from the Red Sea (Schmidt and Müller 1973), but it is rarely seen by casual observers who tend to swim under the sea surface or move too fast over it.

Another arthropod radiation largely confined to land is the Chelicerata, including the very successful scorpions, spiders and mites. Many mites live in the sea, where some are parasitic or commensal on other animals; a favoured site is the bivalve gill. The bizarre pycnogonids or sea spiders are all marine. Sea spiders have very reduced bodies and long legs in which some of the internal organs are situated. All are slow moving predators, often specialising in small colonial organisms such as bryozoans. Pycnogonids are not very conspicuous in the Red Sea and have not received a great deal of attention; a useful recent review was given by Stock (1963).

Crustaceans are highly important marine animals. Although far less numerous than insects (some 31 thousand compared with 750 thousand species) they are much more diverse in body form, life history and ecology. Their exoskeleton is harder and stiffer than that of insects, being reinforced with heavy calcium carbonate, the weight of which is supported by the water. This may explain the much larger size to which many crustaceans can grow. Large spider crabs may have a leg span greater than 2 metres, and some lobsters can weigh several kilograms.

The crustacean fauna of the Red Sea is well documented, but many important references are old or difficult to obtain. Heller (1861) recorded over 100 species including many important new Indian Ocean taxa based on Red Sea material. Paul'son (1875) published (in Russian) a magnificent monograph on Red Sea Crustacea, in an edition limited to only 100 copies and now extremely rare and valuable. Other important early studies include Klunzinger (1906, 1913), who provided translations of some of Paul'son's text, Nobili (1906) and Balss (1929). More recently Guinot (1962, 1966a,b) has reviewed the Red Sea crab fauna in the context of the whole western Indian Ocean, while several papers (in Russian) may be found in Vodyanitzkii (1971). Fishelson (1971) included crustaceans in his account of Red Sea shallow benthic communities. Other relevant papers are mentioned later in the text.

Arguably the most important crustaceans in the Red Sea as in all oceans are the smallest, members of the artificial grouping 'entomostraca' which includes the copepods and barnacles among lesser groups. Copepods are extremely important in the plankton, most are herbivorous and form a major link in the food chain transferring energy from the plant to the animal communities. The planktonic copepods of the Red Sea are well documented, the reader is referred to Chapter 5 of this book and to the review by Halim (1969). Not all copepods are planktonic however, many harpacticoid copepods are important members of the benthic and interstitial fauna. Some harpacticoids are symbiotic with larger benthic animals especially echinoderms (Fishelson, 1974). Humes and Stock (1965) described a highly specialised association of the cyclopoid copepod *Anthessius* species with the giant clam *Tridacna*, undoubtedly careful observation in the Red Sea would reveal many more such associations and several new species.

Barnacles are at their most successful when living intertidally in temperate latitudes, and are not dominant in the Red Sea (Chapter 9), especially in the central region where the tidal range is very small. Several small chthamalid barnacles are known, but the fauna is still rather underinvestigated (Achituv and Safriel, 1980). The large barnacle *Tetraclita squamosa rufotincta* occurs both intertidally and on shallow reef flats. On some wave tossed boulders on exposed barrier reefs *Tetraclita* shells form substantial accumulations. Their attractive rose coloured shells are commonly sold in western curiosity shops.

12.7.2. Peracarids and stomatopods

The remaining Crustacea are grouped within the Malacostraca, and the most familiar members (shrimps, lobsters and crabs) belong to the order Decapoda. Three non-decapod groups will be briefly mentioned here, but others may be significant in the plankton and the benthos. The peracarid orders Isopoda and Amphipoda include the familiar woodlice and fresh water gammarid shrimps. Neither group is of outstanding significance in the Red Sea compared for example with the crabs. Spandl (1924) lists and maps the distribution of 42 species of amphipods, the total Red Sea fauna must be over 70 species. Most live semi-cryptically in submerged rocky and seagrass areas, but a few species are important scavengers on sandy beaches, often associated with mats of stranded algae.

Isopods are also important on sandy beaches, and those of Jiddah and Yanbu are described by Jones (1974). A specialist isopod niche is occupied by the gribble, which bores (often with disasterous consequences) into wooden ships and pilings, digesting the wood with the aid of symbiotic gut bacteria. At Elat, Haderlie (1974) found the gribble *Limnoria tripunctata* reached high densities on submerged experimental panels. The habitat was shared with the amphipod *Chelura terebrans*, although this may not have initiated burrows but merely colonised vacant gribble borings. Isopods of the genus *Sphaeroma* are listed by Por (1971) as part of the special euryhaline Isthmus fauna of Suez.

Mantis shrimps of the order Stomatopoda are among the largest non-decapod crustaceans. They have elongate bodies and are equipped with a pair of very sharp or club-like flexed anterior limbs, used to slash or stun animal prey. Large stomatopods like the Red Sea *Squilla* can be over 13 cm long, and can give a most unpleasant wound if handled. Most stomatopods live in shallow waters and lagoons, burrowing or inhabiting holes in rocks. Several descriptions exist of the Red Sea stomatopod fauna, the most recent general account is that of Holthuis (1967). So far about 29 species are known.

12.7.3. Shrimps and lobsters

These two groups of decapods are characterised by their long muscular abdomens and good swimming ability. They are also very tasty, and some are of economic importance in the Red Sea. Commercially important penaeid shrimps are found in the shallow coastal waters of the Red Sea, but the total shelf area is small, and the only significant shrimp fishery is that of North Yemen which could yield about 800-1000 tonnes annually (Sanders and Kedidi, 1981). Al Kholy and El Hawary (1970) provide a review of Red Sea penaeids. Other shrimp groups are of interest from their associations with other organisms. The beautiful red and white banded cleaner shrimp *Stenopus hispidus* is commonly found in pairs in reef hollows. It makes a living by removing ectoparasites from fish which allow the shrimp to groom them without molestation. Their conspicuous coloration advertises their services in contrast to the rather cryptic coloration and habits of most other shrimp species.

The Alpheidae is the largest and most important family of reef shrimps. They are small, but possess very large chelae which can be used to make surprisingly intense snapping noises which may stun their prey. Some alpheids (e.g. *Alpheus frontalis*) live in tubes built under stones, grazing on algae growing in the tube walls (Fishelson 1966). Others (e.g. *A. sublucanus*) live symbiotically within branching coral colonies or sponges. The large *A. purpuritentacularis* excavates sand burrows which it shares with the goby *Cryptocentrus*. The fish gains a burrow for protection but contributes to the mutual symbiosis by bringing home food to share with the shrimp.

Holthuis (1968) has reviewed the lobsters of the Red Sea, where both spiny lobsters (Palinuridae) and shovel-nosed lobsters (Scyllaridae) can be found. The commonest spiny lobster is *Panulirus penicillatus* which occurs in considerable numbers on rocky shelf and reef areas. *Panulirus* is largely nocturnal, hiding by day in deep clefts or under rocky overhangs. They emerge at night and scavenge over the substrate, often migrating into very shallow water. They are caught at night by divers or by fishermen wading in shallow water and carrying spears and kerosene lamps.

12.7.4. Crabs and hermit crabs

The crabs (Brachyura) are the most diverse decapod group in the Red Sea, and have been described in several monographs. Balss (1929), summarising the results of the German *Pola* investigations mentions 260 species, and Guinot (1966a) considers about 15% of the fauna to be endemic. Recent studies have further increased the known fauna, in particular the work of Griffin and Tranter (1974), Holthuis (1977) and Lewinsohn (1977a,b) should be consulted for species identification.

Crabs are characterised by their highly reduced abdomen, their short head-thorax and their well developed claws on the first pair of legs. The body plan seems very versatile, and the crabs include the most successful large terrestrial crustaceans, but the marine radiation embraces carnivores, scavengers, herbivores, specialist symbionts and filter feeders. The largest single group in the Red Sea is the family Xanthidae or coral crabs, with well over 100 species. Many of these small to medium sized crabs live symbiotically within branching corals, and some are highly host specific. The genus *Tetralia* for example only inhabits corals of the genus *Acropora*, while the large genus *Trapezia* is found in the closely related corals *Pocillopora*, *Stylophora* and *Seriatopora*. Coral tissue forms a major part of the crabs diet, and the larger the coral head, the larger the size to which the pair of crabs inhabiting it can grow. Some crabs may confer some advantage to the corals however, thus *Trapezia cymodoce* uses its claws to protect *Stylophora pistillata* against attack by the coral predator *Acanthaster* (Wolodarsky and Loya 1980). Edwards and Emberton (1980) list the *Stylophora* symbionts from Sudanese waters. Xanthids are relatively conventional crabs in appearance, but another coral inhabiting group, the Hapalocarcinidae are considerably modified, very small and with reduced locomotory ability. These gall crabs are especially common in *Stylophora* and *Pocillopora*, where they become entombed within cavities in the coral tissue and skeleton. They feed on suspended particulate material and seem not to be parasitic, using the coral only as shelter.

Two large swimming crab species are common in coastal lagoons and shallow water areas with muddy or sandy bottoms. *Lupa pelagica* is a particularly aggressive species, threatening any potential attacker with uplifted claws. *Lupa* is fished in trawls and taken using hand spears, especially at night. *Scylla serrata* is generally less common but can be larger and is a highly esteemed food crab.

Many crabs have developed terrestrial tendencies. Species of *Grapsus* and *Geograpsus* are particularly common in the rocky intertidal and littoral fringe (Chapter 9). These attractively patterned small crabs are very nimble and astonishingly difficult to catch intact. The most fully terrestrial crabs in the Red Sea are the ghost and fiddler crabs of the family Ocypodidae. The largest ghost crab *Ocypode saratan* (Fig. 12.5) is common on sandy shores just above the tideline. It digs deep burrows for protection from sun and predators and the excavated sand is piled into characteristic mounds from which the males display to potential mates. *Ocypode* always stays close to the sea, but its relatives, especially fiddler crabs such as *Uca inversa*, may penetrate some distance inland to muddy coastal fields. Other ocypodid and grapsid crabs are characteristic of mangrove communities (Chapter 9).

The final group considered is the Anomura, extensively reviewed by Lewinsohn (1969). These odd decapods are characterised by the loss or reduction of the fifth walking legs, and variable degrees of abdomen reduction. At one extreme some anomurans (e.g. *Galathea*) are rather lobster like, at the other extreme (e.g. *Porcellana*) some are superficially identical to true crabs. The best known anomurans are the hermit crabs or Paguridae, which protect a soft asymmetrical abdomen within an empty gastropod's shell, into which the whole animal can withdraw (Fig. 12.6). There are more than 20 hermit crab species in the Red Sea, and one (*Coenobita scaevola*) is extremely common on all sandy beaches where it lives a semi-terrestrial scavenging existence. When walking along a sandy beach above the tideline it is instructive to stop and stand still for a few minutes, and most of the surrounding 'dead' gastropod shells will begin to start walking away as the hermit crabs emerge from their retreats. The snail shell must be a considerable heat and desiccation shield for *Coenobita*, pre-adapting it to a terrestrial existence. Spaargaren (1977) found it to be less well adapted physiologically to resist water loss or altered salt concentrations than more thoroughly terrestrial crabs such as *Uca*.

12.8. BRYOZOA

The ectoprocts or Bryozoa are tiny animals, living within the protection of variously constructed cases, and feeding by eversible tentacle crowns. They are colonial, and although the individual zooids are less than a millimetre in length, the colonies can spread over quite large areas. Many build their houses

Fig. 12.5. Ghost crab *Ocypode saratan*. The light colour of the carapace and the very large eyes held above the carapace are terrestrial adaptations in this species.

of calcium carbonate, and are sometimes mistaken for corals. Bryozoans are abundant cryptic reef organisms, readily found under plates of coral or in shaded caves. They form part of the 'cryptofauna' which has been much studied for the light it sheds on mechanisms of competition and co-existence in sessile organisms. Several publications deal with Red Sea bryozoans, the most recent is Dumont (1981), based on SCUBA collections to 30 m depth in The Sudan. The fauna includes 86 species, and although the colonies are normally small, Dumont found densities as high as $5,000/m^2$. Surprisingly, marked depth related abundance changes were found for many species, although it is not clear to what environmental parameters the animals were responding. Further studies of the Red Sea Bryozoa and of the cryptofauna in general would be very worthwhile.

In much deeper waters of the Red Sea, below the depth of reef coral growth, Bryozoa become major community components and are responsible with serpulid polychaetes and Foraminifera for reefal carbonate accumulation (Fricke and Schuhmacher, 1983).

12.9. CONCLUSIONS

This chapter has reviewed very briefly and inadequately the biology of certain relatively minor elements of the Red Sea fauna, and the selection of groups for discussion has been biased by the author's interest in sessile benthic organisms. While the total literature on the Red Sea fauna is voluminous,

Fig. 12.6. Large, unidentified hermit crab from the Sudanese coast.

much of it derives from early collections based on brief expeditionary forays in limited varieties of habitat. In all but a few favoured groups such as corals, gastropods and fish, the level of sampling is very inadequate. In particular, many groups have never been surveyed using modern SCUBA techniques, which revolutionise our approach to sampling the marine sub-littoral. There is plenty of scope for further studies on most of the 'minor' invertebrate groups, and this is well illustrated by Dumont's (1981) study of the Bryozoa. Despite very brief time in the field, Dumont's careful SCUBA sampling to 30 m depth added no less than 30 species to the Red Sea list, four of which were completely new to science. Comparable studies on other taxa such as the Foraminifera, Porifera, Ctenophora, Nemertini, Phoronida and Urochordata would probably yield comparably satisfactory results.

REFERENCES

Achituv, Y. and Safriel U. N. (1980). A New *Chthamalus* (Crustacea: Cirripedia) from intertidal rocks of the Red Sea. *Isr. J. Zool.* 29, 99–109.

Al-Kholy, A. A. and El-Hawary, M. M.(1970). Some penaeids of the Red Sea. *Bull. Inst. Oceanogr. Fish. Cairo* 1, 449–77.

Amirthalingam, G. (1970). *Fauna of the Red Sea*. Sawakin Project (No.2), Sudan Research Unit, Faculty of Arts, University of Khartoum.

Amoureux, L. (1983). Note taxonomique et écologique sur une collection d'Annélides Polychètes du Golfe d'Aqaba (Mer Rouge). *Cah. Biol. mar.* 24, 363–9.

Amoureux, L., Rullier, F. and Fishelson, L. (1978). Systématique et écologie d'annélides polychètes de la presqu'il du Sinai. *Israel J. Zool.* 27, 57–163.

Bahafzallah, A. B. K. (1979). Recent benthic Foraminifera from Jiddah Bay, Red Sea (Saudi Arabia). *Neues Jb. Geol. Palaont. Mh.* 1979, 385–98.

Balss, H. (1929). Expedition S. M. Schiff''Pola'' in das Rote Meer, Nordliche und Südliche Hälfte 1895/6 1897/8. Zoologische Ergebnisse Decapoden des Roten Meeres IV. Oxyrhyncha und Schlussbetrachtungen. *Denkschr. Akad. Wiss. Wien Math. naturw. Kl.* 102, 1–30.

Barnes, R. D. (1980). *Invertebrate Zoology*. Saunders College/Holt, Rinehart and Winston, Philadelphia.

Bemert, G. and Ormond, R. F. G. (1981). *Red Sea Coral Reefs*. Keegan Paul International, London.

Bergquist, P. R. (1978). *Sponges*. Hutchinson and Co., London.

Betz, K.-H. and Otte, G. (1980). Species distribution and biomass of the soft bottom faunal macrobenthos in a coral reef (Shaab Baraja, Central Red Sea, Sudan). *Proc. Symp. Coastal and Marine Environment of the Red Sea, Gulf of Aden and Tropical Western Indian Ocean* 3, 13—38.

Benayahu, Y. and Loya, Y. (1981). Competition for space among coral-reef sessile organisms at Eilat, Red Sea. *Bull. mar. Sci.* 31, 514—22.

Boschma, H. (1968). The Milleporina and Stylasterina of the Israel South Red Sea Expedition. *Bull. Sea Fish. Res. Stn. Haifa* 49, 8—14.

Brock, R. E. and Smith, S. V. (1983). Response of coral reef cryptofaunal communities to food and space. *Coral Reefs* 1, 179—83.

Capone, D. G. and Taylor, B. F. (1977). Nitrogen fixation (acetylene reduction) in the phyllosphere of *Thalassia testudinum*. *Mar. Biol.* 40, 19—28.

Chamberlain, J. A. (1978). Mechanical properties of coral skeleton: compressive strength and its adaptive significance. *Paleobiology* 4, 419—35.

Colin, P. L. (1978). *Caribbean Reef Invertebrates and Plants*. T. F. H. Publications Inc., Hong Kong.

Drew, E. A. (1972). The biology and physiology of alga-invertebrate symbiosis. 1. Carbon fixation in *Cassiopea* sp. at Aldabra Atoll. *J. exp. mar. Biol. Ecol.* 9, 65—9.

Ducret, F. (1973). Contribution à l'etude des chaetognathes de la Mer Rouge. *Beaufortia* 20, 135—53.

Dumont, J. P. C. (1981). A report on the cheilostome Bryozoa of the Sudanese Red sea. *J. nat. Hist.* 15, 623—37.

Edwards, A. and Emberton, H. (1980). Crustacea associated with the Scleractinian coral *Stylophora pistillata* (Esper) in the Sudanese Red Sea. *J. exp. mar. Biol. Ecol.* 42, 225—40.

Ferber, I. (1977). *Marine Invertebrate Collections from the Red Sea, the Eastern Mediterranean and Suez Canal and Bibliography of The Hebrew University - Smithsonian Institution Joint Program (1967—1971), Biota of the Red Sea and the Eastern Mediterranean*. Zoological Museum, the Hebrew University of Jerusalem, Israel.

Fishelson, L. (1964). Observations on the biology and behaviour of Red Sea coral fishes. *Bull. Sea Fish. Res. Stn. Haifa* 37, 11—26.

Fishelson, L. (1966). Observations on the littoral fauna of Israel V. On the habitat and behaviour of *Alpheus frontalis* H. Milne Edwards (Decapoda: Alpheidae). *Crustaceana* 11, 98—104.

Fishelson, L. (1970). Littoral fauna of the Red sea: the population of non-scleractinian anthozoans of shallow waters of the Red sea (Eilat). *Mar. Biol.* 6, 106—16.

Fishelson, L. (1971). Ecology and distribution of the benthic fauna in the shallow waters of the Red Sea. *Mar. Biol.* 10, 113—33.

Fishelson, L. (1974). Ecology of the northern Red Sea crinoids and their epi- and endozoic fauna. *Mar. Biol.* 26, 183—92.

Fricke, H. W. and Schuhmacher, H. (1983). The depth limits of Red Sea stony corals, an ecophysiological problem (A deep Diving Survey by Submersible). *P.S.Z.N.I. Marine Ecology* 4, 163—94.

Futterer, D. K. (1974). Significance of the sponge *Cliona* for the origin of fine grained material of carbonate sediments. *J. sedim. Petrol.* 44, 79—84.

Gibson, R. (1974). Two species of *Baseodiscus* (Heteronemertea) from Jidda in the Red sea. *Zoolog. Anz.* 192, 255—70.

Godeaux, J. (1973). A contribution to the knowledge of the Thaliacean faunas of the Eastern Mediterranean and the Red sea. *Israel J. Zool.* 22, 39—50.

Gohar, H. A. F. (1940). Studies on the Xeniidae of the Red Sea. *Publs. Mar. Biol. Stn. Ghardaqa* 2, 25—120.

Gohar, H. A. F. and Eisawy, A. M. (1961). The biology of *Cassiopea andromeda* from the Red Sea (with a note on the species problem). *Publs. Mar. Biol. Stn. Ghardaqa* 11, 3—45.

Griffin, D. J. G. and Tranter H. A. (1974) Spider crabs of the family Majidae (Crustacea: Decapoda: Brachyura) from the Red Sea. *Isr. J. Zool.* 23, 162—98.

Guinot, D. (1962) Sur une collection de Crustacés Décapodes Brachyoures de la Mer Rouge et de Somalie. Remarques sur les genres *Calappa* Weber, *Menaethiops* Alcock, *Tyche* Bell, *Ophthalmius* Rathbun et *Stilbognathus* von Martens. *Boll. Mus. Civ. Stor. Noct. Venezia* 15, 7—63.

Guinot, D. (1966a). La faune carcinologique (Crustacea, Brachyura) de l'Océan Indien occidentale et de la Mer Rouge. Catalogues, remarques biogéographiques et bibliographie. *Mém. Inst. fr. Afr. noire* 77, 235—252.

Guinot, D. (1966b). Les espèces comestibles de crabes dans l'Océan Indien occidental et la Mer Rouge. *Mem. Inst. franc. Afr. noire* 77, 353—94.

Haderlie, E. C. (1974). Wood boring marine animals from the Gulf of Elat. *Isr. J. Zool.* 23, 57—9.

Halim, Y. (1969). Plankton of the Red sea. *Oceanogr. mar. Biol. A. Rev.* 7, 231—75.

Hartman, W. D. and Goreau, R. G. (1970). Jamaican coralline sponges: their morphology, ecology and fossil relatives. *Symp. zool. Soc. Lond.* 25, 205—43.

Hartmann-Schröder, G. (1960). Polychaeten aus dem Roten Meer. *Kiel Meersforsch.* 16, 59—125.

Head, S. M. (1980). *The Ecology of Corals in the Sudanese Red Sea*. Ph.D. Thesis, University of Cambridge.

Heller, C. (1861). Synopsis der im rothen Meere vorkommenden Crustaceen. *Verh. k.k. zool. bot. Gesellsch. Wien* 11, 3—32.

Hiatt, R. W. and Strasburg, D. W. (1960). Ecological relationships of the fish fauna on coral reefs of the Marshall Islands. *Ecol. Monogr.* 30, 65—127.

Hickson, S. J. (1940). The species of the genus *Acabaria* in the Red Sea. *Publs. Mar. Biol. Stn. Ghardaqa* 2, 3—22.

Holthuis, L. B. (1967). The stomatopod crustacea collected by the 1962 and 1965 Israel South Red Sea Expeditions. *Isr. J. Zool.* 16, 1—45.

Holthuis, L. B. (1968). The second Israel South Red Sea Expedition 1965, Report No 7. The Palinuridae and Scyllaridae of the Red Sea. *Zool. Meded. Leiden* 42, 281—301.

Holthuis, L. B. (1977). The Grapsidae, Geocarcinidae and Palicidae of the Red Sea. *Isr. J. Zool.* 26, 141—92.

Humes, A. G. and Stock, J. H. (1965). Three new species of *Anthessius* (Copepoda, Cyclopoida, Myicolidae) associated with *Tridacna* from the Red Sea and Madagascar. *Bull Sea Fish. Res. Stn. Israel* 40, 49—74.

Ivanov, A. V. and Gureeva, M. A. (1980). (A new species of the genus *Oligobrachia* (Pogonophora) from Bab el Mandeb Straits). *Zoologicheskii Zh.* 59, 1587—91. (In Russian).

Jackson, J. B. C., Goreau, F. G. and Harman, W. D. (1971). Recent brachiopod-coralline sponge communities and their paleoecological significance. *Science N.Y.* 173, 623—5.

Jones, D. A. (1974). The systematics and ecology of some sand beach isopods (family Cirolanidae) from the coasts of Saudi Arabia. *Crustaceana* 26, 201—11.

Jones, S. (ed.) (1972). *Bibliography of the Indian Ocean.* Special Publication, Marine Biological Association of India, Mandapam Camp, India.

Kinzie, R. A. III (1973). Coral Reef Project — Papers in memory of Dr Thomas F. Goreau. 5. The zonation of West Indian gorgonians. *Bull. Mar. Sci.* 23, 93—155.

Klunzinger, C. B. (1877). *Die Korallthiere des Rothen Meeres. Erste Theil Die Alcyonarien und Malacodermen.* Gutmann'schen Verlag, Berlin.

Klunzinger, C. B. (1902). Über *Ptychodera erythraea* Spengel aus dem Rothen Meeres. *Verh. Dt. Zool. Ges.* 12, 195—202.

Klunzinger, C. B. (1906). *Die Spitz und Spitzmundkrabben (Oxyrhyncha und Oxystomata) des Rothen Meeres.* Stuttgart.

Klunzinger, C. B. (1913). Die Rundkrabben (Cyclometopa) des Rothen Meeres. *Nova Acta Akad. Nat. Caes. Leop. Car. Germ.* 99, 97—402.

Levi, C. (1965). Spongiaires récoltes par l'expédition Israelienne dans le sud de la Mer Rouge en 1962. *Bull. Sea Fish Res. Stn. Haifa* 40, 3—27.

Lewinsohn, Ch. (1969). The second Israel South Red Sea Expedition 1965, Report No.6. Die Anomuren des Roten Meers (Crustacea, Decapoda, Paguridae, Galatheida, Hippidae). *Zool. Verh.* 104, 1-213.

Lewinsohn, Ch. (1977a). Die Dromiidae des Rothen Meeres (Crustacea, Decapoda, Brachyura). *Zool. Verh.* 151, 3—41.

Lewinsohn, Ch. (1977b). Die Ocypodidae des Rothen Meeres (Crustacea, Decapoda, Brachyura). *Zool. Verh.* 152, 45—84.

Mackenzie, F. T., Kulm, L. D., Cooley, R. L. and Barnhart, J. T. (1965). *Homotrema rubrum* (Lamarck), a sediment transport indicator. *J. sedim. Petrol.* 35, 265—72.

Mergner, H. (1977). Hydroids as indicator species for ecological parameters in Caribbean and Red Sea coral reefs. *Proc. 3rd Int. Coral Reef Symp.* 1, 119—26.

Mergner, H. and Schuhmacher, H. (1981). Quantitative analyse der korallenbesiedlung eines vorriffareals bei Aqaba (Rotes Meer). *Helgoländer wiss. Meeresunters.* 34, 337—54.

Mergner, H. and Wedler, E. (1977). Über die Hydroidenpolypenfauna des Roten Meeres und seiner ausgänge. *Meteor Forsch. Ergebn. (D)* 24, 1—32.

Murina, V. V. (1971). (On the species composition and ecology of sipunculids from the Red Sea). In (*Benthos of the Red Sea Shelf*). Ed. V.A. Vodyanstzkii, pp. 76—88, Naukova, Kiev. (in Russian).

Nobili, G. (1906). Faune carcinologique de la Mer Rouge. Décapodes et Stomatopodes. *Annal. Sci. nat. (Zool.) (9)* 4, 1—347.

Otter, G. W. (1931). Rock-destroying organisms in relation to coral reefs. *Scient. Rep. Gt. Barrier Reef Exped.* 1, 323—52.

Paldi, R. (1969a). *Computer Produced Regional Bibliography: Eastern Mediterranean, Suez Canal and Red Sea.* Sea Fish. Res. Stn., Haifa, Israel.

Paldi, R. (1969b). *List of Articles: Eastern Mediterranean, Red Sea and Suez Canal. Addendum I to the Computer Produced Regional Bibliography.* Sea Fish. Res. Stn., Haifa, Israel.

Paldi, R. (1971). *List of Articles: Eastern Mediterranean, Red Sea and Suez Canal. Addendum II to the Computer Produced Regional Bibliography.* Sea Fish. Res. Stn., Haifa, Israel.

Palombi, A. (1928). Zoological results of the Cambridge expedition to the Suez Canal, 1924. 34, Report on the Turbellaria. *Trans. zool. Soc. Lond.* 22, 579—625.

Paul'son (1875) (Researches on Red Sea Crustacea) Kiev. (In Russian).

Pérès, J.-M. (1960). Sur une collection d'ascidies de la côte israelienne de la Mer Rouge et de la peninsule du Sinai. *Bull, Sea Fish. Res. Stn. Haifa* 30, 39—47.

Por, F. D. (1971). One hundred years of Suez Canal — a century of Lessepsian migration. Retrospect and viewpoints. *Syst. Zool.* 20, 138—51.

Reiss, Z. (1979). Foraminiferal research in the Gulf of Aqaba: a review. *Utrecht micropaleontol. Bull.* 15, 7—25.

Reiswig, H. M. (1971). Particle feeding in natural populations of marine demosponges. *Biol. Bull.* 141, 568—91.

Row, R. W. H. (1909). Reports on the marine biology of the Sudanese Red Sea. 13, Report on the sponges collected by Mr Cyril Crossland in 1904—5. Part I. Calcarea. *J. Linn. Soc. Zool.* 31, 182—214.

Row, R. W. H. (1911). Reports on the marine biology of the Sudanese Red Sea from collections made by Cyril Crossland MA. BSs., FZS. 19. Report on the sponges collected by Mr Cyril Crossland in 1904—5. Part II. Non-Calcarea. *J. Linn. Soc. Zool.* 31, 287—400.

Said, R. (1949). Foraminifera of the northern Red Sea. *Contr. Cushman Lab. Foraminif. Res. Spec. Publ.* 26, 1—44.

Said, R. (1950a,b). Additional Foraminifera from the northern Red Sea. *Contr. Cushman Lab. Foraminif. Res.* 1, 4—9, 9—29.

Sanders, M. J. and Kedidi S. M. (1981). Summary review of Red Sea commercial fisheries catches and stock assessments including maps of actual and potential fishing grounds. *U.N.D.P. United Nations F.A.O. report* RAB 77/008.

Schmidt, G. D. and Paperna, I. (1978). *Sclerocollum rubrimaris* gen. et sp. nov. (Rhabdinorhynchidae: Gorgonorhynchinae) and the Acanthocephala of marine fishes from Israel. *J. Parasit.* 64, 846−50.

Schmidt, H.-E. and Müller, R. (1973). Marine water striders of the genus *Halobates* (Hemiptera, Gerridae) from the Red Sea and Gulf of Aden. *Israel J. Zool.* 22, 1−12.

Schuhmacher, H. (1973). Morphologische und okologische Anpassungen von *Acabaria*-arten (Octocorallia) im Roten Meer an verschieden Formen der Wasserbewegung. *Helgoländer wiss. Meeresunters.* 25, 461−71.

Sebens, K. P. (1977). Autotrophic and heterotrophic nutrition of coral reef Zoanthids. *Proc. 3rd Int. Coral Reef Symp.* 1, 397−404.

Spaargaren, D. H. (1977). On the water and salt economy of some decapod crustaceans from the Gulf of Aqaba (Red Sea). *Neth. J. Sea Res.* 11, 99−106.

Spandl, H. (1924). Expedition S.M. Schiff ''Pola'' in das Rote Meer Nordliche und Südliche Hälfte 1895/6, 1897/8 Zoologische Ergebnisse. Die Amphipoden des Rothen Meeres. *Denkschr. Akad Wiss. Wien math. naturw Kl.* 102, 19−73.

Stephen, A. C. (1965). Echiura and Sipuncula from the Israel South Red Sea Expedition. *Bull Sea Fish. Rẻs. Stn. Haifa* 40, 79−83.

Stiasny, G. (1938). Die Scyphomedusen des Roten Meeres. *Verh. K. Ned. Akad. Wet.* 37, 1−35.

Stiasny, G. (1959). Alcyonaria I. Gorgonaria aus dem Rothen Meere Sammlung Dr R. Ph. Dollfus, Paris aus dem Golf von Suez. *Result. Scient. Miss. R. P. Dollfus, Egypte*, 1−18.

Stock, J. H. (1963). Israel South Red Sea Expedition 1962. Report No.3, Pycnogonida. *Bull Sea Fish. Res. Stn. Haifa* 35, 27−34.

Suchanek, T. H. and Green, D. J. (1981). Interspecific competition between *Palythoa caribaeorum* and other sessile invertebrates on St Croix reefs, U.S. Virgin Islands. *Proc. 4th Int. Coral Reef Symp.* 2, 679−84.

Svoboda, A. (1978). *In situ* monitoring of oxygen production and respiration in Cnidaria with and without zooxanthellae. In *Physiology and Behaviour of Marine Organisms*. Eds. D. S. McLusky and A. J. Berry, pp. 75−82, Pergamon Press, Oxford.

Vasseur, P. (1974). The overhangs, tunnels and dark reef galleries of Tuléar (Madagascar) and their sessile invertebrate communities. *Proc. 2nd Int. Coral Reef Symp.* 2, 143−60.

Verseveldt, J. (1965). Report on the Octocorallia (Stolonifera and Alcyonacea) of the Israel South Red Sea Expedition 1962, with notes on other collections from the Red Sea. *Bull. Sea Fish. Res. Stn. Haifa* 16, 28−48.

Verseveldt, J. (1974). Alcyonacea (Octocorallia) from the Red sea, with a discussion of a new *Sinularia* species from Ceylon. *Isr. J. Zool.* 23, 1−37.

Verseveldt, J. and Benayahu, Y. (1978). Descriptions of one old and five new species of Alcyonacea (Coelenterata: Octocorallia) from the Red Sea. *Zool. Meded.* 53, 57−74.

Vine, P. J. and Bailey-Brock, J. H. (1984). Taxonomy and ecology of coral reef tube worms (Serpulidae, Spirorbidae) in the Sudanese Red Sea. *Zool. J. Linn. Soc.* 80, 135−56.

Vodyanitzkii, V. A. (1971) (ed). (Benthos of the Red Sea's Shelf) Naukova Dumka, Kiev. (In Russian).

Warner, G. F. (1981). Species descriptions and ecological observations of black corals (Antipatharia) from Trinidad. *Bull. mar. Sci.* 31, 147−63.

Waters, A. W. (1910). Reports on the marine biology of the Sudanese Red Sea 12. The Bryozoa, Part 2, Cyclostomata, Ctenostomata and Endoprocta. *J. Linn. Soc. Zool.* 31, 231−56.

Wilkinson, C. R. (1980a). Red Sea sponges of Sudan. In *Final Report, A Description of Studies Conducted in the Sudanese Red Sea*. Ed. S. M. Head, pp. 20−7. Unpublished report to Saudi-Sudanese Joint Red Sea Commission.

Wilkinson, C. R. (1980b). Nutrient translocation from symbiotic cyanobacteria to coral reef sponges. *Coll. Int. C.N.R.S.* 291, 425−47.

Wilkinson, C. R. and Fay, P. (1979). Nitrogen fixation in coral reef sponges with symbiotic cyanobacteria. *Nature* 279, 527−9.

Witzell, W. N. (1983). Synopsis of biological data on the hawksbill turtle *Eretmochelys imbricata* (Linnaeus, 1766). *FAO Fish. Synop.* 137, 1−78.

Wolodarsky, Z. and Loya, Y. (1980). Population dynamics of *Trapezia* crabs inhabiting the coral *Stylophora pistillata* in the north Gulf of Elat (Gulf of Aqaba). *Isr. J. Zool.* 29, 204−5 (abstract).

CHAPTER 13

Red Sea Fishes

RUPERT ORMOND* and ALASDAIR EDWARDS**

*Tropical Marine Research Unit, Department of Biology, University of York, U.K.
**Dove Marine Laboratory, Department of Zoology, University of Newcastle upon Tyne, U.K.

CONTENTS

13.1. INTRODUCTION

The fishes of the Red Sea were amongst the earliest to be studied and named in modern scientific fashion, many having been described as early as 1761 by Peter Forsskål on the ill-fated Danish expedition to Arabia Felix (Figs. 13.1 & 13.2). In the next century several other ichthyologists devoted considerable time and effort to collecting and/or studying specimens from the area — Ehrenberg and Hemprich (1820—1826), Eduard Ruppell (1811—1836) and Carl Klunzinger (1864—1884). And in this century a great number of scientists have made and reported on smaller collections of fish from the Red Sea. Yet it is only recently that attempts have been made to give a comprehensive account of the identification and occurrence of the Red Sea species of which there are about 1000.

A start was made by Botros (1971) who compiled a synopsis of the species recorded by previous authorities. While this presented a major first step, no allowance was made for the considerable uncertainty surrounding many of the names and records in the literature. In addition, the specimens examined by the older scientists were of species which were obtained by fishermen or by using surface based techniques; the records did not necessarily reflect the species that were seen to be the most abundant on coral reefs or in other habitats by snorkellers or SCUBA divers, once diving became a common scientific tool in the late 1960s and 70s.

Ormond (1980a) presented a preliminary list of the common fish species on Red Sea coral reefs, together with notes on their behaviour, based on extensive underwater observation around the Red Sea. Meanwhile a comprehensive checklist of Red Sea species has been published by Menahem Dor (1984) and John Randall (1983) has published his much awaited account of Red Sea reef fishes. This latter book provides not only a beautifully illustrated guide book, but also a serious scientific study of great value. These two works taken together must be regarded as the most significant contribution to the ichthyology of the Red Sea since the work of Forsskål and of Klunzinger.

The fishes are amongst the most colourful and photogenic of Red Sea creatures but unfortunately their beauty cannot be adequately illustrated here so we have confined ourselves to outlining their ecology, behaviour and zoogeography and refer the reader to Randall (1983) for colour illustrations. In our use of common names we have followed Randall so that our text may be complementary to his 'Diver's Guide to Red Sea Reef Fishes' (ISBN 0-907151-05-1) a waterproof guide with colour photographs of 325 species, which is designed to be carried underwater by snorkellers and SCUBA divers.

The greater part of this present chapter is an expansion and revision of Ormond's (1980a) earlier paper, to include more species and fish of habitats other than coral reefs; principally pelagic species, and those associated with inshore sandy and muddy environments, such as are found in particular within the numerous creeks, (known in arabic as *marsas* or *sharms*) which occur on both sides of the Red Sea. The species which are included are those which are common, well-studied in the Red Sea or familiar to the authors. This first part is primarily by Rupert Ormond.

This chapter is thus intended to give an indication of the occurrence and ecology of those species most likely to be encountered in the Red Sea; these are discussed in a series of 24 groups on a family basis. In most groups taxonomically related species are included, but under some other headings are discussed various ·unrelated forms, which have a superficially similar morphology, and/or occur in the same environment. Not only has this approach proved the most convenient, but it is hoped it will provide a sensible framework for the average reader. The chapter finishes with a discussion of the zoogeographical relationships of Red Sea fishes by Alasdair Edwards, with particular emphasis on the impact of its recent geological history and of the opening of the Suez Canal in 1869.

13.2. SHARKS

About 30 shark species have been reported from the Red Sea. The two most frequently seen are the Whitetip reef shark (*Triaenodon obesus*) and the Blacktip reef shark (*Carcharhinus melanopterus*) which are fairly common in shallow water on fringing and offshore reefs throughout the Red Sea. Both these reef sharks seem to establish individual territories on the reef and to be scavengers and predators of small to medium-sized reef fish. They are typically only a metre to a metre and a half long and almost always flee as soon as they see a diver or snorkeller approaching. In slightly deeper water (10—50 m) several other sharks may be regularly seen, especially if bait is put out to attract them. These include the Silvertip (*Carcharhinus albimarginatus*), the Sandbar shark (*C. plumbeus*), and especially the Shortnose blacktail shark (*C. wheeleri*), all of which are typically 1.5—2.0 m long. Until it was distinguished by Garrick (1982) as a separate species, *C. wheeleri* was generally thought to be a form of the Grey reef shark (*C. amblyrhynchos*). It has a very characteristic colour pattern on the caudal fin, consisting of a dark band down the trailing edge and anterior to this a conspicuous white crescent. Other *Carcharhinus* species may also be seen occasionally, especially in or close to deep water (100 m). All the *Carcharhinus* species seem to feed principally on fish.

The four largest predatory species, and the four which can be really dangerous to man, are the Tiger shark (*Galeocerdo cuvier*) which may grow up to 5 m in length, the Mako shark (*Isurus glaucus*), up to 4 m, and the two hammerhead species, (*Sphyrna*) which may grow to 4 or more metres long, although most individuals in each species are probably 3 m or less in length. The Tiger and Mako are again generally associated with deep water. Hammerheads, on the other hand, are more often seen near the seabed, often in groups, either close to offshore reefs or sometimes in shallow water within marsas. Again, all these species feed on bony fish, although the Tiger shark also takes turtles and seabirds and even other sharks which may form a significant part of its diet.

Amongst the smaller sharks are the houndsharks or smoothhounds (Triakidae) which are generally under a metre in length and are found, often two or three together, on or close to the bottom where they feed on small fish and Crustacea.

Other shark families include some specialised feeders. *Alopias pelagicus*, the Pelagic thresher, is the only thresher shark found in the Red Sea. It may be recognised by its extraordinarily long tail, which is approximately as long again as the rest of the body, and according to Smith (1950) is used to strike at and break up schools of sardines or other small fishes. Another comparatively specialised feeder is the nurse shark of which again only one species, *Nebrius ferrugineus* has been found in the Red Sea. This appears to be a bottom feeder, probably mostly active at night, since by day it is generally found resting in caves on the reef. Gohar and Mazhar (1964) found that the stomachs of nurse sharks in the Red Sea contained considerable quantities of coral, especially pieces of *Stylophora*, and also cephalopods, as well as fish. It seems possible that this species of shark may attack reef fish asleep at night among branching corals, ingesting the coral pieces along with the trapped fish.

A highly specialised Red Sea shark, and the largest, is the Whale shark, *Rhincodon typus*, which grows to 12 m or more, and feeds on plankton which are filtered out as water passes through its gills. In consequence, this species, which is not infrequently reported in the Red Sea, is more or less harmless, and from time to time, parties of divers meeting one underwater have been able to hitch rides by hanging on to the fins or tail of one of these impressive beasts.

Fig. 13.1. Remarkably, about 70 of the fishes collected from the Red Sea by Forsskål in 1762–63 still survive as dried skins in the Zoologisk Museum of the University of Copenhagen. These specimens were despatched by Carsten Niebuhr (the sole member of the expedition who was to survive) from Al Mukha, North Yemen, after Forsskål's death in July 1763. Three whole years later, after numerous passages on several ships — during which much of the collections was ruined — the specimens finally reached Copenhagen, having travelled via Bombay, Tranquebar, Calcutta, China and South Africa. As if this wasn't enough the collections then suffered years of neglect in Copenhagen and barely escaped destruction during the British bombardment of the city in 1801.

Two of Forsskål's dried skins are illustrated above: (a) the Yellowbar Angelfish, *Pomacanthus maculosus*, which he collected at Al Luhayya in 1763; this species is known only from the Red Sea, Arabian Gulf and along the coast of East Africa to Zanzibar; (b) the Striped Butterflyfish, *Chaetodon fasciatus*, collected at Jiddah in 1762; this species is endemic to the Red Sea but closely related to *C. lunula* from the Indian Ocean (*see* Table 13.2). (Photo: G. Brovad; courtesy of Zoologisk Museum, Copenhagen)

13.3. RAYS

Much the most frequently encountered ray is the Blue-spotted lagoon ray or Reef stingray (*Taeniura lymma*), which has an ovate body disc, 40—50 cm across and is pale sandy-grey in colour with many distinctive large blue spots. It is found in shallow sandy areas behind or on the reef. Although a stingray, with one to two sharp venomous spines hidden by the base of the tail, this ray will only cause injury if stood on, or intentionally molested. It feeds by excavating sand primarily to search out burrowing polychaete worms (Gohar and Mazhar, 1964). Its digging activities attract many other species of fish, such as wrasses, emperors and triggerfishes, which snatch molluscs, echinoderms, Crustacea, or other invertebrates, which the ray rejects or is too slow to take.

The commonest larger stingray is *Dasyatis sephen* which burrows into muddy sand to feed, principally on Crustacea. This species is typically about 2 m long, including the tail which accounts for about half its length. Immediately recognisable by its much longer tail is the Coachwhip ray (*Himantura uarnak*). Its tail is nearly twice as long as the body, giving an overall length of up to 2.5 m. This distinctive species which is yellowish brown, with many large black spots, is found in comparatively shallow (1—5 m deep) lagoons and sandy areas among reefs; it also seems to feed mainly on small bottom-living Crustacea.

A group of particularly elegant rays are the eagle rays which are represented in the Red Sea by the Spotted eagle ray (*Aetobatus narinari*). This species is readily recognisable by its pointed and domed head and pointed wings, dark on top with many smallish white spots. This species may also grow up to 2.5 m total length and is commonly found in deep lagoons and over shallow sandy areas of sea bed (1—7 m deep) where it probably feeds on molluscs.

The largest and most spectacular of the rays however are the manta rays, which are not infrequently observed in the Red Sea. The Giant manta (*Manta birostris*) though of impressive size swims in a graceful manner, with slow stately beats of its enormous wings, which may grow to 5 m or more across. These mantas are seen in surface waters where like the Whale shark, they strain plankton from the sea with their dense gill-rays. The smaller manta, *Mobula diabolus*, which grows to only about 2 m across, seems to feed by contrast on small schooling fishes. The principal characteristic of these large rays are the two cephalic horns, lobe-like structures projecting forwards on either side of the mouth. In the Giant manta these horns are long, flat and flexible, and used during feeding to channel plankton into the mouth. In *Mobula* the cephalic horns are pointed and shorter, being not much longer than they are broad; yet even in this species they are said to be used in feeding, helping to scoop small fish into the mouth.

In contrast to these two open water species are four other rays specialised for living on the sea bed, mostly in areas of muddy-sand where they dig for food. The sawfish, (*Pristis pectinatus*), can grow to over 4 m in length, the saw itself being up to a metre or more long, with about 30 pairs of teeth projecting from either side. Sawfish may be encountered in shallow muddy areas in marsas where they use the saw for searching through the mud for food, and if necessary, in defence. The shovel-nose ray, (*Rhyncobatus djiddensis*), and the mud skate (*Rhina ancylostoma*) are both half shark-like and half ray-like in appearance. Both have pointed pectoral fins extending back over about a third the length of the body. The shovel nose ray has a more pointed snout and is common in shallow sandy areas; Gohar and Mazhar (1964) found that it feeds almost entirely on crabs. The mud skate, as its name implies, is more characteristic of muddy areas. The fourth species, the guitar-fish (*Rhinobatus halavi*), on the other hand, is not infrequently met with on shallow flat areas of the reef. While its pectoral fins are not much larger than those of the other two species, they are rounded, making the comparison with a guitar, while not very obvious, at least not unreasonable. This species seems to feed mostly on shrimp.

The final species deserving mention is the electric ray (*Torpedo panthera*) which is occasionally found among corals on Red Sea reefs as well as in shallow sandy areas. This species is easily recognised by its short, broad tail and double-disc-like appearance, a smaller disc being formed around the base of the tail, behind the larger body disc formed by the pectoral fins. The electric organs, which are located one on each side of the disc, can deliver quite a powerful shock, and are presumed to be used in helping to stun prey, such as worms, Crustacea and small fish, as well as in defence.

13.4. HERRINGS, SARDINELLAS AND SIMILAR SPECIES

The bonefish (*Albula glossodonta*), Tenpounder (*Elops machnata*), Milkfish (*Chanos chanos*) and wolf-herring (*Chirocentrus dorab*) are all largish herring-like fishes, growing up to a metre or more in length and widespread in the tropical Indo-Pacific. They are all found regularly in small numbers in the catches of local fishermen but are not especially abundant, save that very large schools of milkfish are often seen around the fringing reefs and marsa entrances in spring. Tenpounder and wolf-herring are also open water fishes but the young may be found in the summer in lagoon areas behind the fringing reef. Bonefish, by contrast, are found, even as adults, in sandy and muddy areas inside fringing reefs, marsas and shallow bays.

Herrings and sardinellas (Clupeidae), anchovies (Engraulidae) and silversides (Atherinidae) are all superficially similar small silvery schooling fishes. The atherinids have two dorsal fins whereas the fish of the other two families only have one. The common species are the herrings *Herklotsichthys punctatus* and *H. quadrimaculatus*, the sardinella *Amblygaster sirm*, the round-herrings *Spratelloides delicatulus* and *S. gracilis*, the anchovies *Stolephorus heterolobus* and *Thryssa baelama*, and the silverside *Atherinomorus lacunosus*. Large schools of these fish may be found offshore, or especially within the coastal creeks and marsas, where they are caught by local fishermen with cast nests for use as bait. The herrings and sardinellas apparently make much the best bait and are much preferred by local fishermen, while silversides are only used when herrings or anchovy are unavailable. According to Reed (1964) even the local cats turn their noses up at silversides. Despite this the large silverside schools often encountered around offshore reefs and islands seem to be constantly under attack by jacks and horsemackerel, needlefish and various species of terns, to all of which they appear to be an important source of food.

Though most of the Red Sea herrings are not sufficiently abundant to be exploited commercially in their own right, in the more extensive shallow water areas of the Gulf of Suez and the coast of Eritrea large catches are made using seine nets.

13.5. SURFACE-DWELLING FISHES

For convenience, two unrelated groups of fish are included in this section, firstly the needlefishes, halfbeaks and flyingfishes, which are related to each other, and spend much of their time immediately below the ocean's surface, and secondly the barracuda which, although they may be seen at greater depths, are typically observed within the top few metres. These fish, which tend not to be distinctively coloured, are difficult to identify to species level without close examination.

Needlefishes are predatory, feeding on small schooling fishes such as sardines and silversides. They have a very elongated form with the mouth extended into a long beak bearing numerous fine teeth, and their coloration is blueish or greenish above and silver below. Both form and coloration adapt them for living close to the surface of the sea against which they can be astonishingly well camouflaged, helping them both to approach prey as well as avoid predators. From beneath, the sea's surface often appears as a pattern of horizontal blue/green and silver bands, the result of waves on the surface reflecting or transmitting light. Arranged parallel with this pattern each needlefish seems to be just another wave. Alternatively, needlefishes may slowly approach schools of small fish head-on. When they face their prey all that can be seen of them is a small cross sectional view, suggesting quite a small fish rather than a potential predator.

The halfbeaks resemble the needlefishes in general form, but they lack the upper half of the beak (hence the name) and are rather smaller. They feed on plankton but their form and coloration helps to camouflage them like needlefish against the surface of the sea, reducing their chances of detection by predators. However, when detected and chased by a predator they can jump almost clear of the surface and propel themselves along by rapid beating of their tail which remains in contact with the surface. The halfbeaks are closely related to the flyingfishes and their behaviour seems to represent an intermediate stage through which that of the true flyingfishes must have evolved. The flyingfishes use their tail in the same way to gain speed or sustain flight, and once launched with the wings formed by the spread pectoral fins, they can glide, under favourable conditions, for 200 m or more. Again this behaviour is used to escape predators, although it is often elicited by the approach of a boat which is presumably taken for a particularly voracious predator.

Barracuda resemble needlefish in their elongated body form and predatory habits except, of course, that they are rather larger (some species growing to a length of 1.5—2 m) and will tackle larger prey. There are several species present in the Red Sea, some of which can be difficult to identify with certainty underwater. Some like the Great barracuda (*Sphyraena barracuda*) are usually seen alone, while others such as the Pickhandle barracuda (*Sphyraena jello*) are often seen in large schools. Solitary barracuda may scout along the edge of a reef, often orientated head on to potential prey, thus greatly reducing their apparent size in the same manner as needlefish. Should opportunity present, the barracuda can accelerate with great speed to strike a victim. The elongated muscular form of both barracudas and needlefishes not only assists with disguise but is optimal for such rapid acceleration. Single barracuda are notorious for stalking divers or snorkellers. If one tries to chase the fish away it may swim off briefly but a few minutes later will reappear. Slightly unnerving though this may be, in practice we have never known such a barracuda to attack a diver, and it seems likely that the barracuda is following on the offchance that the diver might disturb some prey which it can then snatch.

Schooling barracuda can show a perhaps more unnerving behaviour in that they may start circling round a diver or group of divers, effectively forming a large wall of perhaps a hundred or more hungry looking fish. According to local fishermen this behaviour is often used when the barracuda encounter a school of herrings. By slowly tightening their revolving circle around the herrings, the herrings are at first forced to bunch, and then to panic and scatter when they can more easily be picked off by barracuda. Again barracuda have not been known to attack divers in this way and almost all human injuries from barracuda have been the result of the fish mistaking the leg of a bather or an arm dangling over the side of a boat for some much smaller prey, often in waters where visibility is poor.

13.6. TUNAS AND MACKERELS

Several species of tuna and mackerel have been reliably recorded for the Red Sea. These are all rapid swimming predators of smaller fishes in the open sea, with the exception of the Indian mackerel, (*Rastrelliger kanagurta*), which is a plankton feeder. This species trawls for plankton with open mouth, plankton being sieved from the water by dense gill-rays, in the same way as the Whale shark and manta rays feed. Groups of Indian mackerel are a common sight just off reefs; they swim in a dense formation, mouths wide open, and in a distinctive irregular manner move upwards and downwards, presumably seeking the densest patches of plankton.

The most important commercial species is the Narrow-barred Spanish mackerel (*Scomberomorus commerson*) which reaches a metre or more in length. This species migrates regularly in schools through inshore waters to reach selected spawning areas in early summer and returns to more extensive feeding grounds during late summer and winter. They are usually caught by trolling, but in certain areas at the right time of year large numbers can be caught with gill nets as they migrate.

The commonest tuna is the Kawakawa (*Euthynnus affinis*) characterised by the pattern of dark wavy lines along the posterior part of its back. According to Reed (1964) they are commonest inshore during December to March when they can be caught around the entrances to various marsas along the coast; these may well be spawning areas. The flesh however is strongly flavoured and so this species is not popular locally.

Of the larger tuna, the Dogtooth tuna (*Gymnosarda unicolor*) is encountered most frequently, but even this species is not common. Unfortunately for local fishermen large pelagic populations of the larger tuna do not occur in the area, probably because of the low productivity of the nutrient-poor open waters of the Red Sea.

Also worth mentioning are the two species of large game fish known to occur in the Red Sea. Both the Sailfish (*Istiophorus platypterus*) and the dolphinfish (*Coryphaena hippurus*) are quite frequently seen in the open sea. The large sail-like dorsal fin of the Sailfish drifting along near the surface can be spotted from some distance, and dolphinfish can be seen as they leap out of the sea often two or three times in quick succession, their long dorsal fins erect. However, neither species is common and they are rarely caught in the area.

13.7. JACKS AND HORSEMACKERELS

The principal species of jack and horsemackerel found in the Red Sea are discussed by Wray (1979) who carried out extensive work on the fisheries of the Saudi Arabian Red Sea coast. Almost all the species are fast swimming predators of herrings, silversides or small reef fish. Together these species represent one of the most important, if not the most important element in Red Sea fisheries.

The species of *Caranx* and *Carangoides* are all very similar in shape, stream-lined, with lunate (forked) tails designed for fast efficient swimming. Near reefs, the most frequently seen species are *Carangoides bajad*, *Carangoides fulvoguttatus*, *Caranx melampygus*, *Caranx sexfasciatus* and *Alepes djedaba*, and they are amongst the most important predators of small reef fish.

A variety of strategies are used by jacks to capture prey. *Carangoides bajad*, *Caranx melampygus* and *C. fulvoguttatus* swim rapidly along the sides of reefs, accelerating around bluffs and outcrops to surprise small fish before they can retreat to hiding places. Often these jacks hunt in small groups in which more than one species may be present. Larger groups of 10—20 or so individuals may form to attack large schools of prey species. We have on many occasions observed large schools of atherinids gathered on the reef top under attack by such groups. The jacks charge at and around the prey school, breaking it up and splitting off small groups and individuals which can then be caught more easily. Similar behaviour by *Caranx* species has been observed in the Indian Ocean by Eibl-Eibesfeldt (1962). The way in which, in the Red Sea *Caranx* gather by the reef edge, out of view of an atherinid school, and then commonly set off simultaneously in two or three groups to approach different sides of the prey school by different routes, suggests a marked degree of active cooperation. Major (1978) has established experimentally that groups of jacks in Hawaii can thus capture more prey from a large school than they would do when hunting alone.

Species of *Caranx* may also make use of the presence of other non-prey fish to facilitate the capture of food. Individuals or small groups may follow other predators which may either disturb prey or increase the effective size of the hunting group. Thus Hobson (1974) describes that in Hawaii *C. melampygus* commonly swim in association with the large piscivorous goatfish, *Parupeneus cyclostomus*. In the Red

Sea, *C. bajad*, *C. melampygus* and *C. fulvoguttatus* are often observed swimming close alongside or behind large non-piscivorous fish, especially the large Bumphead parrotfish, Humphead wrasse and Titan triggerfish, behind which they are able to approach close to potential prey (Ormond, 1980b). Also individual small *Caranx* often swim among schools of the similarly coloured fusilier, *Caesio caeruleus*, and are thus able to approach closely shoals of *Chromis* and other damselfish which never stray far from refuges on the reef; this facultative stalking behaviour is a form of aggressive mimicry.

Other carangids show a variety of forms and feeding behaviours. The queenfish species (*Scomberoides lysan* and *Scomberoides commersonnianus*) and the Rainbow runner (*Elagatis bipinnulata*) are more elongate and spindle shaped than the species so far considered. All three feed on Crustacea and small fish, and are associated with coastal reefs; according to Wray (1979) populations of queenfish may also live in deep lagoons. By contrast, the Snubnose pompano (*Trachinotus blochii*) feeds on molluscs and Crustacea which it crushes with the aid of pharyngeal bones in the throat. Pompanos are often seen in small schools, notably around fringing reefs by the entrance to marsas which they enter in search of shellfish and crabs. Perhaps the most unusually shaped of the Red Sea carangids is the Indian threadfish (*Alectis indicus*), a very deep bodied jack with, at least in the younger fish, the dorsal and ventral fins elongated into plumes that extend beyond the tail.

13.8. COMMERCIAL FISHES FROM LAGOONS AND MARSAS

In this section are included various moderately-common small to medium-sized fish which occupy sandy lagoon or marsa environments, and which tend to get taken in the hauls of commercial fishermen. They belong to eight unrelated families.

Of greater commercial significance are the grey mullet (Mugilidae), of which the four species *Crenimugil crenilabis*, *Mugil cephalus*, *Oedalechilus labiosus* and *Valamugil seheli* are most frequently caught. They are silvery-grey cylindrical fish, 20—40 cms long, which are generally seen in schools near the water's surface within lagoons or marsas or over the reef shelf. They feed on microalgae and detrital matter, often from the surface of the sand or from the material floating on the surface film of the water. Grey mullet are especially favoured as food by the local people in the Red Sea area. They are caught with gill nets, beach seines, or with specially designed verandah nets. With beach seine a pair of fishermen wade out into a fringing reef lagoon where mullet can be seen. On gaining the far side of the shoal the fishermen run as fast as they can through the water, one to either side of the shoal, unravelling the net as they go towards the shore in an attempt to surround the shoal or contain it against the beach. Verandah nets are used especially by Egyptian fishermen who in the winter sail down to the southern coast of Egypt and to Sudan in dhows known as *sambûk*. The verandah nets are set across the lagoons where the mullet occur at this time of year, the particular feature of these nets being an extension of the net from its upper edge back along the surface of the water on the outerside of the net. Mullet can see and avoid ordinary set nets by leaping over the top of them, but using a verandah net, the fish that leap over are caught in the extension or verandah beyond.

The fish in the next four families are often caught in set nets or with beach seines or cast nets within marsas. Mojarras (Gerreidae) are small thin silvery fish, 10 to 25 cm long; they have a curious extensible down-turned mouth and feed by taking in mouthfuls of sediment which they sift through for food. While the young are frequently seen in shallow water, larger individuals occur at depths down to 80 m (Wray, 1979) and are mostly caught by hand-lining. Often occuring in mixed schools with young mojarra are terapons (Teraponidae), particularly *Terapon jarbua*. Both these types of fish are eaten around the Red Sea, mojarra being particularly popular with local fishermen as a convenient and tasty breakfast, half a dozen or so medium-size fish being grilled over an open fire and the flesh picked off and eaten with bread. Larger fish are taken to market while the smaller fish are used for bait. The fish of the next two families though common in marsas and often caught in nets are not of commercial significance. The Mono (*Monodactylus argenteus*) is a laterally-compressed, round silvery fish, 20 cm or so long, with symmetrically pointed dorsal and anal fins, fast swimming schools of which are often conspicuous around quays and harbour works in the Red Sea. Quite dissimilar, the Arabian killifish (*Aphanias dispar*) is a small (5—7 cm) pale cylindrical fish abundant in shallow water on muddy-sandy bottoms within marsas, where it is able to tolerate the hypersaline conditions that occur there.

Flat fish are also not of commercial significance in the Red Sea, but are included in this section for convenience, being also characteristic of sandy bottoms. Two species are regularly recorded. The Moses sole (*Pardachirus marmoratus*) has attracted particular attention since it was discovered that it secretes a

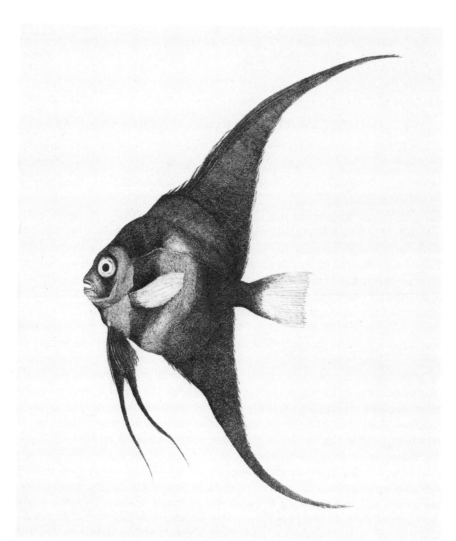

Fig. 13.2. The batfish *Platax teira*, discovered by Forsskål at Al Luhayyah on the coast of North Yemen in early 1763. This is one of only two Red Sea fishes for which drawings by the Danish Expedition's artist, Georg Wilhelm Baurenfeind, survive. First published in 1776. (Photo: G. Brovad; Zoologisk Museum, Copenhagen)

milky white toxic fluid. In experiments it was found that the flesh of this species or of other fish treated with the poisonous fluid was avoided by sharks, which veered away just as they were about to take the bait; the possibility thus arose that this substance might be used as the basis for a first effective shark repellent.

13.9. EELS AND CATFISHES

A large number of species of moray and conger eel are known from the Red Sea, but only the common and readily recognisable ones are discussed here. The largest species and the most frequently seen on reefs through most of the Red Sea is the Giant moray (*Gymnothorax javanicus*) which grows to 2 m or more. Also common are the related Yellowmargin and Yellowmouth morays (*Gymnothorax flavimarginatus* and *G. nudivomer*) which grow to a metre or more. More typical of sandy areas with scattered boulders and corals, or in seagrass beds, are the Snowflake moray (*Echidna nebulosa*) and the Grey moray (*Siderea grisea*) both of which are frequently encountered in the Gulf of Aqaba area.

Conger eels are more typical of deep water but the Moustache conger (*Conger cinereus*) is quite often encountered in shallow water. Much better known is the related garden eel (*Gorgasia sillneri*), colonies of which live in individual burrows in flat sandy areas at depths of 20 m or more, often among seagrasses. These eels hold their bodies upright, two-thirds out of their burrows, and gently waving their heads to and fro appear to feed on larger planktonic organisms drifting by. The larger moray and conger eels are principally predators of fish which are caught while passing the holes in the reef in which the eels are generally found. The eels are not normally aggressive to divers, unless provoked, but occasionally mistake hands for prey and can inflict nasty wounds.

Distantly related to the eels are the eel catfish, one species of which, the Striped eel catfish (*Plotosus lineatus*) is not uncommon in the Red Sea. The adults are up to 25 cm long, dark above with two longitudinal white stripes. In the young, dense schools of which can be found among boulders in sandy reef areas or among seagrass beds, the barbels and fins and front of the stripes are bright yellow. This colouring may be a warning of the venom glands associated with the single sharp spines in the dorsal fin and each pectoral fin. The poison is apparently very toxic and one is advised against handling these attractive fish. The Giant catfish, *Arius thalassinus* is the only other catfish reported from the Red Sea.

13.10. OTHER ELONGATED FISH SPECIES IN THE RED SEA

For convenience various unrelated fish species all of particularly elongated form are discussed together in this section. Despite the similarity of shape these species pursue very different life styles.

The Sharksucker (*Echeneis naucrates*), is the common remora seen clinging to the underside of sharks, rays and turtles in the Red Sea. It has a large sucker (a modified spinous dorsal fin) on the top of its head, two thin white lines down the side, and may grow to a metre in length. In return for free transport and morsels of food left when the larger fish feed, the remoras are believed to tackle parasites on their hosts.

The tilefishes (Malacanthidae) are to some extent reminiscent of the remoras in general proportion and, in the case of the Striped blanquillo (*Malacanthus latovittatus*), in coloration. They are commonly encountered, usually in pairs, near the bottom on sandy or rubble areas below the reef face and when approached quickly retreat into burrows.

The cornetfishes (Fistulariidae) are very elongate cylindrical fish, with tiny dorsal and ventral fins at the base of the pointed tail, large eyes and a long tubular mouth. The common reef species is *Fistularia commersonii*, not *F. petimba*, which occurs in deeper water (Randall, 1983). Cornetfish are normally pale green above and silvery below, but they have a masterful ability to change colour to camouflage themselves against their surroundings. They can hang almost indetectable in the water column, but often align themselves close to gorgonians, anchor ropes or other fishes which they use as camouflage. They thus stalk small fish or Crustacea which are sucked in through the tubular mouth in vacuum cleaner fashion.

The flatheads or crocodile fish (Platycephalidae), as the name suggests, are broad dorso-ventrally flattened fish with crocodile-like heads. They are able to camouflage themselves against sand and typically bury themselves close by coral patches, waiting to ambush passing prey.

The pearlfish (*Carapus variegatus*) has surely found the most extraordinary way of protecting itself. This slender, elver-like fish uses the intestine of a holothurian (sea-cucumber) as a burrow, retreating there through the anus of the sea-cucumber when threatened.

13.11. GROUPERS AND THEIR ALLIES

The main species of grouper and coral trout found in the Red Sea are now discussed along with the species of several closely related groups — soapfish (Grammistidae), jewelfish (Anthiinae), dottybacks (Pseudochromidae) and longfins (Plesiopidae). Randall and Ben-Tuvia (1983) distinguish several species of grouper which had previously been confused. For example the Vermilion (*Cephalopholis oligosticta*) and Sixspot grouper (*C. sexmaculata*) have frequently been mistaken for colour forms of the Coral grouper, *C. miniata*.

As shown by Hiatt and Strasburg (1960) the coral trout and groupers of the family Serranidae are the major predators of smaller fish on reefs and associated habitats. Some species are very common in the Red Sea, particularly the Peacock (*Cephalopholis argus*), Halfspotted (*C. hemistiktos*), Redmouth (*Aethaloperca rogaa*), Lunartail (*Variola louti*) and Roving grouper (*Plectropomus maculatus*). In some parts of the Red Sea, for example along the northern part of the Egyptian coast, Brownspotted grouper (*Epinephelus chlorostigma*) is particularly abundant, whilst in the Gulf of Aqaba, the Blacktip (*Epinephelus fasciatus*) is the most abundant of the smaller groupers.

The smaller grouper species, i.e. the *Cephalopholis* spp. and the Roving and Redmouth groupers are found in shallower parts of the reef and are active by day. Larger species and especially the larger individuals of these occur at greater depth; they are most active around sunrise and sunset, and some may hunt as much by night as by day; some groupers feed on crustaceans as well as fish, especially at night (Hobson, 1974).

All the groupers essentially employ a stalk and/or ambush strategy for hunting, and are equipped with large mouths and deep throats into which large prey can be gulped. Peacock grouper tend to rest on small coral banks and wait for prey, or move up to and wait by crevices or interstices which they have seen potential prey to enter. Coral and Sixpot grouper tend to stalk prey or lie in ambush under overhangs or within large crevices on the reef face. The Redmouth moves along in a head-down position over coral mounds, pausing in the water when prey appears, whilst the Roving grouper lies in ambush on sandy slopes along the reef front, and can undergo marked colour changes to camouflage itself. *Epinephelus fuscoguttatus* rests on its pelvic fins on the sea bed, often behind coral mounds or outcrops, waiting to ambush passing prey; and the very largest groupers such as *E. malabaricus* are reported to lie with mouths open inside caves ready to suck in any small fish which may enter. Some groupers opportunistically follow schools or lone non-prey fish which may incidentally flush out prey as they move along the reef (Fricke, 1972). In a number of species including *C. argus*, *P. maculatus* and *E. fuscoguttatus*, a loose form of group hunting may occur and two or three individuals will lie in wait not far from each other; thus prey which escapes one may be driven towards another. The groupers, like the jacks and snappers, constitute a major portion of the catch of local fishermen throughout the Red Sea.

The other fish families discussed in this section are also predators but of greatly varying sized prey. At one extreme are the jewelfish or *Anthias*, small bright orange-red species, shoals of which add so much colour to the reef edge and reef face zones of steeper more exposed reefs in the Red Sea. These fish are predators, but of zooplankton, and so have evolved to a style of life similar to that of the green and grey damselfish species (*Chromis*) with which they mix. The common species is *Anthias squamipinnis*, but in the Gulf of Aqaba a second slightly larger species (*A. taeniatus*) is also found.

Slightly smaller still are the dottybacks (Pseudochromidae). These are small elongate predators of benthic Crustacea, polychaete worms and zooplankton, each adapted to a particular microhabitat. The three most common and conspicuous species are the Orchid dottyback (*Pseudochromis fridmani*), the Sunrise dottyback (*P. flavivertex*) and the Olive dottyback (*P. olivaceus*). The Orchid dottyback is a brilliant almost luminescent purple colour, and is characteristic of overhangs and cliff-like faces on the reef; it is endemic to the Red Sea. The Olive dottyback is commonly found among the branches of the corals *Stylophora* or *Pocillopora*, and the Sunrise dottyback is common among coral interstices in sand-dominated coral areas, for example among corals within marsas.

The attractive longfins (Plesiopidae) grow up to about 15 cm long and get their name from their particularly long pelvic fins. They are secretive, generally being seen among coral crevices or withdrawn beneath a coral head.

The Serranidae are typically protogynous hermaphrodites, initially maturing as females but, if conditions are right, changing sex to become males later in life. In some species even functional hermaphrodites, that is individuals ripe with both milt and roe, have been found (Reinboth, 1970). The same is probably true of the other serranid related families discussed here. As first recognised by Fishelson (1970), the Scalefin anthias (*Anthias squamipinnis*) is a protogynous hermaphrodite; its social and sexual behaviour has been studied in detail by Shapiro (1978) and social control of sex-change demonstrated. The species live in groups, each group being a harem presided over by a male. By their social interactions the males inhibit the development of the older females into males, but when a male is removed the dominant female rapidly changes into a male and in turn interacts with the other females to prevent them too from changing sex.

13.12. NOCTURNAL SPECIES:
SOLDIERFISHES, SQUIRRELFISHES, CARDINALFISHES AND GOGGLE-EYES

In this section various families are covered which are almost exclusively nocturnal, and which together account for most of the fish which are active on Red Sea reefs by night. By day holocentrids (squirrelfishes and soldierfishes) are usually seen sheltering under overhangs or below corals. Soldierfishes often pass the day at the entrance to a reef crevice or cave which they thus appear to be guarding, hence their common name; while squirrelfishes gain their name from their large eyes, an adaptation to feeding at night. The cardinalfishes (Apogonidae) are much smaller and shelter among or under corals or within the reef, and so the abundance of the 35 or so species known from the Red Sea can be difficult to assess. With perhaps the largest eyes of all these nocturnal species are the Goggle-eyes (Priacanthidae), deep-bodied reddish fish which, like the holocentrids, often shelter under overhangs or in crevices during the day time.

The diet of soldier- and squirrelfishes has been studied by Hobson (1974) and Vivien and Peyrot-Clausaude (1974); all feed on invertebrates. The species of *Sargocentron*, most commonly referred to as squirrelfishes, are bottom feeders, eating Crustacea and polychaetes. The Crown squirrelfish (*Sargocentron diadema*) often feeds in sandy areas, but the others prefer harder substrates. All tend to remain not far from cover even when feeding. By contrast the species of *Myripristis*, (those most frequently referred to as soldierfishes) feed in mid-water on planktonic invertebrates. Eight species of holocentrid are known from the Red Sea; most are relatively common and widespread but the Yellowtip soldierfish (*Myripristis xanthacrus*), which was only separated from the common Blotcheye soldierfish (*M. murdjan*) by Randall and Guézé (1981), is endemic to the central and southern Red Sea and Gulf of Aden.

A distinctive and frequently seen nocturnal fish is the red Goggle-eye (*Priacanthus hamrur*), up to 35 cm long, with a markedly upturned mouth, and, as its name suggests, a conspicuously large eye. The large eye and red coloration are adaptations to nocturnal behaviour, species of *Priacanthus* being predators of Crustacea and cephalopods. Cardinalfishes have also been shown to feed on small Crustacea (Hiatt and Strasburg, 1960; Hobson, 1974). Most feed on planktonic forms in mid-water, but some feed on or near the bottom.

Three further nocturnal species are conspicuous in the Red Sea. Sweepers (Pempheridae) are abundant on some reefs, large schools assembling by day in crevices in the reef. According to Randall (1983) the Red Sea species are *Pempheris varicolensis* and *Parapriacanthus guentheri*. Somewhat similar in habit, but generally occurring at much greater depth is the Flashlight fish (*Photoblepharon palpebratus*), characterised by a luminescent kidney-shaped light organ below each eye. The light is produced by bacteria cultured within the organ, and is apparently used to attract and detect planktonic prey, and also help the fish in a school to keep in contact with one another. The species is especially well-known in the Gulf of Aqaba where, unusually, it is quite common in shallow water, and often observed and photographed by divers.

13.13. BOTTOM-LIVING PREDATORS

Several distantly related families of small species (15—30 cm long) which rely on camouflage and ambush to catch prey (mainly small fish and invertebrates) are now considered.

Both the hawkfishes (Cirrhitidae) and the lizardfishes (Synodontidae) spend much time resting immobile on the sea bottom in a slightly head-up attitude, supported by their pelvic fins; they move only occasionally, to change position, to chase prey, or when threatened. The species of hawkfish are coloured so as to be well camouflaged when positioned on their preferred substrate. Of the Red Sea species the Blackside hawkfish, (*Paracirrhites forsteri*) lies in wait mainly on live coral, especially *Porites* and *Pocillopora*, as described by Hobson (1974): the Pixy hawkfish (*Cirrhitichthys oxycephalus*) waits among rubble and small rocks; the Stocky hawkfish (*Cirrhitus pinnulatus*) ambushes among algal covered stones in shallow water; and the Longnose hawkfish (*Oxycirrhites typus*), an attractive little fish with a pattern of orange vertical and longitudinal stripes, is coloured to blend in perfectly with the orange-red sea-fans on which this species typically settles. Both lizardfish and sandperch (Mugiloididae) are a pale sandy colour to merge in with their background; two common lizardfishes (*Synodus gracilis* and *Synodus variegatus*) and the sandperch (*Parapercis hexophthalma*) are often seen lying in wait on sand not far from coral patches, ready to pounce on unsuspecting prey. Although hawkfishes and lizardfishes are mostly active by day, some species may also feed at night; in particular *S. variegatus* appears to feed regularly at night as well as by day. Most of the above species feed principally on fish except for the Longnose hawkfish which preys on small Crustacea.

The Scorpaenidae (scorpionfishes, stonefishes, and lionfishes) are less mobile than the hawkfishes but have developed an essentially similar strategy to catch prey. Both the false stonefish (*Scorpaenopsis* spp.) and the true stonefish (*Synanceia verrucosa*) lie immobile among corals and rocks, well disguised as algal covered stones or boulders. This mimicry is perfected in *S. verrucosa* which has numerous small growths on the skin resembling strands of filamentous algae. The lionfishes (or turkeyfishes), *Pterois radiata* and

P. volitans are more active hunters, and may be seen cruising or hovering in mid-water. Their distinctive form, with greatly enlarged pectoral fins, and irregular red and white banded coloration, appear to serve several functions. Firstly when waiting motionless in mid-water the lionfish may be mistaken for drifting pieces of algae; thus they may be ignored by prey, and small prey may even try to seek cover among their fins; this disguise may also protect *Pterois* from larger predators. Secondly, as we have observed in the Red Sea, the pectoral fins may be used to herd and drive small fish into a corner or alcove where they can be caught and eaten. Thirdly the elongated spines of the dorsal, anal and pelvic fins are sharp, have poison glands at their bases and can inject a powerful venom; the red and white banding serves as a conventional coloration warning of such defences to potential predators. The poison from some of these *Pterois* can cause agonising pain and even kill — so beware!

13.14. GOATFISHES

The goatfishes (Mullidae) are benthic feeders which have a pair of long submandibular barbels which they use to detect food. Generally they are predators of small soft-bodied invertebrates, but despite similarity in form and size, the species show a marked degree of specialisation as regards habitat and feeding behaviour; and, as described by Hobson (1974), while some species are diurnal others are nocturnal. The three commonest Red Sea species are the Yellowsaddle goatfish (*P. cyclostomus*) which is typical of rich coral areas, Forsskål's goatfish (*P. forsskali*) which feeds over bare rock areas of the reef, and the Longbarbel goatfish (*P. macronema*) which is seen mostly in sandy and lagoon areas. These species feed by day, but the species of *Mulloides*, which by day are seen in stationary schools above sandy bottoms' feed by night.

While most goatfishes feed on small Crustacea and polychaetes, the Yellowsaddle goatfish *P. cyclostomus* takes mostly small benthonic fish (Hiatt and Strasburg, 1960; Hobson, 1974). As described by Ormond (1980b), it is probably to enable it to catch such prey more efficiently that this fish is commonly observed hunting in small groups, or in association with the Bird wrasse (*Gomphosus caeruleus*). The goatfish winkles prey out of coral heads with its barbels, while the Bird wrasse uses its elongated snout. Goatfish and wrasse may follow one another across the reef, and on reaching a coral head one may tackle it from either side, so that prey disturbed by each fish may be driven towards and caught by the other.

13.15. SNAPPERS AND FUSILIERS

The snappers (Lutjanidae) and fusiliers (Caesionidae) are two families which although rather different in behaviour are nevertheless fairly closely related.

The snappers are, as mentioned above, one of the most important families of fish in the area from the commercial fisheries' point of view. The species discussed here fall into two ecological groups; firstly, those species such as the Black-and-white snapper, (*Macolor niger*) and most species of *Lutjanus*, which are fairly commonly seen around coral reefs. Secondly the red snapper (*Etelis carbunculus*) and the jobfishes (*Pristipomoides* spp. and *Aphareus rutilans*) which are all deep water fish typically occurring on rocky or muddy sea beds at 100–200 m depth. Reed (1964) described an interesting method used by local fishermen to catch these deep water species. At the end of a long hand-line several hooks, baited with sardine, are tied together with half a dozen loose sardines and one or two suitable sized stones, using the adjacent part of the line. The line is tied using a loose slip knot. The stones carry the hooks and bait down to the depth where the fish are thought to occur; the jerk as the stones reach the depth to which the line has been paid out, or a sharp pull by the fisherman, releases the knot so that the stones fall away, leaving the loose bait and baited hooks to attract the fish. Some skill is needed to hook the fish since the line needs to be struck as the fish bite, but a skilled fisherman can bring up a line with two or three snappers hooked at a time.

Snappers are essentially predators of medium sized invertebrates, such as Crustacea and small echinoids, and of small fish (Hiatt and Strasburg, 1960; Randall, 1955). The majority, are also nocturnal feeders, occurring by day in relatively inactive schools which disperse at nightfall to forage as solitary individuals. Thus in the Red Sea, Bluestripe (*Lutjanus kasmira*), Blueline (*L. caeruleolineatus*) and Dory snapper (*L. fulviflamma*) form schools which shelter near overhangs or in gullies near the reef crest, often where there is some surf, and wave surge. The movement of these schools amongst the brightly coloured corals is a spectacular and much photographed sight. Other species such as Humpback (*Lutjanus gibbus*) and Onespot snapper (*L. monostigma*) aggregate by day lower down the reef, in inlets on the reef slope or among coral knolls. Yet other species, notably Twinspot (*Lutjanus bohar*) and Black-and-white snapper, form relatively inactive groups in mid-water, 10−20 m away from the reef edge or above the reef slope. These two species take at least some food during the day, and this is presumably an advantage of spending the day in this location; should potential prey be seen abroad on the reef while it is still light, the fish can drop down to snatch it.

The Twin-spot snapper, whose scientific specific name derives from its local arabic name, 'bohar', is commercially probably the single most important species in the whole Red Sea.

The fusiliers (Caesionidae), although closely related to the snapper, are somewhat different in both form and behaviour. They are medium sized (15-30 cm) spindle shaped fish, living in shoals just off the reef front, and feeding on plankton. Most are blue grey in colour, matching the background colour of the shallow sea, and some have longitudinal dark and/or yellow stripes. The Lunar fusilier (*Caesio lunaris*), pale blue-grey above, silvery below and with black tips to each fork of its tail, is particularly common.

13.16. EMPERORS, GRUNTS, BREAMS, CHUBS AND PORGIES

The emperors or emperor-bream (Lethrinidae) are closely related to the snappers, and like them are predators of medium sized invertebrates, including molluscs, crabs and urchins, and many are primarily nocturnal feeders. There are interesting behavioural differences in the ways in which some of the species forage. The Blackspot emperor (*Lethrinus harak*) and the Yellowstripe emperor (*L. ramak*) both search actively for food by day. The former is associated with sandy areas and can be common in lagoons; its pale sandy coloration, save for a black and yellow mark on the posterior flank, blends well with its surrounding, and the fish tends to hover motionless 10−30 cms above the sand waiting for suitable prey to show themselves. Alternatively this species, like a number of wrasse, follows lagoon rays (*Taeniura lymma*) that are digging in the sand, waiting for them to disturb potential prey. By contrast the Yellowstripe frequently hunts around small holes and burrows in the sand. The fish approaches a likely looking burrow or depression in the sand, and comes to rest in a nose-down position with its snout touching the sand just beside the hole. The fish assumes a mottled camouflaged pattern and stays immobile until the prey appears, or until it gives up and moves away.

Of the nocturnally active species some have characteristic day time behaviours. Mahsenas (*Lethrinus mahsena*) tend to hover, perhaps two or three together, a metre or two away from the reef face, especially near hummocks of *Porites* coral, against which the yellowish banded pattern on their flanks provides an element of camouflage. While the Big-eye emperor *Monotaxis grandoculis* spends the day almost motionless in small loose-knit groups in mid-water 10−20 m off the reef slope. This species is common on many Red Sea reefs, and its large eyes and characteristic behaviour make it easy to recognise. Although both species feed mostly at night (Hobson, 1974), they will drop to the sea bed to take prey during the day if the opportunity presents.

A fifth particularly common emperor is the Spangled emperor (*Lethrinus nebulosus*) which can occur in large numbers over extensive areas of shallow sea bed, 10-20 m deep. Where such habitat is predominant, such as to the north of Ghardaqa in the Egyptian Red Sea, this species is of considerable economic importance.

The sweetlips, rubberlips and grunts (Haemulidae) are again predominantly nocturnal in habit, feeding on various medium-sized invertebrates. This family is characterised by thick heavy lips which presumably give some protection from the spines of the animals on which they feed. The Black-spotted grunt (*Plectorhynchus gaterinus*) is especially common and conspicuous, small groups often being encountered by day in gullies along the reef face and being a favourite subject for photography. The Harry hotlips (*Plectorhynchus gibbosus*) is also fairly common, may grow to well over a metre, and can often be approached very closely by divers. This can tempt spearfishermen to tackle it, which is unfortunate since the fish is not particularly good eating.

The sea bream or porgies (Sparidae) are a further group of predators of medium sized invertebrates, principally Crustacea and molluscs, which they crush with molar-like teeth. The Doublebar bream (*Acanthopagrus bifasciatus*) is the most conspicuous species on Red Sea reefs, being readily recognisable by the two vertical black bands across the face. The arabic name of this fish is *abu kohol* which literally translates as 'father-of-the-eye-liner', kohol being the black graphite-like material used by local women to paint their eye lids. The other common species of grunt and bream, the Silver grunt (*Pomadasys argenteus*) and the Yellowfin bream (*Rhabdosargus sarba*) are most usually encountered within marsas.

Two other families will be mentioned briefly. The sea chubs (Kyphosidae) are often encountered in schools, milling about at the ends of deep water reefs. They are omnivorous but feed heavily on benthic algae. Also there is one common species of spinecheek (Nemipteridae), the Dotted spinecheek (*Scolopsis ghanam*), which is common in fringing reef lagoons and on sandy parts of the reef throughout the Red Sea. Particularly conspicuous are a yellowish line over the snout from eye to eye, and its jerky 'pause-and-dart' locomotion as it swims along above the sand searching for food.

13.17. BUTTERFLYFISHES AND ANGELFISHES

Amongst the most beautiful and brightly coloured reef fishes are the butterfly and angelfishes. There are seven species endemic to the Red Sea and Gulf of Aden area (Fig. 13.3). Some such as the Exquisite (*Chaetodon austriacus*) and the Brownface butterflyfish (*C. larvatus*) are common throughout the Red Sea; these two species being the most abundant on almost all Red Sea reefs. Others have a more restricted distribution. The Red Sea bannerfish (*Heniochus intermedius*) is the sole representative of this genus through most of the Red Sea, but it appears to be replaced by the Pennantfish (*H. diphreutes*) in the northern part of the Gulf of Aqaba, and this latter species is also common in the south-eastern Red Sea. The Blackback (*Chaetodon melannotus*) has a very patchy distribution, but is found throughout the Red Sea, while the Crown butterflyfish (*C. paucifasciatus*) seems to occur mostly in the northern half of the Red Sea, and the Paleface (*C. mesoleucos*) mainly in the south. The Yellow-ear angelfish (*Apolemichthys xanthotis*) and the Zebra angelfish (*Genicanthus caudovittatus*) seem both to be restricted to the lower parts of the reef, at 30 m or more, near deep water, except that both occur in comparatively shallow water at the head of the Gulf of Aqaba.

The butterflyfishes (Chaetodontidae) are typically feeders on coral polyps and hydroids and/or other small invertebrates such as polychaete worms. There is probably a marked element of resource

Fig. 13.3. A group of Masked Butterflyfish, *Chaetodon semilarvatus*, with above them a school of fusiliers (Caesionidae) off Dhahrat Miraya Island in the outer Farasan bank, Saudi Arabia. This butterflyfish is one of seven species which are found only in the Red Sea. (Photo: TMRU/Dawson–Shepherd)

partitioning among the Red Sea species. Thus Exquisite, Brownface, and Chevron butterflyfish (*C. trifascialis*) feed predominantly on corals, the first species taking especially *Pocillopora* spp., the second especially *Goniastrea* spp. and the third especially *Acropora* spp. The three corresponding Pacific forms were reported by Reese (1977) to be similarly obligate coral feeders. In addition to taking the polyps themselves, it has been shown by Benson and Muscatine (1974) that the mucus produced by corals has a high nutritional value and may be important to these fishes as food. Whitebrow, Threadfin and Paleface butterflies (*C. fasciatus*, *C. auriga* and *C. mesoleucos*) take a greater proportion of hydroids and small non-coral invertebrates. In the Red Sea each species tends to be associated with a different substrate: Whitebrow feed most frequently in coral areas with coral rock supporting a short filamentous algal lawn; Threadfin are associated with coral rock covered with a fine layer of sand; while Paleface are typically found in the more sandy parts of the reef, e.g. the lagoon. Hobson (1974) describes that in Hawaii *C. lunula* (the Indo-Pacific form of the Whitebrow) and *C. auriga* feed on a wide variety of invertebrates, especially on polychaetes, and that *C. lunula* is a nocturnal feeder. In the Red Sea, Whitebrow, although feeding in part by day, spend much of the daytime inactive beneath coral overhangs, suggesting that they must also feed by night; on similar grounds Masked butterflyfish (*Chaetodon semilarvatus*) would also appear to be a nocturnal feeders.

Of the angelfish (Pomacanthidae), species of *Pomacanthus* have been found by Hiatt and Strasburg (1960), Hobson (1974), and others to feed almost entirely on sponges, while species of *Centropyge* (represented in the Red Sea only by *C. multispinis*) feed almost exclusively on algae and detritus. By contrast the Zebra angelfish (*Genicanthus caudovittatus*) is a mid-water plankton feeder, a habit also shown occasionally by a few butterflyfish.

The butterfly- and angelfishes are often though not always territorial, and in the Red Sea, we have found that most of the smaller butterflyfish species occur in pairs, each occupying a territory from 10–20 m in diameter, while the larger species of butterfly and angelfish have individual or pair territories that may be up to 0.5 km or more across.

13.18. DAMSELFISHES

The damselfishes (Pomacentridae) are one of the most abundant families of fish on coral reefs in the Red Sea. The swarms of small chromis, 5–10 cm long, hovering and feeding just beyond the reef edge, against the backdrop of the blue sea beyond, are typically the strongest impression that a visitor takes away from his first snorkel on a Red Sea reef. There are huge numbers of the Bluegreen chromis (*Chromis caerulea*), the Ternate chromis (*Chromis ternatensis*) and the half black and half white Half-and-half chromis (*Chromis dimidiata*). These are accompanied by shoals of slightly larger (10–20 cm), black and white banded, Sergeant Major and Scissortail sergeant (*Abudefduf saxatilis* and *Abudefduf sexfasciatus*), which rise slightly higher, almost blending with the surface waves. Also present are often schools of the Scalefin anthias (*Anthias squamipinnis*), discussed above in the section on groupers and related species. All these species feed in mid-water on zooplankton brought to the reefs by the ocean currents, but stay close enough to the reef so that they can retreat to shelter when threatened.

The above species are only five of the thirty-five or so species known to occur in the Red Sea. The identification of many of the less abundant species and even of one or two quite common forms had been in doubt until the publication of a major study by Allen and Randall (1980) and a subsequent paper (Randall & Allen, 1982) which cleared up the taxonomy of this group in the Red Sea.

Apart from the group of reef edge plankton feeders, several other ecological groups of damselfish species can be recognised. Some, like those mentioned above, are shoaling planktivorous species, but tend to occur in slightly deeper water, say 10–20 m; these include three other chromis species, the Yellow-Edge, Triplespot and Weber's chromis (*Chromis pembae*, *C. trialpha* and *C. weberi*). Other species while primarily planktivorous occur as scattered individuals over the reef face, for example, the all-yellow Sulphur damselfish (*Pomacentrus sulfureus*) and the greyish Whitebelly damselfish (*Amblyglyphidodon leucogaster*), species which are abundant on almost all Red Sea reefs.

Some species have developed a particular relationship with certain types of coral or other coelenterate. Schools of Bluegreen chromis (*Chromis caerulea*) normally retreat to large tables of *Acropora* coral when chased, and often spawn in dense aggregations over dead *Acropora* tables. But Humbug (*Dascyllus aruanus*) and Blackbordered dascyllus (*D. marginatus*) live in small groups each permanently associated with an individual coral colony of *Stylophora*, *Pocillopora* or *Acropora*. These species of *Dascyllus* feed primarily on plankton, often on larger lagoon dwelling plankton (Emery, 1968); the larger fish of a group may venture a little way (up to a metre) from the colony, but the smallest individuals barely leave the shelter of the coral branches. This close association with coelenterates has been taken furthest in the clownfishes which live among the tentacles of giant sea-anemones. The clownfishes secrete a special mucus which does not stimulate discharge of the stinging cells (nematocytes) of the tentacles of the anemone, while any other fish which comes too close will be immediately bombarded by the stinging cells and severely stung or killed. Thus the clownfish gains protection from the anemone from its predators, while apparently the anemones may benefit from the clownfishes' habit of retrieving items of food, some of which may then be eaten by the clownfish. The common Red Sea species of clownfish is the Twobar clownfish (*Amphiprion bicinctus*), but the Domino Damsel (*Dascyllus trimaculatus*) is also seen in the Red Sea in association with giant anemones, although it doesn't swim among the tentacles to anything like the extent of the clownfish.

A fourth group of damselfish are specialised algal feeding species which feed primarily on the thin coating of filamentous algae with which rocky surfaces on the reef typically become covered. This coating of algae is known as the algal lawn. Most of these damselfish are territorial, and aggressively defend small territories against not only other members of their own species, but against most other algal-feeding fish. As a result the algal lawn within these territories becomes much thicker and taller, perhaps 2–5 cm high, enabling relatively blunt-mouthed damselfishes to browse the algae more easily. Such behaviour by damselfish is widespread (Low, 1971; Myrberg, 1972) and has been studied in the

Red Sea by Vine (1974). The Red Sea species which show this behaviour are Whitebar (*Plectroglyphidon leucozona*) and Jewel damselfish (*P. lacrymatus*), and especially the Black damselfish (*Stegastes nigricans*) which typically occurs in colonies of twenty to fifty or more fish, each pugnaciously defending a small territory (perhaps 0.5 m across) among rubble in the shallow part of the reef against other herbivores. As a result of their defence the rubble becomes thickly coated in green algae (Vine, 1974).

In addition to the feeding behaviour, the social and reproductive behaviour of damselfishes has been studied in some detail in the Red Sea, in particular by Fricke (1973, 1977).

13.19. WRASSES

Apart from semi-cryptic forms, such as gobies and blennies, the wrasses, in the Red Sea at least, are the most diverse family of fishes. Randall (1983) states that sixty species are known from the Red Sea.

The identification of wrasses has been complicated not only by the variety of sometimes similar forms and by the uncertain relationship to Indian Ocean species, but also by the fact that in many species male and female, and in some cases the juvenile too, are so different from each other that they have been described as separate species. Thus male and female Bird wrasse (*Gomphosus caeruleus*), both with an elongated snout, but the former dark blue, and the latter green above and pale yellow below, were associated at a fairly early stage by Strasburg and Hiatt (1957), whereas male and female Bluelined wrasse (*Stethojulis albovittata*) were only associated by Randall and Kay (1974), and male and female Eyefin wrasse (*Halichoeres marginatus*) by Randall (1980), the female having been generally known until then as *H. notopsis*.

The commonest species in the Red Sea are several of the medium-sized forms, 10–20 cm long, including in particular the very abundant common green Klunzinger's wrasse (*Thalassoma klunzingeri*) and also the Moontail (*Thalassoma lunare*), Chequerboard (*Halichoeres hortulanus*) and Bluelined wrasse. Of the larger species the Sling-jaw wrasse (*Epibulus insidiator*) and several species of *Cheilinus* are quite common, including the very largest species, the Humphead (*Cheilinus undulatus*) which grows to over 2 m long. The Cleaner wrasse (*Labroides dimidiatus*) is also very common, as are the two small species of *Pseudocheilinus* which are only 5 to 8 cm long and move cautiously from coral head to coral head.

Most wrasse, as described by Hiatt and Strasburg (1960), and Hobson (1974) are essentially feeders on gastropods and Crustacea which they are able to crush with strong pharyngeal teeth. However the size of the prey and the feeding habits of the different species vary considerably. The species of *Stethojulis* pick small Crustacea and gastropods off the open reef, while the slightly larger *Halichoeres* and *Thalassoma* species feed on a wide variety of smallish benthic invertebrates. The species of *Cheilinus* take correspondingly larger prey, the Broomtail (*Cheilinus lunulatus*) and the Red-breasted wrasse (*Cheilinus fasciatus*) quarter the reef picking medium-sized (0.5–3 cm) gastropods from rubble or sand. The Humphead *C. undulatus* has been found by Randall *et al.* (1978) from analysis of stomach contents to feed on crustaceans and larger molluscs, including full-size cowries, and also on sea-urchins and small fish. Ormond (1980b) has observed how the wrasse obtains this food by upending, and lunging down into large crevices and under coral overhangs when there is often a loud report of the fish striking and sometimes breaking coral or rock; the large hump on the forehead may serve to protect the fish's head during this manoeuvre.

At the other end of the size range are at least two species which feed on zooplankton — the Social wrasse (*Cirrhilabrus rubriventralis*) and Eightline wrasse (*Paracheilinus octotaenia*); and Abel (1960) and Randall and Springer (1975) report that adult Fourline wrasse (*Larabicus quadrilineatus*) feed primarily on coral polyps.

While feeding generally on appropriately sized invertebrates many of the Red Sea wrasse have evolved special behaviour to help obtain their food. Best known is the Cleaner wrasse which picks ectoparasites from client fish which come to the cleaners' feeding station on the reef. Their behaviour has been studied in detail by Eibl-Eibesfeldt (1955) and Potts (1973). Other wrasse species are known to show similar behaviour on either a regular or occasional basis, and Ormond (1980a) has described that, in the Red Sea, juveniles of Klunzinger's, Moontail, and Fourline wrasse will all act as 'cleaners', picking at items on the bodies of larger fish. The latter in particular show such behaviour fairly frequently and it is interesting that their coloration is fairly similar to that of the Cleaner wrasse.

Several species regularly turn over rocks or excavate the sand to find food; these include not only the Rockmover wrasse (*Novaculichthys taeniourus*) but also the larger species of *Coris* (*Coris aygula* and *C. gaimard*). Fricke (1971) describes how at Aqaba, Abudjubbe (*Cheilinus abudjubbe*) crush sea-urchins by carrying them to and striking them against suitable rocks. The Sling-jaw wrasse as its name suggests, has a highly protractile mouth which it can extend suddenly to snatch crustaceans, and probably also small fish from crevices and interstices among coral.

Some species may associate with other quite different fish in order to obtain food. At the simplest level, various Red Sea wrasse are sufficiently opportunistic to follow any larger fish or animal which during its own activities may expose food that can be taken by the wrasse. Checkerboard, Zig-zag (*Halichoeres scapularis*) and Red-breasted wrasse regularly follow lagoon rays or goatfishes in this way, and Klunzinger's and Moontail wrasse are attracted as soon as a large parrotfish or triggerfish, or even a diver, starts turning over or breaking coral.

More specialised adaptations are those in which predatory wrasse will associate with and mimic harmless species in order to approach their own prey in disguise (examples of 'aggressive mimicry'). Ormond (1980b) has described how the Sling-jaw wrasse will turn black to mimic and accompany black herbivorous Sailfin tang (*Zebrasoma veliferum*), while Mental wrasse (*Cheilinus mentalis*) will change colour to mimic non-predatory parrotfish or goatfish.

The social and reproductive behaviour of the wrasses is also of considerable interest, because, as established by Reinboth (1973) and Robertson and Choat (1974), most if not all coral reef wrasses are protogynous hermaphrodites; that is the typical males develop by sex change from the older females. In addition, in quite a number of species there is a second form of male which is born as such, but looks like the typical female. These males are termed primary males, while those that become males on sex change are termed secondary males. As described by Dawson-Shepherd (1981) for Klunzinger's, Moontail and Bluelined wrasse, these two types of male differ not only in coloration, but also in reproductive behaviour. The secondary males are typically territorial and spawn in a pair with one female at a time, while the primary males spawn in a group with numbers of males and several females coming together in a dense shoal.

13.20. PARROTFISHES

The parrotfishes are perhaps the family most clearly associated with shallow tropical waters; they are comparatively large (25—75 cm) torpedo shaped fish, with pronounced beaks formed by the fusion of their anterior teeth. These teeth, and the gaudy coloration of the fish, together with their mode of swimming by beating their pectoral fins, do indeed frequently give them the appearance of parrots flying through an underwater tropical jungle.

The identification of the Red Sea parrotfishes has been even more problematic than that of the wrasses or other families. In almost every case male and female differ in coloration and have originally been described separately in the scientific literature; in addition the characteristic coloration fades almost immediately on death making it difficult to give a scientific description which can be related to the appearance of the fish underwater. Different scientific workers have found it difficult to identify their own fish with descriptions given by others, and so have provided new descriptions and names; as a result some species have been given as many as twenty scientific names at different times. The clarification of the identify and correct scientific name of the different Red Sea species is largely due to the efforts of Randall.

Probably the commonest Red Sea species, numerous on all reefs in the area, is the Bullethead parrotfish (*Scarus sordidus*), the females of which are dull brownish-black with a pattern of eight or so regularly spaced white spots on the body. The males, by contrast, are bluegreen with orange cheeks. Also particularly common around Red Sea reefs are Rusty (*Scarus ferrugineus*) and Longnose parrotfish (*Hipposcarus harid*). The Bluebarred parrotfish (*Scarus ghobban*) is locally abundant and often the main species in shallow sandy reef areas, for example within the Gulf of Suez. Two species belonging to the more primitive sub-family (Sparisomatinae) with much less pronounced beaks — Dotted (*Calotomus viridescens*) and Slender parrotfish (*Leptoscarus vaigensis*) are more characteristic of seagrass beds or areas dominated by algae.

The parrotfishes were for a long time thought to be primarily feeders on coral, their beaks being adapted, it was thought, specially for this task; but while by many authors they were described as feeding on coral, other authors denied this. The situation was reviewed by Randall (1974) and it is now quite clear that parrotfishes are primarily grazers of algae, the beaks being adapted to scraping away at the thin algal lawns of diatoms and green algae that grow on most rocky surfaces of the reef. However, it seems that while in some parts of the world such as the Caribbean and the Great Barrier Reef live coral is rarely if ever taken, in parts of the Pacific (Hiatt and Strasburg, 1960, Bakus, 1969) and in East Africa (Goldman and Talbot, 1976) some species do take significant amounts of coral. This latter situation has been found by Ormond (1980a) to be the case in the Red Sea. While Bullethead, Purple parrotfish (*Scarus niger*), and females (initial phases) of most other species feed entirely on the algal lawn, larger males (secondary phases) of some of the other species will scrape at living coral, particularly the smooth domed colonies of *Porites* on which their teeth marks can frequently be seen. Males of Bicolor (*Cetoscarus bicolor*) and Steephead parrotfishes (*Scarus gibbus*) in particular take significant quantities of coral, while the largest species, the Donkey-fish or Bumphead (*Bolbometopon muricatum*) seems to be primarily a coral feeder. Schools of 50 or so large Bumpheads (up to 120 cm long) may descend on the reef and attack areas of *Porites* or *Acropora*, biting off chunks 2–5 cms in size with their strong jaws — an awesome sight.

The social and reproductive behaviour of parrotfishes, like that of the wrasses to which they are closely related, is complex and interesting. Like the wrasse, the parrotfishes are protogynous hermaphrodites, the colourful typical male form developing by sex change from the drabber female. In some species, as with the wrasse, there are also primary males which resemble females in appearance. In most Red Sea parrotfish species I have observed that individual secondary males hold territories on the reef within which also live from two to ten or more females, effectively the harem of the male. At spawning the male spawns in a pair with one or more of his females in turn.

13.21. SURGEONFISHES, UNICORNFISHES AND RABBITFISHES

The surgeonfishes (Acanthuridae), so-called from the sharp scalpel-like spines placed on either side of the caudal peduncle (Fig. 13.4), are mostly specialised algal browsers feeding, like the parrotfishes, on the filamentous algal lawn which develops on rocky areas on the reef or elsewhere in shallow water. Surgeonfishes have mouthparts specialised to enable them to pluck at algal strands and fronds, but in general, unlike parrotfishes, they cannot rasp at the reef to scrape off the finest algae. Instead they use morphological and behavioural specialisations to gain prior access to algae.

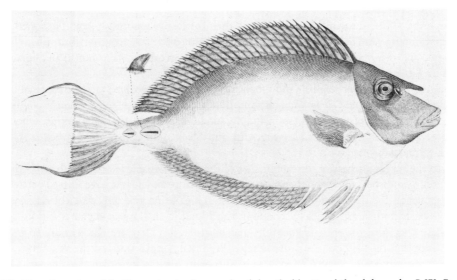

Fig. 13.4. The Bluespine unicornfish, *Naso unicornis*, discovered and described by Forsskål and drawn by G.W. Baurenfeind while the ill-fated Danish Arabia Felix Expedition was at Jiddah in late 1762. A year later both scientist and artist were dead. (Photo: G. Brovad, Zoologisk Museum, Copenhagen)

The most conspicuous of the Red Sea surgeonfishes is the Sohal (*Acanthurus sohal*). This species, endemic to the Red Sea and Arabian Gulf, is brightly coloured with blue fins, blue and white longitudinal stripes along the body, and white cheeks, the latter giving the impression of a fish wearing a surgical mask, as well as carrying scalpels. The fish is abundant in the shallow surf zone on Red Sea reefs where groups may often be seen swirling about and chasing each other. *A. sohal* is able to outcompete the parrotfishes on this part of the reef, partly by an ability to manoeuvre and feed in surge that other fish would find difficult to cope with, and partly by being highly aggressive towards other grazing species which enter their individual territories. Intruders are driven away, the surgeonfish swimming close by and swiping at their opponents with the sharp caudal spines raised. The Sohal thus maintain territories within which the algal lawn grows higher, enabling them to browse more effectively, just as described for several species of damselfish. Interestingly, the surgeonfish nevertheless tolerates within its territory one or two species of herbivorous damselfish, particularly the Onespot damsel (*Chrysiptera unimaculata*). It seems that the small damselfish not only takes but a small proportion of the algae, but also makes a significant contribution to the defence of the area against other herbivores, particularly at dawn and dusk when the Sohal may be travelling to or from a sleeping shelter in deeper water.

Of the other surgeonfishes two brown species are also abundant on Red Sea reefs. The Brown surgeonfish (*Acanthurus nigrofuscus*) is also an algal browser, while the Lined bristletooth (*Ctenochaetus striatus*), similar in overall appearance apart from distinct yellow-orange pectoral fins, apparently feeds on detritus.

Apart from territorial defence, another technique has been evolved by some species of surgeonfish to gain access to denser areas of algal lawn. This is to invade as a school the individual territories defended by the Sohal or by damselfish species which show similar behaviour. Such invading behaviour was first described by Barlow (1974) for the surgeonfish *Acanthurus triostegus* in Hawaii, and in the Red Sea has been described by Ormond (1980b) as being especially characteristic of the Sailfin tang (*Zebrasoma veliferum*). Typically in late afternoon large schools of Sailfin tang may form and invade the algal territories of other surgeonfishes or damselfishes, overcoming the individual territorial holders by sheer weight of numbers. The surgeonfishes will defend their territories vigorously, and Ormond (1980b) has even reported finding dead Sailfin tang left in the water after such an encounter. But usually the Sailfins are successful, and neighbouring Sohal, as well as passing parrotfish, then join in the pillaging of the algal lawn in the lost territory.

Other related species of surgeonfish and unicornfish feed on macroalgae rather than algal lawn, these species including Orangespine (*Naso lituratus*) and Bluespine unicornfish (*N. unicornis*). Yet others feed, at least as adults, on zooplankton, these including species which are typically seen in schools in mid-water off the reef — Longhorn (*Naso brevirostris*) and Sleek unicornfish (*N. hexacanthus*), and Black surgeonfish (*Acanthurus nigricans*).

Related to the surgeonfishes are the rabbitfishes of the family Siganidae. These fish are fairly similar in shape to the surgeonfishes; they lack the spines on the caudal peduncle but have instead particularly well developed spines on the dorsal and ventral fins, these spines being not only particularly sharp but also venomous. The rabbitfish are typically schooling fish found in shallow sandy areas and in lagoons and marsas, where they feed on both filamentous and macroalgae. They are able rapidly to modify their coloration, adopting mottled or sandy camouflage. Presumably both this and the venomous spines help to provide some protection against predators in the open habitat in which they occur. One species however, the Stellate rabbitfish (*Siganus stellatus*), is more brightly coloured and is the one most regularly seen, usually in pairs, in the reef environment.

13.22. BLENNIES

The blennies are a large family of small (2—10 cm long) fishes, approximately forty species of which occur in the Red Sea (Randall, 1983). The majority are semi-cryptic in nature and so less frequently noticed by divers or snorkellers, but Randall describes 16 species which are quite frequently observed.

The majority of the blennies are herbivorous, feeding on the filamentous algal lawn on rocky surfaces on reefs or rocky shores, but some species have become plankton eaters, while one specialised group, the sabre-tooth blennies, have become parasitic on larger fish species.

The herbivorous blennies include several species, notably Chestnut (*Cirripectes castaneus*), Red-Dotted (*Istiblennius periophthalmus*) and Jewelled blenny (*Salarias fasciatus*), which are especially abundant on the reef flats and one-metre terraces of Red Sea reefs. Here they play a major role in restricting the growth of the algal lawn, particularly in situations where this is inaccessible to larger species of fish and sea urchin. Each blenny is closely associated with one or more small holes on the reef, often the empty burrow of a vermetid mollusc; the blenny retreats here frequently, especially when any hint of danger threatens.

Some blennies have successfully invaded the intertidal habitat where they feed on the algal lawn of rocky shores. Rock-skippers (*Istiblennius edentulus*) are common, especially where flat rock, such as a fossil reef flat, is exposed by the fall of the tide. Here they compete for the algal lawn with algal eating grapsid crabs; they move by short jumps and flips of the tail, retreating into the water when they become too dry or are threatened by the approach of a predator from land. More remarkable though is the Leaping blenny (*Alticus kirkii*). This species occurs on rocky shores, and is well adapted to situations where waves regularly break over them. Recently, to the north of Jiddah, this species has colonised the rocks and boulders which form the new shore as a result of the construction of the 'corniche' road. Leaping blennies can cling both to the sides and tops of boulders, even as waves wash over them, probably aided by suction developed on the ventral surface between the pelvic fins. In addition the blenny can leap, up to a metre or more, from boulder to boulder, and can thus escape most potential predators.

The plankton feeding blennies are of particular interest because of the occurrence, in the commonest Red Sea examples, of some well developed cases of mimicry; this allows these species, which are otherwise less agile than many planktivorous fish, to gain some protection from predators under false pretences. The Midas blenny (*Ecsenius midas*), an orange-yellow species with a long forked tail, was first observed by Starck (1969) to join schools of Scalefin anthias (*Anthias squamipinnis*) to feed on zooplankton. By its mimicry of the *Anthias* the blenny avoids being conspicuous, and benefits from the anti-predator advantages of being in a school.

Another planktivore, the Blackline blenny (*Meiacanthus nigrolineatus*) takes more direct action against predators; it possesses a pair of large canine teeth, each with a venom gland at the base, and as shown by Losey (1972), when captured by larger fishes the blenny can bite them in the mouth, thus forcing an escape. The predators thus learn to avoid attacking this species, an effect which is used by a second harmless blenny, the Red Sea Mimic blenny (*Ecsenius gravieri*), which resembles the Blackline not only exactly in coloration, but also in its manner of swimming.

The third group of blennies, the sabre-tooth blennies, feed by attacking passing fish and biting small lumps of scales or skin from their body. There are four moderately common Red Sea species. The Scale-eating blenny (*Plagiotremus tapeinosoma*) is the commonest, and will equally take a nip at a passing diver or snorkeller as at any fish. Another well-known species, the Mimic blenny (*Aspidontus taeniatus*), mimics the Cleaner wrasse, and uses both the characteristic nod-swimming of the cleaner and its resemblance in colour to fool larger fish into approaching it. A third sabre-tooth blenny, the Bluestriped blenny (*Plagiotremus rhinorhynchos*), also has some resemblance to the Cleaner wrasse, while *Plagiotremus townsendi* mimics the Blackline blenny, described above, gaining not only protection from this disguise, but also the chance to approach larger fish more closely by being taken for a 'harmless' planktivore, rather than a parasitic sabre-tooth.

13.23. THE GOBIES

The gobies, like the blennies, are small semi-cryptic species; over 75 species occur in the Red Sea but only the most frequently observed will be considered here.

The gobies are in general predators of small invertebrates, and because of their small size are able to colonise comparatively open habitat by living in burrows or among interstices of the coral or rock. Those typically abundant on reefs and in shallow water around the Red Sea are the Decorated goby (*Istigobius decoratus*) and Eyebar goby (*Gnatholepis anjerensis*), both of which are a mottled sandy colour.

Sand living gobies acquire their burrows in one of two ways. Some species build their own, excavating tunnels by scooping up sand in their mouth which they spit out on returning to the entrance. These species often live in pairs, and include Maiden and Sixspot goby (*Valenciennea puellaris* and *V. sexguttata*) and Tailspot goby (*Amblygobius albimaculatus*). Usually the burrows are either under stones or in sand, where there are sufficient pieces of shell, coral and small stones to support the sand, and prevent collapse of the burrow.

The second approach is to make use of burrows excavated by alpheid shrimps, and there are at least seven species of Red Sea goby which live, typically as pairs, in symbiotic association with alpheid shrimps. The shrimps are usually about 5 cm in size, and have large pincers with which they excavate the burrow, pushing small loads of sand and stone up to the entrance in the fashion of miniaturised bulldozers. The shrimps have a very poor vision and, in return for a home, the gobies stand on guard just in front of the entrance while the shrimps are excavating. The shrimps remain close to the gobies, usually maintaining physical contact with their antennae; should the goby detect a potential predator it moves back to the mouth of the burrow, the movement of the fish's tail sending the shrimps scuttling into hiding. The gobies also tend to carry back larger pieces of food to the mouth of the burrow before consuming them, and it seems likely the shrimps may get some food in this way. The gobies which live in association with these shrimps include three species of *Cryptocentrus* — Ninebar, Luther's and Blue-and-red-spotted gobies; two species of *Amblyeleotris* — Steinitz's and Magnus's, both of which are a pale sandy colour with five or six orange or brown rings around the body; and the little *Lotillia graciliosa*, which is black with a white forehead.

A third group of gobies live between the dense branches of live corals, usually table-form *Acropora*. In the Red Sea these include the Citron goby (*Gobiodon citrinus*), bright yellow with several pale blue lines down the face, and the Lime goby (*Gobiodon rivulatus*), pale green with red lines across the face and down the back.

Finally some gobies, like some blennies, have become plankton feeders. They feed within a metre or two of the substrate, but in the face of any danger retreat into burrows or crevices on the reef. These are the hovergobies, so called from their habit of alternately hovering in mid-water and suddenly darting one way or another. They typically occur in small shoals of from several to 50 or more individuals, a number of which may share the same burrow as a retreat. Several species can be very conspicuous on Red Sea reefs, catching the eye both as a result of their behaviour, and their large widely spread dorsal and anal fins. These include the Smallscale hovergoby (*Ptereleotris microlepis*) and the Bluefin hovergoby (*Pogonoculius zebra*). They usually occur on flat fairly open areas of reef rock near or on a shallow reef-slope.

13.24. TRIGGERFISHES, PUFFERFISHES AND ALLIES

The pufferfish-like group of fishes includes species of very varied form, which tend to use gentle undulations of the dorsal and anal fins to propel themselves forwards (or backwards) through the water. The tail is used only in rapid swimming, for example to escape a predator or threaten an intruder. The use of the dorsal and anal fins for swimming, rather than the body musculature, has enabled the body to become specialised in various ways — to become encased within a bony trunk, as in the boxfishes (Ostraciidae), or to have the ability to swell up by taking in water, as in the pufferfishes (Tetraodontidae).

The most familiar family within the order are the triggerfishes (Balistidae), medium-large diamond-shaped fishes with a conspicuous large dorsal spine, and behind it a second smaller spine, the 'trigger', with which the fish can lock the first spine upright so as to jam itself in a coral crevice or hole in the sea bed. There are four particularly common Red Sea species. Two smaller species are seen on the reef, Orange-striped (*Balistapus undulatus*) and Picasso triggerfish (*Rhinecanthus assasi*), the former more typical of dense coral areas, the latter of sandy patches and shallow lagoons. Two larger species, although they come on to the reef to feed, are much more wary and are usually seen swimming a little way off from the reef slope. These are the Titan (*Balistoides viridescens*) and Yellowmargin triggerfish (*Pseudobalistes flavimarginatus*); again the former species is more typical of coral areas while the latter tends to be associated with deeper lagoons and sandy areas. In the northern Red Sea the Yellowmargin seems to be replaced by the Blue triggerfish (*Pseudobalistes fuscus*).

Triggerfishes have large teeth and powerful jaws which enable them to feed on a variety of armoured invertebrates. The smaller species have been found by Clark and Gohar (1953) to take a variety of Crustacea and molluscs, and also small echinoderms, coralline algae, and in the case of the Orange-striped triggerfish significant amounts of the tips of branches of *Acropora* coral. The larger species can tackle larger or more heavily armoured prey including the large sea urchins *Diadema setosum* and *Echinometra mathaei*. Fricke (1972) has described how the Blue triggerfish lifts *Diadema* and drops it upside down, thus breaking many of the larger spines and exposing the less well protected underside, which has only short spines which the fish can tackle. Ormond *et al.* (1974) found that large triggerfishes, especially Titan triggerfish, are major predators of the Crown-of-thorns starfish in the Red Sea, and they tackle the starfish in essentially the same way, lifting it up and dropping it upside down so that it can be attacked on the unprotected ventral side. In a similar way the large triggerfishes may lift up and turn over pieces of rubble, dead corals, etc. to obtain prey hiding underneath; even quite large lumps, 20 cm or more across, can be turned over. Ormond (1980a) has also described how Titan triggerfishes tackle isolated branching coral colonies, typically *Pocillopora*. Initially they break off the tips of the coral branches, which are consumed; then the lower parts of the branches are broken off, and while these are not eaten, this exposes the various symbiotic crabs and shrimps sheltering within the colony which are caught in their turn.

The large triggerfishes more characteristic of sandy areas have another specialised technique. The Yellowmargin triggerfish feeds principally by excavating deep pits (about 1 metre wide, 20–40 cm deep) in sandy and lagoon areas. This is done by hovering head down over the sand, and blowing away the sand with a stream of water generated with the mouth and pectoral fins, while the dorsal and anal fins are used to maintain the fish's position. Invertebrate life exposed by the excavation is eaten.

Quite different is the feeding behaviour of the Swallowtail triggerfish (*Odonus niger*), which despite its scientific specific name is an attractive deep blue fish. This species occurs in shoals in mid-water over reef slopes at 10–20 m depth where it feeds on zooplankton.

Of the other families in this group, the filefishes (Monacanthidae) are closely related to the triggerfishes, being fairly similar in form but slenderer, with minute scales, and with a single large dorsal spine placed further forward almost above the eyes. Thirteen species have been recorded from the Red Sea (Randall, 1983), but the four discussed here are the only ones at all often observed. The Broom and Wirenet filefishes (*Amanses scopas* and *Cantherhines pardalis*) grow to about 15–20 cm long and feed on coral, by biting at the tips of coral branches, and also on algae and a variety of benthic invertebrates such as sponges, hydroids, amphipods and polychaetes. The much smaller Harlequin filefish (*Oxymonacanthus halli*), green with a regular pattern of large bright orange-yellow spots, is a specialist coral feeder, and is seen in small groups among the branches of arborescent corals, especially *Acropora*, feeding on individual polyps. By contrast Scrawled filefish (*Aluterus scriptus*) grow to over a metre. In the West Indies this species feeds principally on algae and hydroids, including the stinging or fire coral *Millepora*! Observations in the Red Sea suggest that by means of their extraordinary elongated form, and their manoeuvrability, they are able to reach down into coral crevices, for example between *Porites* mounds, to pick at hydroids and algae inaccessible to other large fish.

There are two common boxfishes in the Red Sea. The Bluetail boxfish (*Ostracion cyanurus*), which is typically only 10 cm or so long, has a greenish back and tail; and Cube boxfish (*O. cubicus*) grows to half a metre or so and is yellow-ochre in colour with a regular array of white, black-edged spots. The boxfishes feed on algae and a wide variety of invertebrates, including polychaete worms, sponges and tunicates.

The pufferfishes divide into three morphological and ecological groups — Tetraodontinae, Canthigasterinae, Diodontidae. Of the large pufferfishes three are relatively common. The Masked puffer (*Arothron diadematus*), which is a little smaller than the other two species, is much the most numerous and can be seen in shallow reef areas on almost any snorkel swim in the Red Sea. Randall (1974) found that the very similar *Arothron nigropunctatus* found on Indo-Pacific reefs feeds almost entirely on coral, and it seems that the Masked puffer does the same, nibbling at the tips of coral branches, especially *Acropora*. The other two species, however, the Bristly and Blackspotted puffers (*Arothron hispidus* and *Arothron stellatus*), which are not infrequently encountered either lower down the reef face, or in sandy lagoonal areas, feed on a range of heavily armoured invertebrates which they are able to tackle with their heavy beak-like teeth. In the Pacific the Bristly puffer has been found by Hiatt and Strasburg (1960) and Hobson (1974) to feed especially on sea-squirts, urchins and brittle-stars; in the Red Sea Ormond *et al.* (1974) found that Bristly puffer were, like Titan triggerfish, a major predator of the Crown-of-thorns starfish. Unlike the triggerfish, the pufferfish tackles the starfish from the upper side, but it is able to get at the sides of the bases of the arms, where large spines are absent.

The much smaller sharpnosed pufferfishes or tobies (Canthigasterinae) are delightful little fish, about 10 cm long, which row around among coral heads and crevices and take a wide range of small benthic plants and invertebrates, including coralline algae, sponges and some coral. Three species are known from the Red Sea, all pale with blue spots. The commonly seen species is the Pearl toby (*Canthigaster margaritata*); the Crown toby (*C. coronata*) is apparently seen mostly in deeper water, but like several other deeper water species occurs in shallow water in the Gulf of Aqaba. The Pygmy toby (*C. pygmaea*) is a tiny secretive species only 5 cm long, endemic to the Red Sea, where it was only recently discovered by Allen and Randall (1977).

Finally the porcupinefishes or spiny puffers (Diodontidae) are very similar in size to the large puffers but have the additional feature of spines or burrs arranged over the head and body. The commonest species is the Porcupinefish (*Diodon hystrix*) which, in some parts of the Red Sea, notably along the coast north of Jiddah, is so common that twenty or thirty can be seen in a half-hour swim. Like the large pufferfish, the porcupinefishes feed on a range of large invertebrates, particularly sea-urchins (Starck and Davis, 1966; Hobson, 1974). These they tackle principally at night, their large eyes equipping them for nocturnal activity, while by day they tend to shelter under corals and overhangs, especially under large *Acropora* tables.

13.25. SEAHORSES, PIPEFISHES AND OTHER CURIOUS FORMS

Finally, seahorses, pipefishes and a number of other curiously shaped forms will be considered. Seahorses and pipefishes are closely related, being placed together within the family Syngnathidae. Much the most frequently observed species listed is the Yellow-banded pipefish (*Corythoichthys flavofasciatus*) which is abundant in shallow reef areas, often occurring in a closely associated pair. Other species are usually found in algae or seagrass beds where the only frequently seen seahorse (*Hippocampus histrix*) is also found. Other seahorses from the Red Sea are thought to be mostly deep water species. The Syngnathidae feed by sucking in small planktonic and benthic Crustacea.

The frogfishes (Antennariidae) are curious globular fish whose form is further obscured by various outgrowths of fin and skin and well-developed camouflage. Some species, like anglerfishes, have a false 'bait' that projects forward above the mouth, and is used to attract fish which the frogfishes then gulp down. Others are able to wobble slowly forwards on stump-like pectoral and pelvic fins and surprise prey such as Crustacea or fish. Eleven species are known from the Red Sea (Randall, 1983)

Equally curious is a strange little fish called the Sea moth (*Pegasus draconis*), only about 10 cm long, with an elongated snout, and pectoral fins expanded into transparent semicircular wings. It is not infrequently seen at night in the Gulf of Aqaba, but little is known of the adaptive significance of such a curious form.

The habit of hiding in sand, shown by the flatheads or crocodilefishes, is taken to its extreme in the final species mentioned, the stargazer (*Uranoscopus*), a cylindrical fish with upturned eyes, which buries itself completely in the sand, leaving only a mouthful of teeth visible on the surface. Apparently it detects passing prey electrically.

13.26. ZOOGEOGRAPHY OF RED SEA FISHES

13.26.1. Palaeogeographic background

The present day distributions and affinities of the fish species of the Red Sea are largely a result of the Pleistocene history and unusual oceanographic features of the Red Sea basin (*see* Chapters 2 and 3 respectively). Just as the zoogeography the fishes cannot be understood without reference to the palaeogeography and oceanography of the Red Sea, so the present day relationships of the fish fauna can provide circumstantial evidence as to the likely course of events in the recent history of the region where this is still controversial.

Initially, during the Miocene epoch (from 25 to 5 million years ago), the Red Sea basin was a gulf of the then tropical Palaeo-Mediterranean and had no direct connection with the Indian Ocean. Even as far south as Ethiopia marine fossils from this period are 'Mediterranean' in character. In the latter part of the Miocene links between the Red Sea and Palaeo-Mediterranean basins across the Isthmus of Suez area were tenuous. Periodically inflow from the Atlantic into the Palaeo-Mediterranean and from there to the Red Sea became insufficient to replace water lost by evaporation and the Red Sea and Mediterranean formed a series of hypersaline evaporating basins (*see* Chapter 2) in which few fishes, if any, could have survived.

In the early Pliocene, about 5 million years ago, uplifting and increased deposition of Nile sediments in the Isthmus of Suez region led to a severing of connections between the Mediterranean and Red Sea (Por, 1978). At about the same time the Straits of Bab al Mandab began to open to the south as the Arabian plate rotated away from Africa, and the Red Sea was for the first time colonised by an Indo-Pacific fish fauna. Meanwhile, the Mediterranean was repopulated by tropical and subtropical eastern Atlantic fishes as a deep sea connection with the Atlantic was re-established.

There has been considerable debate as to whether the post-Miocene Red Sea remained connected to the Mediterranean for long enough to be repopulated by eastern Atlantic species. Gohar (1954) among others has suggested that there was a period during which there was an intermingling of Mediterranean and Indo-Pacific faunas in the Red Sea with the gradual replacement of the former by the Indian Ocean invaders. Other workers (*see* review by Por, 1978) consider that the Red Sea basin remained isolated from the Mediterranean from the late Miocene onwards and that it was a hypersaline lake, in which most, if not all, the pre-existing marine fauna had been exterminated, that was colonised by the Indo-Pacific immigrants of the early Pliocene. In either event, despite past claims to the contrary, there is no evidence that the descendants of any fish species of Palaeo-Mediterranean origin (so-called 'Tethys relicts' — Ekman, 1953) survive in the Red Sea today.

Events crucial to the present day zoogeography of the Red Sea occurred during the Pleistocene (*ca.* 2 million to 10,000 years ago). During this epoch global sea-levels rose and fell several times as seawater was periodically released from and trapped in the circumpolar ice caps. At the height of the glaciations sea-level fell as much as 120 m below the present day level, whilst during the warmest interglacial period it was 20 m higher than at present. The sole link between the Red Sea and the Indian Ocean, through the straits of Bab al Mandab, passes over a sill currently a little over a 100 m deep. Thus the Indo-Pacific fishes, which had established themselves in the Red Sea during the Pliocene, now became periodically almost completely isolated from those of the Indian Ocean during each glacial period. Whether complete isolation occurred at any time is still a matter of controversy, because it is unclear what vertical movements of the Earth's crust in the vicinity of the sill have occurred during or since the Pleistocene. Conditions in the Red Sea during the Pleistocene do not appear ever to have been as bad as during the Miocene, but at times of low sea level the considerably reduced inflow of water from the Indian Ocean would have been insufficient to compensate for evaporation, and marked increases in salinity must have occurred (*see* Chapter 2). Concomitant with these would have been falls in water temperature to as low as 13—14°C in the central Red Sea (Por, 1978).

Thus over the last 2 million years the Red Sea fishes have been subjected periodically to harsh conditions quite different from those prevalent simultaneously in the Indian Ocean. Those populations which survived the changing conditions, and persisted through the periods of high salinity and cool temperatures during sea-level falls, would have been both effectively isolated from counterpart populations in the Indian Ocean and subject to differing selection pressures. Thus conditions in the recent history of the Red Sea have been such as to promote speciation.

Even today the marine environment of the Red Sea is in several respects dissimilar to that of the adjacent Indian Ocean (see Chapter 3) and the unique tidal regime and absence of oceanic swell pounding on the coastline contribute to creating a reef and shallow water environment rather different from that in the Indian Ocean. These present day peculiarities of the Red Sea environment and the restricted connection through the Straits of Bab al Mandab between it and the Indian Ocean will help to maintain genetic separation of Indian Ocean and Red Sea sibling species, subspecies or populations. Thus, although now in contact, introgression (merging) of Red Sea forms with their Indian Ocean counterparts will not necessarily occur even though interbreeding may be possible, and further divergence may even be promoted.

Links between the Red Sea and Mediterranean fish faunas severed in the early Pliocene by uplifting of the Isthmus of Suez were re-established in 1869 on the opening of the Suez Canal. Initially a salinity barrier existed, the salinity in the Bitter Lakes being about 70‰. However, at present salinities in the Canal fluctuate around approximately 45‰ and rarely exceed 48‰, levels not dissimilar from those in the northern Gulf of Suez where the salinity may at times reach 44—46‰. Also, minimum sea temperatures at Suez are similar to those in the eastern Mediterranean. Thus species which can survive in the Gulf of Suez are pre-adapted for the high salinities of the Canal and low winter temperatures of the eastern Mediterranean. Several Red Sea fish species have thus been able to colonise both the Suez Canal itself and the eastern Mediterranean.

13.26.2. Zoogeographic relationships of Red Sea fishes

The distributions of 508 species of Red Sea fish in 86 families are analysed. The main references used were Botros (1971), Cohen (1973), Burgess (1978), Allen & Randall (1980), Allen (1981), Garrick (1982), Randall & Ben-Tuvia (1983) and Randall (1983). This analysis is summarised in Figure 13.5 and the ranges of the Red Sea species of 18 fish families which are well represented in the area are examined in more detail in Table 13.1.

70% of the species are widespread in the tropical Indo-Pacific region and are recorded from at least as far east as the western Pacific and often much farther afield. A further 4% of the Red Sea fish species are known from throughout the tropical Indian Ocean but do not occur in the Pacific, whilst 9% are confined to the western Indian Ocean only. A substantial proportion (17%) of the fish are found only in the Red Sea (and in several cases also the Gulf of Aden); these are all considered here as Red Sea endemics. In addition, of the 48 Red Sea species which are also known from the western Indian Ocean but not from farther afield, 7 are recorded from the Arabian Gulf only, and 6 have not been found beyond the Gulf of Oman. Thus 27% of the western Indian Ocean species are confined to the extreme north-west of the area and are probably of Red Sea origin (Fig. 13.5).

It is evident from these distributions that the Straits of Bab al Mandab and Gulf of Aden area is a barrier which cannot be crossed successfully, on the one hand by many species which have evolved to cope with the peculiar environment of the Red Sea, and on the other, by many of the species adapted to the conditions prevalent in the Indian Ocean. The high level of endemism may be ascribed to both the periods of partial isolation from the Indian Ocean during the Pleistocene low sea-level stands and the unusual oceanographic conditions still prevailing.

Many of the endemic Red Sea fishes have close relatives (sibling species) in the Indian Ocean and one can recognise species pairs, one species in each pair being found on either side of the Bab al Mandab/Gulf of Aden barrier (Table 13.2). In these cases the Red Sea species is assumed to have evolved from populations of the ancestors of the Indian Ocean species. In Table 13.1 only distinct species have been considered, however, several quite widespread fishes have forms which are peculiar to the Red Sea. These Red Sea forms often differ slightly in certain characters such as fin ray counts, body proportions or colour patterning. In some instances differences are statistical and only a proportion of individuals in the two populations can be distinguished, but in others there is a clear-cut, consistent difference between the Red Sea form and the Indian Ocean form. Where differences are consistent the Red Sea form has often been designated a subspecies (Table 13.3). Where differences are not clear-cut and merely statistical, such as in fin ray or scale counts, or in the relative proportions of the body, characters which are well known to be affected by environmental changes such as temperature and salinity (Marshall, 1952), there appears to be no justification for formally separating the Red Sea and Indian Ocean populations although these inter-population differences are of interest.

There would seem to be three likely origins for the endemic Red Sea and north-west Indian Ocean fishes. First, they could have evolved during the periods of semi- or complete isolation of the Red Sea during the Pleistocene glaciations when salinities were high and temperatures low. Secondly, they could

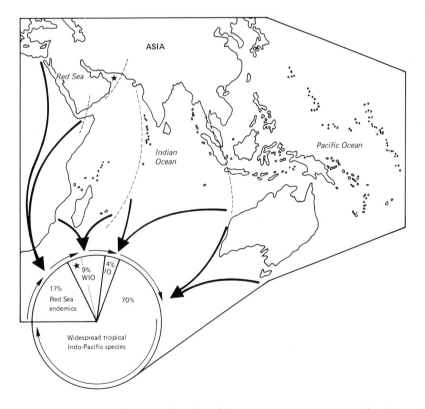

Fig. 13.5. Zoogeographic composition and ranges of Red Sea fish species. WIO = Western Indian Ocean. IO = Indian Ocean. The black star denotes those species endemic to the Western Indian Ocean which, outside of the Red Sea/Gulf of Aden area, are found in only the Gulf of Oman or Arabian Gulf; such species probably originally evolved in the Red Sea.

have evolved in the peculiar Red Sea environment from Indian Ocean species which recolonised the Red Sea after the last (Würm) glaciation (i.e. in the last 10,000 years or so). Thirdly, they may be relict species, that is species once widespread in the Indo-Pacific but now superseded elsewhere in the region and only surviving in this unusual pocket on the edge of their former range. There is no evidence that any are descendants of late Miocene Palaeo-Mediterranean species which have somehow managed to survive the vicissitudes of the Pliocene and Pleistocene.

Certain of the endemic species have no obvious close relatives in the Indian Ocean. In such cases, either the Red Sea species evolved a long time ago, perhaps early in the Pleistocene and has now diverged considerably from the Indian Ocean descendants of its ancestors, or possibly the ancestral species from which it evolved has since become extinct in the Indian Ocean. The existence of endemics without obvious Indian Ocean counterpart (sibling) species may be taken to support the contention that contrary to suggestions by Sewell (1948) the Red Sea fauna was not wiped out in the Pleistocene but, although no doubt decimated, survived, these endemics being the survivors' descendants.

For those species with clear Indian Ocean counterparts, or forms which have only warranted subspecific distinction, it is quite possible that they have evolved from Indian Ocean colonists since the last glaciation. However, rates of evolution vary so much according to circumstances and the taxa involved that no firm conclusions can be drawn.

Although the overall level of endemism in the fishes is approximately 17%, there is enormous variation in the proportion of endemic species in any one family (Table 13.1). This is to a large extent related to the habits and modes of reproduction of the families concerned. Broadly speaking, large strong swimming pelagic fishes which spawn in open water and have long larval lifetimes in the plankton will tend to be more widespread than small benthic territorial species who lay their eggs attached to the substrate and have short larval lifetimes. Families with particularly high levels of endemism are the small triplefins, Tripterygiidae (91% — Clark, 1980) ·and dottybacks, Pseudochromidae (90% — Edwards & Randall, 1983), whilst those with no endemic representatives among the species analysed are the jacks (Carangidae), snappers (Lutjanidae), emperors (Lethrinidae) and

TABLE 13.1. Distribution of the species of 18 major families of Red Sea Fish (Mediterranean immigrants are not considered). The percentage of species in each zoogeographic category is given for each family.

FAMILY	Number species which invaded eastern Mediterranean	Distribution of Red Sea fish species in the Indo-West Pacific				Number of species analysed
		Confined to Red Sea and Gulf of Aden	Confined to western Indian Ocean	Confined to Indian Ocean	Wide ranging Indo-Pacific species	
Carcharhinidae (Requiem sharks)	—	0	7	7	86	14
Serranidae (Groupers and seabasses)	1	12	8	0	80	26
Pseudochromidae (Dottybacks)	—	90	10	0	0	10
Apogonidae (Cardinalfishes)	1	6	12	6	76	17
Carangidae (Jacks)	2	0	0	0	100	11
Lutjanidae (Snappers)	1	0	9	0	91	11
Lethrinidae (Emperors)	—	0	0	0	100	12
Mullidae (Goatfishes)	2	10	0	0	90	10
Chaetodontidae (Butterflyfishes)	—	47	6	0	47	15
Pomacentridae (Damselfishes)	—	21	26	3	50	34
Labridae (Wrasses)	—	14	11	7	68	56
Scaridae (Parrotfishes)	—	33	0	7	60	15
Tripterygiidae (Triplefins)	—	91	9	0	0	11
Blenniidae (Blennies)	—	30	11	4	55	27
Gobiidae (Gobies)	—	13	9	4	74	23
Acanthuridae (Surgeonfishes)	—	9	9	18	64	11
Balistidae (Triggerfishes)	—	0	20	10	70	10
Tetraodontidae (Pufferfishes)	1	45	0	0	55	11

TABLE 13.2. Examples of Red Sea: Indian ocean species pairs.

FAMILY	RED SEA SPECIES	INDIAN OCEAN COUNTERPART
Chaetodontidae	*Chaetodon fasciatus* Forsskål	*C. lunula* (Lacepède)
	Chaetodon paucifasciatus Ahl	*C. mertensii* Cuvier
Pomacanthidae	*Holacanthus xanthotis* Fraser-Brunner	*H. xanthurus* Bennett
Pomacentridae	*Chromis pelloura* Randall & Allen	*C. axillaris* (Bennett)
Labridae	*Cheilinus abudjubbe* Rüppell	*C. trilobatus* Lacepède
	Thalassoma klunzingeri Fowler & Steinitz	*T. quinquevittatum* (Lay & Bennett)
Scaridae	*Scarus fuscopurpureus* (Klunzinger)	*S. russelii* Valenciennes
Balistidae	*Rhinecanthus assasi** (Forsskål)	*R. aculeatus* (Linnaeus)
Monacanthidae	*Oxymonacanthus halli* Marshall	*O. longirostris* (Bloch & Schneider)
Tetraodontidae	*Arothron diadematus* (Rüppell)	*A. nigropunctatus* (Bloch & Schneider)
	Canthigaster margaritata (Rüppell)	*C. solandri* (Richardson)

*occurs to Gulf of Oman

TABLE 13.3. Examples of Indian Ocean: Red Sea sub-species pairs

FAMILY	INDIAN OCEAN SUB-SPECIES	RED SEA SUB-SPECIES
Chaetodontidae	*Chaetodon vagabundus vagabundus* Linnaeus	*C.v. pictus* Forsskål
Pomacentridae	*Plectroglyphidodon leucozona leucozona* (Bleeker)	*P.l. cingulum* (Klunzinger)
Labridae	*Gomphosus caeruleus caeruleus* Lacepède	*G.c. klunzingeri* Klausewitz
	Halichoeres scapularis scapularis (Bennett)	*H.s. coeruleovittatus* Rüppell
	Macropharyngodon bipartitus bipartitus Smith	*M.b. marisrubri* Randall
Balistidae	*Sufflamen chrysopterus chrysopterus* (Bloch & Schneider)	*S.c. albicaudatus* (Rüppell)

sharks (Carcharhinidae). Intermediate levels of endemism are found in such families as the butterflyfishes (47%), blennies (30%), damselfishes (21%) and wrasses (14%). In the last two groups there are also significant components endemic to the western Indian Ocean; 26% for the damselfishes and 11% for the wrasses.

From a zoogeographical viewpoint the Red Sea is by no means homogeneous, nor is the Red Sea: Indian Ocean barrier a sharp boundary. For instance, certain Red Sea endemics survive in the Gulf of Aden, and species of probable Red Sea origin are found as far afield as the Gulf of Oman and the Arabian Gulf (Fig. 13.5). Also certain Indian Ocean or Indo-Pacific species which are found in the Red Sea only survive in the extreme south of the sea where conditions are most similar to those in the Indian Ocean. Examples of the latter are: the butterflyfishes *Chaetodon leucopleura* and *Chaetodon vagabundus*, the angelfish *Pomacanthus striatus*, the damselfish *Pomacentrus leptus*, the surgeonfish *Acanthurus bleekeri* and the triggerfishes *Canthidermis maculatus* and *Melichthys indicus*. Similarly, certain endemic species are not found throughout the Red Sea basin but only in parts of it. For example, the damselfish *Neopomacentrus miryae*, the dottybacks *Pseudochromis pesi*, *P. springeri*, *Pseudoplesiops auratus* and *P. rubiceps*, and the wrasse *Paracheilinus octotaenia* occur in the northern but not the southern Red Sea, whilst the damselfish *Neopomacentrus xanthurus*, the wrasse *Minilabrus striatus*, and the dottybacks *Pseudochromis dixurus* and *P. sankeyi* occur in the southern but not the northern Red Sea. Also, several species are reported commonly to occur more shallowly in the Gulf of Aqaba (where the surface water is cooler) than they do in the Red

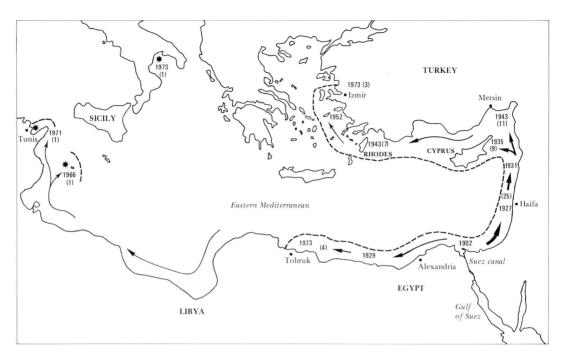

Fig. 13.6. Invasion of the Eastern Mediterranean by Red Sea fish species which have escaped via the Suez Canal. The arrows show the major migrations from the beachhead at Port Said, the dates indicate the advent of the first published records at the various localities, and the figures in parentheses show the numbers of Red Sea species at present recorded from these localities. The dashed outline shows the limits of the area in which the invaders appear reasonably well established.

Sea proper; for example, the dottyback *Pseudochromis fridmani*, the butterflyfish *Chaetodon paucifasciatus*, the angelfish *Apolemichthys xanthotis*, the surgeonfish *Genicanthus caudovittatus*, and the toby *Canthigaster coronata*. The distributions of many fish species thus appear to be strongly influenced by the large scale heterogeneity of environmental conditions within the Red Sea.

13.26.3. Red Sea fishes which have invaded the Mediterranean

Since the opening of the Suez Canal in 1869 about 30 fish species have invaded the eastern Mediterranean (Ben-Tuvia, 1966). The term 'Lessepsian migrants' (after Ferdinand de Lesseps who built the canal) has been coined for these invaders (Por, 1978), many of which are now widespread in the eastern Mediterranean, being found as far west as Sicily and Tunisia and as far north as the southern Aegean Sea (Fig. 13.6). Although few in number these Red Sea escapees have had a profound effect on the ecology of the Levant Basin.

It was not until about 30—40 years after the opening of the Suez Canal that salinities in the Bitter Lakes and Canal had dropped sufficiently to allow Red Sea fishes to survive the crossing of the Isthmus of Suez. The first invaders were spotted near the Canal exit at Port Said in 1902 and by 1930 the invading fishes had pushed northwards to the coasts of Israel and Lebanon. The main thrust continued along the coasts of Syria and Turkey and by the mid-1940s the island of Rhodes had been reached (Fig. 13.6). By contrast, westwards along the Egyptian coast the great freshwater outflow from the Nile delta has proved an inimical and effective barrier to the migration of all but a very few of the invading Red Sea fishes.

One and probably two of the Red Sea invaders managed to cross the Isthmus of Suez before the Suez Canal was built and are thus 'Pre-Lessepsian migrants'. The killyfish *Aphanius dispar* is a euryhaline species (can withstand a wide range of salinities) and was known from the Mediterranean prior to 1869. It may have crossed the Isthmus via the system of artificial and natural waterways that provided a navigational link between the Mediterranean and Red Sea in pharaonic times. The sparid *Crenidens crenidens* is also euryhaline and is found in hypersaline lagoons on either side of the Isthmus of Suez. Paperna and Lahav (1975) found that the parasites of the populations of this species on either side of the isthmus were completely different. This indicates that the populations have been isolated from each other for a considerable time; it thus seems likely that *C. crenidens* was established in Mediterranean lagoons before the building of the Canal. Although two presumed invaders (the herring *Etrumeus teres* and flyingfish *Parexocoetus mento*) are circumtropical in distribution, neither have been found in the western Mediterranean and, in the case of *P. mento*, the eastern Mediterranean specimens appear not to be of the eastern Atlantic subspecies but more closely related to Red Sea specimens (Ben-Tuvia, 1966). They thus seem to have reached the eastern Mediterranean from the Red Sea and not from the Atlantic.

In the Levant Basin almost 12% of the coastal fishes are Red Sea immigrants and several species are now important constituents of local fisheries (Por, 1978). The immigrant goatfish *Upeneus moluccensis* now forms around 30% of mullid catches off Israel (Ben-Tuvia, 1973) whilst the Red Sea lizardfish *Saurida undosquamis*, which was first reported from the Mediterranean in only 1952, now constitutes 20% of trawl catches and 11% of the total landings of the Israel sea-fisheries. The former species seems to compete with the local red mullet, *Mullus barbatus*, and the latter with the local hake, *Merluccius*, in each case displacing the local species into deeper cooler water. The depth limit for the Red Sea species is generally at about 70–80 m, below which temperatures never rise above 18°C. Currently on the coast of Israel about half of the barracuda catch consists of the Red Sea immigrant *Sphyraena chrysotaenia*, and the alien rabbitfish *Siganus rivulatus* may make up as much as 50% of samples of near-shore fishes obtained by poisoning (Ben-Tuvia, 1973). Certain others of the invading species (e.g. the halfbeak *Hemiramphus far*, the silverside *Atherinomorus lacunosus*, the ponyfish *Leiognathus klunzingeri* and the goatfish *Upeneus asymmetricus*) now form a significant part of the diet of commercially important local food fishes such as tuna and hake.

Thus the opening up of the Suez Canal connection between the Indo-Pacific and Atlantic regions has, particularly in the rather species-poor Levant Basin, had profound effects on the local ecology, many of the Lessepsian migrants having been markedly successful. For several invading species there are no known residents with similar ecological requirements (local competitors) so the invasion has, on the whole, enhanced the resident fauna rather than led to local extinction of indigenous species by competitive exclusion.

13.26.4. Mediterranean species which have invaded the Red Sea

There are records of at least five presumed Mediterranean immigrant fish species in the Red Sea, and a further six species have been reported in the immediate vicinity of the southern end of the Suez Canal. However, one of these Mediterranean immigrants is confined to the northern Gulf of Suez and three are found no further than the lagoon of El-Bilaiyim about 200 km south of Suez. Only the sea-bass *Serranus cabrilla* (the only representative of this genus in the Red Sea) has escaped from the Gulf of Suez, reaching the southern Red Sea. *Serranus cabrilla* is frequently caught in the northern Red Sea, and the serranid *Dicentrarchus punctatus* and the mullet *Liza aurata* are quite common at El-Bilaiyim, otherwise the Mediterranean immigrants appear to be uncommon or rare (Por, 1978).

In summary, only a few Mediterranean fishes have managed to invade the Red Sea through the Suez Canal and most of these only survive in a few marginal habitats. This is in marked contrast to the considerable success of Red Sea species in the Levant Basin of the Eastern Mediterranean. Various reasons have been put forward to explain the almost unidirectional movement through the Canal. Three significant factors would appear to be (1) the predominantly northward flow of water in the Suez Canal, (2) the preadaptation of species in the Gulf of Suez to Canal and eastern Mediterranean conditions, and (3) the impoverishment of the indigenous fauna of the Levant Basin (only about 300 fish species as compared to about 1000 in the Red Sea).

REFERENCES

Abel, E. (1960). Zur Kenntnis des Verhaltens und der okologie von Fischen on Korallenriffen bei Ghardaqa (Rotes Meer). *Z. Morph. Oekol. Tiere.* 48, 430–503.

Allen, G. R. (1981). *Butterfly and Angelfishes of the World.* Volume 2. Baensch, Melle.

Allen, G. R. and Randall, J. E. (1977). Review of the sharpnose pufferfishes (subfamily Canthigasterinae) of the Indo-Pacific. *Rec. Aust. Mus.* 30 (17), 475–517.

Allen, G. R. and Randall, J. E. (1980). A review of the damselfishes (Teleostei: Pomacentridae) of the Red Sea. *Israel J. Zool.* 19, 1–98.

Bakus, G. J. (1969). Effects of the feeding habits of coral reef fishes on the benthic biota. *Proc. Symp. Corals Coral Reefs*, 445–8. Mar. Biol. Ass. India, Cochin.

Barlow, G. W. (1974). Extraspecific imposition of social grouping among surgeonfishes (Pisces: Acanthuridae). *J. Zool. Lond.* 174, 333–40.

Benson, A. A. and Muscatine, L. (1974). Wax in coral mucus: energy transfer from corals to reef fishes. *Limnol. Oceanogr.* 19, 810–4.

Ben-Tuvia, A. (1966). Red Sea fishes recently found in the Mediterranean. *Copeia 1966*, 254–75.

Ben-Tuvia, A. (1973). Man-made changes in the eastern Mediterranean Sea and their effects on the fishery resources. *Mar. Biol.* 19, 197–203.

Botros, G. A. (1971). Fishes of the Red Sea. *Oceanogr. mar. Biol. a. Rev.* 9, 221–348.

Burgess, W. E. (1978). *A Monograph of the Butterflyfishes (Family Chaetodontidae).* T.F.H., Neptune.

Choat, J. H. and Robertson, D. R. (1975). Protogynous hermaphroditism in fishes of the family Scaridae. In *Intersexuality in the Animal Kingdom.* Ed. R. Reinboth, pp.263–83, Springer-Verlag, Berlin.

Clark, E. (1980). Red Sea fishes of the family Tripterygiidae with descriptions of eight new species. *Israel J. Zool.* 28, 65–113.

Clark, E. and Gohar, H. A. F. (1953). The fishes of the Red Sea: Order Plectognathi. *Publ. mar. Biol. Stn, Ghardaqa* 8, 1–80.

Cohen, D.M. (1973). Zoogeography of the fishes of the Indian Ocean. In *The Biology of the Indian Ocean.* Eds. B. Zeitzschel and S. A. Gerlach, pp.450–63, Springer-Verlag, Berlin.

Dawson-Shepherd, A. R. (1981). Diandry in the coral reef wrasse, *Thalassoma rueppellii* (Klunzinger, 1871). D. Phil. thesis, University of York.

Dor, M. (1984). *Checklist of the Fishes of the Red Sea.* Israel Academy of Sciences and Humanities, Jerusalem.

Edwards, A. J. and Randall, J. E. (1983). A new dottyback of the genus *Pseudoplesiops* (Teleostei: Perciformes: Pseudochromidae) from the Red Sea. *Rev. fr. Aquariol.* 9, 111–4.

Eibl-Eibesfeldt, I. (1955). Uber symbiosen, parasitismus und andere besondere zwischenartliche Beziehungen tropischer Meeresfische. *Z. Tierpsychol.* 12, 203–19.

Eibl-Eibesfeldt, I. (1962). Freiwasserbeobachtungen zur Deutung der Schwarmverhaltens verschiedener Fische. *Z. Tierpsychol.* 19, 165–82.

Ekman, S. (1953). *Zoogeography of the Sea*, Sidgwick and Jackson, London.

Emery, A. R. (1968). Preliminary observations on coral reef plankton. *Limnol. Oceanogr.* 13, 293–303.

Fishelson, L. (1970). Protogynous sex reversal in the fish *Anthias squamipinnis* (Teleostei, Anthiidae) regulated by the presence or absence of a male fish. *Nature, Lond.* 227, 90–1.

Fricke, H. W. (1971). Fische als Feinde tropischer seeigel. *Mar. Biol.* 9, 328–38.

Fricke, H. W. (1972). *Korallenmeer (The Coral Seas).* Thames and Hudson, London.

Fricke, H. W. (1973). Okologie und Sozialverhalten des Korallenbarsches *Dascyllus trimaculatus. Z. Tierpsychol.* 32, 225–56.

Fricke, H. W. (1977). Community structure, social organisation and ecological requirements of coral reef fish (Pomacentridae). *Helgoländer wiss. Meeresunters.* 30, 412–26.

Garrick, J. A. F. (1982). Sharks of the genus *Carcharhinus. NOAA tech. Rep. NMFS Circular*, 445, i–vii, 1–194.

Gohar, H. A. F. (1954). The place of the Red Sea between the Indian Ocean and the Mediterranean. *Publ. Hydrobiol. res. Inst., Univ. Istanbul, Ser. B*, 2, 1–38.

Gohar, H. A. F. and Mazhar, F. M. (1964). Elasmobranchs of the north-western Red Sea. *Publ. mar. biol. Stn, Ghardaqa* 13, 3–144.

Goldman, B. and Talbot, F. H. (1976). Aspects of the ecology of coral reef fishes. In *Biology and Geology of Coral Reefs* Vol. III. Eds. O. A. Jones and R. Endean, pp.125–54, Academic Press, New York.

Hiatt, R. W. and Strasburg, D. W. (1960). Ecological relationships of the fish fauna on the coral reefs of the Marshall Islands. *Ecol. Monogr.* 30, 65–127.

Hobson, E. S. (1974). Feeding relationships of teleostean fishes on coral reefs in Kona, Hawaii. *Fish. Bull. natn oceanogr. atmos. Admin.* 72, 915–1031.

Losey, G. S. (1972). Predation protection in the poison fang blenny *Meiacanthus astrodorsalis*, and its mimics *Ecsenius bicolor* and *Runula laudandus* (Blenniidae). *Pacif. Sci.* 26, 127–39.

Low, R. M. (1971). Interspecific territoriality in a pomacentrid reef fish *Pomacentrus flavicauda*. *Ecology* 52, 648–54.

Major, P. F. (1978). Predator-prey interactions in two schooling fishes, *Caranx ignobilis* and *Stolephorus purpureus*. *Anim. Behav.* 26, 760–77.

Marshall, N. B. (1952). The 'Manihine' Expedition to the Gulf of Aqaba 1948-1949. IX. Fishes. *Bull. Br. Mus. nat. Hist., Zool.* 1, 221–52.

Myrberg, A. A. (1972). Social dominance and territorially in the bicolor damselfish, *Eupomacentrus partitus* (Poey) (Pisces, Pomacentridae). *Behaviour* 41, 207–31.

Ormond, R. F. G. (1980a). Occurrence and feeding behaviour of Red Sea coral reef fishes. In *Proceedings of Symposium on Coastal and Marine Environment of the Red-Sea, Gulf of Aden and Tropical Western Indian Ocean*, pp.329–71, University of Khartoum, Khartoum.

Ormond, R. F. G. (1980b). Aggressive mimicry and other interspecific feeding associations among Red Sea coral reef predators. *J. Zool. Lond.* 192, 323–50.

Ormond, R. F. G., Campbell, A. C., Head, S. M., Moore, R. J., Rainbow, P. R. and Sanders, A. P. (1974). Formation and breakdown of aggregations of the Crown-of-Thorns Starfish, *Acanthaster planci* (L.). *Nature, Lond.* 246, 167–9.

Paperna, I. and Lahav, M. (1975). Parasites of fish of the hypersaline Bardawil Lagoon, north Sinai — a preliminary communication. *Rapp. Comm. int. Explor. Mer Medit.* 23(3), 127–8.

Por, F. D. (1978). *Lessepsian Migration: the Influx of Red Sea Biota into the Mediterranean by Way of the Suez Canal*, Springer-Verlag, Berlin.

Potts, G. W. (1973). The ethology of *Labroides dimidiatus* (Cuv. and Val.) on Aldabra. *Anim. Behav.* 21, 250–91.

Randall, J. E. (1955). Fishes of the Gilbert Islands. *Atoll Res. Bull.* 47, 1–243.

Randall, J. E. (1974). The effect of fishes on coral reefs. *Proc. 2nd Int. Symp. Coral Reefs* 1, 159–66.

Randall, J. E. (1980). Two new Indo-Pacific labrid fishes of the genus *Halichoeres*, with notes on other species of the genus. *Pacif. Sci.* 34 (4), 415–32.

Randall, J. E. (1983). *Red Sea Fishes*, Immel, London.

Randall, J. E. and Allen, G. R. (1982). *Chromis pelloura* — a new species of damselfish from the northern Red Sea. *Aquarium* 5, 15–19.

Randall, J. E. and Ben-Tuvia, A. (1983). A review of the groupers (Pisces: Serranidae: Epinephelinae) of the Red Sea, with description of a new species of *Cephalopholis*. *Bull. mar. Sci.* 33, 373–426.

Randall, J. E. and Guézé, P. (1981). The holocentrid fishes of the genus *Myripristis* of the Red Sea, with clarification of the *murdjan* and *hexagonus* complexes. *Contr. Sci. Los Angeles*, 334, 1–16.

Randall, J. E., Head, S. M. and Sanders, A. P. L. (1978). Food habits of the giant hump-headed wrasse, *Cheilinus undulatus* (Labridae). *Env. Biol. Fish.* 3, 235–8.

Randall, J. E. and Kay, J. C. (1974). *Stethojulis axillaris*, a junior synonym of the Hawaiian labrid fish *Stethojulis balteata*, with a key to the species of the genus. *Pacif. Sci.* 28, 101–7.

Randall, J. E. and Springer, V. G. (1975). The monotypic Indo-Pacific labrid fish genera *Labrichthys* and *Diproctacanthus* with description of a new related genus, *Larabicus*. *Proc. biol. Soc. Wash.* 86, 279–98.

Reed, W. (1964). *Red Sea Fisheries of Sudan*. Ministry of Animal Resources, Khartoum.

Reese, E. S. (1977). Co-evolution of corals and coral feeding fishes of the family Chaetodontidae. *Proc. 3rd Int. Coral Reef Symp.*, 267-274.

Reinboth, R. (1970). Intersexuality in fishes. In *Hormones and the Environment. Mem. Soc. Endrocr.* 18, 516–43.

Reinboth, R. (1973). Dualistic reproductive behaviour in the protogynous wrasse *Thalassoma bifasciatum* and some observations on its day-night changeover. *Helgoländer Wiss. Meeresunters.* 24, 174–91.

Robertson, D. R. and Choat, J. H. (1974). Protogynous hermaphroditism and social systems in labrid fishes. *Proc. 2nd Int. Symp. Coral Reefs.* 1, 217–25.

Sewell, R. B. S. (1948). The free-swimming planktonic Copepoda: geographical distribution. *J. Murray Exped. 1933–34. Sci. Rep.* 8, 317–592.

Shapiro, D. Y. (1978). The structure and growth of social groups of the hermaphroditic fish *Anthias squamipinnis* (Peters). *Proc. 3rd Int. Coral Reef Symp.*, 571–7.

Smith, J. L. B. (1950). A new dogfish from South Africa with notes on other chondrichthyan fishes. *Ann. Mag. nat. Hist.* (12) 3, 878–87.

Starck, W. A. (1969). *Ecsenius (Anthiiblennius) midas*, a new subgenus and species of mimic blenny from the western Indian Ocean. *Notul. nat.* 419, 1–9.

Strasburg, D. W. and Hiatt, R. W. (1957). Sexual dimorphism in the labrid fish genus *Gomphosus*. *Pacif. Sci.* 11, 133–4.

Vine, P. J. (1974). Effects of algal grazing and aggressive behaviour of the fishes *Pomacentrus lividus* and *Acanthurus sohal* on coral-reef ecology. *Mar. Biol.* 24, 131–6.

Vivien, M. L. and Peyrot-Clausaude, M. (1974). A comparative study of the feeding behaviour of three coral reef fishes (Holocentridae), with special reference to the polychaetes of the reef cryptofauna as prey. *Proc. 2nd Int. Coral Reef Symp.* 1, 179–92.

Wray, T. (1979). *Commercial Fishes of Saudi Arabia*. Ministry of Agriculture and Water Resources, Kingdom of Saudi Arabia: White Fish Authority, United Kingdom.

CHAPTER 14

Turtles and Marine Mammals

JOHN FRAZIER*, G. COLIN BERTRAM** and PETER G. H. EVANS***[1]

*Department of Zoological Research, National Zoological Park,
Smithsonian Institution, Washington, D.C. 20008, U.S.A.
**St. John's College, Cambridge, U.K.
***Edward Grey Institute, Zoology Department, South Parks Road, Oxford OX1 3PS, U.K.

CONTENTS

14.1. MARINE TURTLES

14.1.1. Introduction

The Red Sea presents an uncommon opportunity to understand a group of marine organisms such as marine turtles. Its size, geography, hydrology, and history represent a convenient unit in which to examine this group of enigmatic animals. Yet, despite the many advantages of the Red Sea little is

[1] J. Frazier is responsible for the section on Marine Turtles (14.1), G. C. Bertram for that on the Dugong (14.2), and P. G. H. Evans for that on Cetaceans (14.3).

recorded about its marine turtles, and much basic information is needed for further detailed research and conservation programs.

This report draws mainly from recent studies in the Red Sea including: field work on sea turtles (Urban, 1970; Moore and Balzarotti, 1977; Walczak, 1979; Hirth and Abdel-Latif, 1980; Sella, 1982; Frazier and Salas, in press), a synopsis of turtle information (Frazier, in press (a)), and a multidisciplinary conference on marine sciences in the Red Sea (Thompson, in press). The general and specific characteristics of marine turtles, as they relate to the Red Sea, are described. A discussion of conservation and management issues is also necessary, for Man has created many problems for marine turtles, and some of these are well illustrated by the situation in the Red Sea.

Despite an international community of hundreds of biologists who study sea turtles, and decades of such research, there are many basic questions about these animals that are unanswered: demographic information is poor, and details of feeding biology are scanty — even distributional data are incomplete. In fact, the only aspect of sea turtles that is well understood is that tiny part of their lives when they come ashore to lay eggs. A generalized life history can, however, be inferred from what is known.

Contemporary marine turtles include a group of seven distinctive animals which belong to two taxonomic families: Cheloniidae and Dermochelyidae. Like tortoises and terrapins, they are encased within a box-like shell consisting of a carapace above and plastron below. Most chelonians are able to retract the legs and the head within the shell, thus protecting the soft parts, but the appendages of marine turtles are so enormous that they cannot be accommodated inside. Shells of adults are at least 60 cm long, and in some species they may be well over one metre in length. Like all chelonians, sea turtles not only are air breathers but depend on dry land, for they nest on sandy beaches above the highest tides. Their eggs, which resemble table tennis balls, are normally laid at night in clutches of 100 or more and then buried in the sand and abandoned to incubate with the heat of the sun. Within 6 to 10 weeks the eggs hatch, and the baby turtles dig their way to the surface and scramble to the sea. Both eggs and hatchlings are especially liable to predation. The succeeding stages of the life history are not well understood, but the young of some species may live as macroplankton, floating in oceanic currents. Subadults and adults are common in shallow waters over reefs, in marine pastures of seagrasses and algae, or even in protected bays or estuarine lagoons. The adults often make lengthy migrations from feeding grounds to breeding beaches or 'rookeries'. Each species is distinguished by scales, colour, shape and size (Fig. 14.1) as well as feeding habits.

Marine turtles are most abundant in the tropics, but they are common in some subtropical and even temperate seas. One of the most notable characteristics of their varied habitats is an absence or paucity of humans and human disturbances: Man and turtle are rarely compatible.

The Red Sea is relatively small, covering only 437,900 km^2. Although it is an extension of the tropical Indian Ocean, via the Gulf of Aden, nearly a third of its area lies north of the tropic. The Straits of Bab al Mandab form the southern boundary of the Red Sea; little more than 25 km across, they restrict water exchange with the Indian Ocean. The Suez Canal provides a link between the Red Sea and the Mediterranean, but although numerous animals have taken advantage of this man-made channel there is no evidence that sea turtles have made the passage of 168 km between water bodies (Por, 1978).

Most of the Red Sea is tropical; its coral reefs are extensive, and there appears to be abundant feeding habitat for marine turtles. Yet, winter water temperatures in the north may drop sufficiently low that some tropical organisms at the northern end become scarce or show signs of thermal stress (Pearse, in press).

Mangrove forests, important indicators of coastal conditions, are small and rare along the Egyptian coast, but they are more common toward the south of the Red Sea. Dense patches of seagrasses are offshore of wadis in Sinai, and in localized areas offshore of mainland Egypt and Saudi Arabia. Pastures of these underwater angiosperms, important food for at least one turtle species, are possibly more extensive in the shallow waters at the southern end of the sea.

Guide to identification of Red Sea turtles

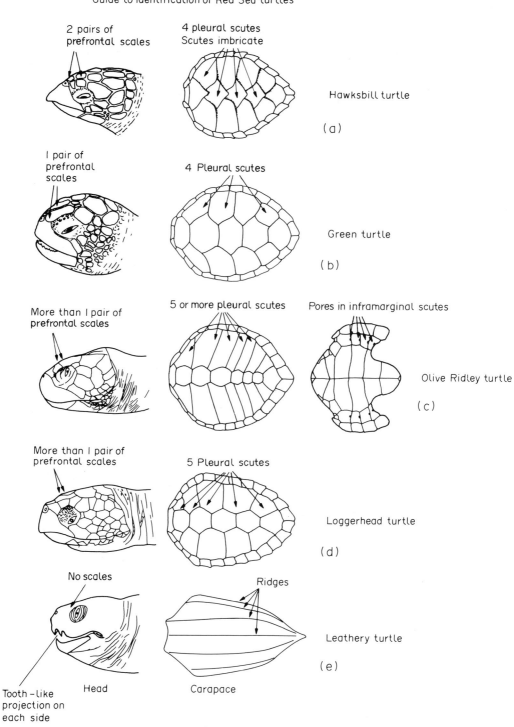

Fig. 14.1. Drawings of heads and dorsal carapaces of 5 turtle species, and plastron of Olive Ridley turtle.

Although the coast line is generally deserted, it is the islands of the Red Sea which are least disturbed by Man. There are hundreds of islands, the major groups being the archipelagoes of Suakin in the Sudan, Dahlak in Eritrea and Farasan in Saudi Arabia. Many of these islands are poorly charted, much less studied, and they provide potentially large and valuable breeding areas for marine turtles.

14.1.2. The status of marine turtles in the Red Sea

All of the five species of pantropical marine turtles occur in the Red Sea (Figs. 14.2 & 14.3). These are listed together with their common names in Table 14.1. The only extant species that are not reported are one found in north Australian waters and one restricted to the Gulf of Mexico and North Atlantic. Sea turtle products were possibly used by Egyptian Pharaohs, and they were unquestionably an important trade item for the Roman Empire (Schoff, 1912). There are signs of ancient and incredibly heavy exploitation of sea turtles on several Red Sea islands.

Several important Red Sea expeditions, that provided a wealth of biological information, were carried out during the last century. Rüppell (1835) and Steindachner (1901) reported on specimens and observations of sea turtles that they collected during these endeavours. Yet, despite this long history of human-turtle interaction, little has been documented about Red Sea turtles, and the first detailed studies were not done until the late 1970's.

Eretmochelys imbricata (L.), the Hawksbill turtle

As the English and Arabic names indicate, this turtle has a well developed beak, resembling that of a hawk; its overlapping, or imbricate, scales are the 'tortoise-shell' of commerce (Fig. 14.1a). Of all the sea turtles, this species is most restricted to tropical waters; its major foods seem to be sponges, soft corals and other sessile soft-bodied invertebrates that grow in the nooks and crannies of coral reefs.

This is evidently the most abundant of the Red Sea turtles, with records from every country but Jordan. Nesting is also widely reported but most commonly from islands. The Dahlak Archipelago, in Eritrea, was said to have nesting (Rüppell, 1835). Sudan's Suakin Archipelago may have some of the most concentrated nesting anywhere; Moore and Balzarotti (1977) estimated that 300 turtles nest there annually, and Hirth and Abdel-Latif (1980) observed 42 females nesting on the island of Seil Ada Kebir during an eight-day period. Nesting occurs along the Egyptian coast from Ras Banas to islands at the mouths of the Gulfs of Suez and Aqaba, such as Gubal el Kebir (Fig. 14.4a-c) and Tiran (Frazier and Salas, in press). Nesting spoor, probably of this species, has been seen on islands along the coast of Saudi Arabia, and there is likely to be nesting on Yemeni islands, for the species is common there (Walczak, 1979).

The nesting season seems to vary latitudinally: it is reported to last from February to March in the Dahlak Archipelago, March to June in the Suakin Archipelago and April to July in Egypt. However, the information is incomplete, and the apparent north-south progression may be incidental.

The most remarkable feature of the Red Sea hawksbill is its low reproductive output. We recently (Frazier and Salas, in press) found the average clutch size in Egypt to be about 113, and 39% of the eggs were small and without yolks. Hirth and Abdel-Latif, (1980) reported that the average clutch in Sudan contained about 100 eggs, 27% of which were yolkless, but in a single clutch as many as half of the eggs might be without yolks. Nests on Jabal Aziz Island in the Gulf of Aden contained an average of 107 eggs, 24% of which were yolkless (Hirth and Carr, 1970). On Masirah Island, Oman, in the Arabian

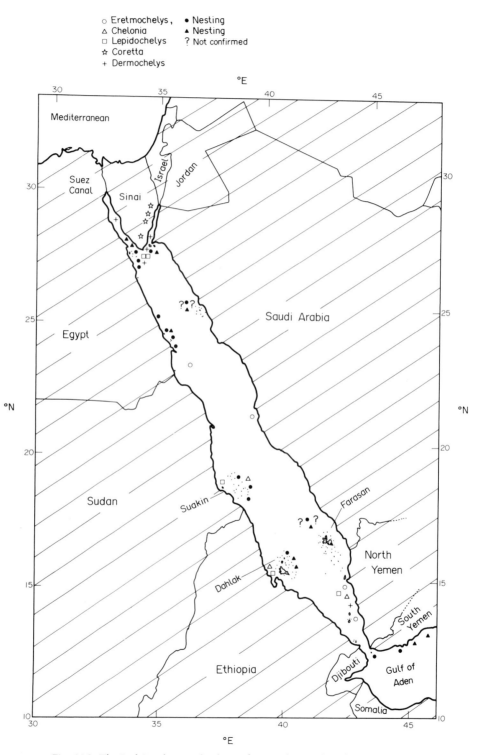

Fig. 14.2. The Red Sea showing localities of sea turtle records and nesting areas.

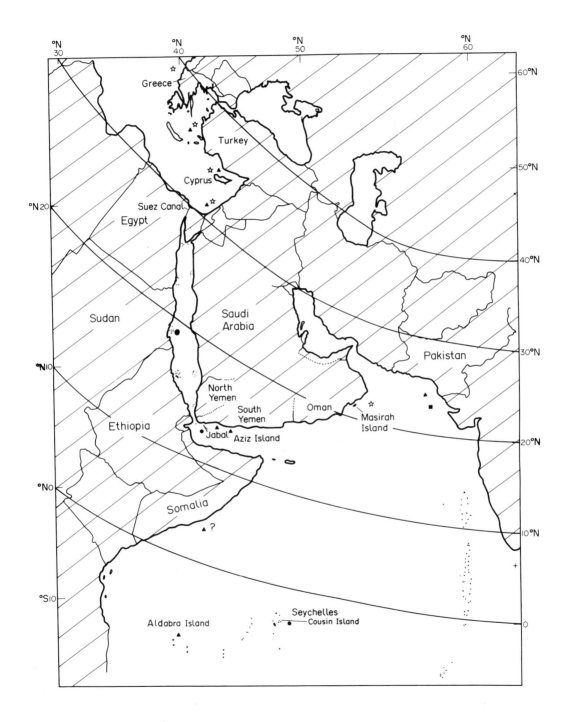

Fig. 14.3. The Red Sea and adjoining seas, showing major nesting areas (symbols as in Fig. 14.2).

Sea the average nest had 97 eggs, 11% without yolk (Ross, 1981). In comparison, the average clutch size on Cousin Island, Seychelles was well over 150, and yolkless eggs have never been seen (Frazier, in press). Low fecundity is unique to hawksbills in the Red Sea and adjoining areas and suggests that special environmental constraints affect the turtle there.

In Egypt most nests are about 5 m inland of the crest of the beach, the highest level the sea reaches. These areas are usually open expanses of sand, fully exposed to the elements. Because hawksbill nests are only about 50 cm deep, the top eggs can be subjected to intense solar heating. Nests are often tucked under small bushes, which provide some shade. An upper layer of infertile, yolkless eggs, that will not rot and 'cost' the female less to make than normal yolked eggs, may increase hatching success by insulating the good eggs beneath from predators and thermal extremes.

The Red Sea is, however, not entirely tropical, and Pearse (in press) found evidence of thermal stress in several invertebrate species, as inferred from lowered reproductive outputs. The winter temperatures that he reported from the Gulf of Suez and northern Red Sea, as low as 10°C, may have drastic effects on animals that are otherwise restricted to the tropics. *E. imbricata* may also be subject to thermal stress, and have lowered fecundity when exposed to subtropical temperatures. The small numbers of fertile eggs in clutches on Masirah Island in the Arabian Sea are thought to be the result of physiological stress from cold water upwellings along the Omani coast (Ross, 1981).

The Hawksbill is thought to be the most sedentary of the sea turtles, and the Red Sea population may have little interchange with those in the Gulf of Aden or other parts of the Indian Ocean. There is no population in the Mediterranean. When Rüppell (1835) described his specimen from Massawa he called it *Caretta bissa* after the local common name. There is no evidence that the Red Sea hawksbill is distinct from those in the Indian Ocean, and because Rüppell's was the first description from the Indo-Pacific, *Eretmochelys imbricata bissa* is now used to refer to hawksbills throughout the Indian and Pacific oceans. There is, however, some dispute whether this is a different subspecies from the Atlantic hawksbill.

Chelonia mydas (L.), the Green turtle (Fig. 14.5a)

This is the turtle of soup fame. It is not entirely green but mottled with a variety of browns on a green background. Two large, elongate scales cover the forehead just above the nostrils where all other sea turtles have two pairs or more (Fig. 14.1b). In all but hatchlings the edges of the beak are serrated, presumably to facilitate grazing on seagrasses and algae. This turtle is unique among reptiles in having a highly efficient mechanism for digesting plant tissues.

C. mydas is the second most common species in the Red Sea. Specimens have been seen or collected in the Dahlak Archipelago, Port Sudan, the Egyptian coast and islands, Sinai, and Yemen. It is the most frequently seen species in eastern Sinai and Yemen.

There is only one confirmed record of observed nestings in the Red Sea: Urban (1970) saw turtles in March on two islands in the Dahlak Archipelago. Old nest spoor, also seen in March, in the Suakin Archipelago indicated that nesting may take place earlier there (Hirth and Abdel-Latif, 1980). In Sinai nests have been found between July and November, but evidently no turtles have ever been seen nesting. Nest spoor has also been found on Egyptian islands. There are unconfirmed reports of *C. mydas* nest spoor on Saudi Arabian islands, and this turtle probably also nests on Yemeni islands.

Each year thousands of Green turtles nest in South Yemen, along the shores of the Gulf of Aden (FAO, 1973); and some tagged females from there have been recaptured in Somalia, more than 2,000 km away (Hirth, 1978). Some of these turtles may move between the Gulf of Aden and the Red Sea, for

Fig. 14.4. a. Female *Eretmochelys imbricata* nesting at Gubal el Kebir, Egypt. (*Photo*. J. Frazier) b. Nest and tracks of *E. imbricata*. Gubal el Kebir. *Photo*. J. Frazier. c. Nest and eggs of *E. imbricata*, Gubal et Kebir. *Photo*. J. Frazier. d. Dead *E. imbricata*, covered in oil, Abu Mingar, Egypt. (*Photo*. J. Frazier)

the distances involved are only a few hundred km, and this species is renowned for making long migrations.

Large populations of *Chelonia* also exist in the eastern Mediterranean, in Turkey (Geldiay *et al.*, 1982), Cyprus (Demetropoulos and Hadjichristophorou, 1982), and Israel (Sella, 1982). There is, however, no evidence that any turtle has ever transited the Suez Canal.

C. mydas is said to be common along eastern Sinai, where there are also luxuriant growths of seagrasses, but in the Egyptian Red Sea both dense marine pastures and the herbivorous sea turtles are uncommon. Although there is an abundance of nesting beaches, and Green turtles are common both to the north and south of the Red Sea (in the Mediterranean Sea and Indian Ocean), this species is less common than one might expect, and this may be related to the lack of adequate feeding habitat. On the other hand, the occurrence of populations of dugongs in the Red Sea shows that there is ample food for at least one large herbivore.

TABLE 14.1. Common names of sea turtles in the Red Sea. Names are Arabic unless indicated* otherwise.

Common Name	Locality (Province or dialect)	Reference
Eretmochelys imbricata (L.)		
Abu Gudr ("the shelled one")	Sudan	Moore and Balzarotti, 1977:12
Baga	Ethiopia (Eritrea)	Rüppell, 1835:6
Bissa	Ethiopia (Eritrea)	Rüppell, 1835:6
Sagr ("hawk")	Egypt	Frazier and Salas, in prep.
Sajlefa (general name)	Egypt	Frazier and Salas, in prep.
Scugar	Sudan (Bejar)	Hirth and Abdel-Latif, 1980:126
Shuker	Sudan (Beja)	Moore and Balzarotti, 1977:13
Sugr	Saudi Arabia and Yemen	Moore and Balzarotti, 1977:13
Chelonia mydas (L.)		
Biswa	Egypt	Frazier and Salas, in prep.
P'saya	Egypt (Abadi)	Frazier and Salas, in prep.
Tersa	Egypt	Frazier and Salas, in prep.
Turas al Abiad ("white sea turtle")	Israel	Sella, *in lit.* 1 March 1976
Lepidochelys olivacea (Eschscholtz)		
Bage	Ethiopia (Eritrea)	Rüppell, 1835:9
Caretta caretta (L.)		
Turas al Asfar ("yellow sea turtle")	Israel	Sella, *in lit.* 1 March 1976
Dermochelys coriacea (L.)		
Na'ama ("ostrich")	Egypt	Frazier and Sales, in prep.
General Name		
Abu Gadah	Egypt (Bish'ari)	Frazier and Salas, in prep.
Sajlefa	Egypt	Frazier and Salas, in prep.
Sajlefa (= *Testudo* sp.)	Israel	Mendelssohn, *in lit.*, 28 Oct. 1982
Salhafa-t	?	Steingass, nd: 423
Sulahfa	?	Steingass, nd: 423
Sulhafah	Saudi Arabia and Yemen	Walczak *in lit.* 20 Jan 1976
Sulkhafa	North Yemen (YAR)	Walczak 1979: 851
Tersa	Egypt	Frazier and Salas, in prep.
Tersa or Tirsa (= *Trionyx* sp.)	Israel	Mendelssohn, *in lit.*, 28 Oct. 1982
Zajlefa Adiyr	Egypt	Frazier and Salas, in prep.
Zugral	North Yemen (YAR)	Walczak, *in lit.* 20 Jan 1976
Ye Woha Ali ("water tortoise") *	Amharinya (Ethiopia)	Largen, *in lit.* 6 March 1976

Lepidochelys olivacea (Eschscholtz), the Olive Ridley turtle (Fig. 14.5b)

Characteristic features of this turtle are: a relatively large head with hooked beak; carapace with thin olive-green scales; highly domed but rounded carapace, nearly as wide as long; and pores in the inframarginal scales, at the left and right sides of the plastron (Fig. 14.1c). This is an uncommon turtle throughout the western Indian Ocean; there are only five records of this species from the Red Sea: one specimen each from Massawa, Eritrea; Suakin, Sudan; and Ras Katib, North Yemen; and also two specimens in southern Sinai. The animal is typically found in water with low salinity, high turbidity, and high organic content where mangroves and shrimp fisheries abound. These features are uncommon in the Red Sea.

Enormous populations in Central America and eastern India have massed nestings in which tens of thousands of females come out to lay eggs on a few kilometres of beach during a period of a day or two. Nesting in the western Indian Ocean is much less concentrated, and the only major rookery is in Pakistan.

There is no record of breeding in the Red Sea. Rüppell recounted that there was nesting in the Dahlak Islands in May, but his informant may have confused this turtle with *E. imbricata*, for both species had nearly the same common name (*see* Table 14.1). The scarcity of *L. olivacea* may reflect its incompatibility with certain hydrographic characteristics of the Red Sea.

Individuals of *L. olivacea* that occur in the Red Sea are apparently waifs from the Indian Ocean. There is no population of this species in the Mediterranean, although strays may occur.

Caretta caretta (L.), the Loggerhead Turtle (Fig. 14.5c)

An enormous head and a large number of scales on the head and carapace are characteristics that this species shares with the former, and this has often resulted in confusion between these turtles. However, the presence of a red-brown colour and a swelling or knob-like projection at the posterior of the carapace are distinctive of the Loggerhead (Fig. 14.1a).

This is the rarest of all the Red Sea turtles. It is known only from the shores of eastern Sinai, where three or four individuals have been accidentally caught in fishermen's nets during the past 30 years. Bones were reportedly found in a cave near Ras Muhammad. There is no evidence of breeding, and the few individuals that occur are evidently strays.

The paucity of records is curious. *C. caretta* is a common species in the Mediterranean, nesting in Greece (Margaritoulis, 1982), Turkey (Geldiay *et al.*, 1982), Cyprus (Demetropoulos and Hadjichristophorou, 1982) and Israel (Sella, 1982). On Masirah Island, in the Arabian Sea, is the world's largest breeding population of this species; tens of thousands of females are estimated to nest each year (Ross and Barwani, 1982). *Caretta* from other populations are known to disperse over thousands of kilometres, and there appears to be every possibility of individuals from Masirah entering the Red Sea. It is thus puzzling that there are not more records of these turtles in the Red Sea, even if they do not breed there.

This turtle feeds especially on molluscs and crustaceans which it crushes with its enormous beak. Both adult and immature Loggerheads frequent estuaries and bays, probably in search of these prey items. The Red Sea is not renowned as a rich area for molluscs, and the scarcity of the turtles may be related to poor feeding habitat.

Dermochelys coriacea (L.), the Leathery turtle (Fig. 14.5d)

The only member of the family Dermochelyidae, this is the largest of all turtles, reaching weights of over half a tonne. Distinctively, it lacks horny scales or claws, but has seven ridges running down its back. The upper beak is unique in having a tooth-like projection on each side (Fig. 14.1e).

There are several records from the Sinai area: south of Abu Rodeis in the Gulf of Suez, and at the southern end of the Gulf of Aqaba, but most of these are sight records. One specimen was stranded off the coast of North Yemen. At least two individuals were brought to the Al Ghardaqa Marine Station: one was of subadult size and a female was of adult size.

There is no evidence of *D. coriacea* nesting in the Red Sea. This species nests in only a few places in the Indian Ocean, and always in relatively small numbers. A small population occurs in the Mediterranean, but it is not known to breed.

This turtle feeds on pelagic, soft-bodied animals and evidently makes migrations of tremendous distances into areas with seasonal concentrations of jellyfish. Swarms of *Aurelia*-like jellyfish in the Egyptian Red Sea may contain dozens of medusae/m^3, and this could be an abundant food source for *D. coriacea*, as well as other turtles. Curiously, there are no reported feeding concentrations of turtles in the Red Sea. Nonetheless, Leathery turtles disperse over enormous areas of ocean and because of this pelagic habit the species is likely to be more common in the Red Sea than the records indicate.

14.1.3. Factors affecting marine turtles

Although chelonians pre-date the dinosaurs, and sea turtles have been on the planet for about 100 million years, the past century or two have seen catastrophic declines in the number of these ancient creatures and several populations have been exterminated. As with other species that have gone extinct during historic times, the faltering status of marine turtles is attributable to Man, but complicated by a multiplicity of factors. These can be grouped into two broad categories: direct predation on the turtles, and disturbances to their environments.

In the life history of a marine turtle the phases most susceptible to predation are the egg and recently hatched turtle. A variety of predators, from ants to dogs, attacks turtle nests, but the only predator on island beaches of the Red Sea that is likely to be significant is the ghost crab, *Ocypode* sp. Mainland nesting beaches are normally subject to many more predators, vertebrate and invertebrate, and it is probably for this reason that islands are so important for nesting. Some sea birds and birds of prey may take hatchling turtles, either from beaches or the surface of the sea, but there are few records of this from anywhere, and none from the Red Sea. Predatory fishes may patrol beaches for hatchlings, and in the Red Sea a small proportion of Tiger sharks (*Galeocerdo cuvier*) have been found with the remains of large turtles in their stomachs.

Without doubt, the most important predator on marine turtles is Man, who exploits them throughout their range for meat, oil and eggs, as well as tortoise-shell and leather. Turtle eggs and meat are consumed by peoples of the Red Sea, but this does not seem to be common. Islamic restrictions on food items, although not specifically prohibiting the consumption of sea turtles, may prevent Muslim peoples from eating turtles with which they are unfamiliar. This seems to be the case in Islamic countries to the south, such as the Comores and Somalia.

Tortoise-shell is, however, another issue: it is an article of commerce, and has been for millennia. The *Periplus of the Erythrean Sea*, written in the First Century AD by a Greek sailor living in Egypt, describes a well organized trade in tortoise-shell focused in Egypt and reaching as far as ports in eastern Africa and Asia (Schoff, 1912).

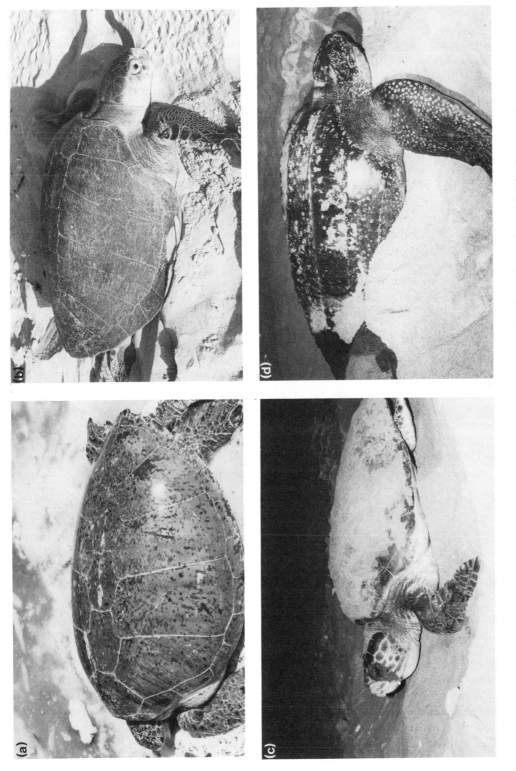

Fig. 14.5. a. Green turtle, *Chelonia mydas*, adult male. (*Photo. J. Frazier*) b. Olive Ridley turtle, *Lepidochelys olivacea*, nesting female. (*Photo. J. Frazier*) c. Loggerhead turtle, *Caretta caretta*, nesting female. (*Photo. J. Frazier*) d. Leathery turtle, *Dermochelys coriacea*, nesting female. (*Photo. J. Frazier*)

There are immense mounds of turtle bones on islands in the Dahlak and Suakin Archipelagoes and also on Egyptian islands. Some mounds with bones of *E. imbricata* were documented at the turn of the century (Steindachner, 1901), but no contemporary fishermen seem to have any knowledge of these remains. These and other mounds are thought to date back well over a century (Moore and Balzarotti, 1977; Hirth and Abdel-Latif, 1980).

This quantity of turtle remains bespeaks an intense exploitation in previous centuries, and the presence of *E. imbricata* suggests an active tortoise-shell fishery. Nowadays there is no significant turtle fishery in the Red Sea, and on the Egyptian coast there is general ignorance of the worth of tortoise-shell, which may fetch as much as US$ 100/kg. The present day absence of a fishery, or any knowledge of its monetary value, indicates that after a period of intense exploitation, the fishery collapsed and has never recovered. We may never know what happened in the Red Sea, but elsewhere heavily exploited turtle populations are known to have declined until they could no longer sustain a significant fishery.

Direct exploitation and predation are relatively easy to identify and comprehend, but other factors, difficult to isolate, can be much more damaging and cause long term disturbances to turtle populations. Many environmental perturbations are out of sight and thus out of mind. Hence, although the Red Sea coast is sparsely populated, parts of it are subject to intense and long lasting disturbances, many of which can be disastrous to marine turtles.

Oil pollution is a major environmental hazard in many parts of the Red Sea. A major part of the Egyptian coast has been contaminated with crude oil; on islands in the north where turtles once nested, I have seen oil several centimetres thick covering the shore and former nesting beaches. In addition to birds and marine invertebrates, turtles have been found dead and enveloped with oil (Fig. 14.4d).

Little is known about the physiological effects of petroleum oil or any other pollutant on turtles, but thick layers on the sea would be fatal to a hatchling and could foul the nose, eyes and throat of a large individual, leading to almost certain death. Some volatile fractions of crude oil can disorient hatchling turtles, causing them to swim aimlessly. Tar balls now occur the length and breadth of most oceans, including the Red Sea. These may be eaten, either intentionally or unintentionally, and are thought to have fatal effects (Frazier, 1980).

Another problem for marine turtles, as well as many other marine organisms, is submarine explosions. The search for oil in the Middle East has gone underwater; exploratory ventures of oil companies may use dozens of explosions in one seismic investigation and there may be hundreds to thousands of these blasts employed in an exploratory block during early investigations. Apparently there are no restrictions on this activity. Not only are animals destroyed or fatally injured by the shock waves, but vast areas of habitat (coral reefs especially) may be denuded and food species destroyed.

In addition to 'legal' blasting, another regular source of marine explosions is illegal. Military personnel in some places use explosives for 'fishing'. Along stretches of the west coast of the Gulf of Suez this is virtually a daily event. Local naturalists who have visited such areas over the past decade(s) describe the transformation of one of the world's richest marine systems into an underwater desert.

14.1.4. Conclusions

The Red Sea provides a multitude of islands and remote beaches for nesting turtles. Its rich coral reefs sustain feeding and resting habitats, especially for *Eretmochelys imbricata*. Marine pastures, which may be less extensive than the reefs, furnish feeding habitats for *Chelonia mydas*. Swarms of jellyfish, which may be seasonal, are a potential food source for *Dermochelys coriacea* and other species. There appears to be little feeding habitat for the other two species which are rare in the Red Sea: *Lepidochelys olivacea* and *Caretta caretta*.

The nesting season of *E. imbricata* lasts from spring through summer, and it may occur latest at the highest latitudes. A characteristic low fecundity, with small clutchs and large numbers of yolkless eggs, is unique to this turtle in the Red Sea and adjacent areas, and this may be a manifestation of physiological stress caused by water temperatures that are far below the norm for this tropical turtle. A paucity of basic information limits any further biological discussion on *E. imbricata* or any other Red Sea species. The status of these animals needs attention, for the Red Sea is in a troublesome state. Surrounded by sparsely populated, desert coastlines of often fast developing countries, the Red Sea is losing its traditional role. Although it was one of the richest marine environments on the planet, with tremendous significance to ancient civilizations, the contemporary value of this sea is as a source of oil and a waterway for international shipping. Unfortunately, the health of this living system is no longer a priority to the majority of economic and political forces exploiting it and shaping its future.

Fundamental information on distributions, nesting beaches and breeding seasons is badly needed. Basic data on major feeding habitats, especially marine pastures and jellyfish swarms, is also required so that the size of these resources can be plotted in time and space. Interesting questions are posed by the singularly low fertility of *E. imbricata* in the Red Sea, and studies of reproductive physiology and nest ecology are needed to understand this situation better.

There is an urgent need for studies of environmental quality, and particularly the effects of oil pollution and underwater explosions on marine turtles and their habitats. There is, however, no doubt that these two factors have caused, and continue to cause, wanton destruction of certain areas of the Red Sea: the need to restrict these perturbations especially when they are unnecessary cannot be overemphasized. It is of the utmost importance that oil pollution be controlled. A regional plan to manage and maintain the Red Sea, recognizing that it is a vital common resource, must become a regulating force.

ACKNOWLEDGEMENTS

My earliest (bibliographic) research on the Red Sea was completed while aided by the East African Wildlife Society and the Fauna and Flora Preservation Society. Work on the Egyptian Red Sea was supported by Friends of the National Zoo, Office of Naval Research, U. S. Fish and Wildlife Service, Egyptian Wildlife Service and Institute of Oceanography. Numerous individuals provided invaluable assistance: D. Ferguson, M. Fouda, I. Helmy, H. Hoogstraal, Said el Gamei, Mary Megaly, Stephanie Sagebiel, and B. Zahuraneck. Susana Salas shared in the research and typed innumerable drafts of Red Sea manuscripts.

14.2. DUGONG

14.2.1. Introduction

The dugong (*Dugong dugon* (Müller)) is a marine mammal belonging to the order Sirenia. This small but interesting group contains the only truly aquatic herbivorous mammals, including the dugong of the Indo-Pacific, the tropical Atlantic Manatee, and the now extinct Steller's Sea-cow of the Bering Sea. Surprising as it may seem, the Sirenia are thought by zoologists to be related to the elephants (order Proboscidea), on the basis of their dentition, general anatomy and fossil record. The orders have however been separate since the early Eocene, some 60 million years ago. The dugong's range extends from the Solomon Islands (or beyond) westwards around the northern coasts of Australia (where the largest remaining populations exist), around Sri Lanka and the tip of the Indian peninsula, southwards along the coasts of East Africa, and north-westwards to the Arabian Gulf and the northernmost parts of the Red Sea (Bertram and Bertram, 1973; Nishiwaki *et al.*, 1979). The Red Sea is therefore the western extremity of the range of this tropical Indo-Pacific species.

14.2.2. Biology and ecology of dugongs

The dugong is a large animal, totally aquatic, with no externally visible back legs. The general appearance can be best appreciated from Figure 14.6. Adult animals may reach a length of about 3 metres, weighing some 320 kg. The detailed anatomy of the dugong, both skeletal and muscular, is conveniently described by Kingdon (1971) in his study of east African mammals. Chiefly notable are the fusiform shape, hind limbs reduced to pelvic traces, immensely long intestine characteristic of a herbivore, very thick hide and down-turned facial disc. The tail is powerful and forms a fluke like that of a whale. The skin is chiefly grey in colour, with scattered single hairs about 2 cm apart.

The evolutionary relationship with elephants is especially indicated by the single pair of pectoral mammary glands close under the fore limbs, and the form and mode of succession of the molar teeth. Small tusks, second upper incisors, are visible in the males; although present in females, they remain unerupted. There are no other permanent incisors in either jaw, their normal function being replaced by horny pads which serve to break and pull aquatic green vegetation. The cheek teeth or molars are sequentially replaced, as in elephants. Teeth move forward in the jaw as they are worn by abrasion, and are replaced from behind by new teeth later in the series, the total being six pairs of teeth in each jaw. The later teeth are larger; finally with replacement, only two remain present and functional on each side above and below.

Fig. 14.6. Dugong (*Dugong dugong*) (Müller). A small individual killed by fishermen.

Dugongs are completely restricted in diet to marine higher plants (the so-called seagrasses) mainly of the families Potamogetonaceae and Hydrocharitaceae (Lipkin, 1975). Anatomical features such as the long intestine and peculiar dentition can be seen as adaptations to this diet, which severely restricts the distribution of the animal to the warm, shallow silty waters where such plants flourish (*see* Chapter 9).

Recent work in Australasia (e.g. Marsh, 1980, 1981) has added greatly to our knowledge of dugong biology. It has been shown that the clear dentine growth layers visible when a tusk is sectioned are almost certainly formed annually, allowing estimation of the age of cadaver material. The unerupted tusks of females are better for this purpose than the functional tusks of males, which are subject to wear. It has been shown that female dugongs may live to 50 or more years of age, while 34 years has been determined from the worn tusk of a male. Puberty appears to occur at about 12—14 years of age, although this may vary on a geographical basis.

Dugongs are rare in captivity, where they have never bred, and there have to date been no successful field studies of particular individuals over time. Nevertheless, it has been possible to determine that the reproduction rate is slow, that gestation takes about 12 months, that a single calf is normally produced, and that it stays with the mother for at least a year. Birth of calves is spaced at a minimum of two year intervals.

Since over most of their range, dugongs are rare and extremely difficult to observe in any deliberate fashion, their detailed habits are poorly known. It is evident however that dugongs are fundamentally gregarious, as are the manatees of the tropical Atlantic, and open in their behaviour. When populations are reduced, the animals seem often to be almost solitary, and markedly timid, and even the tips of their nostrils, when breathing at the surface every few minutes, are rarely seen.

Nowhere have we yet knowledge of important or regular seasonal migrations of dugongs. Some local movement is probable, perhaps in connection with monsoon conditions, and could be looked for in populations along the Queensland coast of Australia, which prior to the commercial fishery of the last century were the largest known.

14.2.3. Dugongs in the Red Sea

There have been suggestions (Gohar, 1979) that the Red Sea dugong may be taxonomically distinct from that of the rest of the Indo-Pacific, but the evidence is scanty. Probably most isolated populations, if sufficiently studied, could be shown to have statistically valid minor morphological differences no more significant than, for example, those separating the races of man.

The physical features of the Red Sea, summarised elsewhere in this volume, are fundamental to the distribution, numbers and prospects for this rare and endangered animal (I.U.C.N. Red Data book). As we have seen, dugongs are restricted to life in shallow silty areas where seagrasses flourish, and coral reefs do not provide a favourable environment. In the largely coral-lined Red Sea, suitable shallow areas of marine vegetation are not abundant, but when occurring are generally landward of offshore reefs. This must isolate feeding grounds, and we have no knowledge of dispersal from one ground to the next, although this must happen on occasion.

Records of dugong sightings exist particularly for the better-studied areas, the Gulfs of Suez and Aqaba, the Port Sudan-Suakin area, the Dahlak Archipelago and elsewhere, but the animals are always in small numbers. The adjacent southern coast of Arabia has almost no record of dugongs, but they do occur on the southern shores of the Gulf of Aden and in the Arabian Gulf (Bertram and Bertram, 1973). When special efforts, using low-flying spotter aircraft are made, doubtless individuals and small groups of dugongs will be recorded along the length of the Red Sea as elsewhere. A similar increase in number of records has happened in recent years for dugongs in Australian waters, and for manatees in the

Caribbean. Study of hydrographic charts of the Red Sea will indicate sites where dugongs are likely to exist, and other areas where their presence is unlikely except for rare wanderers. But one cannot believe that there are more than a few hundred dugongs in the whole of the Red Sea.

14.2.4. Dugongs and man

There seems to be no recorded special hunting of dugongs in the Red Sea, probably because of their natural rarity. But small-scale local predation by man has doubtless been regular, and may have been responsible for further reduction in local populations. Chance has probably been more important than intention in local catching. Until recently there has been no indigenous fishery employing nets large and strong enough to take a dugong, whether by accident or design, as is their frequent fate now in Queensland and Sri Lanka. Little is known of non-human predators of dugongs, sharks at the time of parturition being most probable; so it is difficult to assess the importance of man as a casual predator in comparison with natural enemies.

Whether dugongs are suitable food for the devout of Islam is a local matter. On the south-east Somali coast, where dugongs may possibly still be present in relatively large numbers, it is reported that the permission of the Qadi is required before an animal may be declared fit for human consumption. The supposed likeness to the abominable pig, and the possible use of the dugong as a surrogate woman, are matters reputedly taken into account. The indigenous people of northern Australia and Papua-New Guinea have no such scruples. The big European exploitation of dugongs off the Queensland coast in the last century was based on the extraction of oil, and the use of calcined bone in sugar refining.

In pre-Muslim days, 4,000 years ago, in the Arabian Gulf at Um en Nar, adjacent to modern Abu Dhabi, there was a local culture based on what at the time must have been an abundant population of dugongs (Bibby, 1969). Few survive now in that region, though recently attempts have been made to conserve them. However, the Iraq–Iran war of the early eighties has resulted in extensive oil pollution including the death of dugongs.

Biblical reference to the dugong is slight, but tradition has it that the Ark of the Covenant was covered with a dugong skin during the Wanderings in the Wilderness. This unusual choice was of course based on the immense thickness and toughness of the hide, which is very stiff when dry, and would afford remarkable protection for that which was precious and hidden beneath. In more recent times dugong hide was regularly used for sandal soles and shields by the indigenous people of the Red Sea littoral. This use too, has probably lapsed into history and, for sandals at least, old car tyres have been pressed into service as an equally hard-wearing alternative.

14.2.5. Conclusions

Our knowledge of dugongs in the Red Sea is at present poor. Records of sightings are few, and tend to be associated with centres of communication, probably reflecting principally the distribution of observers. Dugongs are not known to be common in any part of the Red Sea, which is in any case peripheral to the main distribution of the species, and ecologically unsuitable over much of its coastline. However, in view of the somewhat precarious status of the animal worldwide (Bertram, 1974; I.U.C.N. Red Data book) studies of the distribution and abundance of dugongs in the Red Sea should be undertaken, if only by rapid aerial surveys.

Unfortunately, the prospects for the remaining dugongs of the Red Sea area are not good. The inshore habitat makes the animals especially vulnerable to casual human predation. The long prereproductive period and low reproduction rate, make population numbers very sensitive to the effects of local over-exploitation. Since substantial geographical dispersal seems small and irregular, recovery of a local population by immigration from elsewhere is likely to be slow.

14.3. CETACEANS

14.3.1. Introduction

The cetaceans, which include the whales, dolphins and porpoises, are the most strictly aquatic of the mammals, having secondarily returned to living in the sea. They comprise two groups — the Mysticeti or baleen whales, which often attain great size (the Blue whale is the largest animal known ever to have existed) and feed primarily upon plankton, sifting it through their baleen plates; and the Odontoceti or toothed whales, which include the Sperm whale, other smaller whales (such as Killer and Pilot whale), the dolphins and porpoises. These capture their mainly fish or squid prey with the aid of teeth. Although both groups include a number of species with worldwide distributions, the baleen whales generally spend part of the year in areas of high productivity and plankton abundance near the poles before migrating towards the equator where they mate or give birth to their single calves. The toothed whales, on the other hand, include some species which are restricted in their distribution, occurring only in tropical or temperate or arctic seas, though others may have a worldwide distribution. An example of the former is the Plumbeous dolphin, restricted to the Indian Ocean and adjacent enclosed seas, whilst the Killer whale is an example of the latter, occurring in all oceans from the equator to the Arctic. The status as well as distribution of cetaceans varies greatly across species with some, such as the Spotted dolphin, being abundant often in large herds, whilst others, such as the Bowhead whale and N. Atlantic Right whale have been hunted to near extinction.

Despite their relatively large size, our knowledge of cetaceans in the Red Sea is extremely fragmentary. Only seven to eight species have been recorded regularly and it is unlikely that more intensive investigation would reveal many others; now revised to ten to eleven (depending on whether one or two species of *Tursiops* are recognised) — see footnote at end of chapter. A number of reasons probably contribute to the low species diversity compared with the Indian Ocean, with 44 species. These include the enclosed nature of the sea (with a very shallow sill at the southern end and the relatively recent Suez Canal at the northern end), its high salinity, and its relatively low primary productivity. All eight species are odontocetes, and whereas Humpback *Megaptera novaeangliae*, Fin *Balaenopterus physalus*, and Bryde's *B. edeni* whales have been reported at similar latitudes in the Arabian Sea, the narrow straight of Bab al Mandab may inhibit their entry northwards to the Red Sea. The same probably applies to the Blue Whale *B. musculus* and Sperm Whale *Physetermacrocephalus* both of which occur in the Gulf of Aden.

14.3.2. Status and distribution

Grampus griseus Risso's Dolphin (Fig. 14.7)

This species has a worldwide distribution in tropical to temperate seas. It is not uncommon throughout the Red Sea and even into the Gulf of Aqaba where it has been recorded off the south-west point of Tiran Island and the west side of Ras Muhammad (J. Beadon *in litt*). Herds usually number less than twelve individuals although groups of twenty to thirty have been recorded in the Gulf of Aqaba. Sightings have occurred throughout the year.

Fig. 14.7. The Risso's Dolphin occurs in small numbers in the Red Sea. It is readily distinguished from other dolphin species by its rounded snout, absence of a beak, and pencil-like markings on the flanks. (*Photo*. P. G. H. Evans)

Orcinus orca Killer Whale (Fig. 14.8)

Another species with a worldwide distribution extending from the tropics to the arctic seas. It has been recorded occasionally in the Red Sea, mainly in summer, though it is probably more regular than the records suggest. Killer whales generally travel in small herds (a male with a harem of females and some young; a group of young bachelor males; or a group of females) although lone males may sometimes be encountered.

Pseudorca crassidens False Killer Whale

This pelagic deep water species has occasionally been recorded in the Red Sea. There are at least two records of individuals in and around the Gulf of Aqaba, and one of these was captured for the dolphinarium at Tel Aviv. A herd of twenty-five was observed on 2nd July 1984 in association with the Yacht Elendil (V. Papastavrou, *pers. comm.*) in the centre of the Red Sea.

Sousa chinensis Plumbeous Dolphin

Plumbeous dolphin is the local name for the Indo-Pacific hump-backed dolphin and is recognised by some as a local form and others even as a separate species. It is largely confined to the tropics, centred on the Indian Ocean from the Straits of Malacca to Southern India and Sri Lanka to the Arabian Sea, Arabian Gulf, Red Sea and Suez Canal. It is a mainly coastal species and is present in small numbers throughout the Red Sea and Gulf of Suez. It is probably resident all through the year.

Fig. 14.8. A cosmopolitan species, the Killer Whale forms herds usually centred on a group of adult females. It is the only species to regularly take other warm-blooded creatures as prey, and is seen occasionally in the Red sea. (*Photo*. Mark Glover)

Stenella attenuata Spotted Dolphin

This is one of the commonest tropical dolphin species occurring throughout the Pacific and Indian Oceans between the latitudes 22°N and the equator. It is probably the most common dolphin in the Red Sea, and although most records are from the northern end and the Gulf of Aqaba, it is likely to be as abundant in the less well known parts of the Red Sea further south. Commonly, herds number between ten and fifty animals, but large groups of up to two hundred animals may be seen.

Tursiops truncatus aduncus Red Sea Bottle-nosed Dolphin (Fig. 14.9)

The taxonomic status of this form is still in much doubt. Whilst Fraser (In: Ellerman and Morrison-Scott, 1951) regarded it as a separate species, others (e.g. Hershkovitz, 1966; Marcuzzi and Pilleri, 1971) consider it a race of *Tursiops truncatus*, and this tends to be current opinion amongst cetacean taxonomists (though not on any firm evidence). It is clearly very similar to *T. truncatus* and is said to differ by its dark green upper parts, the latter being black or slate grey on the back. It is found throughout the tropical regions of the Pacific and Indian Oceans, south to the Cape of Good Hope, whereas *T. truncatus* occurs in temperate seas with its southern limit being the Mediterranean and the Gulf of Mexico. Marcuzzi and Pilleri (1971) have suggested that the Red Sea form is derived from the typical form, the species having passed down the region of what is now the Suez Canal when still open to the Mediterranean (Chapter 2), to the Indian Ocean and from there to the Pacific. The closing of this passage would have prevented further movements, together with climatic changes (the temperature of the Mediterranean dropping prior to glaciation). In the Red Sea the species appears to be widely distributed with records of small numbers throughout, and herds of up to sixty at Shaab Ali in the Gulf of Suez (J. Beadon *in litt.*). Slijper (1962) considers *aduncus* to be more characteristic of deep waters than *truncatus*, but records of the former in shallow waters of the Gulf of Suez do not support this. Most sightings have been made between February and April, but this may reflect greater coverage by observers at this time of year.

Tursiops truncatus/gilli Gill's Bottle-nosed Dolphin

As with the last form, there is much argument on the taxonomic status of *Tursiops t. gilli* and its relationship to *T. truncatus*. By some authorities, *T. t. gilli* is regarded as a separate species, differing from *T. truncatus* in the size and shape of certain skull characters. Gill's bottle-nosed dolphin occupies a narrow band across the Pacific between latitudes 57°N and 21°N, from Japan to Hawaii, southern California and Mexico. However, it has also been recorded from the Gulf of Suez (Ras Muhammad up to Marsa Bareca) in herds of thirty to forty individuals during the months of August and September (J. Beadon *in litt.*). If these sightings were correctly identified they would represent a very interesting and rather surprising extension of the range of this form. It seems more likely that the animals observed were *T. truncatus* which may derive from the Mediterranean (where the species is now uncommon), although distinguishing between *T. t. truncatus* and *aduncus* on sightings alone is open to doubt. Clearly more work is needed in this region to settle these questions.

Steno rostratus Rough-toothed Dolphin

Widely distributed throughout tropical regions of the Atlantic, Indian Ocean and Pacific, this species has been recorded from the southern part of the Red Sea and Gulf of Aden. It is primarily pelagic, occupying the surface layers of deep water areas, but in the Red Sea may be found in quite shallow waters amidst coral reefs and islands.

Fig. 14.9. The Bottle-nosed Dolphin is generally a temperate species although it is found uncommonly in the Mediterranean and it is from here that records of the species in the Gulf of Suez may derive. The taxonomic status of this genus is still in much doubt, and many authorities consider the Red Sea bottle-nosed dolphin *T. aduncus* a subspecies of the Atlantic form *Tursiops truncatus truncatus*. (Photo. P. G. H. Evans)

TABLE 14.2. Ecological parameters of Red Sea cetaceans.

Species	Habitat	Typical Group Size	Diet	Group Characteristics	Sources
Risso's Dolphin *Grampus griseus*	Coastal	Usually 1–7; ranges 1–300.	Primarily cephalopods; also fish.	Stable with adult male, adult females, and sub-adults.	Mitchell 1975, Evans 1980, unpubl.
Plumbeous Dolphin *Sousa chinensis*	Coastal	Often alone or in pairs; usually less than 10.	Reef fish	Very labile. Larger groups (10) of mixed composition.	Nishiwaki 1972, Saayman & Taylor 1973, 1979.
Spotted Dolphin *Stenella attenuata*	Coastal or pelagic	Often 500–2500 (in pelagic areas); otherwise 1–500.	Fish and cephalopods.	?	Perrin et al. 1973.
Indian Ocean Bottle-nosed Dolphin *Tursiops t. aduncus*	Coastal, sometimes pelagic.	Usually 1–200; ranges 3–1,000.	Benthic and reef fish.	Dominance hierarchy	Tayler & Saayman 1972, Saayman & Taylor 1973, Ross 1977.
Gill's Bottle-nosed Dolphin *Tursiops t. gilli*	Coastal, sometimes pelagic.	Usually 25 or less inshore; occasionally groups of several hundred.	Fish and cephalopods.	Age-sex segregation of sub-groups.	Norris & Prescott 1961, Evans & Bastion 1969, Leatherwood et al. 1972, Leatherwood 1975.
Atlantic Bottle-nosed Dolphin *Tursiops t. truncatus*	Coastal or pelagic.	Usually 1–10; occasionally groups of several hundred.	Primarily benthic fish.	Very labile. Adult males, adult females with offspring; sub-adult males; sub-adult males; mixed females + offspring.	Gunter 1942, Tomilin 1967, Wursig 1978, Wursig & Wursig 1977, 1979, Evans 1980, Wells et al. 1980.
Rough-toothed Dolphin *Steno rostratus*	Pelagic	1–50 occasionally larger.	Fish	?	Layne 1965, Mitchell 1975, Leatherwood et al. 1983.
Killer Whale *Orcinus orca*	Coastal or pelagic.	Usually 4–40; occasionally lone indivs. or small pods of 2–6	Fish, marine mammals, also squid.	Stable with adult male, adult females, and sub-adults; sub-adult males; females and offspring.	Balcomb et al. 1980; Bigg 1981.
False Killer Whale *Pseudorca crassidens*	Pelagic; occas. coastal	1–100 (occasionally larger)	Cephalopods and fish	Some age-sex segregation of sub-groups.	Brown et al. 1966; Pilleri 1967.

14.3.3. Biology and ecology

Our knowledge of the general breeding biology and feeding ecology of the seven/eight dolphin species recorded in the Red Sea is derived exclusively from studies elsewhere. Like other small cetaceans, all the species have an annual cycle, with single young being born during the summer months when sea temperatures are highest. Risso's dolphins are probably polygynous, a male holding a harem of females to which some young males may be attached. The other dolphin species have a fairly open breeding system with promiscuity probably being normal rather than monogamy or polygyny. A review of basic breeding parameters is given for each species in Table 14.2. Although the Risso's dolphin is primarily a cephalopod (squid, cuttlefish) feeder, the diets of the Spotted and Bottle-nosed dolphins are quite diverse, taking a variety of small shoaling fish. Bottle-nosed dolphins tend to be inshore feeders on bottom-living shallow water fish, whereas Spotted dolphins are primarily surface feeders. The Plumbeous and Rough-toothed dolphins have been little studied but their diets are probably rather similar to the other species, the former most closely resembling Bottle-nosed dolphins in their feeding behaviour, and the latter resembling Spotted dolphins.

The two whale species have rather different ecologies. Although both the Killer and False Killer bear single young, these may be at intervals of 3 to 8 years. Both species are generally gregarious. Little is known about the social organisation of False Killer herds, but the Killer whale forms relatively stable groups (known as pods) with one adult male to usually three or four adult females and several sub-adults of both sexes. The pod size of Killers rarely exceeds forty individuals, and often may be less than ten. False Killers, on the other hand, may move in herds numbering fifty or more individuals (though between ten and fifty is more typical). Because they are pelagic and rarely seen, our knowledge of the behaviour and ecology of False Killers is very poorly known, whereas Killer whales have been studied intensively, particularly in N. America, in recent years. Killer whales are the only cetacean to take warm-blooded prey, feeding upon porpoises and seals, often in coastal waters, although their diet is catholic and in many areas fish and squid are the dominant food items. This is probably the case in the Red Sea where there are no marine mammals except sirenians, though seabirds and small dolphins might also be taken. The home ranges of Killer whale pods can be extremely large, extending for 2–300 miles (320–480 km), but when food is concentrated or abundant in one area, this presumably becomes much more restricted. False Killers feed mainly on squid, but also take a variety of fish. They probably have relatively large home ranges.

14.3.4. Threats

Small cetaceans face a number of threats, all associated with human activities. The following are the most important: (1) small whale fisheries; (2) incidental takes in fishing nets; (3) chemical pollutants; (4) disturbance; (5) habitat modification.

There appears to be no large scale dolphin fishery in the Red Sea region (though Plumbeous dolphins have been hunted in the past). There is no information on the extent of incidental takes in fishing nets but the small scale of these activities suggests that the problem is not a serious one for this area. The same is not the case for chemical pollutants. The increased industrialisation of Middle Eastern countries and development of the oil industry has posed serious threats to marine birds and mammals. Unfortunately there is little direct evidence of the effects of pollutants on cetaceans in the Red Sea but in other closed seas such as the Mediterranean and the Baltic, this has been better documented. High residues of chlorinated hydrocarbon insecticides such as DDT have been found in the fat of porpoises from the North Sea and Baltic (Holden and Marsden, 1967) and high levels of heavy metals with apparently associated lesions of body tissues and organs have been found in a variety of small cetacean species in the Mediterranean (Viale, 1974, 1976a). However, as with detailed studies of seabirds, although it is possible to find apparently high levels of pollutants in marine organisms, the consequences of these levels is rarely known. In many cases it has been found that animals can carry large amounts of potentially harmful chemicals with no deleterious effects, either because they are usually in an inactive state (e.g. in the blubber of marine mammals) or because the animals concerned presumably have some physiological mechanism to cope with these chemicals (Osborn, 1978a,b).

Although the effects of oil pollution upon seabirds has been well documented, there is little firm evidence for marine mammals being affected except in cases of chronic pollution in restricted areas. Most of the serious oil pollution incidents have been in temperate regions despite heavy oil traffic elsewhere.

Another potential threat whose direct effect on small cetaceans we know little about is disturbance. This may take the form of military activities, for example target practice (Mitchell, 1975), detonations of explosives in munitions disposal and depth charge practice (van Bree and Kristensen, 1974), or dredging and seismic testing during oil exploration. The latter have apparently caused death directly or induced strandings (FAO, 1976), and represents the potentially more important threat in the Red Sea. A third threat, however, is increased boating activity. There is now quite a lot of indirect evidence to suggest that power boats have a disturbing effect on cetaceans causing them to exhibit avoidance reactions (FAO, 1978).

Our knowledge of the effects of habitat modification upon cetaceans is very limited. Those species occurring mainly in inshore waters are most vulnerable since these are the areas where man's activities tend to be greatest. Sometimes the effects may be beneficial, for example hot water outflow from cooling units of power stations provide conditions that are suitable for planktonic growth and hence attract fish and, in turn, dolphins that feed upon the fish. Most effluents that man deposits in coastal waters have deleterious effects however, such as the cases of pesticides and heavy metals discussed above. Sewage in low concentrations may have a beneficial effect in stimulating growth of food organisms but in excess cause eutrophication, and profound changes in the ecology of the whole water column.

Barriers across coastal areas will regulate the movement of cetaceans and their fish prey. This is a greater problem to river dolphins, for example *Platanista* species, whose living areas are more restricted than those of their marine relatives. These species do not occur in any Red Sea countries.

14.3.5. Information needs

This account has highlighted the scant information we have on the status and distribution of all cetaceans occurring in the Red Sea. Some peripheral areas, such as the Gulfs of Suez and Aqaba, are better known than most parts of the Red Sea proper, for example the Sudanese coast. These latter regions need intensive survey if we are able to make confident statements about the status of any cetacean species in the Red Sea. The production of a small field guide including those species that might conceivably be recorded in the Red Sea would be immensely valuable since identification of cetaceans at sea is very difficult. Similar guides have been produced for other sea areas (Leatherwood *et al.*, 1972, 1976; Evans, 1981; Dugny and Robineau, 1982), and some of the species to be found in the Red Sea are included in these.

None of the species found in the Red Sea has a distribution confined to that region so that as a species, none is threatened by developments in the Red Sea. However, both the Plumbeous dolphin and the Red Sea bottle-nosed dolphin have a fairly restricted distribution, and being inshore species, they are vulnerable to some of the potentially important threats such as pollution and disturbance. Furthermore our knowledge of the biology and ecology of these species (particularly the Plumbeous dolphin) is sparse, though this applies equally to Gill's bottle-nosed dolphin and Rough-toothed dolphin.

Further exploration of the Red Sea for cetaceans will almost certainly add to the number of species known for the area. A short review of those species most likely eventually to be recorded may therefore be useful. It is probable that whales occasionally enter and may strand on the coast. Five species commonly occur in the northern Indian Ocean and have been recorded not infrequently in the Arabian Sea. These are the Blue whale *Balaenoptera musculus*, Fin whale *Balaenoptera physalus*, Bryde's whale *B. edeni*, Humpback whale *Megaptera novaeangliae*, and Sperm whale *Physeter macrocephalus*. All but the Minke whale tend to occur in deep waters but the latter commonly feeds close to the coast, particularly when upwellings around headlands provide rich feeding grounds in the form of plankton concentrations with associated small shoaling fish. The baleen whales feed on plankton such as eupharisiids and with the exception of the Blue Whale on a wide variety of fish. The Sperm whale feeds mainly on squid.

It is perhaps surprising that no porpoise has been recorded for the Red Sea. The Harbour porpoise *Phocoena phocoena* once occurred commonly in the Mediterranean Sea (Barcelo y Combis, 1875), but during this century has become very rare (Viale, 1976b). However, it is primarily a temperate species, its southern limit in the North Atlantic being 10°N and so would be unlikely to occur in the Red Sea unless it had entered from the Mediterranean. In the tropical Indian Ocean and eastern Pacific, its place is taken by the finless porpoise *Neophocaena phocaenoides* which inhabits not only coastal areas but many of the large rivers in S.E. Asia. Its nearest locality to the Red Sea is the south-east coast of the Arabian peninsula and the Arabian Sea, and unless the high salinity limits its distribution, the finless porpoise may sometimes occur in the Red Sea.

Finally, the Common dolphin *Delphinus delphis* and Spinner dolphin *Stenella longinestris* are perhaps surprising absentees from the Red Sea. They occur throughout the world in warm temperate and tropical waters. The former is common in the Mediterranean and like the latter is also found in the northern Indian Ocean, Gulf of Aden and Arabian Sea. They are, however, primarily pelagic species and the enclosed nature of the Red Sea together with its high salinity may discourage them from entering this region.

In conclusion there is a great need for further survey work to be carried out in the Red Sea, particularly in the southern half of the region. Cetaceans must be amongst the most poorly known vertebrate groups, and although their essentially submarine life makes this understandable, great advances are being made in our knowledge of cetaceans elsewhere. Zoological institutes and marine laboratories in the Red Sea countries should take the opportunity to study these creatures in the relatively favourable conditions that an enclosed and generally calm sea provides.

ACKNOWLEDGEMENTS

I should like to thank the many marine observers who have provided me with records for this region and surrounding areas, and particularly Captain P. Chilman, Jonathan Gordon, Paul Newton, Chris Smeenk, and Vassili Papastavrou, and Dr Yoram Yom-Tov who kindly provided some invaluable records from J. Beadon of the Tel Aviv Dolphinarium.

REFERENCES

Balcomb, K. C., Boran, J. R., Osborne, R. W. and Haenel, N. J. (1980). Observations of killer whales (*Orcinus orca*) in greater Puget Sound, State of Washington. *Rep. MMC 78/13 to U.S. Marine Mammal Commission*, Washington D.C., 41pp.

Barcelo y Combis, E. (1875). Apuntes para la fauna balear. *Anales de la sociedad espanola de historia natural* 4, 53–8.

Bertram, G. C. L. (1974). Conservation of Sirenia. *I.U.C.N. Occ. Paper* 12, pp.20.

Bertram, G. C. L. and Ricardo-Bertram, C. K. (1973). The modern Sirenia: their distribution and status. *Biol. J. Linn. Soc.* 5, 297–338.

Bibby, G. (1969). *Looking for Dilmun*. Penguin Books, London.

Bigg, M. (1981). An assessment of Killer whale (*Orcinus orca*) stocks off Vancouver Island, British Columbia. Working Paper submitted to J. W. C. workshop on Killer whales, 23–25 June 1981, Cambridge. SC/JN81/KWA.

Brown, D. H., Caldwell, D. K. and Caldwell, M. C. (1966). Observations on the behaviour of wild and captive false killer whales, with notes on associated behaviour of other genera of captive delphinids. *Los Angeles County Museum Contribs. in Science* 95: 1–32.

Demetropoulos, A. & Hadjichristophorou, M. (1982). Turtle conservation in Cyprus. *Bull. Biol. Soc. Cyprus* 2, 23–6.

Dugny, R. and Robineau, D. (1982). *Guide des mammiferes marins d'Europe*. Delachaux and Niestlé, Neuchâtel and Paris.

Ellerman, J. R. and Morrison-Scott, T. C. S. (1951). *Checklist of Palaearctic and Indian Mammals, 1758 to 1946*. British Museum (Natural History), London.

Evans, P. G. H. (1980). Cetaceans in British waters. *Mammal Review* 10, 1–52.

Evans, P. G. H. (1981). *Guide to Identification of Cetaceans in the North-east Atlantic*. Occasional Publication, The Mammal Society.

Evans, W. E. and Bastian, J. (1969). Marine mammal communication: social and ecological factors. In *The Biology of Marine Mammals*. Ed. H. T. Andersen, pp. 425–76. Academic Press, New York.

Food and Agriculture Organization. (1973). Report to the Government of the People's Democratic Republic of Yemen on marine turtles management, based on the work of H. F. Hirth and S. L. Hollongworth, FAO/TA Marine Turtle Biologists. Rep. FAO/UNDP (TA) (3178).

Food and Agriculture Organization (1976). Ad hoc group 2 on small cetaceans. *Draft report*. ACMRR/MM/SC.3 Scientific Consultation on Marine Mammals, Bergen, Norway, September 1976.

Food and Agriculture Organization (1978). Mammals in the seas. Vol.1. *Report of the FAO ACMRR working party on marine mammals. FAO Fisheries Series no.5*.

Frazier, J. (1980). Marine turtles and problems in coastal management. In *Proceedings of the Second Symposium on Coastal and Ocean Management*. Ed. B.L. Edge, Vol. 3, pp.2395–411.

Frazier, J. (in press (a)). Conservation of sea turtles in the Red Sea. In *Proceedings of the International Conference on Marine Science in the Red Sea*, 24-29 April 1982, Al Ghardaqa, Egypt.

Frazier, J. (in press (b)). Marine Turtles in the Seychelles and Adjacent Territories (Chapter 21). In *Biogeography and Ecology of the Seychelles Islands*. Ed. D.R. Stoddart. Junk, The Hague.

Frazier, J. and Salas, S. (In press). The status of marine turtles in the Egyptian Red Sea. *Biol. Conserv.*

Geldiay, R., Koray, T. and Balik, S. (1982). Status of sea turtle populations (*Caretta c. caretta* and *Chelonia m. mydas*) in the northern Mediterranean Sea, Turkey. In *Biology and Conservation of Sea Turtles*. Ed. K. Bjorndal, pp.425–34, Smithsonian Institution Press, Washington D.C.

Gohar, H. A. F. (1979). Notes on the Red Sea Dugong. Unpublished paper given at Symposium on the Biology of the Dugong, Ocean Research Institute, University of Tokyo, Japan, 6–7 December 1979, pp.8.

Gunter, G. (1942). Contributions to the natural history of the bottlenose dolphin, *Tursiops truncatus* (Montagu), on the Texas coast, with particular reference to food habits. *J. Mammal.* 23, 267–76.

Hershkovitz, P. (1966). Catalog of living whales. *Bulletin of the United States National Museum* 246, 1–259.

Hirth, H. F. (1978). A model of the evolution of the Green turtle (*Chelonis mydas*) remigrations. *Herpetol.* 34(2), 141–7.

Hirth, H.F. and Abdel-Latif, E.M. (1980). A nesting colony of the Hawksbill turtle *Eretmochelys imbricata* on Seil Ada Kebir Island, Suakin Archipelago, Sudan. *Biol. Conserv.* 17, 125–30.

Hirth, H. and Carr, A. (1970). The Green turtle in the Gulf of Aden and the Seychelles Islands. *Verh. K. ned. Acad. Wet. Afd. Natuurk.* 58(5), 1–44, 7 pls.

Holden, A. V. and Marsden, K. (1967). Organochlorine residues in seals and porpoises. *Nature, Lond.* 216, 1274–6.

I.U.C.N. (International Union for Conservation of Nature and Natural Resources). *Mammal Red Data Book* (1981), compiler J. Thornback, pp. 417–42. Sirenia.

Kingdon, J. (1971). *East African Mammals: an Atlas of Evolution in Africa*. Academic Press, London.

Layne, J. N. (1965). Observations on marine mammals in Florida waters. *Bull. Florida St. Mus. Biol. Sci.* 9, 131–81.

Leatherwood, J. S., Evans, W. E. and Rice, D. W. (1972). Whales, dolphins, and porpoises of the eastern North Pacific: a guide to their identification in the water. *Naval Undersea Center Technical Public. no. 282*, San Diego, California.

Leatherwood, J. S., Caldwell, D. K. and Winn, H. E. (1976). Whales, dolphins, and porpoises of the western North Atlantic: a guide to their identification in the water. *NOAA Technical Report, National Marine Fisheries Service, CIRC-396*, Seattle, Washington.

Leatherwood, J. S., Reeves, R. R. and Foster, L. (1985). The Sierra Club Handbook of Whales and Dolphins. Sierra Club Books, San Francisco.

Lipkin, Y. (1975). Food of the Red Sea Dugong (Mammalia: Sirenia) from Sinai. *Israel J. of Zool.* 24, 81–98.

Marcuzzi, G. and Pilleri, G. (1971). On the zoogeography of Cetacea. In *Investigations on Cetacea*, ed. G. Pilleri, vol.3, Part I, pp.101–70.

Margaritoulis, D. (1982). Observations on Loggerhead sea turtle *Caretta caretta* activity during three nesting seasons (1977–1979) in Zakynthos, Greece. *Biol. Conserv.* 24, 193–204.

Marsh, H. (1980). Age determination of the Dugong (*Dugong dugon* (Müller)) in Northern Australia and its biological implications. *Rep. Int. Whal. Comm.* (Special Issue 3), pp.181–201.

Marsh, H. (ed.) (1981). *The Dugong*. Proceedings of a Seminar/Workshop held at James Cook University, 8–13 May 1973. James Cook University, Queensland.

Mitchell, E. (ed.) (1975). Report on the meeting on smaller cetaceans, Montreal, April 1–11, 1974, *Journal of the Fisheries Research Board of Canada* 32, 889–983.

Moore, R. J. and Balzarotti, M. A. (1977). Report of 1976 Expedition to Suakin Archipelago. Mimeographed (Cambridge University) 27 pp.

Nishiwaki, M. (1972). General biology. In *Mammals of the Sea: Biology and Medicine*. Ed. S. H. Ridgway, pp. 3–204. Thomas Press, Springfield, Illinois.

Nishiwaki, M., Kasuya, T., Miyazaki, N., Tobayama, T. and Kataoka, T. (1979). Present distribution of the Dugong in the world. *Sci. Rep. Whales Res. Inst.* No.31, 133–41.

Norris, K. S. and Prescott, J. H. (1961). Observations on Pacific cetaceans of Californian and Mexican waters. *Univ. Calif. Publ. Zool.* 63, 291–402.

Osborn, D. (1978a). A naturally occurring zinc- and cadmium-binding protein from the liver and kidney of *Fulmarus glacialis*: a pelagic North Atlantic seabird. *Biochem. Pharmac.* 27, 822–4.

Osborn, D. (1978b). Toxic and Essential Heavy Metals in Birds. *ITE Annual Reports 1978*, 53–6.

Pearse, J. S. (in press). The Gulf of Suez: signs of stress on a tropical biota. In *Proceedings of the International Conference on Marine Sciences in the Red Sea*, 24-29 April 1982, Al Ghardaqa, Egypt.

Perrin, W. F., Warner, R. R., Fiscus, C. H. and Holts, D. B. (1973). Stomach contents of porpoises, *Senella* spp. and yellow fin tuna, *Thunnus albacares*, in mixed species aggregation. *Fish. Bull.* 71, 1077–92.

Pilleri, G. (1967). Behavior of *Pseudorca crassidens* (Owen) off the Spanish Mediterranean coast. *Rev. Suisse Zool.* 84, 679–83.

Por, F.D. (1978). *Lessepsian Migration — The Influx of Red Sea Biota into the Mediterranean by the Suez Canal*, Springer-Verlag, New York.

Ross, G. J. B. (1977). The taxonomy of bottle-nosed dolphins *Tursiops* sp. in South African waters, with notes on their biology. *Ann. Cape Prov. Mus. Nat. Hist.* 11, 135–94.

Ross, J. P. (1981). Hawksbill Turtle *Eretmochelys imbricata* in the Sultanate of Oman. *Biol. Conserv.* 19, 99–106.

Ross, J. P. and Barwani, M. A. (1982). Review of sea turtles in the Arabian Sea. In *Biology and Conservation of Sea Turtles*. Ed. K. Bjorndal, pp.373–82, Smithsonian Institution Press, Washington D.C.

Rüppell, E. (1835). *Neue Wirbelthiere zu der Fauna von Abyssinien gehörig, Amphibien*, Frankfurt am Main. 18 pp.
Saayman, G. S. and Tayler, C. K. (1973). Social organization of inshore dolphins (*Tursiops aduncus* and *Sousa*) in the Indian Ocean. *J. Mammal.* 54, 993–6.
Saayman, G. S. and Tayler, C. K. (1979). The socioecology of humpback dolphins (*Sousa* sp.). In *The Behavior of Marine Animals, Vol. III.* Ed. H. E. Winn and B. L. Olla, pp. 165–226. Plenum Press, New York.
Schoff, W. H. (1912). *The Periplus of the Erythrean Sea*, Longmans, Green and Co. New York,
Sella, I. (1982). Sea Turtles in the Eastern Mediterranean and northern Red Sea. In *Biology and Conservation of Sea Turtles*. Ed. K. Bjorndal, pp. 417–23, Smithsonian Institution Press, Washington, D.C.
Slijper, E. J. (1962). *Whales*. Hutchinson, London.
Steindachner, F. (1901). Bericht über die herpetologischen Aufsammlungen. *Denkschr. Akad. Wiss., Wien.* 69, 325–35, pls. 2.
Steinglass, F. (n.d.). *A Learner's English-Arabic Dictionary*, Librairie du Liban, Beirut.
Tayler, C. K. and Saayman, G. S. (1972). The social organization and behaviour of dolphins (*Tursiops aduncus*) and baboons (*Papio ursinus*): some comparisons and assessments. *Ann. Cape Prov. Mus. Nat. Hist.* 9, 11–49.
Thompson, M. F. (compiler) (in press). *Proceedings of the International Conference on Marine Science in the Red Sea*, 24–29 April 1982, Al Ghardaqa, Egypt.
Tomilin, A. C. (1967). *Mammals of the U.S.S.R. and Adjacent Countries. IX Cetacea.* Israel Program for Scientific Translations, Israel.
Urban, E. K. (1970). Nesting of the Green turtle (*Chelonia mydas*) in the Dahlak Archipelago, Ethiopia. *Copeia* 1970(2), 393–4.
van Bree, P. J. H. and Kristensen, I. (1974). On the intriguing stranding of four Cuvier's beaked whales *Ziphius curvirostris* G. Cuvier, 1823, on the Lesser Antillean island of Bonaire. *Bijdr. Dierkd.* 44, 235–8.
Viale, D. (1974). Divers aspects de la pollution pour les metaux chez quelques cetacea de Mediterranee occidentale. *Iieme Journee pour L'Etude de la Pollution. Conseil permanent international pour l'exploration de la mer*, Monaco 1974, 183–91.
Viale, D. (1976a). Relation entre les echouages de cetaces et la pollution chimique en mers Ligure et Tyrrhenienne. ACMRR/SC/92 with addendum. Unpublished document for Scientific Consultation on Marine Mammals. 31 August-9 September, 1976, Bergen, Norway.
Viale, D. (1976b). Etude des cetaces en Mediterranee occidentale. ACMRR/SC/122. Unpublished document for Scientific Consultation on Marine Mammals. 31 August–9 September, 1976, Bergen, Norway.
Walczak, P.S. (1979). The status of marine turtles in the waters of the Yemen Arab Republic. *Brit. J. of Herpetol.* 5(12), 851–3.
Wells, R. S., Irvine, A. B. and Scott, M. D. (1980). The social ecology of inshore Odontocetes. In *Cetacean Behavior*. Ed. L. M. Herman. J. Wiley, New York.
Wursig, B. (1978). Occurrence and group organization of Atlantic bottle-nosed porpoises (*Tursiops truncatus*) in an Argentine bay. *Biol. Bull.* 154, 348–59.
Wursig, B. and Wursig, M. (1977). The photographic determination of group size, composition, and stability of coastal porpoises (*Tursiops truncatus*). *Science* 198, 755–6.
Wursig, B. and Wursig, M. (1979). Behavior and ecology of bottlenose porpoises, *Tursiops truncatus*, in the South Atlantic. *Fish. Bull.* 77, 299–442.

FOOTNOTE Since the cetacean section went to press, there have been a number of boat surveys in the Red Sea which have revealed important new information (J. Gordon, C. Smeenk pers. comm. 1986). The Spinner dolphin was recorded commonly in the southern Red Sea, with occasional sightings of False Killer, Common and Striped dolphins (*S. coeruleoalba*). At the entrance to the Red Sea, in the Gulf of Aden, Sperm whales, Killer and Short-finned Pilot whales *Globicephala macrorhynchus* were also recorded. Finally there were some apparently reliable reports of small baleen whales, possibly Minke Whale.

CHAPTER 15

Sea Birds of the Red Sea

PETER G. H. EVANS

Edward Grey Institute, Department of Zoology, Oxford OX1 3PS, U.K.

CONTENTS

15.1. INTRODUCTION

In this chapter we shall be considering those bird species whose families are clearly associated with the sea, although some members may have become secondarily adapted to inland conditions at least for part of the year. Other groups such as waders, herons, and raptors such as osprey, may be seen along Red Sea coasts, in lagoons, mangrove swamps or upon islets. However, these are not true seabirds and consequently lie outside the scope of this review, although a short section is included, summarising their status and distribution for the interested reader.

TABLE 15.1. Relative importance of particular seabird families in the Red Sea and nearby areas.

Seabird Family	No. of breeding species per family as a percentage of total number of species for that region				
	Red Sea	Arabian Gulf	Indian Ocean	Mediterranean	World
Petrels/Shearwaters	6.25	8.3	8.1	12.5	20.7
Storm-petrels	0	0	0	6.25	7.4
Tropic-birds	6.25	8.3	8.1	0	1.05
Pelicans	6.25	8.3	5.4	0	2.8
Gannets/Boobies	12.5	0	8.1	0	3.2
Cormorants	6.25	8.3	16.2	12.5	10.2
Frigate-birds	0	0	5.4	0	1.8
Gulls	12.5	8.3	8.1	18.75	15.8
Terns/Noddies	50.0	58.3	37.8	50.0	14.7
Skimmers	0	0	2.7	0	0.35
rest	0	0	0	0	21.4
Total No. of species	16	12	37	16	285

TABLE 15.2. Number of breeding species per seabird family for different regions of the Red Sea

Seabird Family	Region					No. of species per family in each region as percentage of total no. of species in world
	Gulf of Suez	Gulf of Aqaba	Northern Red Sea	Southern Red Sea	Total for Red Sea	
Petrels/ Shearwaters	0	0	0	1	1	1.7
Tropic-birds	1	0	1	0	1	33.0
Pelicans	0	0	0	1	1	12.5
Gannets/Boobies	0	0	0	2	2	22.2
Cormorants	0	0	0	1	1	3.4
Gulls	1	1	1	2	2	4.4
Terns/Noddies	2	1	6	8	8	19.0
TOTAL	4	2	8	15	16	55.6

A number of features of the Red Sea determine its seabird fauna. Perhaps most important is its partial isolation from the open ocean; the sill at the southern entrance is only about 100 m deep so that the deep waters of the Indian Ocean are prevented from entering. At the northern end, the Gulf of Suez has only recently (since the opening of the Suez Canal in 1869) formed a connection with the Mediterranean. These two features provide barriers to a number of marine organisms, including the pelagic species upon which certain seabirds feed. For this reason the pelagic shearwaters and petrels are poorly represented. With its high coast length to sea area ratio, the long and slender Red Sea has instead a seabird fauna dominated by coastal species, notably terns (Table 15.1).

The Red Sea, lying as it does between the Mediterranean and the Indian Ocean may receive immigration from both, although because the latter is much larger, extending over many degrees of latitude, and is less isolated from the Red Sea, it is not surprising that a high proportion of seabird species occurring in the Red Sea are found only in the southern part (Table 15.2). With the continents of Africa and Asia on either side, and Europe and the Middle East to the north, the Red Sea is also an important migration route for birds (both seabirds and land birds) moving south to Africa in autumn and returning the next spring.

15.2. DISTRIBUTION AND ECOLOGY OF RED SEA SEABIRDS

Lying entirely within an arid tropical zone, the Red Sea experiences high evaporation which largely exceeds precipitation so that salinity and temperature are comparatively high (*see* Chapter 3 for details). These climatic conditions have important consequences upon the physical oceanography of the region which in turn affects the distribution and ecology of its seabirds.

15.2.1. Factors controlling the distribution of seabirds

Physical oceanography may affect seabirds in a number of ways. Upwellings to the surface of nutrient rich water arise where there are irregularities of the undersea topography or where current systems meet, and these often result in plankton fronts and associated concentrations of predatory fish, marine birds and mammals.

Seaward from the coast lies a coral reef zone, with depths of less than 50 m, then shelves which vary in depth between 300 and 600 m; most of the seabirds live in these regions. Seaward of these is an irregular broken floor to depths of more than 1100 m and then a deep continuous axial trough. These conditions do not form rich feeding grounds for seabirds and together with the lack of breeding places nearby, result in the low densities of birds witnessed by most observers when travelling by ship down the Red Sea. To the north lies the Gulf of Suez, flat-bottomed and shallow with depths never exceeding ninety metres, contrasted with the Gulf of Aqaba, which is a deep basin with narrow shelves and then an abrupt descent to more than 1,000 m. It is probably the shallow depth of the Gulf of Suez that makes it relatively richer in seabirds, particularly the surface-feeding terns that take small coastal fish (Table 15.2). The entrances to the Gulfs of Suez and Aqaba have an intricate maze of reefs and islands as does the southern half of the Red Sea, and it is in such areas that most of the seabird colonies are concentrated. Tidal streams which may cause upwellings occur at the two extremities of the Red Sea, the Gulf of Suez and the strait of Bab al Mandab, and also in shallow areas, particularly among the extensive coastal archipelagoes in the south. These areas therefore have relatively high levels of nutrients as do also the surface waters at the southern entrance to the Red Sea, caused by upwellings particularly during the winter inflow of water from the Gulf of Aden (Morcos, 1970). This winter inflow brings with it a number of pelagic planktonic organisms (Halim, 1969), and together with the relatively low temperatures further north, may account for the southerly movements of seabird species in autumn. Nutrients and plankton distributions are discussed in Chapter 5.

Where nutrients are readily available, one may expect concentrations of plankton and fish, and if suitable breeding sites are available nearby, these provide ideal conditions for seabird colonies, the sites of which are shown in Figure 15.1.

15.2.2. Seabirds of the Red Sea in a world context

Sixteen species of seabirds have been recorded breeding in the Red Sea (although three of these may not be presently breeding). These are listed in Table 15.3 and described further below. Between 11 and 13 of these are probably of Indo-Pacific origin, only two derive from the Western Palaearctic or western part of the European/Asian land mass through the Mediterranean. One species, the White-eyed Gull, is endemic to the region, and is probably derived from the same stock as the Sooty gull which it closely resembles both morphologically and behaviourally (Fogden, 1964). Half of the breeding species are found throughout the tropics, but three species, the Socotra cormorant, Sooty gull, and White-cheeked

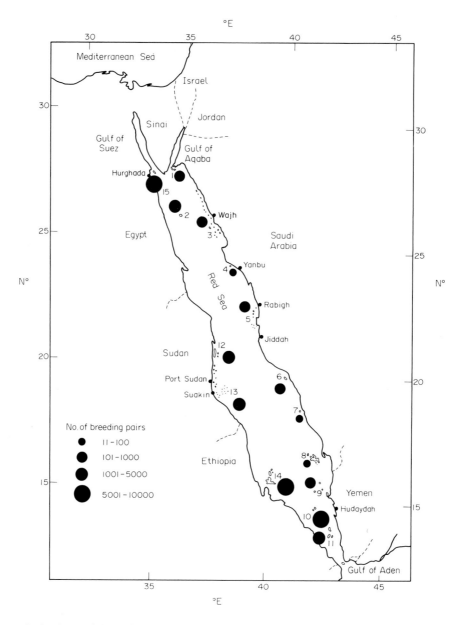

Fig. 15.1. Seabird colonies of the Red Sea, amalgamating data on all breeding species, from several sources. The numbered islands are 1 – Tiran Island, 2 – Perim and Brothers Islands, 3 – Islands on the Wajh Bank, 4 – Islet north of Yanbu al Bahr, 5 – Jiddah and Rabigh, 6 – Janabiyat, 7 – Kutunbol, 8 – Farasan Islands, 9 – Islands near Al Hudaydah, 10 – Zubayr Island, 11 – Hanish Islands, 12 – Muhammad Qôl and associated islands, Mukawwar, Mayetib and Taila, 13 – Suakin Archipelago, 14 – Dahlak Archipelago, 15 – Islands near Hurghada (Mugawish, Om Qamar, Abu Mingar and Shadwan).

TABLE 15.3. Postulated origin of Red Sea breeding species.

(1) Indo-Pacific origin	Audubon's Shearwater
	Red-billed Tropic-bird
	Brown Booby
	Blue-faced Booby
	Pink-backed Pelican
	Socotra Cormorant
	Sooty Gull
	White-cheeked Tern
	Bridled Tern
	Swift Tern
	Lesser-Crested Tern
	Brown Noddy
(possibly)	Little Tern (*saundersi* race)
(2) Palaearctic origin	Caspian Tern
	Roseate Tern
(3) Endemic Red Sea	White-eyed Gull

tern, though probably deriving from an Indian Ocean fauna, have a more restricted distribution, confined to the Red Sea, Arabian Peninsula and Arabian Gulf. It should be noted that the Socotra cormorant is a doubtful breeder in the Red Sea and probably should be excluded from the list.

The seabird fauna of the Red Sea clearly has stronger affinity to the Indo-Pacific rather than Mediterranean fauna. This is also the case with the rest of the marine fauna, which probably derives from the Indian Ocean. The latter has a larger diversity of organisms and is less isolated from the Red Sea, than is the Mediterranean. Seabirds of course do not have the same barriers to movement, but nevertheless depend upon these organisms as prey and so associate with them. This relative ease of movement amongst seabirds will account for the low endemism, since the Red Sea is a small and not very isolated area bordered by large land masses with their associated coastlines that act as sources of immigration.

Compared with other regions of comparable area in the world, the Red Sea is fairly rich in seabird species, and contains some large breeding colonies, though numbers are much smaller than some colonies in the central Pacific, Arctic and Antarctic. For species with a broad tropical distribution (e.g. Brown booby, Bridled tern and Brown noddy) the region is of minor importance. However, for species such as the Sooty gull, White-cheeked tern, and White-eyed gull, the Red Sea forms an important centre of their world distribution.

15.2.3. Biology of breeding and feeding

A review of aspects of breeding biology and feeding ecology for all sixteen breeding species is given in Table 15.4. Most are shallow-water fish feeders (reflecting the availability of shallow reefs with associated high concentrations of small shoaling fish), nesting colonially on low flat islands, often in association with related species. Breeding occurs in all species (with the possible exception of the Audubon's shearwater) during the summer months, particularly between June and August when sea surface and air temperatures are highest.

Many species, particularly open-nesters such as the terns are vulnerable to both aerial and ground predators. Nest dispersion and behaviour are modified according to the breeding ecology of the species.

Thus the Lesser-crested tern nests in very dense colonies which afford partial protection simply by force of numbers, but they will also aggressively drive away any aerial predators if necessary. Its close relative the Swift tern is less aggressive and so more vulnerable to predation. It also gains protection by nesting in dense colonies, but when its numbers are low it may be found associating with Lesser-crested terns,

TABLE 15.4. Biology and Ecology of seabirds breeding in the Red Sea.

Species	Main Breeding Season	Clutch Size	Incub. Period	Fledging Period	Breeding habitat and coloniality
Audubon's shearwater (*Puffinus lherminieri*)	poss. Nov–Feb	1	44–60	62–100	Semi-colonial; holes, crevices, in rocks, amongst boulders.
Red-billed Tropic-bird (*Phaethon aethereus*)	April–August	1	42–44	80–90	Colonial; holes, crevices in rocks, boulders.
Brown Booby (*Sula leucogaster*)	March–August	1–3	39–48	85–103	Semi-colonial; steep sites, often on bare rock, also on slopes, flat ground, sandy beaches, sometimes flanked by dense vegetation
Blue-faced Booby (*Sula dactylatra*)	March–October	1–3	38–49	c.120	Colonial; scrape of bare sand, gravel or rock decorated with gravel
Pink-backed Pelican (*Pelecanus rufescens*)	August–November	1–3	30	c.85	Colonial; well-built nest in tall trees beside rivers, lakes, sandy islands.
Socotra Cormorant (*Phalacrocorax nigrogularis*)	May–October	2–3	?	?	Colonial; scrape of bare sand or gravel, often on islands.
White-eyed Gull (*Larus leucopthalmus*)	June–September	(1–3)	?	?	Semi-colonial; scrape on bare sand, often surrounded by seaweed.
Sooty Gull (*Larus hemprichii*)	June–September	(2–3)	?	?	Solitary; scrape of bare sand or gravel, usually under bushes, on islands or promontories.
Swift Tern (*Sterna bergii*)	June–August	1–2	25–30	38–40	Colonial; scrape of bare sand, gravel, coral or rock, often without shelter on low-lying islands.
Roseate Tern (*Sterna dougalli*)	June–August	1–2	21–26	27–30	Semi-colonial; scrape of bare sand or gravel, on sand-dunes or spits, shingle, coral or low rocky islands.
White-cheeked Tern (*Sterna repressa*)	June–August	1–3	?	c10	Semi-colonial; scrape of bare sand on coral-girt islands, sometimes bare sand-flats.
Caspian Tern (*Sterna caspia*)	March–May	1–3	20–22	30–35	Semi-colonial; scrape of bare sand, gravel, or rock, on sand-dunes, reefs, islets, or skerries.
Lesser-Crested Tern (*Sterna bengalensis*)	July–October	1–2	21–26	32–35	Colonial; scrape of bare sand among dwarf, sparse vegetation, on sand-spits, coral reefs.
Bridled Tern (*Sterna anaethetus*)	June–September	1–2	28–30	55–63	Colonial/semi-colonial; scrape in crevice under bush, ledge, protected by low scrub on sandy and coral islands.
Little Tern (*Sterna albifrons*)	May–June	1–3	18–22	19–20	Semi-colonial; bare scrape on sand, shingle, or shell-beach.
Brown Noddy (*Anous stolidus*)	June–July	1	32–35	c.42	Semi-colonial; well-built nest in scrub, low trees, also on rock shelves, crevices or bare shingle.

nesting around the periphery of their colonies. Likewise the White-cheeked tern may associate with Lesser-crested terns; otherwise it shows much greater nest dispersion, relying upon this together with camouflage of eggs and young to escape the attentions of ground predators. White-cheeked terns are relatively non-aggressive but they will drive off aerial predators should occasion demand. The Caspian

Feeding Methods and Habitat	Food	Sources
Surface-dipping or diving; coastal.	Small fish and crustaceans.	Snow (1965a), Harris (1969).
Plunge-diving; usually pelagic.	Fish and squid.	Stonehouse (1962b), Snow (1965b), Harris (1969b).
Plunge-diving; usually inshore in v. shallow water.	Fish (+ squid and prawns).	Kepler (1969), Nelson (1978).
Plunge-diving; usually inshore.	Fish (+ squid).	Kepler (1969), Nelson (1978), Gallagher & Woodcock (1980).
Surface-dipping; inshore shallow water.	Fish	Din & Eltringham (1974a,b); Gallagher & Woodcock (1980).
Surface-diving; inshore.	Fish	Meinertzhagen (1954), Gallagher & Woodcock (1980).
Plunge-diving; inshore and offshore.	Mainly fish; also crustaceans, molluscs and annelids.	Su-Aretz, S. (1980); Cramp & Simmons (1983).
Scavenging or plunge-diving, predatory or kleptoparasitic; inshore.	Fish, offal, birds' eggs, molluscs, and crustaceans.	Archer & Godman (1937), Fogden (1964).
Plunge-diving; usually pelagic.	Fish (+ crustaceans).	Archer & Godman (1937); Ali & Ripley (1969); Cramp & Simmons (1985).
Plunge-diving; inshore shallow water.	Fish	Dunn (1972); Nisbet in Cramp & Simmons (1985).
Surface-dipping; usually inshore over coral reefs.	Small fish and invertebrates.	Archer & Godman (1937); Clapham (1964); Cramp & Simmons (1985), Nikolaus (pers. comm.).
Plunge-diving; inshore near sand-spits.	Fish (+ invertebrates).	Su-Aretz, S. (1979); Cramp & Simmons (1985).
Plunge-diving or surface-dipping; often pelagic.	Small fish (+ crustaceans).	Archer & Godman (1937); Clapham (1964); Nikolaus (per. comm.).
Surface-hovering and dipping; pelagic often away from reefs.	Small fish, + planktonic crustaceans, molluscs.	Meinertzhargen (1954), Ali & Ripley (1969), Hulsman (1974, 1977), Diamond (1976), Trott (1952).
Plunge-diving; usually inshore shallow water.	Small fish and invertebrates (crustaceans, insects).	Cramp & Simmons (1985).
Surface-dipping; usually inshore.	Small fish	Dorward & Ashmole (1963); Cramp & Simmons (1985); Watson (1908).

tern has very similar habits to the White-cheeked, and nests at low densities. Both the Roseate and Little terns tend to hide their nests from aerial predators and their nests are well dispersed to reduce conspicuousness. Finally both Bridled tern and Brown noddy have relatively dispersed nests that are hidden amongst vegetation, in the former case often under a bush and in the latter case, in the branches of a tree. Possibly because its nest is not so easily concealed, the Brown noddy tends to be the more aggressive, but Bridled terns will also mob gulls. However, the latter sometimes also adopt distraction display where they feign injury (by trailing a wing and moving away in an erratic manner along the ground) and in this way, lead a potential predator away from the nest.

Most of the seabird species feed by scooping fish from or close to the surface, usually in shallow waters, though some species (notably the Red-billed tropic bird and Swift tern) feed further offshore in deeper waters. Terns and gulls will follow shoals of fish, which when driven to the surface by predatory fish, are taken by surface-dipping or plunge-diving. Often these fish may be momentarily stunned or confused and probably this eases their capture.

15.3. STATUS AND DISTRIBUTION OF SPECIES

We may consider the seabirds of the Red Sea under two headings — those that actually breed (or at one time have been recorded breeding) in the area, and those which spend part of the year there, but without breeding. Since our knowledge of the seabirds of the Red Sea is so fragmentary, it is possible that one or two species reviewed below, are considered in the wrong category; and three species — Audubon's shearwater, Socotra cormorant and Roseate tern have not recently been recorded breeding in the Red Sea.

15.3.1. Breeding species

Puffinus lherminieri Audubon's shearwater (Fig. 15.2)

The typical race (*P. lherminieri lherminieri*) of this species has a pan-tropical distribution, occurring in the Atlantic, Pacific, and Indian Oceans. The Persian race, *P.l. persicus*, however, is restricted to a small area along the Arabian Peninsula, in the Arabian Gulf and the southern part of the Red Sea where it may breed on islets off the Yemen coast (Brown *et al.*, 1982; Cornwallis and Porter, 1982) although good evidence for this is lacking. From April and May, numbers begin to move towards the coast of Dhufar where large concentrations occur over the cool waters of the upwelling area during the peak months of the South-east Monsoon from June to August (Bailey, 1966). Its status as a breeding species remains unproven.

Phaethon aethereus Red-billed Tropic-bird

This species is widely distributed in equatorial regions of the world, breeding on islands along both the Pacific and Atlantic coasts of central America, Brazil, the Atlantic islands of Cape Verde, Ascension and St. Helena, in the Red Sea, Arabian Sea and Arabian Gulf.

In the Red Sea, the species breeds on islets in the Gulf of Suez (Etchecopar and Hüe, 1967), the Brothers Islands (Jones, 1946), almost certainly in the Dahlak Archipelago (Clapham, 1964), the islands

Fig 15.2. The Audubon's shearwater is restricted in the Red Sea to breeding on islets off the Yemen coast. (*Photo* M. de L. Brooke)

near Al Hudaydah off the coast of North Yemen (Jennings, 1981) and on the Zubayr (Jazair az Zubayr) and Hanish groups in the southern Red Sea (Morris, 1962). It probably also occurs elsewhere in the Red Sea. Although resident in the region, numbers probably disperse over a wide area outside the breeding season.

Sula leucogaster Brown Booby (Fig. 15.3)

The species has a pan-tropical distribution, breeding in all the major oceans. Its range extends into the Red Sea with the species recorded breeding in small numbers in the north on the islet of Om Qamar, Hurghada group and on Ashfrafi islands (Al-Hussaini, 1939), the Farasan and Suakin islands, Sudanese coast (Mackworth-Praed and Grant, 1957), Byreema islands near Al Wajh (Moore and Ormond, 1982), and the Dahlak archipelago (Clapham, 1964). At least one hundred pairs breed on Byreema Island, and a large colony of several hundred probably breed on Dahurt Ghab (Seven Islets), southern Red Sea (Moore, 1977). Boobies, singly or in pairs, were also observed at a number of islands along the Sudanese coast though breeding was not proved. Tuck (1980) mentions the species nesting in considerable numbers on the Zubayr, Abu Ail, and Hanish island groups (*see also* Phillips, 1947; Morris, 1962) and 6—10,000 birds were counted on or around the Zubayr group in April 1964, including about 250 pairs with well-feathered young on Quoin (Bourne, 1966). Small numbers have also been recorded in the Gulf of Aqaba, at least between February and September (Safriel, 1968; Jennings *pers. comm.*). The species is resident in the Red Sea but with some evidence of dispersal outside the breeding season.

Large guano deposits and old references to colonies of boobies, terns and gulls traceable to the Dahlak Archipelago and the Hanish and Mohabbakah islands suggest that large seabird populations have existed on these islands for some considerable time. Although the Hanish islands still contain large numbers of Brown booby (and Red-billed tropic-birds and Bridled terns), the Dahlak Archipelago held only about 120 pairs of Brown booby in August—September 1962 (Clapham, 1964). Although this was towards the end of the breeding season, other species (notably White-eyed gulls, Lesser-crested and Bridled terns) with similar breeding seasons, were present in large numbers.

Fig. 15.3. The Brown Booby is widely distributed as a breeding species in the Red Sea, often nesting in large numbers, although the species has almost certainly suffered from human exploitation. (*Photo* M. de L. Brooke)

Sula dactylatra Blue-faced Booby (Fig. 15.4)

The world distribution of this species is similar to the Brown booby, occurring throughout the tropical zone. Blue-faced boobies nest in small numbers alongside Brown boobies on the Zubayr, Abu Ail and Hanish islets in the southern Red Sea (Morris, 1962). They had young at all stages on November 4th 1961 but in May of the following year there were no breeding birds at all. The species has also been recorded on Kutunbol Island, Saudi Arabia (Gallagher *et al.*), although it has not been recorded further north than here (Moore, 1977).

Pelecanus rufescens Pink-backed Pelican

The species breeds in much of Africa, south of the Sahara, and in small numbers in the Red Sea north to about 20°N (Etchecopar & Hüe, 1967) where it may be found on a few islands near Al Qunfidhah and Al Lith, Saudi Arabia. Clapham (1964) recorded the species as present but not breeding in the Dahlak Archipelago, and Meinertzhagen (1954) noted it as a summer visitor to the Arabian coast from Aden to Jiddah. It is partially migratory, dispersing widely outside the breeding season, some moving inland from colonies along the coast of Eritrea ·(Smith, 1957). Similar movements have been recorded elsewhere in the Red Sea region (Cramp & Simmons, 1977).

Phalacrocorax nigrogularis Socotra Cormorant

This species has a very restricted distribution, occurring in the Arabian Gulf and possibly along the Arabian Peninsula into the southern Red Sea, where it is suspected to breed on islets off Dhufar, Aden and Socotra (Bailey, 1966). One was seen in April 1928 a little to the north of Perim island (Alexander, 1929) though it may have only been a vagrant. Tuck (1978) records the species in the southern half of the Red Sea although I cannot find any evidence to substantiate this. In the Arabian Gulf it may be seen all the year round but will also disperse over the north-west part of the Arabian Sea around the Gulf of Aden. At present its status as a breeding species in the Red Sea must remain doubtful.

Larus leucopthalmus White-eyed Gull (Fig. 15.5)

Otherwise known as the Red Sea Black-headed gull, this species is confined to the Red Sea where it is the only endemic seabird. Although generally a common resident, its distribution appears to be patchy. On the Sudanese coast, Moore (1977) recorded it in large numbers on Mukawwar, but did not see it south of Eitwid islet, although Mackworth-Praed and Grant (1955) reported it nesting in very large numbers on the Somalia coast. Jennings (1981) describes the species as common north to Al Wajh where small numbers bred on the Wajh Bank in summer 1982 (Moore and Ormond, 1982), and Su-Aretz (1980) records it as resident along the shores of northern Sinai with 30–50 pairs also nesting on the island of Tiran. Jones (1946) found birds with eggs on Perim and Brothers islands, and it is also recorded from Hurghada on the Egyptian coast. Although the species does not breed on Mukawar Island, 36 pairs were found breeding in August 1982 on Deganub Island, between Muhammad Qôl and Mukawar Island (Nikolaus pers. comm.). Further south, about 1,400 adults were observed on the islands in the Dahlak peninsula in 1962 (Clapham, 1964). The species tends to migrate south in winter, particularly from the northern Red Sea and Gulf of Aqaba, though birds apparently remain in the Gulf of Suez throughout the year.

Fig. 15.4. The Blue-faced Booby is restricted in the Red Sea to the southern parts, where it may nest in small numbers alongside its relative the Brown Booby. (*Photo* M. de L. Brooke)

Larus hemprichii Sooty or Aden Gull (Fig. 15.6)

The Sooty gull is found only in the Red Sea, the Arabian Gulf and the coasts of Somalia and Iran. In the Red Sea, Moore (1977) recorded it present in small numbers on almost every island he visited along the Sudanese coast. Although nesting mainly in isolated pairs, he observed large numbers, which he presumed were nesting, on Mukawwar Island. The species also breeds in the Dahlak Archipelago (Smith, 1957; Clapham, 1964). Its northerly breeding limit appears to be about 26°N; Jennings (1981) records it as an uncommon resident on the Arabian coast north to Jiddah, and small numbers have been found breeding on the Wajh Bank (Moore and Ormond, 1982), whilst it was noted as very plentiful in June 1945 on the Brothers Islands (Phillips, 1947). It also occurs infrequently in winter in the Gulfs of Suez and Aqaba.

Sterna caspia Caspian Tern

This species has a very broad distribution. Although mainly warm temperate or sub-tropical, outlying populations exist in tropical regions of South and West Africa, north Australia, and cold temperate regions of the north-east United States and Canada, and the Baltic. In the Red Sea region, it is frequently observed in the Gulf of Suez and the Sinai peninsula during most months of the year although proof of breeding has rarely been confirmed. Small numbers were found probably breeding along the coast of the southern part of the Gulf of Aqaba in February 1973, and surveys in the same area between March and July of that year revealed one small colony of about 25—30 pairs on the island of Tiran (Su-Aretz, 1973). Borman (1929) also found nests with eggs on Abu Mingar and Um-el-Meimab (27°N 34°E), and Nikolaus (pers. comm.) found one pair nesting at Suakin on a very small coral island. A few pairs probably nest elsewhere along the Sudanese coast.

Fig. 15.5. Two White-eyed Gulls with a group of Black-headed Gulls on the Egyptian Red Sea coast in winter. (*Photo* P. J. Ewins)

Fig. 15.6. A mixed flock of White-eyed Gulls, a juvenile Sooty Gull and behind this, a Spoonbill feeding in the winter on the Red Sea coast of Egypt. (*Photo* P. J. Ewins)

Fig. 15.7. A White-cheeked Tern courtship-feeding its mate, sitting on a motor boat in the Sudanese Red Sea. (*Photo* S.M. Head)

Sterna repressa White-cheeked Tern (Fig. 15.7)

This tern has a rather narrow distribution embracing the Red Sea, coast of Somalia and the Arabian peninsula, the Arabian Gulf and the coasts of Iran, Pakistan and western India. Both Smith (1957) and Clapham (1964) record it breeding in numbers in the Dahlak archipelago. In the Sudanese Red Sea,

Moore (1977) recorded the species as a common breeder on most islands with colonies varying in size from twenty to three hundred pairs and similar numbers probably exist along the coast of Eritrea and Yemen. Along the Arabian coast it breeds commonly on offshore islands north to Jiddah (Jennings, 1981), on the Wajh Bank (Moore and Ormond, 1982), and on Abu Mingar and Om Qamar near Hurghada in the north of the Red Sea (Al-Hussaini, 1939). About fifty pairs have been recorded breeding on the east coast of Sinai (Cramp & Simmons, Vol.4, 1985). On the south coast of Sinai, Su-Aretz (1979) records the species returning to breed on Tiran Island in a flock of about 150 pairs, at the end of June or in July, usually together with Lesser-crested terns. Although mainly a summer visitor to the Red Sea, occurring between April and September/October, some flocks remain in the southern Red Sea off Eritrea, feeding offshore in winter (Smith, 1957). Nikolaus (pers. comm.) has never found the species breeding along the Sudanese coast, although he has observed a few with dependent young in August, 10km south of Suakin. At the end of August and beginning of September, a regular passage (with a few hundred individuals roosting) occurs here.

Sterna albifrons Little Tern

Mainly a north temperate species, the Little tern does, however, have a very wide distribution, extending across southern Asia, the east Indies, north and east coasts of Australia, North and West Africa and the southern Red Sea. In the latter region, the typical race breeds in the Nile Delta and Lake Bardawil, Egypt, and is a regular and common migrant to the Egyptian coast. The smaller race *S.a. saundersi* is a resident breeder in Sudan and on the Arabian coast north to Jiddah where it is often confused with the typical race which migrates through the region (Gallagher *et al.*, 1984). There are some old records of *S.a. saundersi* breeding on the Brothers and Om Qamar islands in the northern Red Sea (Ticehurst, 1924; Al-Hussaini, 1939) and between Jiddah and Rabigh on the Arabian coast (Trott, 1952), but no recent information is available from here, except for individuals observed at Yanbu al Bahr and Al Wajh, and collected from Mugawish Island (Hurghada) in 1982 (Moore and Ormond, 1982).

Sterna anaethetus Bridled Tern (Fig. 15.8)

This species has a pan-tropical distribution, its northern limit occurring in the Caribbean, Red Sea and Arabian Gulf. It is a common breeding summer visitor to offshore islands along the Sudanese (Moore, 1977) and Arabian coasts north to Jiddah (Jennings, 1981), reaching its northerly limit on the Wajh Bank (Moore and Ormond, 1982), where it is relatively uncommon compared with further south on the islets near Yanbu al Bahr. It is also recorded from Om Qamar (Al-Hussaini, 1939) and near Hurghada, south of Suez (Etchecopar and Hüe, 1967). Thousands were recorded nesting on the Brothers in the north of the Red Sea but there is no recent evidence of breeding here (Ticehurst, 1924; Cramp and Simmons, Vol.4, 1985). Moore (1977) saw Bridled terns in fairly large numbers (with colonies of between 200 and 500 pairs) nearly everywhere in the Suakin islands and on the islands in the north of Sudan, although he noted fewer in the southernmost parts of the Sudanese coast. About thirty pairs were nesting at Muhammed Qôl (two of the Deganub Islands) in August 1982 (Nikolaus pers. comm.). About one thousand birds were counted on the islands of the Dahlak Archipelago where Clapham (1964) found birds breeding, and a flock of about two thousand was seen feeding on a shoal of fish off Massawa in August. Outside the breeding season the species disperses over a large area, occurring mainly some distance offshore.

Sterna bergii Swift or Great Crested Tern (Fig. 15.9)

A pan-tropical species, although it also breeds around much of the coast of Australia, islands of the south-east Pacific, and southern Africa. It occurs along the coasts of the Red Sea from Suez south, breeding on the Brothers Islands (Jones, 1946) though no nests have yet been found along the Egyptian

Fig. 15.8. The Bridled Tern is one of the commonest breeding species in the Red Sea. It is a summer visitor, the population dispersing over a wide region in winter, becoming mainly pelagic in habit. (*Photo* P. G. H. Evans)

Fig. 15.9. The Crested or Swift Tern breeds throughout much of the tropics. In the Red Sea it is a widespread breeder, though mainly in the southern part, with large numbers on the Sudanese coast.

or Arabian coasts north of 23°N (Meinertzhagen, 1930, 1954; Etchecopar and Hüe, 1967). It is fairly common on the Egyptian coast near Hurghada although no signs of breeding have been found (Al-Hussaini, 1938), and it possibly breeds on Ras Gemsa, at the mouth of the Gulf of Suez (Marchant, 1941). The species nests mainly in small numbers along the Sudanese coast, particularly in the south of the Suakin peninsula, where the greatest numbers (about three hundred pairs) were recorded on one of the 'Two Islets' (Moore, 1977) and on the coast of Eritrea (Cramp and Simmons, Vol.4, 1985). The species has bred in the Dahlak archipelago (Smith, 1957). Breeding probably also takes place along the coasts of Yemen, and at least the southern part of the Arabian coast of the Red Sea. It disperses mainly northwards outside the breeding season, although along the Egyptian coast it appears to be a resident non-breeder, with numbers occurring between April and August (Al-Hussaini, 1938, 1939). There is limited dispersal south of the Gulf of Aden.

Sterna bengalensis Lesser-Crested Tern

This smaller relative of the Swift tern occurs almost exclusively within the tropical zone between the Red Sea and the coast of Somalia in the west and the north of Australia in the east. In the northern Red Sea it breeds on islands off Hurghada (Meinertzhagen, 1954), the Brothers Islands (Jones, 1946), Safaga (Bulman, 1944), on Tiran Island (60 pairs in September to October 1975 and 1979), on the Wajh Bank and islets off the coast between Jiddah and Rabigh (Moore and Ormond, 1982). It was also described as abundant on Abu Mingar island by Al-Hussaini (1939). Meinertzhagen (1930) suggested that it probably bred in the Gulf of Suez though this has not been confirmed recently. In the central Red Sea, Moore (1977) found large numbers on many islands along the Sudanese coast, and although he visited many of the more southerly islands before breeding had begun, many birds were clearly preparing to nest on Seil Ada and Eitwid islets in the Suakin archipelago and at the latter site Nikolaus (pers. comm.) has counted 100–150 pairs breeding in 1980–82. By the time Moore reached the Muhammad Qôl area further north, large numbers were nesting and he recorded 900 pairs on the Taila islands and 800 pairs on Mukawwar. Nikolaus (pers. comm.) recorded 50 pairs on one of the Deganub Islands (Muhammed Qôl) in August 1982. The species was also breeding on islands at the southern end of the Red Sea (Meinertzhagen, 1954) but these islands were not identified and the author may have been referring to the Gulf of Aden. Clapham (1964) counted about 2,300 adults in the Dahlak Archipelago in August to September 1962, and found newly hatched young. The species appears to be at least partially migratory since numbers throughout the Red Sea are reduced in winter, and uncommon in the Gulf of Suez.

Anous stolidus Common or Brown Noddy (Fig. 15.10)

An almost exclusively tropical species, the Brown noddy breeds in all the major oceans of the world. In the Red Sea, it is apparently confined to the southern part where it is locally common in spring and autumn (Gallagher *et al.*, 1984). Moore (1977) found only one colony along the entire Sudanese coast — Barra Musa Saghir, in the Suakin archipelago, where perhaps three hundred were nesting in 3 m high shrub vegetation in the centre of the island. Although birds were seen by Moore and Ormond (1982) on the Wajh Bank, no sign of breeding was found. The species migrates southward outside the breeding season, presumably dispersing over the Arabian Sea and the Indian Ocean.

Sterna dougalli Roseate Tern

Although primarily tropical in its distribution, this species also breeds in north temperate regions of western Europe and eastern United States. However, the Roseate tern has been undergoing a worldwide decline in recent years and many colonies have become extinct. In the Red Sea, it was said to

Fig. 15.10. In the Red Sea, the Common or Brown Noddy is restricted to the southern parts where it nests in isolated colonies. In winter it migrates south, probably dispersing over the Indian Ocean and Arabian Sea. (*Photo* P. G. H. Evans)

breed on the Egyptian coast (Etchecopar and Hüe, 1967), but this seems unlikely since it is a vagrant to Sinai (Meininger and Mullié, 1981); there are no recent records, and it has not been recorded elsewhere in the region. Thus whilst it may still occur in a few localities, its status as a breeding species remains unconfirmed. Populations in western Europe migrate south after the breeding season along the coast of West Africa.

15.3.2. Non-breeding species

A number of seabird species occur in the Red Sea region without breeding. These are reviewed in Table 15.5 although it should be noted that the distribution of records probably reflects the distribution of observers more accurately than that of the birds. Most species are of palaeantic origin although some vagrants to Elat (e.g. Shy Albatross, Pale-footed Shearwater and Swinhoe's Storm Petrel) almost certainly derive from the Indian Ocean, and have traversed the length of the Red Sea. Unless otherwise stated, a summary of the status is derived from Etchecopar and Hüe (1967) for the western Red Sea, Jennings (1981) for the Arabian coast, Meininger and Mullié for Egyptian coast, or records provided by U. Paz for the Gulfs of Suez and Aqaba.

15.3.3. Coastal non-seabird species

Many shorebirds breed on the offshore islands and coastal lagoons and mangrove swamps that fringe areas of the Red Sea. These include Western reef herons (*Egretta gularis*), Green (*Butorides striatus*) and Goliath herons (*Ardea goliath*), Spoonbills (*Platalea leucorodia*), Greater flamingoes (*Phoenicopterus ruber*),

TABLE 15.5. Non-breeding seabird species

Calonectris diomedea	Cory's Shearwater.	Occasional autumn visitor to Elat.
Puffinus carneipes	Pale-footed Shearwater.	One site record from Elat, August 1980.
Puffinus griseus	Sooty Shearwater.	Regular spring visitor at the head of the Gulf of Aqaba.
Oceanites oceanites	Wilson's Petrel	Small numbers visit from the south, mainly in June and July (Alexander 1929, Maclaren 1946).
Hydrobates pelagicus	Storm Petrel	Occasional visitor (Alexander, 1929)
Oceanodroma leucorhoa	Leach's Petrel	Sometimes present from October to March.
Oceanodroma monorhis	Swinhoe's Storm Petrel	One found dead at Elat in January 1958 (Safriel 1968)
Phalacrocorax carbo	Cormorant	Twelve seen on islands at Ras Jemsah, early 1940 (Marchant 1941).
Diomedea cauta	Shy Albatross	One bird seen at Elat, February 1981, later found dead.
Fregata minor	Great Frigate-Bird	Frigate birds, probably of this species have been seen along the southern Sinai coast particularly in April/May.
Pelecanus onocrotalus	White Pelican	Once common at Suez and on Nile sandbanks, now winters only in small numbers (Meininger and Mullié, 1981). Occurs as soaring migrant at Suez and along the southwest coast of Saudi Arabia in winter.
Stercorarius parasiticus	Arctic Skua	Common at Elat January to August, also in Gulf of Suez and along the Egyptian Red Sea coast as a passage migrant and winter visitor.
S. pomarinus	Pomarine Skua	Common passage migrant February to July in northern Red Sea and Gulfs.
S. longicaudus	Long-tailed Skua	Rare spring passage migrant at Elat.
Larus audouini	Audouin's Gull	One collected in Gulf of Aqaba (Phillips 1915)
L. canus	Common Gull	Common winter visitor and passage migrant at Elat, particularly in spring (Safriel 1968)
L. argentatus	Herring Gull	Winter visitor to east Red Sea and Gulf of Suez, rare in Sudan.
L. fuscus	Lesser Black-backed Gull.	Uncommon winter visitor to Arabian coast north to Suez. More common as autumn passage migrant in Sudan, commonest in southern Red Sea and Aden. Strong spring passage at Elat (Safriel 1968).
L. hyperboreus	Glaucous Gull	One taken in Gulf of Aqaba (Phillips 1915)
L. ichthyaetus	Great Black-headed Gull.	Regular winter visitor to northern Red Sea and Aqaba, uncommon migrant and winter visitor in Sudan and southern area (King, 1978)
L. minutus	Little Gull	Rare migrant to Gulf of Aqaba.
L. ridibundus	Black-headed Gull	Common migrant and winter visitor to all coasts.
L. genei	Slender-billed Gull	Common passage migrant at Elat, February to May. Regular in autumn to spring on Sudan coast. Recorded in August off Hurghada and Suez.
Chlidonius hybridus	Whiskered Tern	Common winter visitor to Nile delta, may occur in Red Sea.
C. leucoptera	White-winged Black Tern	Infrequent passage migrant on Arabian and Egyptian coasts. Regular migrant in Sudan, large numbers pass through Elat in spring.
Gelochelidon nilotica	Gull-billed Tern	Common passage migrant at Elat, March–May. Also common in winter along Arabian and Sudanese coasts.
Sterna sandvicensis	Sandwich Tern	Rarely on Sudanese and Arabian coasts.
S. hirundo	Common Tern	Spring and autumn passage migrant in Eritraea, Sudan, and at Hurghada; also common late spring migrant at Elat. (Safriel 1968)
Sterna fuscata	Sooty Tern	Two seen in August 1982 at Muhammad Qôl (Nikolaus pers. comm.)
Rhynchops flavirostris	African Skimmer	Regular in small numbers between Port Sudan and Suakin in September and October. (Nikolaus, pers comm)

and Crab plovers (*Dromas ardeola*). In spring and autumn, enormous numbers of storks, cranes, pelicans and raptors pass through the area of the Gulfs of Suez and Aqaba (particularly Elat) on migration between their breeding grounds in Europe and Asia and their winter quarters in Africa (*see* Safriel, 1968; Meininger and Mullié, 1981, for reviews). The Bay of Suez is also an important wintering area for waders (Meininger and Mullié, 1981). Most rocky islands have breeding Osprey (*Pandion haliaetus*) and Sooty falcon (*Falco concolor*), the former feeding upon coastal reef fish and the latter preying upon insects, small birds and bats.

15.4. POSSIBLE THREATS TO SEABIRDS

Before reviewing the overall future prospects for Red Sea seabirds, six major threats to seabird populations may be recognised. These are: direct exploitation by humans, incidental takes in fishing nets, competition with fisheries, pollution and poisoning, habitat disturbance and destruction, and introduced predators. I shall deal with each in turn, first introducing the problem in a world context and then dealing specifically with the situation in the Red Sea.

15.4.1. Direct exploitation

For many centuries seabirds have been of importance to local communities for their flesh, fat, eggs or feathers, and a number of cultures, such as the eskimo of Greenland and Alaska, have evolved a close relationship with large neighbouring seabird populations which they regularly harvest. Many tropical island cultures, though less dependent upon seabirds, have nevertheless a strong tradition in collecting their eggs. This occurs on many islands in the Pacific and Indian Oceans but is also important in the Arabian Gulf (Gallagher *et al.*, 1984) and to a lesser extent on the islands in the Red Sea. Where there is a strong dependence upon seabirds for survival, human populations have tended to develop rules which allow a sustained harvest over a long period of time. As dependence becomes less a vital consideration, the taking of seabirds becomes more of a sport, and this together with the increased impact provided by improved technology including firearms and motor boats, may lead to over-exploitation of the resource and its subsequent demise.

To what extent man has directly affected the status of breeding seabirds in the Red Sea is difficult to determine. There are few historical records to provide any idea of the past numbers of birds present upon the islands and where large numbers are indicated in sites where only small numbers now exist, it is impossible to attribute this change of status to a particular cause. Many of the islands are still very remote from human habitation but may nevertheless be vulnerable to egg-collecting which almost certainly still takes place. Although egg-collecting may not be directly responsible for changes in status of particular species, it may well be an important contributory factor and attention should be paid to its possible effects. The large and dense colonies of Lesser-crested and Bridled terns, the Brown booby, and the large but diffuse colonies of the Sooty gull are perhaps the most vulnerable.

15.4.2. Incidental takes in fishing nets

The use of long-lines of gill-nets to entrap shoaling fish such as Salmon, Tuna or Halibut has in recent years resulted in heavy mortality amongst surface-diving seabirds, notably auks, shearwaters and cormorants. The areas worst affected appear to be the north-west Pacific, Davis Strait west of Greenland

and east of Canada, and Monterey Bay in the Gulf of California. There is little information available for the Red Sea at present on the extent of mortality due to this factor but gill-nets are used quite extensively to catch shark (Ormond, 1976) and for more general fishing at Jiddah and Jizan. These could pose a threat to seabirds such as the Socotra cormorant, although most Red Sea species are likely to be little affected.

15.4.3. Competition with fisheries

As human populations rise and pressures for food resources increase, there has been a greater tendency to turn towards the sea for food. There is a long history of fisheries being over-exploitative of their resource, leading to the decline and often collapse of fish stocks. Since many commercial food fishes are important components of the diet of seabirds, at least in juvenile stages, the possibility exists of competition between man and seabirds for food. As fish stocks come under increasing pressure, the threat to seabird populations may become significant. So far, the evidence for declines in seabirds attributable to this factor is scanty, involving mainly North Atlantic auks (notably the Atlantic puffin *Fratercula arctica* in west Norway and Newfoundland) and we have no information from the Red Sea, but it is another potential threat which needs consideration, particularly for species like the terns with restricted foraging ranges during the breeding season. Red Sea fishermen regard fish-eating birds as competitors and many destroy the nests and eggs of terns, boobies and cormorants for this reason. The Brown booby with its extended breeding season and apparently limited foraging range, is perhaps most vulnerable to food scarcity. Breeding failure in this species and fluctuations in numbers breeding have in the past been attributed to lack of food (Dorward, 1962) and may account for possible changes in status of this species in the Red Sea and Arabian Peninsula. Such changes, if real, may of course be due to natural changes in population size or distribution of their food prey or to over-exploitation by man.

15.4.4. Pollution and poisoning

This factor has received considerable attention in the popular press. Birds are light and durable and tend to float after death. Large numbers washed ashore are conspicuous evidence of mortality resulting for example from an oil spill. The increasing dependence upon oil as a fuel for transport has resulted in a disturbing increase in the number of oil pollution incidents, particularly along the coasts of Europe and North America. As with drowning of seabirds in fishing nets, those groups most vulnerable are the auks, cormorants, and divers. The Red Sea carries a heavy traffic of oil tankers and so the chances of oil spills are high. However, fortunately there have been few major incidents so far and those that have taken place have apparently involved relatively few birds. This is almost certainly in part due to the careful precautions taken by local authorities and the oil industry in the region but may also be due to the effect of high temperatures upon oil which rapidly loses its volatile constituents and becomes reduced to solid residues which are harmless to birds (Bourne, 1976). Notwithstanding this, there is much oil pollution along the coast of Egypt and the Gulf of Suez with a band of tar noted almost everywhere in the former case (Ormond, 1981). The Socotra cormorant is probably one of the most vulnerable species to oil pollution and suffered heavy mortality near Aden in July 1960 and at Bahrain in the Arabian Gulf in August–November 1980 (Gallagher *et al.*, 1984).

Oil is not the only pollutant which can have important negative effects on bird populations. Toxic chemicals are often discharged into the sea either as industrial waste or in the form of pesticides from agricultural activities. They may enter the sea directly from factories or via river systems. Enclosed areas such as the Red Sea are particularly vulnerable to pollution since there is little through-put in the system and chemicals can readily build up concentrations in confined areas. This has occurred in particular in the Baltic and the North Sea. In the 1960s evidence accumulated to show the effects of persistent organochlorine pesticide residues on breeding and survival of the top predators in food chains, notably birds of prey and freshwater fish-eating birds. Improved analytical techniques have provided more sensitive methods of detecting levels of sub-lethal chemicals. In the last decade, with tighter controls on these pollutants in Europe and North America, their ecological effects have generally been reduced. In the Red Sea region industrial developments have occurred only rather recently. The very low rainfall in the region restricts the effects of inland terrestrial pollution by limiting freshwater input to the Red Sea from rivers, so that pollution from agricultural pesticides is unlikely to be important. However, pollution from new industries such as fertiliser plants, and the building of ports and installations for phosphate mining are increasingly serious threats, particularly in areas like the Gulf of Aqaba. Inshore fish feeders or scavengers such as the Socotra cormorant, Sooty and White-eyed gulls, and some of the terns are probably the most vulnerable species. The problem will be to interpret levels of pollutants contained in seabirds and to identify any sub-lethal effects, which although falling short of killing adults may profoundly interfere with breeding (*see also* Chapter 18).

15.4.5. Habitat destruction and disturbance

In most cases the effects of habitat destruction upon seabirds relate to destruction of their terrestrial breeding sites rather than their feeding grounds. The building of industrial plants, hotels and holiday homes when sited along the shore or upon islands may destroy important breeding habitat for seabirds. These activities are rapidly increasing in many parts of the Red Sea, for example around Jiddah and near Port Sudan, and large seabird colonies could readily become threatened. Human disturbance both from egg-collecting expeditions and by tourists can also lead to increased predation when parent birds have been disturbed from their nests leaving them unguarded, and even lead to nest desertion.

15.4.6. Predators

Predation by humans has been considered under the heading of direct exploitation. However, other animals may be important predators of seabirds or their eggs and their effects can be considerably increased by man. Natural predators such as gulls may increase in numbers due to an increase in food provided by man and so have a greater negative effect upon other seabirds. Other predators such as small rodents, or domestic animals, may be introduced by man and cause widespread damage. In the Red Sea the most important natural predators are raptors such as the Osprey and Sooty falcon, and gulls, particularly the Sooty gull. Raptors may prey quite heavily on birds nesting on islands but their numbers are usually not large enough to make a significant impact. Sooty gulls are present in larger numbers and can cause greater damage, taking the eggs or young of terns, boobies and the Socotra cormorant. They will also take eggs of their own species, and chase White-eyed gulls forcing them to disgorge food. White-eyed gulls in turn may also prey upon the eggs and young of terns.

In many parts of the world, introduced mammals have wrought havoc amongst seabird colonies. Rats and cats are the most serious predators although other mammals such as goats can have important indirect effects by damaging nesting habitats. In the Red Sea, Black rats *Rattus rattus* are present on a number of islands and are known to feed upon birds' eggs and young. However, Clapham (1964) noted that they did not seem to affect Brown boobies nesting in exposed positions on islands in the Suakin Archipelago, and postulated that nesting gulls were more likely to be affected. Feral cats from nearby habitations can be serious predators on ground-nesting birds but the extent to which they affect seabird populations is not known. It is unlikely that they are at present more than a local problem, restricted largely to coastal sites on the mainland, but with increasing pressures upon once remote areas, introduced predators such as these may become important.

15.5. CONCLUSIONS, FUTURE INFORMATION NEEDS

The seabirds of the Red Sea are really rather poorly known. Although expeditions have taken place to the Suakin and Dahlak archipelagos, the islets on the Wajh Bank, and other islands along the Sudanese coast, such visits have usually only occurred once and a proper evaluation of the status of breeding species can rarely be made. It is therefore impossible to state for the Red Sea as a whole whether seabird populations are increasing, stable or decreasing. Only the Gulf of Aqaba appears to have received more regular attention and our knowledge of the seabirds of Tiran Island is probably better than any other area in the Red Sea. Our knowledge of the seabirds of the Dahlak Archipelago and of the Hanish and Zubayr groups rely upon visits made some twenty years ago, and some other islands have not been visited by ornithologists for a half a century or more. The two areas most needing attention are the coasts and islands at the northern end of the Red Sea (Barqan and Naman on the Arabian coast, and Shadwan island off the coast of Egypt) and those at the southern end, notably the Farasan group, the Hanish and Zubayr groups and neighbouring islands. There is also a need for surveys of the Arabian coast south of Jiddah.

Three developments in the Red Sea countries are likely to affect seabirds. These are the oil industry, tourism and fisheries. The threats they may pose have been discussed above and there is a need for a monitoring programme to determine the impact of each of these. At the same time, it is important that sites are set aside as reserves where development and disturbance is controlled. Already coastal and marine parks are being established in various parts of the Red Sea, notably over part of the Dahlak Archipelago, Hurghada, and Port Sudan (*see* Ormond, 1976 and chapter 19 of this book). Other areas where reserve development has been considered include Al Hudaydah (in the Yemen), near Aqaba (Jordan) and at Jiddah.

The most important seabird localities presently known to us are the Brothers Islands in the northern Red Sea, the Zubayr and Hanish groups off the Yemen coast, the outer western islands of the Dahlak Archipelago, a number of islands in the Suakin archipelago, and further north, the islands in the Muhammad Qôl area. It is recommended that measures should be taken to establish at least some of these as reserves or marine coastal parks, and where mainland seabird colonies exist near human habitation, that tighter controls are introduced to minimise disturbance and exploitation.

ACKNOWLEDGEMENTS

I should like to thank the following for providing much useful unpublished information: Michael Jennings, G. Nikolaus, Dr. Uzi Paz, Shalom Su-Aretz, Glen Tyler and Dr. Yoram Yom-Tov.

Dr. Euan Dunn and Michael Jennings kindly criticised an earlier draft of this chapter, and Drs. P. J. Ewins and M. de L. Brooke generously provided photographs.

REFERENCES

Alexander, H. G. (1929). Some Birds seen in the Indian Ocean and the Mediterranean. *Ibis* (12) 5, 41–53.
Al-Hussaini, A. H. (1938). Some birds observed in Ghardaqa (Hurghada), Red Sea Coast. *Ibis* (14) 2, 541–4.
Al-Hussaini, A. H. (1939). Further notes on the birds of Ghardaqa (Hurghada), Red Sea. *Ibis* (14) 3, 343–7.
Ali, S. and Ripley, S. D. (1969). *Handbook of the Birds of India and Pakistan*. Vol.3, Bombay.

Archer, G. and Godman, E. M. (1937–61). *The Birds of British Somaliland and the Gulf of Aden*. 4 vols. Oliver and Boyd, London.

Bailey, R. S. (1966). The sea-birds of the southeast coast of Arabia. *Ibis* 108, 224–64.

Borman, F.W. (1929). An ornithological trip in the Gulf of Suez and Red Sea. *Ibis* (12) 5, 639–50.

Bourne, W. R. P. (1966). Observations of seabirds. *Sea Swallow* 18, 9–39.

Bourne, W. R. P. (1976). Seabirds and Pollution. In *Marine Pollution*. Ed. R. Johnson, pp.403–502. Academic Press, London.

Brown, L., Urban, E. K. and Newman, K. (1982). *The Birds of Africa*, Vol. 1. Academic Press, London.

Bulman, J. F. H. (1944). Notes on the birds of Safaga. *Ibis* 86, 480–92.

Clapham, C. S. (1964). The Birds of the Dahlac Archipelago. *Ibis* 106, 376-88.

Cornwallis, R. K. and Porter, R. (1982). Spring observations on the birds of North Yemen. *Sandgrouse* 4, 1–36.

Cramp, S. and Simmons, K. E. L. (1977-continuing). *Handbook of the Birds of Europe, the Middle East and North Africa*, Vol.1 (1977); Vol. 2 (1980); Vol. 3 (1983); 4 (1985). Oxford University Press, Oxford.

Croxall, J. P., Evans, P. G. H. and Schreiber, R. (eds.) (1984). *The Status and Conservation of the World's Seabirds*. ICBP Publication.

Diamond, A. W. (1976). Sub-annual breeding and moult cycles in the Bridled Tern *Sterna anaethetus* in the Seychelles. *Ibis* 118, 414–9.

Din, N. A. and Eltringham, S. K. (1974a). Ecological separation between White and Pink-backed Pelicans in the Ruwenzori National Park, Uganda. *Ibis* 116, 28–43.

Din, N. A. and Eltringham, S. K. (1974b). Breeding of the Pink-backed Pelican *Felecanus rufescens* in Ruwenzori National Park, Uganda; with notes on a colony of Marabou Storks *Leptoptilus crumeniferus*. *Ibis* 116, 477–93.

Dorward, D. F. (1962). Comparative biology of the White Booby and the Brown Booby *Sula* spp. at Ascension. *Ibis* 103b, 174–220.

Dorward, D. F. and Ashmole, N. P. (1963). Notes on the Biology of the Brown Noddy *Anous stolidus* on Ascension Island. *Ibis* 103b, 447–57.

Dunn, E. K. (1972). *Studies on terns with particular reference to feeding ecology*, Ph.D. thesis, University of Durham.

Etchecopar, R. D. and Hüe, F. (1967). *The Birds of North Africa*. Oliver and Boyd, Edinburgh.

Fogden, M. P. L. (1964). Notes on the reproductive behaviour and taxonomy of Hemprich's Gull *Larus hemprichi*. *Ibis* 106, 299–320.

Gallagher, M. D. and Woodcock, M. W. (1980). *The Birds of Oman*. Quartet Books, London.

Gallagher, M. D., Scott, D. A., Connor, H. J. and Jennings, M. C. (1984). Seabirds breeding in the Arabian Gulf and Arabia. pp. 421–456. In *The Status and Conservation of the World's Seabirds*. Eds. Croxall, J. P., Evans, P. G. H. and Schreiber, H., ICBP Publication (1984).

Halim, Y. (1969). Plankton of the Red Sea. *Oceanogr. mar. Biol. Ann. Rev.* 7, 231–75.

Harris, M. P. (1969a). Breeding seasons of seabirds in the Galapagos. *J. Zool. Lond.* 9, 145–65.

Harris, M. P. (1969b). Factors affecting the breeding cycle of the Red-billed Tropicbird in the Galapagos Islands. *Ardea* 57, 149–57.

Hulsman, K. (1974). Notes on the behaviour of terns at One Tree Island. *Sunbird* 5, 44–9.

Hulsman, K. (1977). Feeding and breeding biology of six sympatric species of tern (Laridae) at One Tree Island, Great Barrier Reef. Ph.D. thesis, Univ. of Queensland.

Jennings, M. C. (1981). *The Birds of Saudi Arabia: a check-list*. Jennings, Cambridge

Jones, H. B. (1946). An account of a visit to the Brothers (Jebel Teir) Islands in the Gulf of Aden. *Ibis* 88, 228–32.

Kepler, C. (1969). The breeding biology of the blue-faced booby (*Sula dactylatra personata*) on Green Island, Kure. *Publs. Nuttall Orn. Club* 8, 1–97.

King, B. (1978). April bird observations in Saudi Arabia. With Addendum 1, January 1977 observations: Addendum 2, Arabian Bustard Expedition April 1977, and Jebal Sawdah — a National Park? *J. Saudi Arab. nat. Hist. Soc.* 21, 3–24. (see also Errata and Addendum) *J. Saudi Arab. nat. Hist. Soc.* 24, 35–6).

Mackworth-Praed, C. W. and Grant, C. H. B. (1955). *African Handbook of Birds, Series 1. East and North Africa*. Vol.2, Longmans, London.

Marchant, S. (1941). Notes on the Birds of the Gulf of Suez. Parts I and II. *Ibis* (14) 5, 265–95.

Meinertzhagen, R. (1930). *Nicoll's Birds of Egypt*, 2 Vols, Hugh Rees, London.

Meinertzhagen, R. (1954). *Birds of Arabia*. Oliver and Boyd, London.

Meininger, P. L. and Mullie, W. C. (1981). *The Significance of Egyptian Wetlands for Wintering Waterbirds*. The Holy land Conservation Fund, New York.

Moore, R. J. (1977). Notes on Birds Nesting in the Islands of the Sudanese Red Sea. In *Report of 1976 Expedition to Suakin Archipelago (Sudanese Red Sea). Results of marine and bird life*. Eds. Moore, R.J. and Balzarotti, M.A. Mimeographed. iv + 27pp.

Moore, R. J. and Ormond, R. F. C. (1982). Sea-bird Survey: Northern Saudi Coast. August/September 1982. Unpublished MS. 8pp.

Morcos, S. A. (1970). Physical and Chemical Oceanography of the Red Sea. *Oceanogr. mar. Biol. Ann. Rev.* 8, 73–202.

Morris, R. O. (1962). Two visits to the Haycocks (Hanish Islands, southern Red Sea). *Sea Swallow* 15, 57–8.

Nelson, J. B. (1978). *The Sulidae. Gannets and Boobies*. Oxford University Press (for University of Aberdeen).

Ormond, R. F. C. (1976). The Red Sea. In *Promotion of the Establishment of Marine Parks and Reserves in the Northern Indian Ocean including the Red Sea and the Persian Gulf*. Papers and Proceedings of the Regional Meeting held at Tehran, Iran, 6–10 March 1975. IUCN Publics. New Series No. 35, pp.115–23, Morges, Switzerland.

Ormond, R. F. C. (1981). Report on the need for management and conservation in the northern Egyptian Red Sea. Report Biol. Dept., University of York. 35pp.

Phillips, J. C. (1915). Some Birds from Sinai and Palestine. *Auk* 32, 273–89.

Phillips, W. W. A. (1947). The ornithological voyage of S.S. Samluzan. *J. Bombay nat. Hist. Soc.* 46, 593–615.

Safriel, U. (1968). Bird migration on Eilat, Israel. *Ibis* 110, 283–320.

Smith, K. D. (1957). An annotated check-list of the birds of Eritrea. *Ibis* 99, 1–26, 307–77.

Snow, D. W. (1965a). The breeding of Audubon's Shearwater *Puffinus lherminieri*) in the Galapagos. *Auk* 82, 591–7.

Snow, D. W. (1965b). The breeding of the Red-billed Tropic Bird in the Galapagos Islands. *Condor* 67, 210–14.

Stonehouse, B. (1962). The Tropic Birds (Genus *Phaethon*) of Ascension Island. *Ibis* 103b, 124–61.

Su-Aretz, S. (1973). Survey of Breeding Birds on the Island of Tiran and the coast of Southern Sinai. Unpublished MS. 11pp.

Su-Aretz, S. (1979). White-cheeked Tern *Sterna repressa*. Unpubl. MS for *Handbook of the Birds of Europe, Middle East and North Africa* (ed. by S. Cramp and K. E. L. Simmons) Vol. 4 (1985).

Su-Aretz, S. (1980). White-eyed Gull *Larus leucopthalmus*. Unpubl. MS for *Handbook of the Birds of Europe, Middle East and North Africa* (ed. S. Cramp and K. E. L. Simmons) Vol. 3 (1983).

Ticehurst, C. B. (1924). Birds from the Red Sea lights. *Ibis* 6, 282–3.

Trott, A. C. (1952). Notes on the breeding of the Bridled Tern (*Sterna anaethetus* Scop.). *Proc. Zool. Soc. Lond.* 122, 537–9.

Tuck, G. S. (1980). *A Guide to Sea-birds of the Ocean Routes.* Collins, London.

Watson, J. B. (1908). The behaviour of Noddy and Sooty Terns. *Pap. Tortugas Lab. Carnegie Inst. Washington* 2, 185–255.

CHAPTER 16

The Human Settlement of the Red Sea

MARK HORTON

St Hugh's College, Oxford, U.K.

CONTENTS

16.1. INTRODUCTION

For three million years, man and his immediate ancestors have occupied the shores of the Red Sea, leaving behind a long and complex sequence of archaeological sites. The Red Sea is sited at the junction between Africa, Asia and maritime Europe, and the development of occupation is of considerable historical interest. Here we may seek evidence for the migrations of early man, and for the dissemination of food crops, technology, language and culture between the continents. For the last three millenia, the Red Sea has been a major trade route and commercial link between the countries that border the Indian Ocean and the world of the Mediterranean.

Despite its importance, the Red Sea region never spawned civilisations of its own, those that developed nearby were based inland, in milder climates along rivers to ensure an adequate water supply, or where valuable raw materials were to be found. The civilisations of Egypt, of Meroe, Axum and southern Arabia were located in such favoured areas. All however, except possibly Meroe (Shinnie, 1967), established outpost trading communities on the shores of the Red Sea. Between these outposts on both shores of the Red Sea, lived scattered traditional societies. These pastoral or fishing groups never reached high population densities, and the archaeological evidence suggests their way of life survived unchanged for thousands of years. Only recently has political strife, and the pervasive spread of Western influence begun to threaten the character and very existence of these societies.

The Red Sea region has been poorly researched by historians or archaeologists, who tend to focus on the more accessible inland or riverine civilisations. Many of the famous ports of antiquity have still to be precisely located on the ground. Few of the coastal traditional societies have been documented by modern anthropologists. Evidently, any account of human settlement in the Red Sea region will be incomplete and in places speculative. The opportunities for exploratory field research along the coasts of the Red Sea are immense.

16.1.1. The climatic background

Of the various ecological factors that influence human settlement patterns, climate is one of the most important. Reconstruction of past climates is especially relevant when understanding settlement in a region like the Red Sea, which has been one of considerable variation of temperature and rainfall, although very hot and arid today. During the Pliocene and Pleistocene periods (see Chapter 2), the faunas of North and East Africa were similar, suggesting the Saharan desert belt was much less extensive than today.

Study of the Nile terraces indicates that open savannah lands associated with wetter climates were formerly much more extensive, and both north-east Africa and the Arabian peninsula enjoyed higher rainfall than now. Evidence for rainfall patterns in the late Pleistocene and Holocene have come from study of lake sediments (Street and Grove, 1976), suggesting that on the African side, rainfall was high from 24,000 to 18,000 years B.P. (before present), and around 12,000 B.P. and 8,000 B.P. On the Arabian side, the sequence in the Rub' al Khali depression indicates that there were wet phases between 37,000–36,000 B.P., 30,000–21,000 B.P. and 9,000–6,000 B.P. The period around 15,000 B.P. was very arid (McClure, 1976).

The modern, very arid climate on both sides of the Red Sea, with generally much less than 17 cm annual rainfall (Chapter 3), appears to have set in about 2,500 B.C. although with some local variation. Surviving records of the height of the annual Nile flood in the Old Kingdom and in the Islamic period, indicate considerable short term variation.

16.2. THE EMERGENCE OF TRADITIONAL SOCIETIES

16.2.1. Early beginnings — the Lower Palaeolithic

Evidence for the earliest occupation of this region has recently come from the Afar Depression in Ethiopia, a northern extension of the East African Rift Valley complex. Here, richly fossiliferous deposits have yielded hominid material from 4.0 to 2.4 million years old (Johanson et al.,1982). The most publicised find is a partially complete skeleton nick-named 'Lucy', but much other material has come to light. Evidence from these finds, and from similar material collected in Tanzania, suggests that this hominid, (*Australopithecus afarensis*) may have walked upright. There is still much controversy over the status of 'Lucy' and her relatives (Johanson, 1980), especially as to whether they were directly ancestral to *Homo* and modern man. These early hominids seem to have been confined to African tropical savannah, and probably lived by scavenging and fruit gathering rather than by hunting and organised exploitation of the environment (Isaac, 1982). It is not clear whether pebble tools found in some of the *A. afarensis* sites (Corvinus and Roche, 1980) were used by the early hominids.

Coincidental with the movement of early hominids from savanna to arid, temperate and forest zones was the appearance of a new hominid species, *Homo erectus*, and the development of more sophisticated stone tools, such as bifacial axes. This change happened about 1 million years B.P., the earliest hand axes and fossils have been found in Africa, but with later examples in Europe and Asia. As yet no fossils of *Homo erectus* have been found around the Red Sea, but their hand axes are widely distributed. Acheulian industries, as they are termed, have been located in the Afar depression (Corvinus, 1976) and Melka Kunture in Ethiopia, at sites along the Nile valley and adjacent oases in the Sudan and Egypt (Fig. 16.1). The most famous site is at Kharga (El Kharga), where 350 hand axes were found in a single mound (Caton-Thompson, 1952), but other nearby sites include Khor Abu Anga, Ternifine, Sidi Zin, Amanzi and Arkin 8. Excavations at Arkin 8 found traces of an early Stone Age hut structure (Chimielewski, 1968).

Early man probably moved into Arabia via the Isthmus of Suez, as the archaeological evidence suggests connections with stone tool industries of the Levant and North Africa. Among important sites surveyed in the Sakaka basin of northern Arabia is a stone tool 'factory' (Parr *et al.*, 1978). Hand axes have been found in southern Arabia and Qatar, but their date and significance are uncertain. It is likely however, that by about 100,000 years B.P., both shores of the Red Sea were inhabited by early man, hunting and gathering in the savannah in low densities and with minimal social organisation.

16.2.2. Middle and Upper Palaeolithic

Melting of land ice during the last interglacial period caused a global rise in sea level, amounting to some 8 m above present levels, about 125,000 years B.P. (*see* Chapter 2). The Afar depression was flooded by the Red Sea, and the lower Awash river became a fresh water lake. Coinciding with this was a major change in the range of stone tool industries of the area (Clark, 1982), marking the beginning of the Middle Paleolithic. The clumsy hand axes were abandoned in favour of lighter, more versatile tools struck from cores. This change ocurred generally at this time in East Africa, but stone axes continued in use in some areas, blurring the boundary between the Lower and Middle Paleolithic.

The best sequence for the Middle Palaeolithic sites comes from Porc Epic Cave, near Dire Dawa in Ethiopia (Clark, 1982) where an important collection of points and blades were found associated with a jaw fragment of 'Neanderthal Man', *Homo sapiens neanderthalis*. It is interesting that this Ethiopian material resembles that from other East African sites (Clark, 1954), but differs from material from the Nile valley, the Levant and North Africa. Little is known of contemporary industries in Arabia, but they seem to constitute a third tradition (Caton-Thompson and Gardner, 1939; Parr *et al.*, 1978). It would seem, that as modern *Homo sapiens* evolved from *H. erectus*, regional or cultural differences arose which had been absent in the Lower Palaeolithic. Other changes appear to have occurred at the same time. We find the first evidence of communal hunting for 'big game', careful disposal of the dead, and rudimentary art. During the Middle Palaeolithic the first use began to be made of the maritime food resources of the Red Sea. Regional diversity and specialisation increased between 100,000 and 10,000

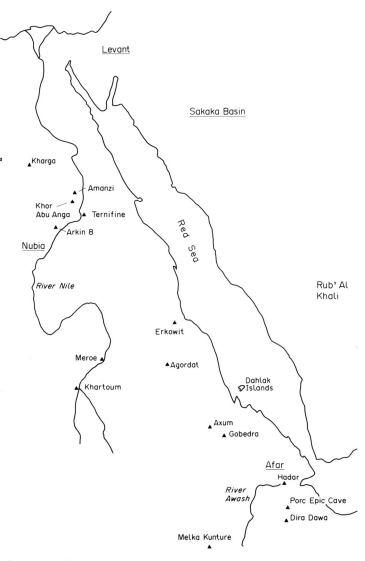

Fig. 16.1. The Red Sea region showing pre-historic sites mentioned in the text. Historical sites are shown in Figure 16.3.

years B.P. Along the Nile are found communities of settled hunter-gatherers, living after about 16,000 B.P, in what may be called villages. Fishing was widespread, and in Nubia and at the site of early Khartoum (Arkell, 1955) formal cemeteries are found for the first time. These developments seem not to have occurred at the southern end of the Red Sea, where much of the material belongs to a more conservative tradition, although sophisticated tools have been found at Gobedra near Axum (Phillipson, 1977) and at Melka Kunture (Isaac, 1982).

16.2.3. The Neolithic

The transition from hunter-gathering to the deliberate cultivation of crops and domestication of animals — the 'neolithic revolution' appears to have taken place independently in several parts of the world at different times. The regional diversity that began in the Palaeolithic became even more marked, as some groups adopted agriculture while others remained hunter-gatherers. Bordering the Red Sea region the Levant, the Nile valley and highland Ethiopia adopted agriculture independantly, and eventually became centralised states with complex social systems.

Along the shores of the Red Sea itself, the poor rainfall and increasing desiccation discouraged such developments. Although agricultural communities stretched down the Nile by 5,000 B.C. there is little evidence of them to the east. Two exceptions are the undated sites at Agordat (Arkell, 1954) and Erkowit in the Red Sea Hills. A similar picture appears in Arabia, where in the north, material related to the pre-pottery Neolithic of the Levant has been found, but further south and west, no traces of Neolithic material of this date are known (Masry, 1977, McAdams et al., 1977). Despite the increasing aridity, the coastal semi-desert areas were occupied by nomadic pastoralist groups who must have obtained their stock initially from the mixed agricultural communities of the Nile valley and the Levant. Unfortunately, nomadic groups leave few traces for archaeologists, and their origins remain obscure. In north-west Arabia, stone circles dating from 3,000 B.C. and probably used for stock keeping have been found (Parr et al., 1978). Stone burial cairns are found on both Red Sea coasts, and over much of Somalia. While some may be ancient, others date to the recent past, and are still being made by pastoralist groups today. The elaborate stone monuments of Erkowit in the Red Sea Hills (Fig. 16.2), described by Seligman (1915) await excavation to establish their date and function, and by whom they were built.

16.2.4. Rock Art

Numerous examples of rock art can be found on both sides of the Red Sea. Two types are known, pigments painted onto rock surfaces and only surviving in sheltered sites, and markings carved or pecked into stone. Some examples, such as at Porc Epic Cave belong to the Palaeolithic, but others depict domestic animals such as long and short horned cattle, and so must belong to the Neolithic. Rock art is widely distributed over western Arabia (Parr et al., 1978), north-east Sudan and Eritrea as far as north-east Somalia (Clark, 1954).

Fig. 16.2. Prehistoric tombs of unknown age near Erkowit, Red Sea Hills.

16.2.5. The Beja and the Bedouin

The present inhabitants of the desert lands bordering the western shore of the Red Sea are the Beja people, who are Northern Cushitic speaking camel pastoralists. They are divided into many groups, of which the main ones are the Ababda, Amanar, Beni Amer, Bisharin, Bongo, Hadendowa and the Tigre. All herd cattle and camels except the Ababda, who mainly fish. As a people they are related to the Afar of Ethiopia, who are Eastern Cushitic speakers. Today, the Beja and Afar are either Muslim or pagan, but some of the Beja were once Christian.

It is generally assumed that the Beja are the direct descendants of the early pastoral groups and have been resident from at least the first century A.D. in the area, where there is the first evidence for camel domestication in north-east Africa. Only then would habitation in the eastern desert of Egypt be feasible on any substantial scale. Writers such as Procopius (circa 6th century A.D.) describe the Blemyes in terms that closely fit the Beja today (MacMichael, 1922). The Beja are noted by name in treaties of A.D. 651 and A.D. 831, and controlled part of Upper Egypt between A.D. 268 and 451 (Paul, 1954). The origin of these Northern Cushitic speakers and of the Beja from Nubia has been of considerable interest to African historians and linguists, who were keen to see Egypt as an origin of cultures and people. Accordingly, the 'Hamitic' people of East Africa were thought to be derived from an Egyptian cradle. While there is a little evidence from both pottery styles and rock art to suggest connections between Nubia and the pastoralists of the northern Ethiopian plateau (Crowfoot, 1928; Arkell, 1954), it can now be shown that there were independant Cushitic speakers in northern Ethiopia (Fattovich, 1978).

On the eastern side of the Red Sea, the Bedouin society resembles that of the Beja, based on camel and cattle pastoralism, and organised into nomadic confederations (Murray, 1935; Naval Intelligence, 1946). These groups are probably older than the Beja; within Arabia there is evidence for camel domestication from at least the 2nd millenium B.C. The Bedouin are Semitic speakers, their language closely related to classical Arabic, and their way of life has changed little in many thousands of years.

16.2.6. Southern Arabia and Axum

From about 1000 B.C. a number of complex societies developed, first in southern Arabia, and later in the highlands of Ethiopia. In both areas by the beginning of the Christian era, there were considerable cultural achievements, including monumental architecture and inscriptions. The oldest were probably the southern Arabian communities (Doe, 1971), which developed in areas of very low rainfall, building large scale irrigation works to facilitate agriculture. The wealth of these kingdoms was partly based on the incense trade, for these aromatic gums grew principally in southern Arabia, and were much in demand in the Mediterranean world. A series of caravan routes were established across the Arabian peninsula. The archaeology of these societies is still very undeveloped, and we know little beyond what is contained in the classical authors such as Pliny and Ptolemy, and from study of numerous surviving inscriptions.

The origins of the Ethiopian kingdoms are equally obscure. At the Gobedra rock shelter, domesticated finger millet (*Eleusine coracana*) has been found in levels dated from the third to the fourth millenium B.C. (Phillipson, 1977). Other crops such as *tef* (*Eragrostis tef*) and *ensete* (*Ensete ventricosum*) may be indigenous domesticates with the same origin. Although the region was a very early centre of crop domestication, state-like societies did not emerge until later, possibly as late as the third century A.D. (Chittick, 1978). To what extent these societies were influenced by contact across the Red Sea with Southern Arabia is still a matter of controversy. Some of the dialects spoken in Ethiopia are Semitic, close to those of Southern Arabia.

During the flourishing of the Ethiopian kingdom of Axum, from the third to the seventh century A.D. there was little evidence of Arabian influence despite communication between the two areas (Doresse, 1957). The society erected large monumental stelae (inscribed upright pillars) with elaborate carving, as well as massive temples and tomb complexes (Kobishchanov, 1979). A detailed account of a visit in the sixth century A.D. by Cosmas Indicopleustes, an Egyptian merchant turned monk, gives a picture of society that can be supplemented by coins, inscriptions and excavations (Chittick, 1974; Butzer, 1981). But much relating to the social organisation and economic basis of Axumite society remains a mystery.

16.2.7. The emergence of medieval society in the Red Sea region

Although our understanding of the history of the area will certainly improve with further research, the following outline may be sketched. During the Palaeolithic, in both the Arabian peninsula and north-east Africa, the extensive moist savannah grasslands were inhabited by groups of hunter-gatherers using stone tools. As the climate became drier, Early Man was forced to adopt specialist strategies including the exploitation of maritime and riverine resources. Population began to concentrate in areas of good food supply.

The adoption of agriculture and the domestication of animals from the sixth millenium B.C. marked another specialisation. The use of cattle and subsequently, camels, allowed groups of these neolithic populations to move into the arid desert and become nomadic pastoralists. These groups were the ancestors of today's Beja and Bedouin.

We know little of the populations that lived off the sea. Language studies suggest they were derived from the nomadic pastoralists, and a close relationship still exists today along the coastal area between fishing and herding groups. This is illustrated by such practices as feeding camels and cattle on dried fish. However, it may also be that some of the fishing groups are directly descended from late Palaeolithic populations of the Red Sea coast, who intermarried and adopted Cushitic languages.

It is clear that by the beginning of the Christian Era, the foundations had been laid of the traditional societies which occupy the region today. These societies were well adapted to live off a harsh and difficult environment, and to retain their ecological stability over long time periods.

16.3. THE MARITIME TRADE OF THE RED SEA

16.3.1. Introduction

Since prehistoric times the Red Sea (Fig. 16.3) has been one of the busiest and most important sea routes of the world. Despite the treacherous sea, difficult winds and currents, and the lack of water and food along its shores the central geographic position of the Red Sea has guaranteed its importance, long before the construction of the Suez Canal.

The reason for this is two-fold. On the one hand, the Red Sea contained important natural resources for which there was demand outside the region; on the other, it links the two great maritime trading systems of the classical and medieval worlds: the Mediterranean Sea and the Indian Ocean. For this reason the fortunes and prosperity of the region, were determined by political and economic changes occurring in other regions often far away. Of the greatest economic importance was the finely balanced system of trade operating in the Indian Ocean. Across this ocean, ships travelled great distances, following the monsoon winds and equatorial currents. As the monsoon blew half-yearly in opposing

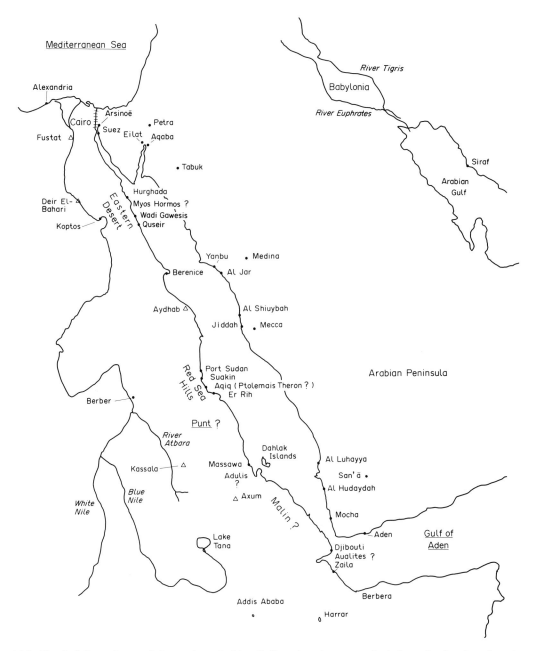

Fig. 16.3. The Red Sea and part of the northern Arabian Gulf to show important classical, medieval and modern sites mentioned in the text. Archaeological sites are marked with open triangles. The exact positions of many ports and regional areas are still uncertain.

directions, it was possible to make the return journey between the Red Sea ports and southern and South-east Asia in a single year. Thus the products of a catchment area covering one fifth of the globe could easily be brought to Europe.

However the Red Sea was not the only route linking these two areas, and goods of the East were also brought over land from China (the silk route), via the Arabian Gulf, and then by caravan to the Levant. Alternatively, they came by sea to southern Arabia, and then overland by caravan, or finally, by sea around the Cape of Good Hope, a passage opened up by the Portuguese in the 16th century A.D.

It was the competition between these routes, and between the political powers that controlled them that coloured the prosperity of all the maritime settlements in western Asia.

16.3.2. The trade goods

What the West demanded of the East and *vice-versa* varied considerably over time. The first large scale maritime trade along the Red Sea was by the Egyptians trading with Punt, during the 2nd millenium B.C. They brought back wild animal skins, gold, ivory and ebony, but chiefly *cntyw* incense (O'Connor, 1982). During the Classical and Medieval periods, the spice trade was probably the most important (Miller, 1969). Many such products heavily demanded in Europe were grown only in India and South-east Asia. Ivory was obtained from India and East Africa. Gold came from several areas including the Red Sea Hills themselves, but by the 13th century, was chiefly coming from the Sofala coast (now Tanzania and Mozambique). Precious stones came from India and Ceylon, and rock crystal from East Africa. The only manufactured goods seem to have been textiles, muslin cloth and especially silk. Silk was first carried overland from China, but soon the maritime trade became important. By the 6th century A.D. it was bought from India and Ceylon.

Manufactured goods seem to have been the main trade item from the West, although Mediterranean coral and gold coins were also used; large numbers of Roman coins have been found in southern Asia. *The Periplus of the Erythraean Sea*, a trade manual written during the first century A.D., frequently mentions manufactured iron goods and glassware (Huntingford, 1980). During the medieval period, pottery and glassware seem to have been very important trade goods and today are found in the excavations of all the Indian Ocean port towns.

The Red Sea also supplied local products into the trading system. During the early period these include ivory and tortoiseshell as well as a little rhinoceros horn. Gold and gems could be obtained from Upper Egypt, the Blemyes in particular are noted as supplying these products, but were probably exploited through a land route. Cosmas Indicopleustes leaves a description of the gold trade with Axum into the far interior of Africa. The gold was later exported through Adulis on the Red Sea (McCrindle, 1897).

Southern Arabia was always famous for its aromatic products notably frankincense and myrrh, but despite occasional references in the *Periplus*, most of the incense was not exported by sea but by long distance caravan routes to Petra and Tabuk. Later the Yemen coast of the Red Sea became famous for its coffee, which brought prosperity to the port of Mocha (Al Mukha) during the seventeenth century.

16.3.3. The available routes

The Red Sea did not physically link the Indian Ocean with the Mediterranean, for ocean-going ships until construction of the Suez Canal in 1869. Therefore ports of embarkation and transhipment had to be set up along the African shore of the Red Sea (Fig. 16.4). The Egyptians were the first to try to establish a direct link between the Nile and the Red Sea. Later traditions attributed the first attempt to the reign of Rameses II (1289–1284 B.C.), but the first definite evidence comes only in the 6th century B.C. when work was suspended on the canal, because it was choked with sand. Around 274 B.C., the canal was completed only as far as Arsinoe, on the Bitter Lakes. Strabo (64 B.C.—21 A.D.) however maintained that it was possible to sail right through to the Red Sea, but his view is not supported by any other authors. Around A.D. 104, Trajan definitely did complete a canal and road link between the Nile and the Red Sea. It continued in use, and was redug by the Fatimids during the tenth century, but thereafter this canal seems to have been of little importance (Huntingford, 1980).

This route to the very head of the Red Sea had a number of serious drawbacks. The traders faced fickle winds, very shallow seas and numerous pirates. As a result a number of land routes developed, linking the Red Sea to the Nile much further south. During the classical period the port of Myos Hormos was the most important — it was the starting point of the *Periplus* voyage. The *Antonine Itinerary* gives details of a road linking Koptos on the Nile with Berenice, a little to the south of Myos Hormos. The Islamic successors to these towns were Quseir and Aydhab, also linked by a route to the Nile. Some ships preferred not to sail even that far north, and ports were established in what is now modern Sudan, in particular at Suakin and settlements further south, linked by caravans to the Nile valley.

16.3.4. The Egyptians in Punt

During the Eighteenth Dynasty (1552—1305 B.C.), the Egyptian kings first developed maritime trade along the Red Sea and one of their ports, at Wadi Gawesis, has recently been excavated (Sayed, 1977). Accounts of these Egyptian expeditions are preserved in both inscriptions and mural paintings, the most famous being the reliefs at the temple of Queen Hatsheput at Deir el-Bahari.

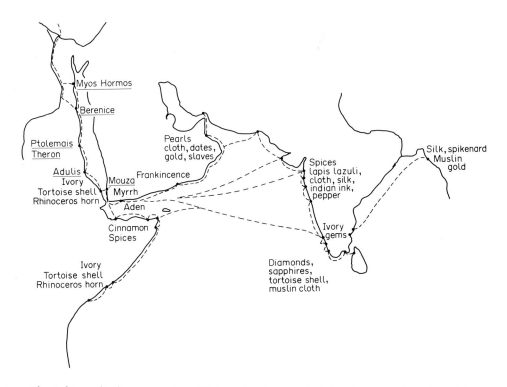

Fig 16.4. The Red Sea and Indian Ocean, about 100 A.D. showing the principal trade routes linking the Red Sea ports with India, the Persian Gulf and East Africa. The main trade goods mentioned in the *Periplus* are also indicated.

The Egyptians gave the name *Punt* to the place they traded with. Disagreement among historians has placed Punt in many different locations, from India to southern Arabia and northern Somalia; however opinion now favours northern Eritrea and the Sudanese coast (O'Connor, 1982).

The inhabitants of Punt are shown on the reliefs as non-negroid, with reddish skin; both men and women have long hair, and the men have goatee beards. Their houses are beehive shaped huts set on piles, very similar to the reed huts in North Yemen today. The Egyptians were mainly interested in the *cntyw* incense, that they used in temple ritual, but Punt also supplied them with ivory, ebony and short-horned cattle. Significantly, the cattle resembled those depicted on what may be contemporaneous rock paintings in Eritrea. Contact with Punt seems to have ceased suddenly, around 1150 B.C.

16.3.5. Greek and Roman trade with India and East Africa

Herodotus states that during the reign of the Egyptian king Nekho II in the 6th century B.C. an attempt was made to circumnavigate Africa, and that the voyage was apparently a success. Such expeditions, if they did indeed occur, were probably isolated incidents, and only in the 3rd century B.C. can a revival of maritime interest be noted. Ptolemy II Philadelphos (282–247 B.C.) needed war elephants, and established hunting stations along the African coast. Strabo (64 B.C.–21 A.D.) noted that the chief place founded by Philadelphos was Ptolemais Theron, but also several other 'elephant hunts and less known towns and islands along the coast' (Strabo, *Geography* XVI). One such place may have been Adulis, where in the sixth century Cosmas Indicopleustes recorded a dedicatory inscription set up by Ptolemy III Euergetes (247–221 B.C.) describing his campaigns using war elephants. Unfortunately no archaeological trace of these sites has yet come to light, although Ptolemais Theron was possibly near Aqiq on the Sudanese coast where some classical architectural pieces have been found, and Adulis was probably at the modern village of Zula (Crowfoot, 1911; Hibbert, 1936; Munro-Hay, 1983).

These elephant hunting stations along the Red Sea developed into trading ports that dealt with a variety of products. The first century A.D. saw an enormous growth in the Roman spice trade with southern and South east Asia (Warmington, 1928). To reach the Indian Ocean, the ships sailed down the Red Sea. According to Pliny the Elder (23 A.D.–79 A.D.) and the *Periplus of the Erythraean Sea* (*circa* 90 A.D.), many different spices and products were carried from India, ivory from East Africa, and cinnamon from northern Somalia. The latter may itself have been a staging post for an Indonesian trade from South east Asia (Miller, 1969). The Red Sea ports were important stopping-off points. Mouza (?Al Mukha) according to the *Periplus*, was humming with business, and the inhabitants used their own boats. Ptolemais Theron was less important as it had no harbour, but none the less supplied true 'tortoiseshell' (from sea turtle), land tortoise and ivory. Mouza supplied these, as well as incense and marble or alabaster. Aden, known as Eudamon Arabia, was another important port, at the southern end of the Red Sea. There is archaeological evidence to indicate the importance of the trade. The chief Roman ports in the northern Red Sea can be identified on the ground; Myos Hormos (Abu Sha'ar), Philoterus (Wadi Gawesis), Albus Portus (Quseir) and Berenice (Ababda Medinat al Harrar). There are also considerable Roman remains in the Eastern Desert (Murray, 1925). Further to the south, the evidence is less good and none of the *Periplus* ports have been positively identified on the ground, although there is a cistern complex on Condenser Island at Suakin, that is constructed in burnt brick that may be classical in date.

Perhaps the most interesting aspect of the Greek and Roman trade along the Red Sea is that it has left us with the first ethnographic descriptions of the people living there. The African shore was inhabited by fish eaters, the *Ichthyophagi*, and by pastoralists, the *Troglodytes*. The fish eaters lived on the shores, were naked and indulged in the common procreation of children. They fished by capturing fish stranded by the tide; eating it as sun-dried cakes. They also ate whales if they became stranded on the shore. Some lived in caves, others in huts constructed from the ribs of fish (Agatharkhides, quoted in Huntingford, 1980). Pliny adds some details as to how turtles were caught in the Red Sea. They were captured either while basking in the sea at midday, or very early in the morning 'the noise of their snoring betrays them to the fisherman'. When captured the turtles are turned on their back and towed ashore (Pliny, *Natural History*, XI). The shells were so big that they formed the roof of the huts of the inhabitants, as well as boats.

The *Troglodytes*, according to Strabo, led a pastoral life, with many chiefs, and were often in dispute over grazing lands. Their food consisted of meat and bone crushed together, or blood mixed with milk. They circumcised themselves, and buried their dead in cairns, often laughing at the funeral (Strabo *Geography*, XVI). Many of these practises can be found among the pastoral groups still living in eastern Africa today.

We can conclude that both the fishing and pastoralist groups played an important role in supplying the Red Sea trade. Of the natural products, the ivory is most likely to have been brought to the shore by the pastoralists. The fisherman provided the small boats for the traders to land. Procopius (6th century A.D.) stated that these boats were fastened with cords, not nails, and not covered with pitch. These sewn boats were common over much of the Indian Ocean, and are noted by Pliny from Ceylon, and the *Periplus* from East Africa. They survived even into the 20th century in some areas of southern Arabia, northern Somalia and East Africa.

A second type of boat operating in the region was described as a raft. The *Periplus* states that the market of Aualites was reached by such vessels, and Pliny adds that the pirates that lived at the southern end of the Red Sea sailed on rafts made of inflated skins (Pliny, *Natural History* VI). He later described how cinnamon is traded in the Gulf of Aden, 'on rafts which have no rudders to steer them or oars to push them, or sails, or other aids to navigation (Pliny, *Natural History* XII). These vessels have normally been taken as 'outrigger canoe' types that are still to be found along the coasts of East Africa and Somalia (Drake-Brockman, 1912) and the type may originally have derived from South-east Asia and the Pacific Ocean.

16.3.6. Medieval trade and the rise of Islam

The Red Sea route was at its most important during the first to fifth centuries A.D. after which a very marked decline in fortune can be noted. The reasons for this are probably threefold; the opening up of alternative routes by the Sassanians through the Arabian Gulf, the decline in demand for eastern luxuries in the West and a great increase in uncontrolled piracy. The Sassanian or New Persian Empire (A.D. 224—636) greatly developed the Indian Ocean trade, and this is illustrated in a series of disputes between competing Axumite and Sassanian merchants in India and Ceylon during the fifth and sixth centuries. Sassanian ports like Siraf, situated on the north-east shore of the Arabian Gulf, were important trading centres, and their growth had by the seventh century eclipsed the Red Sea settlements. Only with the rise of the Fatimid Dynasty in Egypt in A.D. 969, did the Red Sea again become important, but by then a general realignment of trade had taken place (Whitehouse and Williamson, 1973).

The limited prosperity of Axum, and its port Adulis (Fig. 16.5) at the southern end of the Red Sea can be assessed from the account of Cosmas Indicopleustes, written between 535 and 547 A.D. (McCrindle, 1897). From Adulis, merchants sailed to southern Arabia, India and Ceylon, bringing back silks, peppers, spices and incense. However from Cosmas' account the route from the Red Sea to East

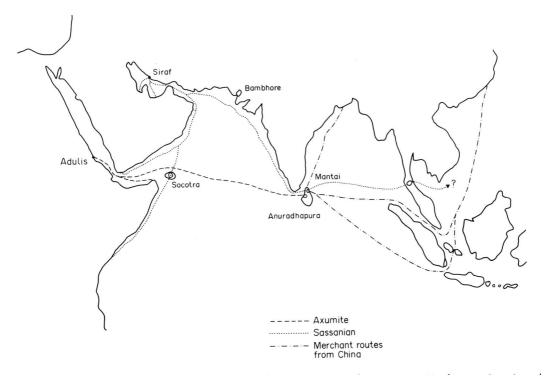

--- --- Axumite
.... Sassanian
--- - --- - Merchant routes
from China

Fig. 16.5. Trading routes from the Red Sea and Arabian Gulf at about 600 A.D. showing competition between Axumite and Sassanian merchants.

Africa was no longer followed, at least by the Axumite merchants. The importance of the trade of Axum can be judged from the *Alexandrian Tariff*, a list of goods subject to important duties at the time of the Roman emperor Justinian, who was a contemporary of Cosmas. Fifty four different items are listed, including twenty three spices, linens, furs, ivory and iron, silk, gems, lions, panthers and even Indian eunuchs (Miller, 1969).

In A.D. 570, the prophet Muhammad was born, and the Islamic faith that he founded in western Arabia was to play a major role in the development of trade and settlement in the area. Muhammad was born in Mecca (Watt, 1970), one of the caravan towns on the incense route, between southern Arabia and the Levant. He was part of an elite class that lived in these permanent settlements, formed from a fusion of nomadic and urban society. Around the age of 40, Muhammad received the first of a series of Divine Revelations, and began to preach against the current religious practices. But his activities were not well received and he and his followers fled from Mecca in A.D. 622, to Medina. Others fled to Ethiopia and some may have settled in the Dahlak Islands. It was this event that marked the beginning of the Muslim calendar, lunar years being counted after the *Hijva* or *Haj,* and designated with the initials A.H.

Muhammad's authority continued to grow, first in Medina, and later in Mecca where he returned in triumph in A.D. 630. By the time of his death, a series of alliances had been established over most of Arabia with the new faith based in Mecca and Medina. His successors, Abu Bakr and Omar, initiated a series of campaigns that carried the Muslim faith to Spain in the west and the Sassanian empire in the east. The first Caliphs maintained their capital at Mecca, but with the election of Mu'awiya, and the start of the Umayyad dynasty (661–750) the capital was moved away from western Arabia, and the holy cities of Mecca and Medina became only centres of pilgrimage and learning. But this brought new prosperity to the Red Sea, as ports sprang up to transport pilgrims from the newly converted countries. Places such as Aydhab and Suakin on the African side, Jiddah and Al Jar on the Arabian side grew prosperous in this new trade in people, a trade that only began to decline with the completion of the Hijāz railway in 1908.

Although Islam spread quickly across North Africa and Egypt, it appears that the African side of the Red Sea (that part of the continent closest to Mecca) was hardly penetrated. The conquest of Egypt by Amr ibn al 'As in 640 A.D. was limited to the area north of the First Cataract, and the Christian kingdom of Nubia retained its independence until the 13th century A.D. and was only fully converted in the 16th century (Hasan, 1967; Hrbeb, 1977). Ethiopia also remained independent, although its Christianity became isolated as the tribal peoples of eastern Africa were converted to Islam (Trimingham, 1952).

New Muslim settlements were established along the Red Sea in order to remove the control of trade from the Christian Ethiopian kingdom. In 702 A.D. Abyssinian 'pirates' burnt Jiddah, and in retaliation the Muslims captured the Dahlak Islands, and destroyed Adulis. Muslim ports were set up at Suakin and Badi, although there were Christian merchants living in Suakin at late as 1270 A.D.

On the Dahlak Islands, a Muslim sultan reigned until the tenth century, when he was replaced by a Christian king, but a Muslim sultanate had again been established by the eleventh century A.D. In the Afar region during the tenth century Muslim states were being established, often by Sharifs. These communities gained considerable wealth through slave trading, and ruined stone towns survive in eastern Ethiopia and northern Somalia. Harrar, the 'forbidden city' visited by Richard Burton during the mid nineteenth century was the most famous of these. Lesser known states included 'Adal, Saho, 'Afar and Ifat. Their ports were Massawa, Zaila and Berbera (Lewis, 1955; Trimingham, 1952).

One description of the peoples of the Red Sea and their trade during the eighth century comes from an unlikely source, the *Ching-hsing Chi* of Tu Huan. Although only preserved in fragments (Wheatley, 1975) this Chinese account relates Tu Huan's captivity for two years by the Arabs (after the battle of Talas in 751 A.D.) during which he travelled from Nubia to the Arabian Gulf, via the Red Sea. He described a place called Ma - lin, that is probably a coastal port in Eritrea, where the 'people are black and their nature is fierce... they feed their horses on dried fish... the people eat the hu - mang (the date). They are not ashamed of debauching the wives of their fathers and chiefs, they are in this respect the worst of the barbarians....' (Duyvendak, 1947). Tu Huan states that in this area there were three religions, Islam, Christian and Zamzam — probably pagan.

Archaeological evidence supports some of Tu Huan's observations. Most of the coastal strip was inhabited by Beja pastoralists, who still feed their cattle on sun-dried fish today. Muslim trading settlements were also being established on narrow peninsulas, or on islands securely cut off from the coastal hinterland. The Dahlak settlement was very important, judging from the wealth of tenth to twelfth century inscriptions, numerous buildings and cistern complexes preserved there. Further north, on the island of Er Rih (Fig. 16.6), there are extensive middens, stone buildings, cisterns, and a tenth century cemetery of 10 hectares (Fig. 16.7) (Crowfoot, 1911; Combe, 1930). The earliest Arabic inscription from this coast was found at Knor Nubt nearby, it dates from 861–863 A.D. (Glidden, 1954).

A similar dual political system operated on the Arabian coast between the ports and the mainland tribes. The hill peoples of the Yemen retained their identity after the coming of Islam, the modern names of their tribal confederations can be traced to the classical pre-Islamic authors. They constructed spectacular settlements in the interior using stone, daub and mud brick in ways still employed today (Varanda, 1981; Serjeant and Lewcock, 1983). The coastal settlements were very different, based on trade and fishing, not agriculture and pastoralism. This is still reflected in the coastal architecture, houses are open to the air, decorated with stucco and contain elaborate woodwork screens.

Fig. 16.6. Subterranean cisterns on the island of Er Rih, probably the port of Badi, dating to the 9th–11th century.

16.3.7. Later medieval trade

In 961 A.D. the powerful Fatimid Dynasty was established in Egypt, and the Red Sea became once more an important trade route (Taurat, 1977). For three centuries eastern trade had been directed through the Arabian Gulf (Guillian, 1856), but during the tenth century merchants returned to the Red

Fig. 16.7. Arabic tombstone of Al-Walid ibn Ahmad ibn Al-Walid ibn Abun found at Er Rih, dated to 997 A.D., one of the early inscriptions known from the African coast of the Red Sea.

Sea. Both Muslim and Jewish traders were important in this revival (Chandhuri, 1985; Goitein, 1973). Fustat and Alexandria became trading centres of immense importance, new ports were founded along the Red Sea and others, including Quseir and Aydhab, experienced a revival in prosperity. Both have recently been investigated by archaeologists and the wealth of finds have suggested trading connections that ranged from China, to the Gulf, Aden and Yemen (Figs. 16.8–16.10). Wealthy houses have been uncovered, and complex water storage cisterns investigated (Whitcomb and Johnson, 1982).

Aydhab relied on the pilgrim trade for a sizable part of its income. It was a notorious stopping off place for the pilgrimage and the inhabitants, according to Makrisi, were beasts rather than men. Ibn Jubayr tells us the pilgrims were fleeced for everything they were worth and that the poorer pilgrims were not uncommonly suspended by their testicles, in order to induce additional payments. They were then carried across the sea in overloaded boats, and while travelling across the desert, the Aydhab guides took them along waterless tracks, and stole the goods of those that perished. The Fatimid officials appointed to the town condoned these unpleasant activities, and claimed one third of the revenues collected. The excesses continued until 1426 when the Mamluks (a dynasty that succeeded the Fatimids in Egypt in 1390) sacked the port, and the inhabitants that fled were later massacred at Suakin.

Suakin (Fig. 16.11) controlled a more southerly route to the Nile, now followed by the Berber Railway, and is beautifully set on a circular island in a superb natural harbour south of Port Sudan. Suakin is mentioned first in 969 A.D. and again during the eleventh and fourteenth centuries, but its prosperity began with the sack of Aydhab, after which it became the main pilgrim port and the centre of commerce on the Red Sea. The site has never been investigated archaeologically, but sixteenth century drawings shows that it was extensively built up with numerous streets and houses by this time (Crossland, 1913; Greenlaw, 1976).

Jiddah, founded in 646 A.D. lies almost opposite Suakin, but remained only a small settlement until the 11th century, when a population of 5,000 is reported. By the twelfth century, there was a stone mosque but most of the houses were made of reeds. Jiddah was not a trading centre but mainly the port of disembarkation for pilgrims. The town was sacked by the Mamluks in 1425, but only received a town wall in 1511 as a protection against the Portuguese.

Much less is known of the other ports on the Arabian coast, such as Al Jar, Al Shinybah and Al Mukha (Mocha), which were by then minor stopping places. At the head of the Red Sea, Aqaba was also a trading place, but eclipsed by Suakin and Aydhab to the south. Aqaba was taken by Renaud de

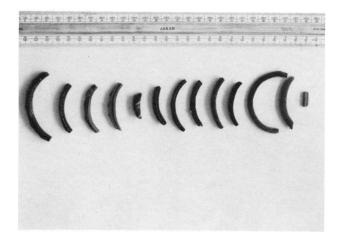

Fig. 16.8. Glass bracelets made in Southern Arabia and widely traded throughout the Red Sea and Indian Ocean. These were found at the site of Aydhab (Khartoum National Museum Collection).

Fig. 16.9. Chinese porcelain of about 14th century date, found at Aydhab. This material was traded through India and shipped to the Red Sea and Egypt (Khartoum National Museum Collection).

Fig. 16.10. Islamic glazed pottery of 12th–14th century date from Aydhab (Khartoum National Museum Collection).

Chatillon during his expedition down the Red Sea in 1183. The prosperity of Kulzum (Suez), the town that gave the Red Sea its Arabic name of Bahr al Kulzum, had greatly declined in the later medieval period. Al Idrisi in the twelfth century stated that the place was in ruins.

The fortunes of the Ethiopian ports collapsed at the end of the Axumite period, and they never regained their importance. Adulis was abandoned in favour of Massawa, but most of the trade passed through the Dahlak Islands. The African products that formerly travelled overland through Ethiopia, were now being shipped directly from the Swahili coast of East Africa, to the Arabian Gulf and southern Arabia (Chittick, 1977). It was in southern Arabia that the strategic port of Aden was located. The *Periplus* noted it as Eudaimon Arabia — the 'entrepôt of Arabia', but by the fifth century it was already known as Adane. Aden (Fig. 16.12) was the terminus of the trans-Arabian caravan route and the western Indian Ocean trade, as well as a stopping off point for ships proceeding up the Red Sea. However, excessive duties often deterred traders from using Aden during the medieval period, and the Portuguese noted it as a very small place by the sixteenth century.

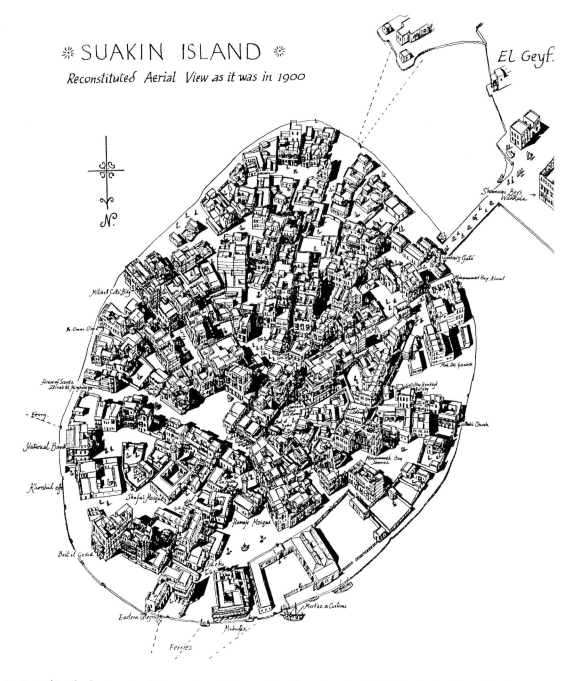

Fig. 16.11. Suakin Island as it was in 1900, reproduced with permission from Greenlaw (1976), *The Coral Buildings of Suakin*, Oriel Press.

16.4 THE MODERN HISTORY OF THE RED SEA

16.4.1. The Ottoman conquests and the Portuguese

In 1498, Vasco de Gama discovered a sea route to India around the Cape of Good Hope that involved neither the Red Sea nor the Arabian Gulf. By outflanking the Muslim stranglehold on the spice trade, goods could be brought to Europe much more cheaply, on Portuguese (and later Dutch) ships. Egypt rapidly lost its prosperity, and by 1517, the Mamluk sultans had been deposed by the Ottoman Turks. The Red Sea itself became a trade backwater. Over the next 400 years attempts were made by the Turks to establish and secure their influence in the Red Sea. Their fortunes were mixed; some towns such as Medina, Suakin and Yanbu (Yanbu al Bahr) could be nominally controlled, but in other areas, such as in the Yemen, no headway was made at all, and Turkish armies were periodically massacred.

The Portuguese also tried to exclude the Turks from the Red Sea. In 1490, Pedro de Covilham was the first modern European to reach the Ethiopian Court. Stories began to be circulated in Europe of a mythical Christian kingdom of 'Prester John', besieged by its Muslim neighbours. At the same time the Portuguese were keen to lessen the threat of Turkish maritime activities in the western Indian Ocean, and a series of sea campaigns were launched in order to open the Red Sea to Portuguese shipping. The first was in 1513 by Alfonso de Albuquerque, who attacked Aden without success, then sailed into the Red Sea, but never reached Jiddah, his destination. Further campaigns in 1517, 1520 and 1547 all ended without success, and left no permanent influence in the area (Cooke, 1933; Kramnerar, 1929).

16.4.2. The English and the Dutch in the Red Sea

Next to come were the English. An attempt was made to set up a 'factory' or trading station in Aden in 1608, but it came to nothing. Later that year, an East Indian Company ship, *Ascension* reached Aden, and two of its company sailed on to Al Mukha. In 1610 Sir Henry Middleton sailed to Al Mukha, where he was imprisoned. When later set free, he blockaded the town.

Interest in the area was resumed after 1660, when Al Mukha became a place of importance, supplying coffee to the Indies trade, chiefly to the English. An account of this trade by Rev. John Ovington, chaplain to the East India Company in Surat, between 1690–1693 reports English, Dutch, Portuguese, French, Indian and Danish ships, as well as several from the Arabian Gulf. They brought cloth and piece goods, but the main export was coffee (Foster, 1949).

Ovington's contemporaries, Joseph Pitts and William Daniels, who travelled the Red Sea also indicate that apart from Al Mukha and its nearby coffee ports, the ancient ports of the Red Sea were either abandoned, or could rely only on the pilgrim trade for their prosperity. Suakin still exported a little ivory and gold dust (Bloss, 1936–7). Jiddah, although 'in a constant traffic' had only a mud wall for fortification. Yanbu 'had lost abundance of its former glory'.

In 1664, the Dutch East India Company established a factory at Al Mukha, and they were soon joined by the French. However it seems that most of the trade was left in the hands of Indian merchants who acted as middle-men. The Indians were of some importance as late as the mid nineteenth century, returning every year to the annual fair at Berbera (Burton, 1856). They were also important in the Yemen capital of San'ā, (Serjeant and Lewcock, 1983).

In 1712, the coffee plant was smuggled out of the Yemen and carried by the Dutch to Java and Surinam, from where it was extensively planted over both the New and the Old Worlds. This broke the monopoly of Al Mukha on the coffee trade, and the prosperity of the coffee producing areas rapidly declined.

Fig. 16.12. The town of Aden in the sixteenth century, after de Braun 1577.

Diplomatic interest on the Red Sea continued even though the trading significance of the region was now completely lost. France saw the control of the Red Sea as a way of influencing the naval situation in the Indian Ocean (Marston, 1961). This was countered by Britain in 1801 when an expedition sailed up the Red Sea under General Baird, and attacked the French in Egypt, landing at the ancient port of Quseir. In 1839, the British took Aden by force to establish a coking station on the sea route to India.

16.4.3. The 'Red Sea Style'

Although the conditions of trade seem to have appreciably declined during the early part of the nineteenth century, the same period sees the development of a unique style of architecture at the port towns on either side of the Red Sea. This became known as the Red Sea Style, formed out of an amalgam of local forms, together with Turkish, Egyptian and Arabian types (Fig. 16.13). Buildings in this style are found in such towns as Suakin, Jiddah, Al Luhayyah, Al Hudaydah, Al Mukha and Massawa. Many of these beautiful houses survive, although an increasing number are under threat from decay and dilapidation, (Greenlaw, 1976; Varanda, 1981).

The houses were constructed of the coral *Porites solida* cut while still wet from the sea bed into square or rectangular blocks. Left in the sun the coral will harden into a very durable building material. Some very high buildings were constructed, occasionally up to four stories. Typically the ground floor was occupied by the store and shop, since most of these houses were the property of merchants. The house was entered through the *dihlis*, often flanked by a *mastaba* or stone seat. Nearby would be a guest room, for guests were generally not admitted into the upper floors. The remainder of the ground floor was filled with the *diwan* which served as a reception room. The *diwan* was always one of the most decorated rooms in the house, heavily carpeted and with stucco friezes along the walls.

The houses each had a second entrance which led by an internal stairway to the upper floors. This part of the house was known as the *harem*, and here the women and children, as well as the many members of the extended family lived. Some of these rooms were allocated as *majlis* or private living rooms for

Fig. 16.13. Ruins of a substantial house in the Red Sea Style on the Island of Suakin. Note the surviving *roshans* or wooden casement windows, and the thinness of the flooring, which with the irregularity of the building blocks has contributed to the very rapid decay of these noble buildings.

married couples, others were communal space. The top floor contained the great *majlis*, the suite used by the head of the family, often lavishly decorated. The rest of the area might be filled by an open terrace or *kharjas* and kitchen. Deep pit latrines led down from the *majlis*. Water was stored in a cistern on the ground floor and also in porous pots at various points in the *harem*.

The most characteristic feature of the Red Sea House were the *roshans* of casement windows. They jutted out from virtually every upper floor *majlis*, were often built in timber, finely carved and acted both as balconies and as openings to receive a cooling breeze. The *roshans* were the most frequently lived in part of the house — coffee was drunk here, or a hookah was smoked. The balcony was wide enough to lie down fully on, so at night they were used for sleeping.

These very complex houses that acted as both shop and office, warehouse and harem achieved their most splendid development in the middle part of the nineteenth century, although some may date from as early as the seventeenth. The reason for the sudden late flowering of these opulent residences may be sought in an economic upturn in the fortunes of the region, associated with the opening of the Suez Canal.

16.4.4. The Suez Canal

In 1869, the Red Sea was suddenly transformed into one of the most important seaways of the world, and the lifeline of the British Empire in the East. Two related factors brought this about; the opening of the Suez Canal and the appearance of steam navigation. Powered ships were able to sail the narrow seas more easily, and the canal at last allowed ocean going vessels to pass directly between the Red Sea and the Mediterranean. The old ports, such as Yanbu al Bahr, Jiddah and Massawa became thriving centres of mercantile activity. Suakin, which in 1805 was 'nearly in ruins', and in 1869, had a population of only 8,000, suddenly became the main port for the Sudan and upper Egypt. Caravans left every three months for Kassala and Berber, carrying manufactured European goods. Each caravan contained up to 1000 camels (Greenlaw, 1976), and brought back gold ivory, hides and cotton in return.

Railways soon became important in the development of the ports and their hinterland. The first was the Berber line running between Suakin and the Nile Valley. Later Djibouti was connected by railway to Addis Ababa, and the Hijāz railway was laid from Damascus to Mecca. These new routes all had their effect on the pattern of trade and the prosperity of the sea ports. One result was the disappearance of the long distance caravan everywhere except western Arabia. Another was the decline of the maritime pilgrim trade, and of those medieval ports that no longer lay directly on the new maritime or railway network.

One such port, was Suakin itself. A new port had been built further north shortly before the First World War, to accommodate the larger modern ships that could no longer use the narrow *marsa* at Suakin. This new harbour, Port Sudan, fast eclipsed the picturesque Suakin and the merchants and government offices had all moved there by 1936. The fine merchant houses, jetties, mosques and custom houses were all left to ruin, and today the island is deserted. Such is the transitory nature of commerce that a port that had been one of the most prosperous in the Red Sea for over 1000 years, could, within a single decade become of heap of ruins. (Roden, 1970).

16.4.5. The present day settlement of the Red Sea

Nine modern nations border the Red Sea; Egypt, Sudan, Ethiopia, Djibouti, North and South Yemen, Saudi Arabia, Jordan and Israel. Each has established important interests along the Red Sea shores. In 1980 Egypt earned $US660 million in dues collected from ships using the Suez Canal and the Egyptian economy suffered badly when war and its aftermath left the canal closed in 1956—7 and 1967—75. The reopened canal retains its importance as an international shipping route (Lapidoth-Eschelbacher, 1982).

Egypt's major Red Sea port is Suez, and lesser ones are at Hurghada and Safaga. Sudan has only Port Sudan, which today is the terminus of the railway and road link with Khartoum, while Suakin is now little more than a fishing port. Ethiopia relies on Massawa, but the area is much disturbed by civil war. Djibouti, connected by railway to Addis Ababa, has a population of only 320,000 and relies heavily on trade and foreign aid.

On the eastern side, North Yemen has three ports, Al Hudaydah, the most important, Al Mukha and Salif which serve an impoverished country. In Saudi Arabia, 60% of the population live along the Red Sea shores and the immediate hinterland. The annual pilgrimage still remains an important source of employment. In 1967, 226,000 pilgrims visited the Holy Cities; in 1977 this figure had grown to 709,000. Jiddah has become an important communications centre, with a new airport, and increased harbour facilities. In 1972 it handled 1,022,000 tons of shipping freight; a two-fold increase in ten years. The only other major Saudi Arabian port on the Red Sea is Yanbu al Bahr where considerable expansion is currently under way.

Aqaba is Jordan's only port on the sea, and therefore very important especially in the export of bulky materials, phosphates account for 95% of the exports of Aqaba, representing in 1974, 1,116,000 tons. Elat, next to Aqaba, is Israel's only port on the Red Sea, and was formerly very important when the Suez Canal was closed to Israeli ships and cargoes. After the Camp David agreement and the reopening of the Suez Canal, Israel's Mediterranean ports, closer to the centre of industry, now take most of the shipping.

REFERENCES

Arkell, A. J. (1954). Four occupation sites at Agordat. *Kush II* 33—63.
Arkell, A. J. (1955). *A History of the Sudan to 1812*, London.
Bloss J. F. E. (1936—7). The story of Suakin, *Sudan Notes Rec.* XIX, 271—281; XX, 247—70.
Burton, R. (1856). *First Footsteps in East Africa*, London.
Butzer, K. R. (1981). The rise and fall of Azum, Ethiopia, a geo-archaeological interpretation. *Am. Antiq.* 46.3, 471—95.
Caton-Thompson, G. (1952). *The Kharga Oasis in Prehistory*. Oxford.
Caton-Thompson, G. (1953). Some Paleoliths from South Arabia. *Proc. prehist. Soc.* 19, 189—218.
Caton-Thompson, G. and Gardner, E. W. (1939). Climate, irrigation and early man in Hadhramaut. *Geogr. J.* XLIII, 18—38.
Chaudhuri, K. N. (1985) *Trade and Civilisation in the Indian Ocean*. CUP Cambridge.
Chimielewski, W. (1968). Early and Middle Palaeolithic sites near Arkin, Sudan. In *The Prehistory of Nubia*. Ed. F. Wenday, pp.110—93, Dallas.
Chittick, H. N. (1974). Excavations at Aksum 1973—4: a preliminary report. *Azania* IX, 159—206.
Chittick, H. N. (1977). The East Coast, Madagascar and the Indian Ocean. In *Cambridge History of Africa*, Vol III. Ed. R. Oliver, pp.183—231, C.U.P. Cambridge.
Chittick, H. N. (1978). Notes on the archaeology of Northern Ethiopia. *Abbey* 9, 15—20.
Clark, J. D. (1954) *The Prehistoric Cultures of the Horn of Africa*. C.U.P., Cambridge.
Clark, J. D. (1982). The cultures or the Middle Palaeolithic/Middle Stone Age. In *Cambridge History of Africa* Vol 1. Ed. J. D. Clark pp.248—341, C.U.P. Cambridge.
Cooke, B. K. (1933). The Red Sea coast 1540. *Sudan Notes Rec.* XVI, 151—-9.
Coombe, E. T. (1930). Four Arabic inscriptions from the Red Sea. *Sudan Notes Rec.* XIII, 288—91.
Corvinus, G. (1976). Prehistoric exploration at Hadar, Ethiopia. *Nature, Lond.* 261, 571—2.
Corvinus, G. and Roche, H. (1980). Prehistoric exploration at Hadar in the Afar, Ethiopia in 1973, 1974 and 1976. *Proc. VII Pan African Cong. Prehistory and Quaternary Studies* 156—88.
Crossland, C. (1913). *Desert and Water Gardens of the Red Sea*. Cambridge.
Crowfoot, J. W. (1911). Some Red Sea ports of the Anglo-Egyptian Sudan. *Geogr. J.* 37, 523—50.
Crowfoot, J. W. (1928). Some potsherds from Kassala. *J. Egypt. Archaeol.* 14, 112—6.
Doe, B. (1971). *Southern Arabia*, London, Thames and Hudson.
Doresse, J. (1957). L'Ethiopie et l'Arabie meridionale aux IIIe et IVe siècles A.D. d'après les descouvertes recentes. *Kush* 5, 49—60.
Drake-Brockman, R. E. (1912). *British Somaliland*, London.
Duyvendak, J. J. L. (1947). *China's Discovery of Africa*. School of Oriental and African Studies, London.

Fattovich, R. (1978). Traces of a possible African component in the Pre-Aksumite culture of Northern Ethiopia. *Abbay* 9, 25–30.

Foster, Sir W. (1949). *The Red Sea and adjacent countries at the close of the seventeenth century, as described by Joseph Pitts, William Daniel and Charles Jacques Poncet*, Hakluyt Society, London.

Glidden, H. (1954). The Khor Nubt Tombstones. *Kush* 2, 63–5.

Goitein, S. D. (1973). *Letters of Medieval Jewish Traders Translated from the Arabic*, Princeton.

Greenlaw, J. P. (1976). *The Coral Buildings of Suakin*, Oriel Press, London.

Guillain, C. (1856). *Documents sur l'histoire, la géographie et la commerce de l'Afrique orientale*, Paris.

Hasan, Y. F. (1967). *The Arabs in the Sudan*, Edinburgh.

Hibbert, H. E. (1936). Relics at Agig. *Sudan Notes Rec.* 19, 193.

Hrbeb, I. (1977). Egypt, Nubia and the Eastern Desert. In *Cambridge History of Africa* Vol. III. Ed. R. Oliver, pp.10–97, C.U.P., Cambridge.

Huntingford, G. W. B. (1980). *The Periplus of the Erythraean Sea*. Hakluyt Society, London.

Isaac, G. L. (1982). The earliest archaeological traces. In *Cambridge History of Africa* Vol. I. Ed. J. D. Clark, pp.157–247, C. U. P., Cambridge.

Johanson, D. C. (1980). Early African hominid phylogenesis. In *Current Arguments on Early Man*. Ed. L. K. Konigsson and S. Sundstrom, pp.31–69, Pergamon Press, Oxford.

Johanson, D. C., Taieb, M. and Coppens, Y. (1982). Pliocene hominids from the Hadar Formation, Ethiopia (1973-1977). *Am. J. Phys. Anthrop.* 57, 373–719.

Kobishchanov, Y. M. (1979) *Axum*. University of Pennsylvania.

Kramnerar, A. (1929). *La Mer Rouge*. Cairo.

Lapidoth-Eschelbacher, R. (1982). *The Red Sea and the Gulf of Aden*. The Hague.

Lewis, I. M. (1955). *Peoples of the Horn of Africa*. London.

McAdams, R. and others. (1977). Saudi Arabian archaeological reconnaissance. *Atlal*, 21–41.

McClure, H. A. (1976). Radio Carbon chronology of the late Quaternary lakes in the Arabian Desert. *Nature*, London. 263, 755–6.

McCrindle, J. W. (1897). *The Christian Topography of Cosmas*. Hakluyt Society, London.

MacMichael, H. A. (1922). *A History of the Arabs in the Sudan*. C. U. P., Cambridge.

Marston, T. E. (1961). *Britain's Imperial Role in the Red Sea Area*. Hamden, Connecticut.

Masry, A. H. (1977). Introduction: The historical legacy of Saudi Arabia. *Atlal* I, 9–20.

Miller, J. I. (1969). *The Spice Trade of the Roman Empire*. Oxford University Press, Oxford.

Munro-Hay, S. (1982). The foreign trade of the Aksumitic port of Adulis. *Azania* 17, 107–26.

Murray, G. W. (1925). The Roman roads and stations in the Eastern Desert of Egypt. *J. Egypt. Archaeol.* 11, 138–50.

Murray, G.W. (1935). *Sons of Ishmael*, London.

Naval Intelligence Division (1946). *Western Arabia and the Red Sea*. H.M.S.O. London.

O'Connor, D. (1982). Egypt 1577–664 B.C. In *Cambridge History of Africa* Vol I. Ed. J. D. Clark, pp.830–925, C. U. P., Cambridge.

Parr, P. and others (1978). Preliminary report of the second phase of the Northern Province Survey 1397/1977. *Atlal* 2, 29–50.

Paul, A. (1954) . *A History of the Beja Tribes of The Sudan*. C. U. P., Cambridge.

Phillipson, D. W. (1977). The excavation of Gobedra Rock-Shelter, Axum. *Azania* 12, 50–82.

Roden, D. (1970). The twentieth century decline of Suakin, *Sudan Notes Rec* 51, 1–21.

Sayed, Abdel Monem A. H. (1977). Discovery of the site of the 12th Dynasty Port at Wadi Gawesis on the Red Sea Shore. *Revue Égyptol.* 29, 138–77.

Seligman, C. G. (1915). An undescribed type of building in the Eastern Province of the Anglo-Egyptian Sudan. *J. Egypt. Archaeol.* 2, 178–83.

Serjeant, R. B. and Lewcock, R. (1983). *San'ā', an Arabian Islamic City*. World of Islam Festival Trust, London.

Shinnie, P. L. (1967). *Meroe*. Thames and Hudson, London.

Street, A. F. and Grove, A. T. (1976). Environmental and climatic implications of late Quaternary lake-level fluctuations in Africa. *Nature*, London. 261, 385–90.

Taurat, T. (1977). Ethiopia, the Red Sea and the Horn. In *Cambridge History of Africa* Vol. III. Ed. R. Oliver, pp.98–177, C. U. P., Cambridge.

Varanda, F. (1981). *The Art of Building in The Yemen*. Art and Archaeology Research Papers, London.

Warmington, E. H. (1928). *The Commerce Between the Roman Empire and India*. London.

Watt, W. M. (1970). Muhammad. In *Cambridge History of Islam* Vol 1. Ed. P. M. Holt *et al.*, pp.30-56, C. U. P., Cambridge.

Wheatley, P. (1975). Analecta Sino-Africana Recensa. In *East Africa and The Orient*. Ed. H. N. Chittick and R. I. Rotberg, Africana Publishing Company, New York & London, pp.76–114, 284–90.

Whitcomb, D. S. and Johnson, J. H. (1982). Quseir al-Qadim 1978, preliminary report. *American Research Centre in Egypt Reports* v.7, Malibu.

Whitehouse, D. and Williamson, A. (1973). Sassanian maritime trade. *Iran* XI, 29–50.

CHAPTER 17

Red Sea Fisheries

STEPHEN M. HEAD

Zoology Department, University of the West Indies, Kingston 7, Jamaica

CONTENTS

17.1. GENERAL INTRODUCTION

Fisheries are among the most important of the world's food resources. In 1982, the last year for which United Nations Food and Agriculture Organisation (FAO) data are presently available, global fish production had reached 76,772,800 metric tonnes, of which nearly 90% came from marine rather than freshwater fisheries (FAO, 1984a). The annual catch has been steadily increasing and has almost exactly doubled since 1960.

Fisheries are important to national economies in two ways. Fish (including crustacean and molluscan shellfish) are an excellent source of high grade protein, and are eaten directly by humans, as well as being extensively used when processed for livestock feed and fertiliser. Fisheries may also represent an important source of employment in countries with a small industrial base; and the pattern of employment often evolves in a characteristic way as the nation develops. Most fisheries start on a simple scale, with relatively large numbers of self-employed fishermen, working close inshore from small boats, and supplying a largely local market. As internal communications improve and demand increases, the scale of exploitation tends to rise and fishermen enter cooperatives or become employees on medium sized boats with extended working ranges. Highly populated and developed countries are usually unable to supply their demand with locally caught fish, and their industry becomes international in scope, while *per capita* consumption often declines in favour of meat and poultry products unless maintained at a high level (as in Japan) for cultural reasons.

Fishery production from the Red Sea is small, accounting for only about 0.07% of the world total, although the surface area of the Red Sea (400,000 km^2 according to Morcos (1970)) represents about 0.123% of the total area of the oceans. Historically, fishing has been an important element in the economy of the countries bordering the Red Sea. The earliest accounts of the area indicate a division of settlement between nomadic pastoralists and fishing communities on the coastal plain (*see* Chapter 16), and a fascinating picture of an equivalent modern Arabian culture is painted by Bertram (1959). The historical artisanal fisheries of the Red Sea were limited in scope by the sparseness of the population, poor communications and water supplies, and by the primitive methods employed. These fisheries seem to have been perfectly in harmony with their fishing grounds, removing insignificant proportions of the fish stocks, and scarcely influencing the ecology of the area. Today, communications have changed beyond recognition, large centres of population have sprung up along the coasts, and the fishing industry is moving into a more sophisticated phase, raising the need for scientific assessment and management of the limited fish stocks.

17.1.1. Factors governing the productivity of fisheries

The study of fisheries and their yields is becoming a sophisticated and exact science, of which no more than an outline can be presented here. For a general introduction to the subject, the reader is recommended to try Cushing (1975a) or Odum (1980), or one of the many introductory texts on Marine Zoology.

Marine production starts with plants, the primary producers, including both phytoplankton and benthic algae and higher plants. The majority of phytoplankton are extremely small, and few if any adult fish feed directly on them. Instead, the phytoplankton are grazed by zooplankton, and the resulting secondary production passed on to fish through complex pelagic food webs. In most of these there are several intermediate levels, for example, herbivorous copepods are eaten by carnivorous copepods, which are eaten in turn by medium to large plankton such as chaetognaths, which are preyed on by small or larval fish which in their turn are prey for adult commercial species. Each feeding animal consumes a set quantity of food, of which it is able to assimilate a varying proportion, averaging between 50 and 95% (Tranter, 1976) of the total consumed. Part of the assimilated food must be respired for maintenance metabolism and the animal's activity; the remainder is available for growth and reproduction. Overall, copepods may convert between 10 and 50% of their food input into copepod biomass available for consumption by the next level in the food chain. At each intermediate consumption stage between primary production and commercial fish, about 90 to 50% of the available food is therefore lost, so that the eventual productivity of the fishery may be orders-of-magnitude less

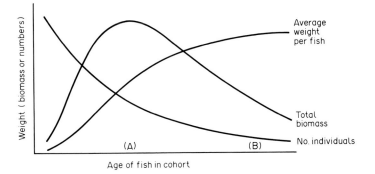

Fig. 17.1. Relation between age and the average weight, survivorship and total biomass for a cohort of fish. The cohort is a group of fish of the same age, the progeny of a single season's spawning. Note that while the average weight of a fish rises, the number surviving in the cohort falls, and the biomass available for capture declines. Catching fish at age 'A' results in a much greater catch biomass than at age 'B', although the older fish will be much larger. Adapted from Clark (1978).

than that of the phytoplankton. Margalef (1967) has provided an analysis of a Mediterranean fishery, suggesting that of 120mg C m^{-2} day^{-1} fixed by primary producers, only 8.0 mg was incorporated into larger zooplankton, and about 2.0 mg into fish. The fisheries yield of 1.2 mg C m^{-2} day^{-1} was only 1% of the primary production.

Given that fishery production can never be more than a small fraction of primary production, the fishery production of an area will nevertheless be controlled partly by the level of primary productivity, which is in turn largely controlled by nutrient availability. Open ocean waters, especially in the tropics, have very low nutrient levels, because of the low rate of return of sunken nutrients from deep to shallow water. Tropical oceanic waters support very low fishery yields, and the fish that are caught in them are largely the big, long-lived species such as tuna, which when caught may represent decades of productivity collected over a wide area of ocean. The majority of fisheries operate in coastal waters where nutrients are continually or seasonally returned to the productive upper layers to be incorporated into organic material by phytoplankton. Areas where nutrient-rich bottom waters continually upwell can have primary production levels in excess of 1g C m^{-2} day^{-1} (Raymont, 1980). Upwelling areas support the most productive fisheries, such as the anchoveta fishery of Peru, which in 1970 yielded the staggering total of 13.1 million tonnes, or 21% of the (then) world total catch.

So far we have considered pelagic productivity and pelagic fish stocks. Most demersal or bottom living fish depend on pelagic primary productivity to sustain them, but in shallow areas (such as coral reefs or seagrass beds) benthic primary production by algae and higher plants forms an important additional food source. As we will find in the next section, primary productivity levels in such areas can be as high or higher than in upwelling pelagic systems, and can support a high fish production.

The actual yield of a fishery is dependent on many factors in addition to the ambient productivity levels, and is critically dependent on the methods and intensity of fishing. As fish grow, their efficiency of food conversion declines (as with humans) and old fish may consume large quantities of food but grow very little if at all. The food consumed by these old fish could be used to bring more small fish to adult size. Likewise, the older the fish at capture, the greater the proportion of the original stock which will have been lost to natural predators. Catching small fish will therefore tend to maximise the catch biomass. On the other hand, it is essential to allow a proportion of the new recruits to grow to reach sexual maturity and breed, and fish only become acceptable to the consumer above a certain size. The age/yield relationship is illustrated graphically in Figure 17.1. The total yield of any fishery is highly dependent on the 'effort' put in by the industry. Effort can be measured by the number of boats per unit area or by the man-hours spent in fishing. Assuming an initially untapped resource, the all-important relationship between effort and yield is shown in Figure 17.2. Initially, as fishing effort increases so will the yield, but as stock numbers are reduced, the yield stabilises and eventually declines as the stock reduces faster than it can be replenished. Figure 17.2 also shows the maximum sustainable yield (MSY) of the hypothetical fishery; increasing or decreasing effort away from this position on the curve would tend to reduce yields. In practice, the optimal yield for the fishery may not be at the MSY, for as Figure 17.2 shows, the return per unit effort will already have reduced by this point. The price of fuel sets a constant real cost for each fishing trip, and if the catch value is not substantially greater than the cost, the fisherman is forced to abandon his trade. Fishing at the MSY may also result in an unacceptably low average size for the catch. These factors form an important negative feedback system for fisheries, helping to prevent the complete extinction of stocks, but often taking effect too late to prevent the collapse of an economic fishery.

17.2. PROBLEMS OF FISHERIES IN CORAL SEAS

17.2.1. Limitations on productivity

Tropical waters are paradoxically among the most and least productive of any in the world. Production in warm oceanic waters is very low. Lewis (1977) has collected data on primary production for many tropical marine environments, and gives estimates between 20 and 50 g C $m^{-2}yr^{-1}$ for oceanic waters in the vicinity of reefs. Such low production sustains very small fishery yields, although some of the world's most spectacular fish such as marlin, sailfish and tuna are found in such waters. Except for these oceanic giants, tropical fisheries are limited to shallow shelf areas, especially in water depths of less than about 30m, where reefs and associated communities thrive. Primary productivity of coral reefs and seagrass beds is very high indeed, Lewis (1977) suggests between 300 and 5000 g C $m^{-2}yr^{-1}$, or about

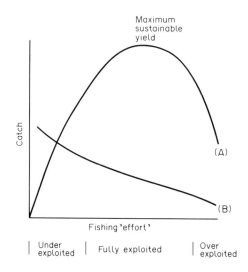

Fig. 17.2. Catch versus fishing effort for a hypothetical fishery. Curve A shows the total catch for the fishery, with a maximum sustainable yield at intermediate fishing effort. Curve B shows the same catch rates expressed as catch per unit effort, a measure of which could be the number of boats/km². Curve B shows that the return per boat declines steadily as the effort increases. Adapted from Odum (1980).

15 to 100 times the values in adjacent waters. Some of the reasons for this high production in the midst of such poverty have been outlined in Chapter 7, the key factors being the ability of the reef community to trap and recycle nutrients, and as has recently been recognised, to fix inorganic nitrogen in significant quantities.

Smith (1978) has estimated the potential global fish production on reefs and closely associated shelves to be about 6 million metric tonnes per year, or about 9% of the world fishery total. Reefs occupy only about 0.17% of the total ocean area (or about 15% of the area of shelf less than 30 m deep), so Smith concluded that reefs are indeed important fishery resources. Stevenson and Marshall (1974) noted that many reef areas support fish stocks 10—15 times greater per unit area than in the commercially important North Atlantic, and suggested that annual catches could approach 5 tonnes km^{-2}. In an important review, Munro (1983b) suggested that catches averaging 4—6 tonnes $km^{-2}yr^{-1}$ are possible over coralline shelves to 200m depth, or as much as 8—18 tonnes $km^{-2}yr^{-1}$ over actively growing coral at depths less than 8 m. The first figure is comparable to the world average potential production from coastal waters (3.3 tonnes $km^{-2}yr^{-1}$) calculated by Ryther (1969), the second is substantially higher. The actual yield for the heavily fished reefs of Apo Island in the Philippines average 11.4 tonnes $km^{-2}yr^{-1}$, but the income of the individual fisherman is very low (Alcala and Luchavez, 1981). Estimates for the Jamaican shelf and offshore banks (Munro, 1983a) suggest yields of 3.7 tonnes $km^{-2}yr^{-1}$ for the highly overfished north coast and only 0.76 tonnes $km^{-2}yr^{-1}$ for the underfished offshore Pedro Bank.

Apart from the factor of fishing effort, the achievable production of any tropical shelf area is heavily dependent on the nature of the substrate, especially the proportion of shallow water and the proportion covered by living coral. The use of an 'edaphomorphic index' (Saila and Roedel, 1980; Marshall, 1981) has been proposed to enable the potential yield of an area to be estimated from a study of the substrate character, this approach has successfuly been applied in lakes.

In many reef areas, including the Red Sea, reefs are narrow linear structures often plunging to abyssal depths on their windward slopes, with very little shallow shelf developed. While the productivity per unit area of reef and shelf is high, the total area available for fishing can be very low. To take a Red Sea example, the Sudan has a total potential fishing area of 9000 km^2 (Sanders and Kedidi, 1981), but has a coastline 700 km long, so the shelf averages only 12.8 km wide. The country has a population of 20.6 million, so the *per capita* fishing area is only 0.04 hectares, for which Sanders and Kedidi's production potential estimates suggest a potential annual *per capita* yield of only 305 g. Present production figures are only one sixth of this.

17.2.2. Overfishing and its prevention

Overfishing is recognisable in the small average size of fish in the catch, with many specimens sexually immature, in the scarcity of predators and the high proportion of herbivorous species, and by a decreasing or stable yield despite increasing fishing effort. The problem is compounded in reef areas by the nature of the environment and the fisheries. Coral reefs and adjoining shelves are morphologically complex, with rapidly changing depths, and are quite inappropriate for exploitation by large vessels using trawls or nets. Reef areas are ideal for use by one- or two-man boats equipped with sails or outboard engines, fishing with handlines, pots or spearguns. Small boats with shallow draught can safely penetrate deeply into reef systems, and fish in shallow waters. Such small boats are however very restricted by weather and wave conditions, forcing fishermen to concentrate on relatively few safe sites and further restricting the available fishery area during at least part of the year. Heavy fishing pressure on one site can seriously deplete stocks of key species, because many reef fish are territorial and do not disperse rapidly outside a small area (Russell, 1977). This seems particularly characteristic of the groupers (Serranidae), amongst the best eating of reef fish, but is also well marked in the snappers, wrasse and triggerfish. Even apparently pelagic fish such as jacks (*Caranx* spp.) seem to haunt individual small reefs for long periods rather than dispersing freely over large areas. Heavily fished reefs can often by recognised immediately by experienced divers from the scarcity of large snapper and medium sized grouper (Craik, 1981). If fishing pressure is reduced, individuals of these species will be replaced by juveniles, but the process is relatively slow.

The effects of overfishing on reef communities may be considerable but are not clearly documented as yet. Polunin and Frazier (1974) suggest that removal of large predators may lead to a general reduction in species diversity. As overfishing becomes more serious, other types of fish will become increasingly scarce. Woodley (1979) has suggested that reduction of part of the fish fauna could lead to important changes of abundance in other taxa, including territorial damselfish, algae, corals and echinoids. Fish have recently been identified as important controllers of coral zonation on some reefs (Wellington, 1982). It could be predicted that reduction of numbers of algal-grazing parrotfish and surgeonfish could seriously affect the balance between coral and algal cover, especially on reefs also subject to some degree of urban pollution and eutrophication.

Overfishing is in theory preventable by voluntary or legal restriction of fishing effort. In practice, conservation of fish stocks has had a chequered career throughout the world (Cushing, 1975b), especially when the regulatory bodies are intergovernmental. Managing reef fisheries is essential to keep healthy stock levels (Munro, 1983b) but is particularly difficult for two reasons. Firstly, reef fisheries are multi-species, not single-species industries. As the stocks of prime large predatory fish such as snappers and groupers are reduced to uneconomic levels, instead of causing abandonment of fishing as would happen in a single-species fishery, fishermen continue to derive a useful income from catches of other species, while still landing the now rare prime fish. Vulnerable fish species are therefore denied the built-in negative feedback protection of a single-species fishery and can be fished-out to oblivion long before the overall fishery goes into decline. Secondly, most reef fisheries are artisanal in nature, the fishermen are poor, many work only part time at the industry and are unregistered. In these conditions legal enforcement of catch or gear restrictions is almost impossible. Viable conservation options include prohibition of fishing in nursery areas and provision of tempting financial subsidy of approved fishing gear. Monitoring of catches and observance of size limits can be assisted by providing centralised marketing facilities and controlling market prices. In some small Pacific island communities where fishing provides the main source of food, complex taboo systems evolved which very effectively regulated the exploitation of stocks and prevented overfishing (Fosberg, 1973). Such cultural controls seem never to have existed in the Red Sea, where the low density of habitation served adequately to prevent overfishing.

17.2.3. Problems of marketing

Reef fisheries are necessarily situated in hot tropical countries most of which are classed as developing nations. In small island nations, where settlement is mainly coastal and fish is sold locally, transport and marketing present no problems. This is not the case in the Red Sea, where the majority of the inhabitants of the bordering countries live inland at considerable distance from the coast. Getting fresh fish to the consumer in good condition can present almost insuperable difficulties. Increasing the consumption of marine fish is therefore not merely a question of increasing the catch. Much more importance lies in the logistic requirements of ice factories, adequate cold storage, and efficient communication routes.

Traditionally a proportion of the fish catch from the Red Sea has been sun-dried, especially in the case of small 'sardines' in the south-eastern coastal region used as a source of human and camel food (Bertram, 1959). Dried fish is not dependent on efficient fast transport for marketing and offers important possibilities for supplying often poorly nourished inland communities. Mastaller (1981) has reviewed the possibilities and problems of low-technology fish drying in the Sudan. At present the quality is low as a result of massive maggot and dermestid beetle attack on the drying flesh. Mastaller designed and tested some very simple solar fish-driers made from locally available materials and demonstrated that considerable increases in efficiency of drying could be achieved, resulting in much reduced contamination and a more acceptable product. Such cheap and simple contrivances could revolutionise the fishing industry in inaccessible coastal areas.

17.2.4. Ciguatera and poisonous fish species

Toxicity in its various forms seems to be a relatively common occurrence in reef inhabitants, and may be related to a comparatively sedentary and territorial way of life (Cameron, 1976). A small number of reef fish species are always dangerously toxic when eaten, other species may be harmless or toxic on occasion through the poisoning phenomenon called ciguatera. Several common Red Sea fish contain toxins in their skin or tissues, and the subject has been extensively reviewed by Halstead (1967). Tissue poisons are best known in puffer and porcupine fish of the family Tetraodontidae, which contain the poison Tetrodotoxin in their liver, gonad and nervous tissue. Remarkably, puffer fish are eaten as a delicacy in Japan, where they are known as *fugu*. *Fugu* can only legally be prepared by highly trained *fugu* chefs skilled at the removal of all dangerous tissues, but adventurous Japanese frequently elect to consume small quantities of liver to experience mild degrees of poisoning. Other poisonous types include filefish and boxfish, which, in common with other species, produce a poisonous secretion from the skin. The larger poisonous fish are well known and avoided by fishermen, but as Cameron (1976) points out, the toxicity of some small types makes the factory processing of large quantities of indiscriminately collected reef fish for meal or paste a potentially dangerous practice.

Banner (1976) has provided an interesting review of the problem of ciguatera poisoning, in which normally wholesome species become toxic on an apparently random basis. Ciguatera poisoning results from the toxic secretions of the dinoflagellate *Gambierdiscus toxicus*, cells of which can be found on the benthic red alga *Spyridia filamentosa* (Shimizu et al., 1982). Toxicity is therefore commonly found in herbivorous fish, especially parrotfish and surgeonfish, but the toxin can be accumulated by predators such as snapper, barracuda, jacks and groupers, unfortunately the best eating of the reef fish. Ciguatera poisoning, which can be lethal in severe cases, is widespread in the Pacific and Caribbean, but seems not to occur in the Red Sea, although the alga *Spyridia* occurs there.

17.3. THE RED SEA FOOD FISHERIES

17.3.1. The artisanal fisheries

At present the Red Sea fisheries are overwhelmingly artisanal in nature, conducted by individuals or small groups of fishermen, working in shallow waters close to shore using small boats. 70% of the present catch comes from such fisheries, and even with the full development of potential trawling and purse seining grounds artisanal reef fisheries would account for more than half of the projected total yield (Sanders and Kedidi, 1981).

Fig. 17.3. Artisanal spear-fisherman in Sudanese waters, using a sharpened steel reinforcing rod to catch reef fish, in this case the abundant surgeonfish *Acanthurus sohal*.

Artisanal fishing methods vary somewhat from country to country, but are dominated by handlining which is cheap and effective. Sardines and sardinellas are used as bait, and are caught in coastal lagoons using hand cast-nets. Skilled hand-line fishermen can land more than 8 kg fish per hour. If a shoal of snapper can be located, catch rates can become prodigious, and I have assisted a team of 6 people to fill a one tonne ice chest in only two hours of fishing in southern Sudanese waters. When the catch rate is high however, fishing at any one site is usually eventually terminated by the arrival of shark which steal the fish as they are being hauled in (Reed, 1964).

Other methods used by artisanal fishermen include trolling and various forms of net fishing. Trolling involves towing a large artificial lure through the water behind a boat, and is almost always used during journeys to and from the hand-lining sites. Trolling catches relatively few fish, but these are usually large and command a high price; trolling is a particularly good way of catching jacks and barracuda. In the Red Sea, nets are used on a relatively small scale because of their high cost and the problems of snagging and tearing over coral bottoms. The types of gear employed are well described by Neve and Al Aiidy (1973) for the Saudi Arabian fishery. The main type used is the gill net, with a large mesh in which fish become entangled in the gill area. Gill netting is probably the second most important technique after hand-lining, and the big nets, up to 50 m long and 2 m deep, are set across reefs or the entrances of creeks and marsas, and are often left overnight. Beach seining was once used very successfully in Ethiopia; one end of a long net is carried out by boat from a gently shelving shore, looped round to enclose a volume of water, and brought back to shore. Both ends are hauled in on land, an extremely arduous task requiring the efforts of a large number of fishermen working as a team. The Ethiopian fishery once yielded some 25,000 tonnes annually by this method (Ben Yami, 1964). Other special net types are used for special purposes, such as catching shrimp or mullet. Limited use is made (Fig. 17.3) of hand spears, usually made of concrete-reinforcing steel with sharpened ends, and a few fishermen have equipped themselves with rubber-powered spearguns. In either case, considerable expertise is called for, and as a minimum the fishermen needs a face mask, preferably with snorkel and fins, items difficult to obtain in many Red Sea countries. Spiny lobsters (*Panulirus* spp.) are caught by snorkel divers, or more commonly are speared at night by fishermen wading over shallow reef flats, and

TABLE 17.1. Main genera of fish caught by artisanal techniques.

English	Genus	Arabic	Catch %
Horse mackerel	*Trachurus, Decapterus*	—	
Indian mackerel	*Rastrelliger*	Bagha	21.7
Spanish mackerel	*Scomberomorus*	Derak	
Jacks and	*Caranx*	Bayad	
Trevally	*Trachinotus*	Teeman	4.79
	Alectis	Shawish	
	Epinephelus	Tauwina	
Groupers and	*Variola*	Louti	
coral trout	*Cephalopholis*	Kut rouban	4.04
	Plectropomus	Najil	
	Lutjanus	Bohar, safin	
Snapper	*Pristipomoides*	Koreib	3.72
	Aprion	Farsi	
Emperor	*Lethrinus*	Sha'oor	3.53
	Valamugil		
Mullet	*Mugil*	Arabi	1.66
	Crenimugil		
	Argyrops	Fofal	
Bream	*Sparus*	—	
	Acanthopagrus	Abu kohol	0.27
	Mylio	—	

Data summarised from Sanders and Kedidi (1981). Arabic names (Sudanese) from Reed (1964). Percentages refer to the whole catch, the categories above account for nearly 40% of the total catch.

carrying paraffin lamps. Small numbers of crab, octopus and cuttlefish are caught in the same way. Fish potting, the dominant form of fishing in many reef areas seems to be little used in the Red Sea, perhaps because of the limited durability of traditionally available materials.

Table 17.1 lists the main types of fish caught by the artisanal industry, and some representative types are illustrated in Figures 17.4 and 17.5. Reed's (1964) account of the Sudanese industry makes good reading and provides useful information on size ranges, price and palatability, and capture techniques. The pelagic species are dominated in the catch by small sardines, anchovies and sardinellas, and the large horse and spanish mackerel. The most important fish caught by handlining are the groupers (mainly species of *Epinephelus*), snapper (*Lutjanus* and other genera), emperors (*Lethrinus*) and jacks (*Caranx*). All are excellent eating and command good prices in local markets. Table 17.1 lists only the most abundant categories of fish recorded by Sanders and Kedidi (1981), but together these add up to only about 40% of the total catch. The rest is made up of a plethora of large and small species of lesser individual abundance, some of which are highly esteemed and expensive.

17.3.2. Purse seining

Purse seining is an increasingly popular form of fishing in which a long, deep net is paid out behind a vessel which steams in a circle. Both ends of the net are recovered, and the circular form of the set net encloses a large volume of water and fish. A rope runs along the bottom of the net, and when this is tightened, the net closes like a purse, trapping the fish inside. The net is slowly hauled in with the fish, or until the fish are trapped in a small volume of water and can be gaffed. Purse seining is widely used to trap schools of large pelagic fish, and relies on the skill of the skipper in locating and surrounding large schools.

At present purse seining in the Red Sea is only conducted by the Egyptians, who have a large industry based in the Gulf of Suez, and a smaller one in Foul Bay south of Berenice, close to the Sudanese border. According to Sanders and Kedidi (1981), annual landings (1979–1980) are about 16,500 tons, although this does not seem consistent with FAO data quoted later in this chapter. Awadallah (1982) has completed a detailed study of the Egyptian purse seine industry. A total of 84 boats are employed, mostly operating in the Gulf of Suez. The fishery is very seasonal, and lasts from September to May, with maximum catches in November, after which the fishery rapidly declines as stocks become exhausted.

In the Gulf of Suez the catch consists mainly of Horse mackerel and Indian mackerel (65%), with some 20% round herring and 5% sardines. Most of the Foul Bay catch consists of sardinella (55%), and mackerel species make up the remainder.

17.3.3. Trawling

Trawling requires a powerful vessel capable of pulling a heavy weighted net across the sea bottom. Trawlers require fairly shallow sea bottoms which are flat and offer no obstacles in which the net can become entangled. These restrictions rule out much of the Red Sea shelf over which a rough and

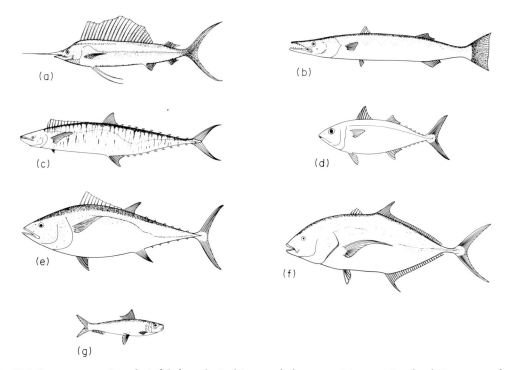

Fig. 17.4. Some representative pelagic fish from the Red Sea, caught by purse seining or artisanal techniques except for the sailfish which is a popular sport fish. A. Sailfish *Istiophorus*, B. Barracuda *Sphyraena*, C. Spanish Mackerel *Scomberomorus*. D. Indian Mackerel *Rastrelliger*, E. Tuna *Gymnosarda*, F. Jack *Caranx*, G. Sardinella *Sardinella*. After Reed (1964).

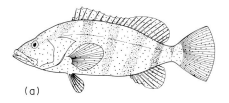

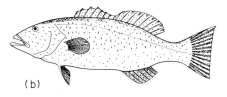

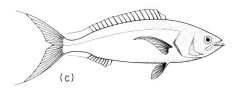

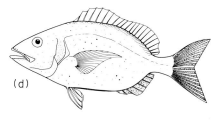

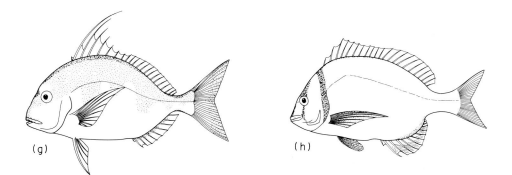

Fig. 17.5. Representative demersal reef fish caught by artisanal techniques A. Grouper *Epinephelus*, B. Coral Trout *Plectropomus*, C and D Snapper *Aprion* and *Lutjanus*, E. Emperor *Lethrinus*, F. Mullet *Mugil*, G and H Bream *Argyrops* and *Acanthopagrus*. After Reed (1964).

irregular reef dominated terrain is developed. At present only the Egyptians operate a trawling industry in the Red Sea, although in the 1970s a small industry existed in Ethiopia. The Egyptian industry operates within the same grounds as the purse seiners, and during the same times of year. Annual landings are about 5,500 tonnes from the Gulf of Suez, and 600 tonnes from Foul Bay (Awadallah, 1983). Most of the catch consists of lizard fish (*Saurida*), snapper and shrimp, the latter making up 13% by weight of the catch (Sanders and Kedidi, 1981). Surveys suggest that the limited trawling area in the northern Red Sea is already fished to close to its maximum capacity, but that there are considerable untapped resources in the southern Red Sea.

17.3.4. Production by the fisheries

An analysis of annual catch by the three methods is given in Table 17.2 for 6 Red Sea nations. Data are included for Ethiopian fisheries before their recent collapse, which has been caused by political rather than biological events. More recent information can be obtained from the invaluable Yearbooks of Fishery Statistics, published annually by the FAO. For each year two volumes are published, the first (e.g. FAO, 1984a) detailing catches and landings, the second (e.g. FAO, 1984b) covering international trade in fish products. Table 17.3 summarises data for 1982, and shows striking differences in production, trade and consumption by the seven Red Sea nations.

The Egyptian fishery production is more than four times larger than that of any other Red Sea nation, but this is made up largely of catches within the Mediterranean, and the prodigious inland fisheries of the Nile. The total Sudanese production is second largest, but again all but a trivial proportion comes from inland fisheries, as is the case with Israel. The Saudi Arabian Indian Ocean fishery is the largest, but more than half the catch comes from the more fertile waters of the Arabian Gulf. At present the largest Red Sea fishery is that of North Yemen, although Ethiopia until recently exceeded this total.

Egypt, Saudi Arabia, Jordan and Israel are major importers of fish products, purchasing as much or more tonnage than their own fleets produce, and no country yet exports significant quantities. Even the Sudanese, who export 50% more fish than they import, make a net loss of US $270,000 on the exchange. Only North Yemen presently satisfies a high proportion of its fish consumption from Red Sea sources.

Net consumption figures shown in Table 17.3 range from 3850 tonnes in Ethiopia to 227,467 tonnes in Egypt, but net *per capita* consumption is more relevant. It must be noted that the data quoted include fish meals used in animal feedstock, so the *per capita* figures may not exactly reflect direct human consumption. The stark poverty of Ethiopia is demonstrated by the present consumption of only 100 g

TABLE 17.2. Red Sea fishery production data

Country	Artisanal	Purse seining	Trawling
Egypt	2,300	16,500	4,500
Sudan	600	—	—
Ethiopia	(25,000)		(1,500)
North Yemen	13,500	—	—
Saudi Arabia	10,200	—	—
Jordan	100	—	—
TOTAL	51,700	16,500	6,000

Summary of production by class of fishery for the six Red Sea countries surveyed by Sanders and Kedidi (1981). Note that the data are averaged over 'recent years' and so do not agree exactly with the FAO data of Table 3. The data for Ethiopia refer to the late sixties before the collapse of the Ethiopian fishing industry. Data in metric tonnes.

fish per person in 1982, while Saudi Arabia and Israel consumed 7.0 and 11.2 kg per person during the same year, sustained by massive importation. These figures are still relatively low by world standards. Nearby South Yemen and Oman, which enjoy an important upwelling system off their coast, consumed 32.3 and 77.9 kg fish per person in 1982, almost all from their own production. The British *per capita* consumption in 1982 was 25.2 kg, that of the USA 19.8 kg, in both countries butcher's meat is now more highly regarded than fish. The Japanese consumption was 93.8 kg per person, mainly caught by their huge international fleet, the largest in the world.

Figure 17.6 shows trends in Red Sea fishery production over the last ten years for the main fishing nations, excluding Saudi Arabia, whose Red Sea and Arabian Gulf catches are not separated in FAO statistics. During the period in question, Saudi Indian Ocean landings have been fairly stable at about 23—26,000 tonnes, while in 1978 the Red Sea catch was 10,200 tonnes or about 40% of the total. It can be seen from Figure 17.6 that production by North Yemen and Egypt has risen fairly steadily, although Egyptian production has only recently recovered to the levels enjoyed in 1965 (14,800 tonnes: FAO, 1973). Ethiopian production collapsed from 25,800 tonnes in 1972 (FAO, 1973), falling to zero in 1979, and now very slowly climbing back. Data for Ethiopia are only FAO estimates however, and the real production levels are unknown.

17.3.5. Potential for increased production

The present and potential Red Sea fisheries have been reviewed by Sanders and Kedidi (1981), and a summary of their conclusions is given in Table 17.4. The biggest potential lies in the more productive southern Red Sea waters, where North Yemen could probably triple its present catch, and Ethiopia could more than double its high production of the late 1960's and early 1970's. The artisanal fishery potential of Saudi Arabian waters is already well exploited, but there appear to be considerable resources for trawling and purse seining, at present commercially untouched. The country also has the material

TABLE 17.3. 1982 catch, import and export data for Red Sea countries, from FAO (1984a,b).

		Egypt	Sudan	Ethiopia	North Yemen	Saudi Arabia	Jordan	Israel
Catch	Red Sea and western Indian Ocean	13,386	1,050	350	22,000	26,425	19	76
"	Other marine areas	11,208	—	—	—	—	—	10,118
"	Inland	112,614	28,660	3,400	—	—	—	13,485
"	Total	137,208	29,710	3,750	22,000	26,425	19	23,679
Imports		90,642	400	100	3,130	48,752	4,630	19,670
Total input		227,850	30,110	3,850	25,130	75,177	4,649	43,349
Total exports		383	600	—	—	2,682	—	460
Net consumption		227,467	29,510	3,850	25,130	72,495	4,649	42,889
Population (millions)		49.5	20.6	31.3	5.7	10.4	3.6	4.1
Per capita consumption (kilograms per year)		4.6	1.4	0.1	4.4	7.0	1.3	11.2
% consumption local origin		55.4	100.0	97.4	87.5	36.5	0.4	31.6
% consumption from Red Sea		5.9	3.6	9.1	87.5	14.1	0.4	0.2

All data are in metric tonnes except where stated. Catch figures for Saudi Arabia include both Red Sea and western Indian Ocean (Arabian Gulf) tonnage, the Red Sea catch has been estimated at 10,200 tonnes (Sanders and Kedidi, 1981), and this figure is used to calculate the percentage of consumption derived from Red Sea fisheries.

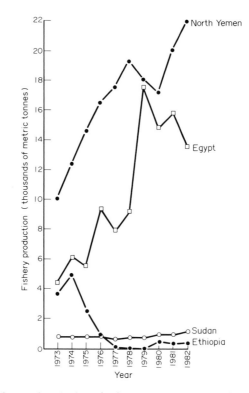

Fig. 17.6. Recent trends in Red Sea Fishery landings for four major nations. Data from FAO (1984a). Ethiopian figures are FAO estimates.

wealth to invest in the relatively expensive boats and gear necessary for these types of fishing. Shrimp trawling probably represents the greatest potential untapped fishery resource in North Yemen. Sustainable catch rates as high as 93 tonnes shrimp per vessel per year have been predicted, with the bonus of a by-catch of some 73 tonnes of saleable fish (Campleman et al., 1977; Walczak, 1977).

Sudan could expand its present very small production ten fold, but as we have already seen, is held back by communication difficulties. Even realising its full production potential from the sea could only add about 280 g fish to the national per capita consumption, the Nile fisheries will always be of greater significance. Egypt, more than any other Red Sea nation has experience in modern fisheries, but would appear from the data of Table 17.4 to be already very close to its potential production limits. The future for both Egypt and Sudan, which have excellent internal water supplies, must lie with further development of inland fisheries and aquaculture. Finally, neither Israel nor Jordan have sufficient coastline of their own to constitute a significant resource. Israel has instead invested heavily in inland fisheries production and large fishing vessels operating in the Mediterranean and south-east Atlantic.

17.4. MINOR NON-FOOD FISHERIES

17.4.1. The aquarium fish trade

Tropical marine aquarium keeping has become a popular if expensive pastime in the wealthier western nations, and a substantial industry has grown up in these countries and in tropical nations to supply the trade with living organisms. Almost all fish offered for sale are wild-caught, the breeding success in captivity is still low. While the larger and prettier fish command the highest retail prices, there is also a substantial demand for living coral, gastropods, tube worms, echinoderms and even 'live' rock and

TABLE 17.4. Potential fisheries production in the Red Sea.

	Artisanal	Purse Seine	Trawling	Total
Egypt	3.3 (70%)	18.7 (88%)	4.6 (98%)	26.6 (88%)
Sudan	4.6 (13%)	1.5 (0%)	0.5 (0%)	6.6 (9%)
Ethiopia	50.0 (50%)	1.6 (0%)	7.4 (20%)	59.0 (45%)
North Yemen	22.0 (61%)	2.3 (0%)	7.4 (0%)	31.7 (43%)
Saudi Arabia	15.2 (67%)	11.0 (0%)	19.0 (0%)	45.2 (23%)
Jordan	negligible	0.0 —	0.0 —	negligible

Sanders and Kedidi (1981). Percentages in brackets indicate the proportion of the potential realised within 'recent years'. These catch estimates differ somewhat from the 1982 FAO dated quoted in Table 3, the catch data for Ethiopia in particular refer to 1960's levels. All figures are thousands of metric tonnes, and where Sanders and Kedidi indicated a range of potential catch estimates, only the highest estimate is included here. Israeli fish production from the Red Sea is negligible.

sand. Retail prices of larger reef fish can be very high, exceeding US $100, although the return to the original fisherman is only a minute proportion of this. In some tropical countries the aquarium trade has become a very big business. Albaladejo and Corpuz (1981) have provided an interesting review of the trade in the Philippines. From its beginnings in 1970, the trade has grown to an annual (1979) value of 20 million pesos or about US $1.85 million, and is within the top ten exported fishery products.

The aquarium industry in the Red Sea offers considerable potential, for suitable reef fish abound in numbers and diversity, but there are several problems, practical and environmental. The essential pre-requisite is an efficient international airport within rapid access of the sea, so that the fish, individually packed in plastic bags under a layer of pure oxygen, and sealed in insulated boxes, can be rushed to their destination with minimal delay. Unfortunately, even with experienced operators and good communications, a high proportion of fish arrive dead at their destination. Communications are excellent from Jiddah, but from elsewhere around the coast the connections to potential markets are still rather indirect.

A serious concern is the ecological damage which can be caused by heavy collection, which naturally tends to concentrate on reefs close to airports, the same reefs most in demand for the tourist industry. Gill nets are often used to catch the fish, and these tangle with and break coral, and often kill more fish than they capture alive. Cyanide is still (illegally) used in the Philippines, and is a serious source of damage to reefs and fishermen alike. Another damaging technique is to frighten small fish into the shelter of a branching coral colony, then detach the coral and break it up within a bucket to release the fish. These collecting methods damage reefs directly, while the removal of certain key species may have serious long term effects on reef ecology (Lubbock and Polunin, 1975). We still know relatively little of the recruitment rates of Red Sea reef fish. Gundermann and Popper (1975) followed fish recolonisation after a poison spill killed all resident fish over a 600m strip of reef at Elat, northern Gulf of Aqaba. This study suggested complete community re-establishment within 12 months, probably because the affected area was small relative to the potential recruitment stock. While this augurs well for reef recovery after heavy fish collection, the latter process would be likely to take place continuously over a large area of reef, and to concentrate on a few, often rarer species. The steady attrition of stocks of these species may have serious consequences on population regeneration.

It may be concluded that the aquarium fish business offers a substantial revenue in foreign exchange for a small capital outlay, but it is a sophisticated and potentially damaging exercise that would need tight controls to prevent deterioration of reef communities.

17.4.2. Coral and black coral

Wells (1981) has reviewed the world trade in coral and shells, which she estimates reached in 1978 no less than 1216 tonnes of coral and over 35,000 tonnes of shells, including mother of pearl. Again, the largest single supplier country has been the Philippines, prior to an export ban on corals introduced in

1977, and already being broken. The trade in coral is for jewelry (especially precious and black coral) and curios to gather dust on the collective mantlepieces of wealthy nations. The trade represents a welcome source of hard currency to developing tropical nations, although the return to the country of origin is a very small proportion of the retail price. Unfortunately, coral growth is slow, and a moderate sized brain coral taken for the curio market may be ten or more years old, and would take a comparable time to replace. Although branching corals grow more quickly, they afford important refuges to fish and invertebrates and so their removal from reefs is correspondingly serious to the community. Trade in precious and black corals has already seriously depleted stocks in certain areas (Wells, 1981) and could be very serious in the Red Sea where black coral exists, but is not abundant (*pers. obs.*). A management plan has recently been drafted for the coral reserves of Florida (Gulf of Mexico and South Atlantic Fishery Management Councils, 1982), and this estimates a maximum sustainable yield of only 140 kg of scleractinian corals per year (for scientific and educational use) and approximately 1400 gorgonian colonies from an area including most of the extensive Florida reef tract. While this yield may be regarded as conservative relative to the natural mortality rates on reefs, it emphasises the vulnerability of such habitats to exploitation.

There is at present no evidence of any large scale international trade in Red Sea corals, but collection by and for sale to tourists is common in many areas. Further discussion of this problem is found in Chapter 19.

17.4.3. Pearl shell fisheries

For many years there has been a minor trade in pearl shell from the Red Sea, based on the gastropods *Trochus dentatus* and *T. niloticus* and the bivalve *Pinctada margaritifera* or pearly oyster. The world trochus industry is probably worth US $6 million annually, with a turnover of about 6 thousand tonnes, and the trade has increased in volume and revenue considerably in recent years (Heslinga, 1981). The shell is exported to Europe and Asia for manufacture of pearl buttons. Serious stock depletion has occurred in heavily exploited areas like Palau, and artificial rearing and release of juveniles is being attempted (Heslinga, 1981).

No recent data are available on the Red Sea trochus trade, which seems to be concentrated on the western shores. Astonishingly malodorous warehouses exist on the Sudanese coast where shells are amassed and cleaned before export. The warehouses probably serve a small local trade, but shells are also collected from large vessels called *sambûk* (Fig. 17.7) using dug-out canoes or *houri* (Fig. 17.8) to cover the reefs. The vessels set out from North Yemen or from ports outside the Red Sea laden with bread and cooking oil, which, supplemented with fish, is slowly consumed by the fishermen as they forage for shell, often for several weeks or months. As the food is consumed, so the empty space becomes filled with decaying uncleaned gastropods, the undelicate scent of which is detectable several miles downwind of an approaching *sambûk*.

The pearl oyster occurs naturally but in small numbers along the Red Sea, where in historical times they were a source of pearls, replaced eventually by better quality pearls from the Arabian Gulf. In 1904, Cyril Crossland, under the direction of Professor Herdman, visited several Red Sea sites before returning temporarily to Britain in 1905 with an extensive and important collection of organisms eventually described by several authors. Crossland observed the small native pearl shell fishery in Dungunab Bay, Sudan, and returned there later in 1905 to initiate an oysterculture industry which occupied him for 12 years, and survived fairly successfully until 1969. Crossland published a very complete account of the physical conditions in Dungunab Bay (Crossland, 1911), and his manuscript notes on the fishery were eventually published posthumously (Crossland, 1957).

Oyster culture at Dungunab was conducted conventionally using separate spat collector rafts to collect settling larvae, nursery trays and growing trays made of wire mesh. In 1969, a massive mortality hit the industry, followed by a second collapse in 1973, from which the industry has not fully recovered. Nasr (1982) studied the many possible causal factors for the catastrophic mortality, and was unable to come to any definite conclusion. Unidentified spherical bodies in the tissues of dying oysters pointed to a possible parasite, and research into the problem is continuing.

The larger gastropods are used by fishermen as bait, and some are eaten as food, although conch is not as popular in the Red Sea as in the Caribbean, where it is highly prized as a delicacy. The opercula of conch, especially of *Strombus tricornis* are used as the base for perfume manufacture, and are sold as *dufra*. According to Schroeder (1981) sales of *dufra* in the Port Sudan market average about 13 pounds per day, and each pound contains the opercula of 1000–2000 animals. Most of the *dufra* is imported from the Arabian Gulf or from India, but depletion of local stocks must be ocurring to some extent.

Fig. 17.7. *Sambûk* in full sail, Sudanese Red Sea. Note the dug-out canoe or *houri* tied to the deck.

Fig 17.8. Dug-out canoe or *houri* used by individual fishermen to penetrate deep into reef complexes. The craft is moored by a stone in very shallow water while the owner collects trochus shell.

17.5. CONCLUSIONS

The Red Sea does not, and can never support a major fishing industry like that of the Arabian Sea and Gulf of Aden exploited by South Yemen and Oman. The fisheries are however relatively underexploited at present, except in Egypt, and could eventually yield up to 3.5 times the present catch of edible fish. The impact of realising the full marine fishery potential of Egypt or Sudan would be almost imperceptible, because both countries have vastly more important inland fisheries. Likewise, neither Jordan nor Israel could benefit significantly, because their coastlines are too small. Production in Saudi Arabia, already the second highest consumer of fish in the region, could be increased by some 45%, but the proportion of fish derived from the Red Sea would still be small compared with that from the Arabian Gulf and from importation. North Yemen currently enjoys a very successful Red Sea fishing industry, with a high *per capita* consumption and a low (12%) import rate. North Yemeni production could probably be doubled, and fish exports could contribute to reducing her current serious balance of trade deficit. At present however, the country that stands to benefit most from increased Red Sea fishery production is Ethiopia, currently torn by internal conflict, drought and famine. The Ethiopian marine fishery reserves are probably the largest in the Red Sea, and if fully exploited could produce more than twice their previous maximum annual output, providing about 1.8 kg fish *per capita* every year, a small but significant quantity of great value to the undernourished population.

REFERENCES

Albaladejo, V. P. and Corpuz, V. T. (1981). A market study of the aquarium fish industry of the Philippines : an assessment of the growth and the mechanics of the trade. *Proc. 4th Int. Coral Reef Symp.* 1, 75—81.
Alcala, A. C. and Luchavez, T. (1981). Fish yield of the coral reef surrounding Apo Island, Negros Oriental, central Visayas, Philippines. *Proc. 4th Int. Coral Reef Symp.* 1, 69—73.
Awadallah, M. W. (1982). *An incomes and costs study of the Egyptian Seine Fishery operated in the Gulf of Suez and off the Southern Red Sea coast during 1980-1981.* Project for the development of fisheries in areas of the Red Sea and Gulf of Aden. RAB/81.002/2.

Awadallah, M. W. (1983). *An incomes and costs study of the Egyptian Trawl Fishery operated in the Gulf of Suez and off the Southern Red Sea coast during 1980-1981.* U.N.D.P./F.A.O. Project for the development of fisheries in areas of the Red Sea and Gulf of Aden. RAB/81/002/4.

Banner, A. H. (1976). Ciguatera : a disease from coral reef fish. In *Biology and Geology of Coral Reefs*. Ed. O. A. Jones and R. Endean, Vol III, Biology 2, pp.177–213, Academic Press, New York.

Ben Yami, M. (1964). *Report on the Fisheries of Ethiopia*. Report to the Ministry of Foreign Affairs, Israel.

Bertram, C. (1959). *Adam's Brood. Hopes and Fears of a Biologist*, Peter Davis, London.

Cameron, A. M. (1976). Toxicity of coral reef fishes. In *Biology and Geology of Coral Reefs*. Ed. O. A. Jones and R. Endean. Vol III, Biology 2, pp.155–76, Academic Press, New York.

Campleman, G., Perovic, Y. and Simons, B. (1977). *Fisheries and marketing in the Yemen Arab Republic.* IOP/TECH/77/13. F.A.O./U.N.D.P. Rome.

Clark W. G. (1978). Dynamic pool models. In *Models for Fish Stock Assessment*. FAO Fisheries Circular 701.

Craik, G. J. S. (1981). Underwater survey of coral trout *Plectropomus leopardus* (Serranidae) populations in the Capricornia section of the Great Barrier Reef Marine Park. *Proc. 4th Int. Coral Reef Symp.* 1, 53–8.

Crossland, C. (1911). A physical description of Khor Dongonab, Red Sea. *J. Linn. Soc.* 31, 265–86.

Crossland, C. (1957). The cultivation of the mother-of-pearl oyster in the Red Sea. *Austr. J. mar. Freshwater Res.* 8, 111–13.

Cushing, D. H. (1975a). *Marine Ecology and Fisheries*. Cambridge University Press. Cambridge.

Cushing, D. H. (1975b). *Fisheries Resources of the Sea and Their Management*, Oxford University Press, Oxford.

FAO (1973). *Yearbook of Fishery Statistics. Catches and Landings*. Vol. 34. Food and Agriculture Organisation of the United Nations, Rome.

FAO (1984a). *Yearbook of Fishery Statistics. Catches and Landings*, Vol. 54, Food and Agriculture Organisation of the United Nations, Rome.

FAO (1984b). *Yearbook of Fishery Statistics. Fishery Commodities*. Vol. 55, Food and Agriculture Organisation of the United Nations, Rome.

Fosberg, F. R. (1973). Past, present and future problems of Oceanic Islands. In *Nature Conservation in the Pacific*. Ed. A. B. Costin and R. H. Groves, pp. 209–15, Australian National University Press, Canberra.

Gulf of Mexico and South Atlantic Fishery Management Councils (1982). *Fishery Management Plan for Coral and Coral Reefs of the Gulf of Mexico and South Atlantic.*

Gundermann, N. and Popper, D. (1975). Some aspects of recolonisation of coral rocks in Eilat (Gulf of Aqaba) by fish populations after poisoning. *Mar. Biol.* 33, 109–17.

Halstead, B. W. (1967). *Poisonous and Venomous Marine Animals of the World*, Vol. 2, U.S. Government Printing Office, Washington.

Heslinga, G. A. (1981). Growth and maturity of *Trochus niloticus* in the laboratory. *Proc. 4th Int. Coral Reef Symp.* 1, 39–45.

Lewis, J. B. (1977). Processes of organic production on coral reefs. *Biol. Rev.* 52, 305–48.

Lubbock, H. R. and Polunin, N. V. C. (1975). Conservation and the tropical marine aquarium trade. *Environ. Conserv.* 2, 229–32.

Margalef, R. (1967). El ecosistema. *Ecologia marina, Monografia 14. Fund. La Salle de Ciencias Naturales Caracas* 377–453.

Marshall, N. (1981). Exploring the applicability of the edaphomorph index concept to estimating the fisheries potential of coral reef environments. *Proc. 4th Int. Coral Reef Symp.* 1, 112.

Mastaller, M. (1981). Feasibility and evaluation of methods for drying reef fishes. *Proc. 4th Int. Coral Reef Symp.* 8, 105–9.

Morcos, S. A. (1970). Physical and chemical oceanography of the Red Sea. *Oceanogr. mar. Biol. A. Rev.* 8, 73–202.

Munro, J. L. (1983a). Coral reef fish and fisheries of the Caribbean Sea. In *Caribbean Coral Reef Fishery Resources*. Ed. J. L. Munro, pp.1–9, ICLARM, Manila.

Munro, J. L. (1983b). Epilogue: progress in coral reef fisheries research 1973-1982. In *Caribbean Coral Reef Fishery Resources*. Ed. J. L. Munro, pp. 249–65, ICLARM, Manila.

Nasr, D. H. (1982). Observations on the mortality of the pearl oyster *Pinctada margaritifera* in Dongonab Bay, Red Sea. *Aquaculture* 28, 271–81.

Neve, P. and Al-Aiidy, H. (1973). The Red Sea fisheries of Saudi Arabia. *Bull Mar. Res. Centre Saudi Arabia* 3. Marine Research Centee, Jeddah.

Odum, W. E. (1980). Utilisation of aquatic productivity by Man. In *Fundamentals of Aquatic Ecosystems*. Ed. R. S. K. Barnes and K. H. Mann, pp. 143–61, Blackwell, Oxford.

Polunin, N. V. C. and Frazier, J. G. (1974). Diving reconnaisance of 27 Western Indian Ocean coral reefs. *Environ. Conserv.* 1, 71–2.

Raymont, J. E. G. (1980). *Plankton and Productivity in the Oceans*, Vol. 1, *Phytoplankton*, 2nd Ed., Pergamon Press, Oxford.

Reed. W. (1964). *Red Sea Fisheries of Sudan*. Government Printing press, Khartoum.

Russell, B. C. (1977). Population and standing crop estimates for rock reef fishes of north-eastern New Zealand. *N.Z. J. Mar. Freshwater Res.* 11, 23–36.

Ryther, J. H. (1969). Photosynthesis and fish production in the Sea. *Science N.Y.* 166, 72–6.

Saila, S. B. and Roedel, P. M. (1980). *Stock Assessment for Tropical Small Scale Fisheries*. Int. Cent. Mar. Res. Dev., University af Rhode Island, Kingston.

Sanders, M. D. and Kedidi, S. M. (1981). *Summary review of Red Sea commercial fisheries catches and stock assessments including maps of actual and potential fishing ground.* U.N.D.P/F.A.O. Project for the development of fisheries in areas of the Red Sea and Gulf of Aden. RAB/77/008/19.

Schroeder, J. H. (1981). Man versus reef in the Sudan: threats, destruction, protection. *Proc. 4th Int. Coral Reef Symp.* 1,

253–7.

Shimizu, Y., Shimizu, H., Scheuer, P. J., Hokama, Y., Oyama, M. and Miyahara, J. T. (1982). *Gambierdiscus toxicus*, a ciguatera-causing dinoflagellate from Hawaii (USA). *Bull. Jap. Soc. Scient. Fish.* 48, 811–4.

Smith, S. V. (1978). Coral-reef area and the contributions of reefs to processes and resources of the world's oceans. *Nature* 273, 225–6.

Stevenson, D. and Marshall, N. (1974). Generalisations on the fisheries potential of coral reefs and adjacent shallow water environments. *Proc. 2nd Int. Coral Reef Symp.* 1, 147–56.

Tranter, D. J. (1976). Herbivore production. In *The Ecology of the Seas*. Ed. D. H. Cushing and J. J. Walsh, pp.186–224, Blackwell, Oxford.

Walczak, P. (1977). *A Study of the Marine Resources of the Yemen Arab Republic*. F.A.O. F.I. DP/YEM/74/003/5, Rome.

Wellington, G. M. (1982). Depth zonation of corals in the Gulf of Panama: control and facilitation by resident reef fishes. *Ecol. Monogr.* 52, 223–41.

Wells, S. M. (1981). International trade in ornanental corals and shells. *Proc. 4th Int. Coral Reef Symp.* 1, 323–30.

Woodley, J. D. (1979). The effects of trap-fishing on reef communities in Jamaica. *Proc. Is. Mar. Labs. Carib.* 13, 27.

CHAPTER 18

Pollution

BRIAN DICKS

Oil Pollution Research Unit, Orielton Field Centre, Pembroke, Dyfed SA71 5EZ, U.K.

CONTENTS

18.1. INTRODUCTION

The Red Sea comprises a wide range of tropical marine habitats, some of considerable conservation, scientific, economic or recreational value. These receive, either locally or more widely, a variety of stresses as a result of man's activities. Considerable industrial and urban developments are currently taking place at many locations both onshore and offshore around the Red Sea, the largest and most widespread of which are concerned with oil exploration and production. In this brief contribution, pollution effects in key Red Sea habitats are reviewed and simple guidelines provided for monitoring strategies. However, published literature on the effects of pollutants on Red Sea organisms or habitats is limited. Consequently publications from other similar tropical areas and temperate seas as well as the results of studies on the Red Sea have been relied on. Inevitably, pollution may affect a wide range of habitats, many of which are dealt with in detail in other parts of this volume. Overlap has been avoided wherever possible.

18.2. NON-LIVING RESOURCES AND SOURCES OF CONTAMINATION

Sources of contaminants to the Red Sea arising from the exploitation of non-living resources and the presence of man can be conveniently listed under three headings: urbanisation and tourism, oil, and other industrial inputs. Oil has been singled out from industrial inputs in view of the extensive and increasing oil production in the northern Red Sea. The nature and magnitude of pollution problems in the Red Sea do not necessarily follow trends elsewhere in the world. In particular, the enclosed nature of the Red Sea in conjunction with the limited water exchange with the Indian Ocean (Chapter 2) considerably reduces the potential for the dispersion of pollutants. This is especially so in the Gulfs of Suez and Aqaba which are relatively shallow compared to the main body of the Red Sea which is very deep along most of its length.

Much of the input of contaminants is to geographically localised areas around urban and industrial developments. The most notable of these are in the Gulf of Suez (onshore and offshore oilfields plus discharges from refineries and industries in the Suez area — *see* Fig. 18.1), the northern end of the Gulf of Aqaba (Elat and Aqaba; oil, phosphates and industrial discharges), and near ports, industrial developments, refineries, petrochemical plants, and oil production areas in Saudi Arabia (Yanbu, Jiddah, Jizan), Yemen (Al Hudaydah, Al Mukha), Ethiopia (Massawa), Egypt (Hurghada) and Sudan (Port Sudan).

More widespread and general contamination may be expected from the considerable and probably increasing ship traffic through the Red Sea to and from the Suez Canal, or from supertanker traffic to and from oil terminals. The fate of pollutants in marine systems has been reviewed in a number of recent publications and is not repeated here. Particularly useful publications are Bryan (1976), Topping (1976), Malins (1977), Jordan and Payne (1980) and Royal Commission for Environmental Pollution (RCEP) (1981).

In addition to the above-noted pollution possibilities, the potential mineral wealth of deep-water sedimentary areas has been recognised and may be exploited in the future. This would involve considerable sediment disturbance, and if minerals are processed locally, the development of further coastal industry.

18.2.1. Urban inputs, tourism and shipping

The greatest single input from urban conurbations to the sea is of sewage. Sewage (treated and untreated) is usually discharged to, or just below, the intertidal zone via pipelines, and is thus mostly a coastal problem. The volume (and thus area of effect) depends on the number of people involved and considerable inputs may occur around cities and large towns (e.g. Suez, Jiddah), but no figures are available. The composition of sewage varies considerably (Topping, 1976) but major effects reported in the scientific literature result from increased nutrient and suspended solid loading or from human health problems associated with coliform bacteria on recreational beaches. It is not possible to accurately predict zones of damage in relation to levels of input because local topography and hydrography considerably influence rates of dispersion, fate and effects. Biological effects are inevitable but can be considerably reduced by the installation of adequate treatment facilities. More widespread and probably insignificant contributions of sewage result from ship traffic and offshore platforms.

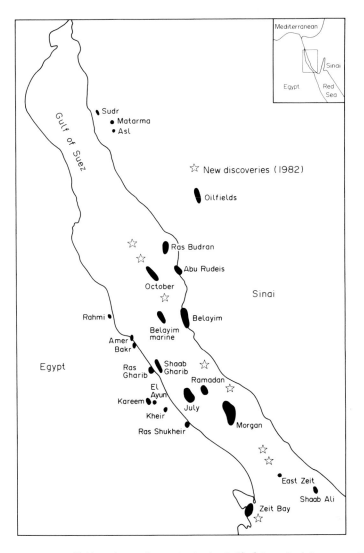

Fig. 18.1. Existing oilfields and new discoveries in the Gulf of Suez, Red Sea, up to 1982.

In addition to sewage, considerable amounts of garbage (especially plastic containers) also enter the sea from urban and recreational areas and from ship traffic and offshore platforms (Fig. 18.2). Wenninck and Nelson-Smith (1977) report that 'the entire coast of Saudi Arabia (which receives onshore winds for much of the year) is "seriously polluted" by plastic deposits, occasionally covered in oil. They are a considerable nuisance and aesthetically unpleasant but probably do little biological damage.' Such deposits occur in the strand zone on many Red Sea beaches, and shores in the Gulf of Suez are particularly affected.

Land reclamation and coastal road construction also affect shore zones and nearshore waters. Apart from areas of the sea which are lost, sediment loading of the water increases and may affect coastal habitats in a similar manner to dredging (*see* below). At least part of a motorway construction project just completed to the north of Jiddah has been built on a reef flat, and any further building in this manner could cause serious losses to coastal habitats.

Tourism may also produce both direct and indirect effects on coastal resources and is currently growing rapidly on Red Sea coasts. Construction of coastal hotels and roads may result in similar problems to those noted immediately above. The increased numbers of people produce more sewage requiring disposal, and the discharges deplete the quality of the very amenities which attract the tourism in the first place, i.e., the clean beaches and spectacular underwater reefs. A classic case was the destruction of the Coral Gardens of Kaneohe Bay in Hawaii as a result of increasing sewage discharge (Smith, 1977). Collection of biological specimens by divers (either for souvenirs or for sale), especially corals, sea fans, starfish and urchins, crustaceans and molluscs, may cause loss of diversity in marine habitats and community changes and degradation of coral reefs. Small boat activities also produce anchor damage to corals.

18.2.2. Oil

The Middle East is currently the world's largest oil production area (ca. 30%) and is likely to remain so with an estimated 55% of the world's 'proven' oil reserves (approximately five times more than the nearest rival, Latin America, with 10.6%). The majority of new exploration and production is offshore,

Fig. 18.2. Garbage on the coast at Sharm Yanbu, Saudi Arabia. The Sharm is used for recreation by both Yanbu residents and expatriate workers from the Yanbu industrial complex. (*Photo*: Dawson-Shepherd/TMRU).

Fig. 18.3. Nodding donkeys near Ras Gharib, Gulf of Suez. Repeated small spills have resulted in severe oiling of nearby coastal reefs and shores. (*Photo*: OPRU)

and one of the fastest-developing areas with enormous production potential is in the Gulf of Suez in the northern Red Sea (Figs. 18.3 and 18.4).

A substantial proportion of Middle East oil is transported by pipeline and/or ship to other parts of the world, with local consumption and refining utilising only about 9% of production. Refining and local consumption are growing rapidly with urban and industrial development programmes, but the emphasis is likely to remain on oil production and shipping, in the short-term at least, as the major cause of oil pollution. Although figures are not available for the Red Sea, the pattern of oil inputs is likely to be similar to that of the Arabian Gulf, which is compared with world figures below (Table 18.1).

Increases in refining and urban and industrial development throughout the Middle East will lead to an increase in chronic discharges (refinery effluents and offshore production water discharges) and suggest that in the future more attention should be directed to their environmental effects.

Although mainly marine habitats are affected, it should be remembered that some oil pollution of terrestrial and freshwater habitats also occurs through blowouts, pipeline breaks, storage tank leaks, road or rail accidents, boat traffic on inland waterways and by general industrial discharges. The degree of oil pollution of shorelines around the Red Sea (with the exception of the Sudanese coast) has been well-summarised by Wenninck and Nelson-Smith (1977, 1979). Quoting these authors:

> Coastal pollution by oil was found to be a very serious problem along major sections of the Gulf of Suez and Egyptian Red Sea coast and it is necessary at the outset of this study, to dispel the notion that the Red Sea as a whole is a relatively unpolluted sea, since we found this to be a total fallacy. The degree of pollution encountered is obviously not uniform for the long coastline involved and there are many sections, even in the North, which are relatively free from the effects of oil. However, at the same time, a number of areas must rank with some of the worst polluted coastlines of the world.

On recent visits to the Gulf of Suez (1980, 1981 and 1983) observations showed that this situation persists (Fig. 18.5). Indeed, the Gulf of Suez is undoubtedly the most-affected area, and many kilometres of coastline are severely oiled intertidally from spills from oil rigs and ships. Weathered oil pavements many centimetres thick blanket rocky promontories, sandy beaches, and in a few locations shallow patch and fringing coral reefs. Some sandy beaches in heavily-oiled areas appear relatively clean, probably as a result of self-cleaning by wave action, but also because oil may be buried and the beach skimmed-over with clean, wind-blown sand.

I support the view expressed by Wenninck and Nelson-Smith that in some heavily-oiled areas in the Gulf of Suez clean-up is impracticable, and the primary concern should be to prevent further areas being damaged rather than cleaning those already polluted by oil.

Outside of the Gulf of Suez and the Hurghada area, oil pollution is less of a problem except locally around oil-handling ports such as Elat, Aqaba and Jiddah (Wenninck and Nelson-Smith, 1977, 1979). Only minor oil pollution was observed around smaller ports (Yanbu, Jizan and Tuwwal) by these authors. However, extensive marine terminal developments are underway at Yanbu and may be a source of future concern. Contingency plans for control of pollution and spill clean-up would reduce the potential for spills and the likely extent of damage.

Wenninck and Nelson-Smith also report most of the Red Sea coast to be patchily subjected to 'smudging' by oil, mostly as soft tarry patches and globules resulting from passing ship traffic. Tank

TABLE 18.1. Estimated inputs of oil to the marine environment (as a percentage of total), Arabian Gulf vs rest of the world.

	Arabian Gulf		World
	1977 Neumann (1979)	1980 Golob (1980)	Cowell (1978)
1. Tanker and ship traffic (spills and routine discharges)	86%	58%	23%
2. Offshore production and natural seepages	14%	22%	14%
3. Refining, industrial and urban	<0.12%	20%	63%

Fig. 18.4. Oil production in the Gulf of Suez. (a) Drilling from a jack-up rig. (b) Tankers moored ready for filling, near the coastal development at Ras Shukheir. (*Photo*: OPRU)

Fig. 18.5. Coastal oiling: the most obvious threat to the northern Red Sea. (a) Spectacular cliff scenery on the coast north of Ras Zeit, Gulf of Suez. The black band along the shoreline is caused by oil from coastal oilfields further north. (b) A closer view of the upper shore; unfortunately, such accumulations of oil, weed, seagrass and rubbish are now common on north-western Red Sea shores, not only damaging the upper-shore communities but making access to the sea both messy and difficult. (*Photos*: OPRU)

washings and uncontrolled discharges produce this type of pollution, and although it is unlikely to cause widespread harm to marine life, serious accumulations may result in some areas (dependent on wind, tide and current) and spoil otherwise beautiful recreational beaches.

Following oil pollution, the use of inappropriate clean-up techniques or extensive use of dispersant chemicals may produce additional damage to resources, and should be carefully controlled.

Aside from the direct effects of spills and routine refinery effluent and oil rig discharges, oil developments can produce environmental stress during the construction and operation phases, notably around offshore platforms (Dicks, 1982) and shore terminals. Effects may result from a combination of oil discharges, sediment disturbance during construction of platforms, pipelines and jetties, coastal blasting and dredging, an 'artificial reef' effect from marine fouling communities on structures, discharges of drill muds and rock cuttings, discharges of process and anti-corrosion chemicals and of sewage and garbage. The discharge of mud and cuttings is worth considering further.

TABLE 18.2.

Depth (m)	Diameter (cm)	Approx. vol. cuttings (m^3)
0— 150	46.5	26
150—1,500	33.5	120
1,500—3,000	24.0	75
3,000—4,500	17.5	39
	Total volume of cuttings	260 m^3

The drilling of oil wells results in production of sediments as rock cuttings, the volume of which depends on the diameter of the hole drilled. Typical values are: for a 30 cm hole about 7.6 m^3/100 m and for a 24 cm hole about 4.7 m^3/100 m. In general the diameter of a well decreases with depth. For example, a well 4,500 m deep may comprise a series of hole diameters (Table 18.2).

The cuttings are normally discharged to the sea immediately around the platforms. During drilling, fluids known as drilling muds are used to lubricate the drill bit and counteract formation pressures. These muds are usually oil (diesel, kerosene or refined de-aromatised diesel) or water-based and contain a wide variety of additives which control lubrication and density properties. The muds are usually reclaimed, but a proportion (approximately 10%) is lost with the cuttings. A single offshore platform may drill up to 40 or more wells, and thus produce of the order of 10,000 m^3 of cuttings and lost mud, which would be deposited immediately around the platform and continuously distributed by water movements. It is more usual for a platform to drill ten or less, however, and volume would be proportionally smaller. Wells to 4,500 m are common in the Gulf or Suez, and there are currently more than 20 platforms in the area. Mud and cuttings may have various biological effects ranging from direct toxicity to simple smothering (Szmant-Froelich et al., 1981).

18.2.3. Other industrial inputs to coastal waters

Many of the observations on the scope of the oil problem apply to all coastal industrial inputs to the Red Sea. Contamination or disturbance may arise from the construction and operation of many large and small coastal industries. Dredging, land reclamation, blasting and jetty construction may all cause localised effects, and subsequently during operation the discharge of effluents may cause local or more widespread problems. Major contributions of contaminants reported for the Red Sea arise from phosphate and mineral (manganese and bauxite) loading in the Gulf of Aqaba (Fishelson, 1973; Walker and Ormond, 1982) and at Quseir.

Seawater desalination plants are common in association with urban and industrial development. Sardar (1981) reported 20 in operation on the Saudi Arabian coast of the Red Sea in 1981, but some 30 more are planned for construction by the year 2000. These plants discharge hot brine, which may contain significant heavy-metal loading from condenser coils, back into the sea. The impact of desalination plants was discussed by Chesher (1975).

18.2.4. Deep-sea mining

The deep trenches of the central and southern Red Sea contain thick deposits of metal-bearing muds in association with hot (60°C) brine pools at depths of 2000 m. The muds are rich in silver, copper, zinc and a number of other metals (Georghiou and Ford, 1981). The German company Preussag, in conjunction with the Saudi Arabian and Sudanese governments, has completed a small pilot scheme for extraction of these deposits by use of a long pipe and suction head from a surface ship, subsequently separating metal-bearing fractions onboard and disposing of waste sediment (mostly silts and clays) to the water. In commercial production this venture would produce *ca*. 270,000 tonnes (= 310,000 m^3) of waste tailings per day, a mixture of mud and warm brine, plus treatment chemicals used in the metal extraction process. This resource and its problems are discussed fully in Chapter 4.

18.3. LIVING RESOURCES OF THE RED SEA AND THE BIOLOGICAL EFFECTS OF POLLUTANTS

The living resources of the Red Sea are described in detail in many of the other contributions to this volume. All that is necessary here is a brief summary of the resources which are of value and are most 'at risk' from pollution. Much of the intertidal and shallow coastal water zone is at risk from floating pollutants (e.g. oil) and coastal discharges (e.g. industrial effluents, sewage) and from disturbance. The coastal zone comprises extensive sand and mud areas, rocky and stony outcrops, shallow coastal lagoons with seagrass beds, fringing and patch coral reefs, and offshore of such reef systems, sediments, coral knolls and seagrass beds. Typical nearshore profiles are given in Figure 18.6. South of Sinai mangroves grow locally. These habitats together form a rich and productive coastal ecosystem on which many fisheries depend, which are of considerable scientific and educational value, in which coral reefs produce a coastal barrier against wave action and erosion, and which are a potential tourist attraction. In deeper water, seabed sediments support a variety of communities, but the relationships between deep- and shallow-water habitats are poorly known.

Some living resources have direct commercial value, e.g. certain fishes and shellfish, and the fishing industry is described in Chapter 17. Both pelagic and lagoon or shallow reef-associated fish which form the major part of existing fisheries are at risk from coastal and offshore contaminants.

Although not a major bird migration route on the scale of, for example, north-west Europe, the area supports resident and breeding populations of waders, terns and boobies and, at least in the northern part, is on the migration route of many species (*see* Chapter 15).

The susceptibility of communities to pollutant damage may vary with geographical location within the Red Sea. Natural gradients in, for example, water salinity and temperature, limit the distribution of species, and it is likely that those species approaching the natural limits to their range may be particularly susceptible to additional pollution stress. A summary of available information regarding effects of pollutants in the major Red Sea habitats is given below. Naturally enough, with the emphasis on oil production, much of the published material concerns oil.

18.3.1. Coral reefs

Much of the available information discusses the effects of oil on coral, although, for the Red Sea, reports are available for phosphate dust pollution (Fishelson, 1973; Walker and Ormond, 1982); dredging and siltation (Johannes, 1970); and sewage (Johannes, 1971; Walker and Ormond, 1982). In addition a certain amount of anecdotal information can be obtained which concerns the effects of diving activities and tourism at some locations.

The effects of oil pollution on coral reefs has been reviewed recently by Ray (1981) and Loya and Rinkevich (1980). There have been few detailed quantitative field studies of oil pollution and laboratory studies are difficult to compare because of the wide range of techniques used.

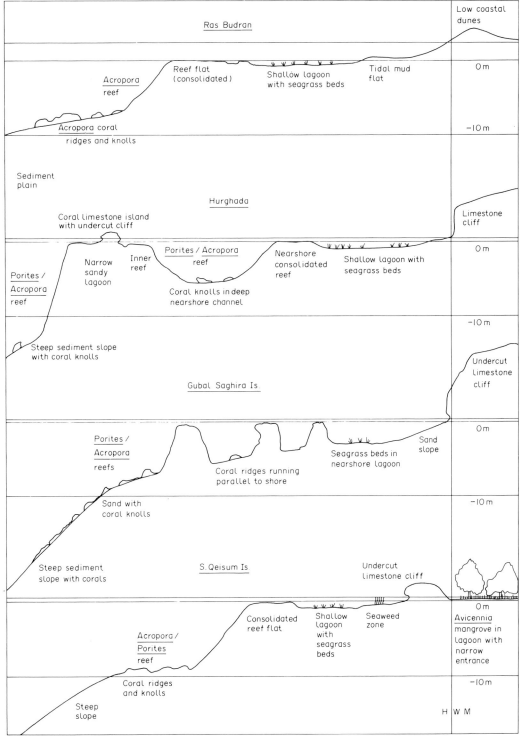

Fig. 18.6. Sketch profiles of nearshore habitats in the northern Red Sea and Gulf of Suez. The distribution of sediments, seagrasses, coral and mangroves are indicated.

Oil has been implicated in a recolonisation failure in the Gulf of Aqaba. Following a 90% mortality of corals in shallow water due to unusually low tides in 1970, recolonisation of two areas was studied: *1)* a reef exposed to large inputs of oil, located 3 km south of an oil terminal from which numerous spills occurred; and *2)* a relatively unpolluted reef 5 km south of the terminal. Recolonisation was normal on the unpolluted reef but there was a failure of opportunistic species such as *Stylophora pistillata* to recolonise the oiled reef. This reef is probably the most badly-oiled of coral reefs that have been studied. Only counting those spills large enough to 'blacken' the reef, 95 spills occurred from 1971 to 1973, at a frequency ranging from one to seven spills per month. In addition to the oil input, there was also phosphate and sewage eutrophication of the shallow lagoon behind the fringing reef due to nearby fertiliser plants and sewage lines (which makes definitive conclusions regarding cause and effect relationships rather difficult). However, Loya (1975) inferred that oil may damage the reproductive system of corals, interfere with production of larvae or reduce larval viability and inhibit their normal settling, and Rinkevich and Loya (1979) have confirmed that oil affects coral reproduction in laboratory experiments.

Several field experiments are of particular interest. Kinsey (1973) floated crude oil above coral and found no toxicity effects or abnormal behavioural patterns. Legore *et al.* (1983) in similar experiments which included dispersant application found that, although growth remained unaffected, some behavioural changes and delayed effects of oil and dispersant were noted, especially during a cold winter season. In contrast, Johannes *et al.* (1972) found that oiled corals which were partially exposed to air were damaged. Knapp *et al.* (1983) found temporary behavioural effects in *Diploria strigosa* under the influence of oil and dispersant chemicals, but no long-term damage.

These experiments were necessarily carried out in still conditions. In practice it is highly probable that wave action would drive oil on to the reef structure near the water surface and expose both coral and associated faunas to high concentrations (Fig. 18.8). None of the reported experimental or field survey data have addressed this possibility. My own observations in chronically-polluted areas in the Gulf of Suez indicate that oil has been driven into the reef structure by wave action, and appropriate scientific studies would provide much useful information on the effects of oil in reef systems (Fig. 18.7). These should be a priority for future studies.

Laboratory experiments reviewed by Ray (1981) and Loya and Rinkevich (1980) indicate a range of possible responses to oil, including abnormal mouth-opening and feeding behaviour, mucus secretion, decreased growth rates and increased tissue death rates. Ray (1981) has described mucus secretion as the

Fig. 18.7. Tarballs were stuck within the branches of this head of living coral (*Montipora* sp.) on the shallow reef flat near Ras Gharib, Gulf of Suez. Severe and chronic oiling from coastal operations is a feature of this shoreline and has caused extensive environmental damage. (*Photo*: OPRU)

single most important defence mechanism because the mucus tends to either repel or trap the oil contacting the surface. However, mucus-consuming fish such as butterflyfish (Chaetodontidae) could become contaminated as a result. Chronic (three-month) exposure of the coral *Manicina areolata* to water-accommodated fractions of No. 2 fuel oil (prepared by contact between diesel fuel and the water phase) led to atrophy of mucus-secreting cells and other pathological responses (Peters *et al.*, 1981).

Death of corals as a result of sewage and phosphate pollution has been reported by Johannes (1971), Fishelson (1973) and by Walker and Ormond (1982) in the Gulf of Aqaba. Quoting the latter authors:

> Localised pollution of coral reef areas is occurring at Aqaba, Red Sea, as a result of sewage discharge, and as a result of spillage of phosphate dust during loading of phosphate mineral onto ships. The rate of death of colonies of the coral *Stylophora pistillata* was found to be 4–5 times as great in the polluted area as in a control area. Coral damage in the control area is generally caused by grazing or by extreme low tide, but the cause of coral death in the polluted area was not readily apparent. The growth of algae, both on the damaged corals, and on glass slides placed out in the reef, was greatly stimulated in the polluted area, but it appeared that such algal growth was not the direct cause of coral death. Corals in the polluted area may be under stress because of reduced light intensity, inhibition of calcification by excess phosphate, and increased sediment load. It was found that in the polluted area there was a greater weight of sediment settling on the glass slides for a given weight of algae. But in addition, since algal growth was faster in the polluted area, the sediment load was increased by the sediment trapping capacity of the enhanced algal growth. Thus it is suggested that increased algal growth stimulated by increased nutrient concentrations may be important in greatly increasing the sediment load experienced by corals.

Dredging has affected reefs in Hawaii, where Johannes (1971) reported *ca.* 2800 ha of reef damaged or killed by dredging and fishing operations. Damage results from physical disturbance or sediment redistribution. Effects of dredging, pipeline route blasting and excavation have not been reported in the literature for the Red Sea (Fig. 18.9).

Drilling muds and rock cuttings also affect corals. Thompson and Bright (1977) report that sedimentation from drilling mud and cuttings discharges killed hermatypic corals. None of the species (*Diploria strigosa*, *Montastrea cavernosa*, *M. annularis*) were able to clear the muds, and died under even fairly light sediment loadings. In the North Sea oil-based muds and cuttings have been demonstrated to produce zones of severe effects up to one kilometre from platforms. The persistence of these zones of effect is not currently known as the use of oil-based muds is relatively recent. They are being used in the Red Sea, but as yet their effects have not been studied. However, the current increase in nearshore drilling for oil in the Red Sea may produce local problems and deserves immediate attention.

Damage to reefs from tourism is poorly documented, but there is much anecdotal evidence which suggests that prolonged disturbance from intensive diving activities, as well as souvenir collections, can cause reef depletion. Many underwater reserve management plans recognise the potential for damage

Fig. 18.8. Chronically oiled reefs like this one near Ras Gharib have been severely damaged. No hard corals survive in shallow water and the reef rock is dominated by encrusting algae and soft corals such as *Litophyton*. (*Photo:* OPRU)

Fig. 18.9. Damage to reefs caused by pipeline laying and seismic exploration at Ras Rahmi, Gulf of Suez. (*Photo*: OPRU)

and specifically exclude any form of collection by visitors and demand care in anchoring and whilst swimming to avoid physical damage to reefs. In areas of the Red Sea where tourism is being encouraged, diving activities should be carefully controlled and the opportunity taken to monitor the effects.

18.3.2. Seagrass beds

Seagrass beds are a widespread and highly productive habitat in coastal areas in the Red Sea. Although undoubtedly of great importance to primary productivity in the Red Sea, the effects of pollutants on these beds are little described in the scientific literature (Fig. 18.10). Two examples reported from European waters are as follows.

The effects of the 'Amoco Cadiz' oil spill on the seagrass *Zostera marina* were studied by Jacobs (1982). He gave data on changes in shoot density, length of internodes and biomass production, and concluded that the effects were short-term and local (evidenced by blackened leaves). Baker *et al.* (1982) are currently using field experiments to study the effects of oil and dispersants on the intertidal seagrass *Zostera noltii*; a preliminary conclusion is that oil/dispersant mixtures are particularly damaging.

The lack of information on seagrasses suggests an urgent need for study, particularly of tropical species, and the presence of seagrasses in areas of the Red Sea already affected by industrial and urban inputs, as well as new developments, offer many opportunities for coordinated study.

Fig. 18.10. Weathered crude oil adhering to seagrass leaves (*Halophila ovata*) from the seabed at a depth of 5 m near Ras Gharib, Gulf of Suez. (*Photo*: OPRU)

18.3.3. Mangroves

The term mangrove refers to approximately 70 species of tree or bush which occur on sheltered shores and in estuaries or nearshore waters in the tropics and some sub-tropical regions. Recent reviews of mangrove ecology include Macnae (1968), Lugo and Snedaker (1974) and the IUCN (1982) Commission on Ecology report on the global status of mangrove ecosystems.

Although mangroves are relatively poorly-developed in the Red Sea (*see* Chapter 9), they form a community which is particularly vulnerable to the effects of pollution, and reports have been made of damage from oil, industrial effluents and sewage (Fig. 18.11). Mangrove roots typically grow in fine anaerobic muds and receive oxygen through aerating tissue which communicates to the air through small pores (lenticels) on the stilt roots or special 'breathing' roots (pneumatophores). Thus any contaminants which interfere with aerating tissue (e.g. oil deposits on aerial roots) may reduce oxygen diffusion to the underground root system. Likewise, pollutants incorporated into sediments may damage root systems and so interfere with the ultrafiltration process. Such effects are usually followed rapidly by defoliation and death. The slow growth of trees to maturity means that recovery of these systems, following damage, takes a considerable time (50+ years).

Much of the published information or damage to mangrove concerns oil pollution, the acute short-term effects of which are likely to be defoliation, death of seedlings and high mortalities of associated invertebrates. Superficial oil is likely to weather comparatively quickly, but little is known about oil penetration and its fate deeper in the sediments, or whether the presence of either fresh or weathered oil affects recolonisation processes. Mangrove swamps are probably the most difficult type of habitat to clean, and spill response should concentrate on trying to prevent oil entering such areas in the first place. Oil has been reported on mangrove areas at Ras Muhammad (Hiscock, *pers. comm.*) and the Red Sea provides opportunities for mangrove studies at the limits of their biological range. Further studies in the

Fig. 18.11. (a) A heavily oiled stand of mangrove (*Avicennia marina*) on South Geisum Island, northern Red Sea. The oil has completely coated the breathing roots and killed the mangrove. The fine carbonate muds at this site are also heavily oiled and anaerobic. (b) At this site on the Egyptian Red Sea coast the oil has only coated the lower half of the breathing roots and the mangrove has survived. Remarkably, in areas with coarse, well-drained and oxygenated sediments, mangrove root systems remain alive and the mangrove can survive even when the breathing roots and sand surface are completely coated in oil (*Photos*: OPRU and Dawson-Shepherd/TMRU)

Red Sea should be directed towards effects of new or existing developments on mangrove areas where appropriate.

18.3.4. Intertidal mud and sand flats and shallow nearshore lagoons

From the ecological point of view, pollution of sheltered sediments causes great concern, firstly because they are more likely to retain oil and other pollutants, and secondly because they often have a rich fauna, including a great variety of polychaete worms, bivalves and crustaceans, which are important as demersal fish food. Shallow lagoons and intertidal areas may also be spawning or nursery grounds for some marine species, especially commercial species and those of conservation value (e.g. turtles). Such animals may be killed in large numbers either if toxicants penetrate the sediment, or if massive algal growth covers the mud surface as a result of eutrophication caused by sewage. Wenninck and Nelson-Smith (1979) report such filamentous algal growth around many sewage discharges to the Red Sea. Penetration by pollutants is of course, more likely if the sediments are perforated with burrows. Rich intertidal flats often form important bird and fish feeding grounds and could be rendered useless for this purpose following a large spill of fresh oil or other pollutant damage.

Of some concern has been the use of dispersant chemicals in attempts to clean up oiled intertidal sediments. Recent field experimental work (*see*, for example, Little *et al.*, 1981) has been aimed at answering the question: how would modern dispersant use at the time of spilt oil coming ashore influence penetration and retention or oil in sediments? Preliminary findings indicate that:

> 1. On waterlogged mud flats, dispersant had no significant effect on oil removal (oil disappeared from the experimental plots regardless of whether dispersants were used or not.)
>
> 2. On well-drained, fine-sand flats, there was some marginal evidence that dispersant alone could reduce worm cast production, and evidence that some dispersant treatments enhanced penetration of oil into sediments, where it was retained at greater concentrations than where no treatment had occurred.
>
> 3. None of the experiments carried out indicate that dispersant application is likely to be a useful cleaning method for relatively sheltered intertidal sediments.

Although information that is available worldwide may be incorporated into contingency plans and be used for relatively crude predictions of likely environmental effects in the Red Sea, the lack of specific information suggests an urgent need for study. This should be focussed on existing areas where effects are known to have occurred (e.g around Suez), as well as clean areas for comparison, and in particular areas of future development.

18.3.5. The seabed sediments

Contaminants may reach the seabed from effluents and routine discharges, oil spills or other accidents. Studies indicate that seabed sediments form a 'sink' for a number of pollutants, notably heavy metals, organic wastes, suspended sediment (e.g. minerals and drill muds) and oil, either from refinery effluents or spills. No information is currently available for the Red Sea, but a few examples which illustrate effects elsewhere are given below.

Severe macrofaunal depletion in combination with opportunist expansion has been demonstrated around sewage discharges (Read *et al.*, 1983), refinery effluents (Dicks and Hartley,1982), drilling mud discharges (American Petroleum Institute, 1980) and in ports and harbours (Reish *et al.*, 1980). In each case the area of effect has been determined by a combination of discharge factors and natural dispersion characteristics of the receiving waters. A zone characterised by opportunistic species, often in extremely high numbers, frequently occurs around the zone of severe damage. Outside of this region lies a zone in which more subtle biological changes may be observed which grades into apparently unaffected areas. This often-reported pattern is well-summarised by Pearson and Rosenberg (1978).

Eisler (1973, 1975) has reported effects of crude oil and dispersant on a variety of seabed and intertidal macrofauna, including Red Sea species of coelenterates, molluscs, crustaceans, echinoderms and fish. Organisms were tested in laboratory tank experiments and showed a variety of lethal and sub-lethal responses including behavioural and metabolic changes, but likely effects cannot be predicted from these findings.

Offshore oilfields are considered in some detail here because of the extensive developments currently underway in the Gulf of Suez. Many long-term studies have used the relatively immobile macrobenthic community as an indicator of the overall biological health of the area. In some cases biological effects have been observed; in others, not. The size and form of effects are largely dependent on the environmental conditions at a site (water depth and currents, community type, prevailing winds) as well as the size and type of the platform and the processes being carried out. Where effects have occurred, it has not so far been possible to completely separate those which result from oil from other influences. For example, physical disturbance during construction, discharges of drilling muds and cuttings, redistribution of sediments as a result of pipelaying and anchorage, discharges of cooling water, sewage and garbage from platforms and ships may all make a contribution (see Dicks, 1982). Of particular recent concern has been the discharge of oil-based muds which are increasingly being used worldwide, including in the Red Sea. Although severe but localised effects on seabed communities have been noted in offshore fields (mostly in unpublished reports), these have usually been restricted to within ca. 1000 m of the platform, and there is no evidence that oil developments have resulted in widespread environmental decline. However, the current oil developments in the Red Sea are so intensive that monitoring is essential. Our current research programmes include studies and monitoring schemes for both offshore platforms and onshore terminals in the Gulf of Suez (see Dicks, 1983), and further studies in a range of key habitats are vital to assessing the total effect, as well as making recommendations for acceptable levels of discharges and for environmental protection and management.

18.3.6. Open waters offshore, including fisheries

The most likely contaminants of offshore waters in the Red Sea are oil and, in the event that deep-sea mining becomes viable, mining operation wastes. Two cases in which oil was released in large volumes at sea and left untreated were the 'Argo Merchant' (NOAA, 1979) and 'Ekofisk Bravo' spills (AIBS, 1978). In the former case, fouled zooplankton, moribund fish eggs and a relatively high percentage of oiled birds were observed near the spill site. It was concluded, however, that abundance studies in benthic and pelagic communities, including commercial fish species, did not suggest an overall major adverse impact. NOAA (1979) concluded that the outcome of the spill was fortunate in that 1) the density of the oil was low enough to prevent wholesale sinking and subsequent contamination of the bottom, and 2) the spill occurred in the winter when biological activity, productivity and fishing activities were relatively low. Following the Ekofisk blowout, there was some evidence that fish absorbed Ekofisk oil via the digestive tract. However, the concentrations of analysed hydrocarbons in the liver and muscle tissues of fish caught soon after the oil flow was stopped, and again some two months later, were within the range found in the past from samples caught in the open seas around the U.K. Changes in the fauna of the bottom sediments could not be specifically attributed to the blowout, although oil from the blowout did reach the seabed.

The potential for environmental damage as a result of the discharge of sediment, brine and chemicals from mining operations undoubtedly exists, but the magnitude is difficult to predict and will depend greatly on dispersion and biological characteristics of the discharge area. In the absence of published data, it is clear that mining activities deserve further study.

Fisheries constitute an important open-water resource, and effects of contaminants on a wide range of fish species have been demonstrated; these range from lethal to subtle and include the uptake and, in some cases, depuration of contaminants. Oil may produce particular problems in the Red Sea, and a summary of the effects of oil on fishes can be found in Royal Commission on Environmental Pollution (1981) and Malins and Hodgins (1981). Main concerns arise from potential mortality to eggs, larvae and juveniles (many of which can be found in coastal waters), the mortality of adults, or the tainting of commercial species which renders the catch unpalatable. Larval stages have been shown to be particularly vulnerable.

It is clear that under laboratory conditions chronic exposure to relatively low levels of petroleum hydrocarbons (0.1–1 ppm) can have marked effects on the proportion of eggs that hatch and the subsequent growth of larvae. Developmental abnormalities have also been noted. According to the limited data available, such concentrations are likely to be maintained only in confined areas such as industrialised estuaries or close to the outfalls of industrial discharges into the sea.

Although there is no definitive evidence which suggests that oil or any other form of pollution has significant effects on adult fish populations in the open sea, this does not mean that effects do not exist. Sampling problems result in such large confidence limits on field survey estimates that mortality below an order of magnitude greater than normal would be virtually impossible to detect.

As in many other habitats, few examples of effects on fish species are available for the Red Sea. Potential problems exist from a range of contaminants (e.g. oil, heavy metals, organochlorines, sewage components), which may produce human health risks or tainting and loss of palatability. Wenninck and Nelson-Smith (1979) report that in spite of sewage and industrial pollution, fisheries near the city of Suez have not declined, but they also note that, as a result of military activity, fishing stopped in some areas and may have resulted in natural restocking.

18.3.7. Birds

A host of factors can influence the survival and distribution of birds ranging from loss of, or damage to feeding area, to disturbance at nesting sites or mortality from a variety of pollutants or hunting. Little information specific to the effects of pollution on birds can be found for the Red Sea. An expedition is currently underway surveying bird oiling along the Egyptian coast from Suez to Quseir (Jennings, *pers. comm.*).

Floating oil is undoubtedly one of the largest threats to birds on the water surface in the Red Sea. The Royal Commission for Environmental Pollution (1981) provides an excellent summary of the mechanisms of oil pollution damage to birds.

The most obvious effect is to damage the plumage on which they depend for their insulation and waterproofing. The matting of the plumage by oil allows water to penetrate the air spaces between the feathers and the skin, with the result that birds lose buoyancy and may sink and drown. In cold conditions they rapidly lose heat, and in an attempt to counteract this the metabolism of food reserves is increased. Oiled birds may attempt to preen themselves and thus ingest significant amounts of toxic material, leading to internal damage. Stress and shock enhance the effects of exposure and poisoning, and the chances of survival are small.

Although effective cleaning methods for oiled birds exist, only a very small proportion of oiled seabirds can usually be rescued, cleaned and returned to the sea, even in the most favourable circumstances. Consequently some organisations such as the Royal Society for the Protection of Birds (RSPB) of the United Kingdom recommend that it is generally more humane to destroy oiled birds.

18.4. RESEARCH, SURVEILLANCE, MONITORING AND MANAGEMENT

It is clear from the preceding sections that there are considerable gaps in knowledge of pollution effects in the Red Sea, but it is also true that some habitats have already been identified as being both valuable and susceptible, notably coral reefs, seagrass beds and mangroves. Certainly the area offers substantial opportunities for further study of both acute and chronic pollution problems, and these opportunities must be taken if a comprehensive database is to be established on which future management of resources and control of industrial activity can be undertaken. In acquiring such a database, field studies have the considerable advantage over laboratory studies of providing descriptions of real field effects which are useful for making predictions in new areas.

There is no easy distinction between research, surveillance and monitoring. The terms surveillance and monitoring are widely used, but precise definitions have not been unanimously agreed. Monitoring implies measurements repeated at intervals and compared with a standard, while surveillance suggests repeated observation without a definite comparative standard. Both may have elements of research included, and vice-versa.

In the study of marine systems as well as in the recording and understanding of pollution effects, many approaches have been used, and at present there is no universally-applicable approach or method. The methods selected depend largely upon the aims and objectives of the work, and in practice many studies have very simple aims such as:
 — to identify and describe the distribution of the main community/habitat types.
 — to quantify natural variability in a system over a period of time.
 — to identify the causes of observed variation.
 — to assess the conservation, scientific or economic value of an area.
Pollution studies may be conducted to meet one or more of the following additional aims:
 — to assess the extent and form of biological change in relation to a particular development, discharge or operation.
 — to measure how effects change with time.
 — to assess whether changes become sufficiently serious to warrant taking corrective action.
 — to define areas of effect for compensation.
 — to monitor recovery of damaged resources to a normal or usable condition.
 — to assess the effectiveness of clean-up techniques or preventive measures.
 — to improve our understanding of the way living systems work and to allow better protection and management.

It is not appropriate to provide an exhaustive summary of suitable survey methods for Red Sea habitats as they have been reviewed and described in numerous publications. Some particularly useful publications are as follows: *Corals*: Stoddart and Johannes (1978); *Sediments and Macrobenthos*: Holme and McIntyre (1981), Stirn (1981); *Intertidal and Subtidal Rocks*: Price et al. (1980), Hiscock (1979).

It is obviously impractical to provide a complete biological description of coastal communities and the fate of pollutants in them, even for the most easily accessible and well-studied parts of the marine environment. Species, populations or sections of communities, habitats or defined resources must be selected for study on the basis of knowledge of their usefulness as indicators of pollution effects, or for their economic or biological value. Where this is not known, more general studies of patterns of organism distribution may be appropriate. Wherever possible sampling should be aimed at those organisms which show the least spatial and temporal variation and can be sampled quantitatively. Such organisms are usually sedentary or have low mobility and so cannot avoid pollution.

In the Red Sea these constraints naturally focus attention on coral reefs, seagrass beds, shallow coastal lagoons and mangrove areas. However, it is sometimes necessary to assess or monitor populations or communities which are difficult to quantify but are either commercially useful (e.g. fish and shellfish or their eggs and larvae) or of conservation value (e.g. marine birds and mammals). In such cases the 'best practical means' must be employed, but the data obtained are likely to be equivocal and would need cautious interpretation.

Where pollution arises from a point source (e.g. effluent discharges, sewage discharges), a particularly useful approach is to define gradients of effect away from sources, with the most remote sites being sufficiently far away to be outside areas of effect and thus act as reference or control sites. Although it is desirable to obtain accurate quantitative data, rapid semiquantitative or qualitative methods can provide valuable information. Financial constraints or simply the size of the area in question may also mean that only simple, rapid descriptive approaches and photography are all that are practicable, and in such cases it may be useful to select carefully a few 'representative' sites for more detailed study. In the Red Sea, fringing coral reefs are extremely extensive, and such an approach is particularly appropriate.

Some examples of methods used in the Red Sea illustrate the range of approaches which may be employed. Wenninck and Nelson-Smith (1977, 1979) have studied coastal oil pollution on the Red Sea coasts of Egypt, Saudi Arabia and Jordan. Their field surveys were carried out by visits to selected areas and by overflights of extensive sections of coastline in helicopters and light aircraft. In the field visits the biologists examined shorelines, provided brief descriptions and maps of habitat types, and obtained some biological samples for laboratory inspection. The degree of oil pollution was assessed visually. Information on fisheries, commercial interests, tourism and industrial developments was obtained from fishermen, appropriate government ministries and departments, and industrial companies or groups. Where necessary field observations were backed-up by photography and by study of available scientific literature. Their approach provided a wealth of general observations for a very large coastline, but ecological and analytical information at specific sites was almost completely absent. Nevertheless, their reports provide an excellent basis for identifying useful future studies and some broader aspects of environmental management.

A second, and again relatively simple scheme has recently begun to obtain specific information on Red Sea coral reefs. Entitled 'Reefwatch', its objectives are set out as follows (Dawson-Shepherd, *pers. comm.*).

> Reefwatch is an underwater project designed to enable divers or expeditions visiting coral reefs on holiday or for more serious purposes to contribute to our knowledge of coral reefs. We badly need detailed information about the present condition of coral reefs in many areas, and to learn more about various processes which can destroy the reef communities, so that coral reefs can be conserved more effectively. Divers are asked to complete a card asking for basic information on the form of the reef, its degree of coral cover and the extent of any pollution or other damage. The card also asks for supplementary information on the number of sea urchin, Crown-of-Thorns starfish, and groupers seen, and on the species of butterflyfish present, and if divers can complete one or more of these sections this too could be of great value. Full guidance on how to record the information is given in a project booklet, which also includes diagrams of the species involved. These species are also illustrated in a mini-print set. The project also aims to provide more general advice for those taking part in the project on planning visits to, and diving in, coral reef areas.

A combination of simple photographic and descriptive studies with detailed quantitative scientific study has been used at an oilfield and marine terminal development site in the Gulf of Suez at Ras Budran (Dicks, 1984). The study area was *ca.* 10 km square, encompassing the proposed location of an offshore oilfield, pipeline routes, a coastal oil-processing terminal and offshore oil loading buoy system for tankers.

The main survey was preceded by discussions with management, engineers and planners as to the form of developments, and preliminary site visits were made. The overall purpose of the environmental programme was to provide information to the oil company management to allow minimisation of environmental damage. The specific aims, defined following the preliminary discussions, were *a*) to describe the main biological communities and types of habitat of the Ras Budran oilfield and terminal and to identify those particularly at risk from developments; *b*) to establish quantitative sampling stations at selected locations as the first stage of a long-term monitoring programme which aims to

Fig. 18.12. Development of an oil terminal at Zeit Bay. Environmental impact assessments and dialogue between scientists and development engineers can keep damage to a minimum during both the construction and operation of the plant.
(*Photo*: OPRU)

identify and assess the impact of the oil company operations on marine communities; *c*) to identify existing pollution of the area, particularly by oil; and *d*) to incorporate biological information into future planning and operation (including preparation of a response plan to deal with oil spills).

By careful selection of habitats and methods of study it was possible to provide adequate data for decision-making within a practical and cost-effective framework. Surveys and sampling were carried out on seabed sediments, patch reefs and a fringing reef system, nearshore lagoons, intertidal rock and sand (Fig. 2). Sediments were also collected for analysis of oil and organic carbon content and for particle size analysis. Sampling strategies included *a*) descriptive surveys of belt transects, both intertidally and subtidally, *b*) preparation of a detailed photographic record of nearshore areas, *c*) grab sampling and trawling offshore, and *d*) the collection of core samples in intertidal and nearshore sediments. A wide range of habitats and communities was described, the richest and potentially most susceptible being the fringing reef and inshore lagoon system at Ras Budran. As a result of the survey a series of recommendations was prepared and acted on by the Suez Oil Company management. These included proposals for the location of effluent discharges, pipeline routes, the construction of a jetty, and an oil spill contingency plan. An identical approach has been used for the Zeit Bay Field further south in the Red Sea. (Fig. 18.12).

18.5. CONCLUSION

I have attempted to identify and summarise the most likely extent of current and future pollution problems in the Red Sea, as well as outlining some of the practical aspects of pollution surveys. There is obviously no single approach or technique which is applicable in all situations, and decisions must be made at every stage of planning and execution of surveys as to the most appropriate strategy, the correct types of sampling equipment and the best ways to treat and interpret data. Although standardisation may be desirable and appropriate to particular types of investigation or, in some cases, geographical areas, it is evident that rigid standards are not widely applicable in view of the site-specific nature of most pollution incidents, industrial developments or discharges. In practice, it is often useful to keep a degree of flexibility in a programme to allow easy adaptation to field conditions or to the findings of previous surveys.

Two related questions are frequently posed about biological monitoring of polluted areas. At what point does damage to marine systems become significant, and how long must monitoring continue? There are no simple answers or magic numbers that can be applied at present. Where very little or no effect is found in studies of particular habitats over a period of years, it may not be unreasonable to expect that little significant damage will have occurred to the system as a whole. As the database for a particular sea area grows it should be reviewed by scientists, industrialists and government and where a consensus of opinion finds little cause for concern, monitoring studies may then be reduced in frequency or stopped. But what if a clear area of effect, slowly spreading form a source, is found? Many industrial developments produce areas of biological damage, and these are part of the price paid for development. In most instances, decisions about levels of acceptability of the damage, the need for remedial action and the form such action should take, are made on a case-by-case basis, taking into account local conditions, amenities, resources and habitat values, and there is little fault in this approach provided decisions are made on the basis of suitable data. At present, the database is somewhat limited in the Red Sea. Governments and industry have a responsible role to fulfil, in co-operation with the scientific community, to design and carry out further studies and to interpret data on which decisions can be made. An excellent starting point would be the preparation of a general assessment of the coastal resources of the Red Sea, much of which could be prepared from existing marine charts and publications without need for extensive field survey. Given that the Gulf of Suez is one of the largest and most rapidly-developing offshore oil production areas of the world, as well as one of the most chronically oil-polluted areas, it is an ideal location for detailed studies of effects of oil in seagrass beds and coral systems, which are currently poorly known and understood. No rigorous comparisons have been made between existing clean and polluted areas for most of the Red Sea, and should be a priority for future studies.

ACKNOWLEDGEMENTS

I am particularly grateful to the following individuals for providing information or critical review of the manuscript: Drs. Jenifer Baker, Keith Hiscock, Alasdair Edwards, John Hartley, Alec Dawson-Shepherd. I am also indebted to Lucille Evans for typing of the manuscript.

REFERENCES

American Institute of Biological Sciences (AIBS) (1978). *Proc. of Conference on Assessment of Ecological Impact of Oil Spills.* June 1978, Keystone, Colorado, U.S.A.

American Petroleum Institute (1980). *Proc. of Symposium on Research on Environmental Fate and Effects of Drilling Fluids and Cuttings.* January 1980. Lake Buena Vista, Florida, U.S.A. Two volumes.

Baker, J. M., Little, D. and Oldham, J. (1982). Comparison of the fate and ecological effects of dispersed and non-dispersed oil in a variety of marine habitats. In *Proc. of Symposium, Oil Spill Dispersants: Five Years of Research.* 12-13 October 1982, West Palm Beach, Florida, U.S.A. ASTM.

Bryan, G. W. (1976). Some aspects of heavy metal tolerance in aquatic organisms. In *The Effects of Pollutants on Aquatic Organisms.* Ed. A. P. M. Lockwood, pp.7–34. Cambridge University Press.

Chesher, R. H. (1975). Biological impact of a large scale desalination plant at Key West, Florida. In *Tropical Marine Pollution.* Ed. E. J. Ferguson Wood and R. E. Johannes. Elsevier Oceanography Series 12, pp. 99–153.

Cowell, E. B. (1978). Pollution of coastal zones by hydrocarbons. Oral submission to the European Parliament. European Parliamentary Hearings, 4 July 1978, pp.14–15.

Dicks, B. (1982). Monitoring the biological effects of North Sea platforms. *Mar. Pollut. Bull.* 13, 221–7.

Dicks, B. (1984). Oil pollution in the Red Sea — environmental monitoring of an oilfield in a coral area, Gulf of Suez. *Proceedings of Mabahiss/John Murray Symposium: Marine Science of the Northwest Indian Ocean and Adjacent Waters.* September 1983, Alexandria, Egypt, U.A.R.

Dicks, B. and Hartley, J. P. (1982). The effects of repeated small oil spillages and chronic discharges. *Phil. Trans. R. Soc. Lond.* B 297, 285–307.

Eisler, R. (1973). Latent effects of Iranian crude oil and a chemical oil dispersant on Red Sea molluscs. *Israel J. Zool.* 22, 97–105.

Eisler, R. (1975). Toxic, sublethal, and latent effects of petroleum on Red Sea macrofauna. *Proc. of 1975 Conference on Prevention and Control of Oil Pollution.* 25–27 March 1975. San Francisco, California, U.S.A. EPA/API/USCG, pp.535–40.

Fishelson, L. (1973). Ecology of coral reefs in the Gulf of Aqaba (Red Sea) influenced by pollution. *Oecologia (Berl.)* 12, 55–67.

Georghiou, L. and Ford, G. (1981). Arab silver from the Red Sea mud. *New Scientist* 19 February 1981, pp. 470–2.

Golob, R. (1980). Statistical analysis of oil pollution in the Kuwait Action Plan Region, and the implications of selected oil spills worldwide to the Region. In *Proc. of International Workshop on Combating Marine Pollution from Oil Exploration, Exploitation and Transportation in the Kuwait Action Plan Region*, December 1980, Manama, Bahrain. IMCO/UNEP.

Hiscock, K. (1979). Systematic surveys and monitoring in nearshore sublittoral areas using diving. In *Monitoring and Marine Environment*. Ed. D. Nicholls, pp. 55–74. Institute of Biology, London.

Holme, N. A. and McIntyre, A. D. (eds.) (1981). *Methods for the Study of Marine Benthos*. IBP Handbook no. 16. Blackwell Scientific Press, Oxford.

International Union for the Conservation of Nature and Natural Resources (1982). *Report on the Global Status of Mangrove Ecosystems*. IUCN Commission on Ecology. Gland, Switzerland.

Jacobs, R. P. W. M. (1982). *Component Studies in Seagrass Ecosystems along West European Coasts*. Ph.D. thesis, University of Nijmegan. Drukkerij Verweij B.V., Mijdrecht, 216pp.

Johannes, R. E. (1970). How to kill a coral reef I. *Mar. Pollut. Bull.* 1, 186–7.

Johannes, R. E. (1971). How to kill a coral reef II. *Mar. Pollut. Bull.* 2, 9–10.

Johannes, R. E., Maragos, J. and Coles, S. L. (1972). Oil damages corals exposed to air. *Mar. Pollut. Bull.* 3, 29–30.

Jordan, R. E. and Payne, J. R. (1980). *Fate and Weathering of Petroleum Spills in the Marine Environment*. Ann Arbor Science, 174pp.

Kinsey, D. W. (1973). Small-scale experiments to determine the effect of crude oil films on gas exchange over the coral back reef at Heron Island. *Environ. Pollut.* 4, 167–82.

Knapp, A. H., Sleeter, T. D., Wyers, S. C., Frith, H. R., Smith, S. R. and Dodge, R. E. (1983). The effects of oil spills and dispersant use on corals: a review and multidisciplinary experimental approach. *Oil Petrochem. Pollut.* 1, 157–70.

Legore, R. S. (1983). A field experiment to assess impact of chemically dispersed oil on Arabian Gulf corals. In *Proc. of Middle East Technical Conference of the Society of Petroleum Engineers*. Manam Bahrain, 14–17 March 1983, pp. 51–9.

Little, D. I., Baker, J. M., Abbiss, T. P., Rowland, S. J. and Tibbetts, P. J. C. (1981). The fate and effects of dispersant-treated compared with untreated crude oil, with particular reference to sheltered intertidal sediments. In *Proc. of 1981 Oil Spill Conference (Prevention, Behavior, Control, Cleanup*. 2–5 March 1981, Atlanta, Georgia, U.S.A. EPA/API.USCG, pp. 283–96.

Loya, Y. (1975). Possible effects of water pollution on the community structure of Red Sea corals. *Mar. Biol.* 29, 177–85.

Loya, Y. and Rinkevich, B. (1980). Effects of oil pollution on coral reef communities. *Mar. Ecol. Prog. Ser.* 3, 167–80.

Lugo, A. E. and Snedaker, S. C. (1974). The ecology of mangroves. *Ann. Rev. Ecol. System* 5, 39–64.

Macnae, W. (1968). A general account of the fauna and flora of mangrove swamps and forests in the Indo-West Pacific region. *Adv. Mar. Biol.* 6, 73–270.

Malins, D. C. (ed.) (1977). *Effects of Petroleum on Arctic and Subarctic marine Environments and Organisms. Vol. I. Nature and fate of Petroleum*. 321 pp. *Vol. II. Biological Effects*, 500 pp. Academic Press, New York.

Malins, D. C. and Hodgins, H. O. (1981). Petroleum and marine fishes: a review of uptake, disposition and effects. *Environ. Sci. Techn.* 15(11), 1272–80.

National Oceanic and Atmospheric Administration (NOAA) (1979). *The 'Argo Merchant' oil spill: a preliminary report*. National Oceanic and Atmospheric Administration, U.S. Department of Commerce.

Neuman, L. D. (1979). The protection and development of the marine environment and coastal areas of the Kuwait Conference Region: the program of the United Nations System. In *Proc. of 1979 Oil Spill Conference (Prevention, Behavior, Control, Cleanup)*. 19–22 March 1979, Los Angeles, California, U.S.A. EPA/API/ISCG. pp. 287–91.

Pearson, T. H. and Rosenberg, R. (1978). Macrobenthic succession in relation to organic enrichment and pollution of the marine environment. *Oceanogr. mar. Biol. A. Rev.* 16, 229–311.

Peters, E. C., Meyers, P. A., Yevich, P. O. and Blake, N. J. (1981). Bioaccumulation and histopathological effects of oil on a stony coral. *Mar. Pollut. Bull.* 12(10), 333–9.

Price, J. H., Irvine, D. E. G. and Farnham, W. F. *The Shore Environment. Vol. 1. Methods*. Systematic Association special volume 17(a). Academic Press, 321pp.

Ray, J. P. (1981). The effects of petroleum hydrocarbons on corals. *Proc. of Petromar 80 Conference, Petroleum and the Marine Environment*. 27–30 May 1980, Monaco. Graham & Trotman Ltd., pp. 705–26.

Read, P. A., Anderson, K. J., Matthews, J. E., Watson, P. G., Halliday, M. C. and Shiells, G. M. (1983). Effects of pollution on the benthos of the Firth of Forth. *Mar. Pollut. Bull.* 14(1), 12–16.

Reish, D. J., Soule, D. F. and Soule, J. D. (1980). The benthic biological conditions of Los Angeles-Long Beach harbors: results of 28 years of investigations and monitoring. *Helgoländer Meeresunters.* 34, 193–205.

Rinkevich, B. and Loya, Y. (1979). Laboratory experiments on the effects of crude oil on the Red Sea coral *Stylophora pistillata*. *Mar. Pollut. Bull.* 10, 328–30.

Royal Commission on Environmental Pollution (1981). *Eighth Report. Oil Pollution of the Sea*. H.M.S.O., London, 307 pp.

Sardar, Z. (1981). Red Sea states unite against pollution. *New Scientist* 19, February 1981, p.472.

Smith, S. V. (1977). Kaneohe Bay: a preliminary report on the responses of a coral reef/estuary ecosystem to relaxation of sewage stress. *Proc. 3rd Int. Coral Reef Symp.* 2, 578–83.

Stirn, J. (1981). *Manual of Methods in Aquatic Environment Research. Part 8. Ecological Assessment of Pollution Effects*. Guidelines for the FAO (GFCM)/UNEP Joint Coordinated Project on Pollution in the Mediterranean. FAO, Rome, 70 pp.

Stoddart, D. R. and Johannes, R. E. (1978). *Coral Reefs: Research Methods*. UNESCO, Paris, 581 pp.

Szmant-Froelich, A., Johnson, V., Hoehn, T., Battey, J., Smith, G. J., Fleischman, E., Porter, J. and Dallmeyer, D. (1981). The physiological effects of oil-drilling muds on the Caribbean coral *Montastrea annularis*. *Proc. 4th Int. Coral Reef Symp.* Manila 1, 163–8.

Thompson, J. H. and Bright, T. J. (1977). Effects of drill mud on sediment clearing rates of certain hermatypic corals. In *Proc. of Oil Spill Conference* 8–10 March 1977, New Orleans, Louisiana, U.S.A. EPA/API/USCG, pp. 495–8.

Topping, G. (1976). Sewage and the sea. In *Marine Pollution*. Ed. R. Johnstone, pp. 301–51. Academic Press, London.

Walker, D. I. and Ormond, R. F. G. (1982). Coral death from sewage and phosphate pollution at Aqaba, Red Sea. *Mar. Pollut. Bull.* 13(1), 21–5.

Wenninck, C. J. and Nelson-Smith, A. (1977). *Coastal Oil Pollution Study for the Kingdom of Saudi Arabia*. Vol.1. Red Sea Coast; Vol.2. Gulf Coast. IMCO, London.

Wenninck, C. J. and Nelson-Smith, A. (1979). *Coastal Oil Pollution Study for the Gulf of Suez and the Red Sea coast of the Republic of Egypt*. IMCO, London.

CHAPTER 19

Conservation and Management

RUPERT ORMOND

Tropical Marine Research Unit, Biology Department, University of York, U.K.

CONTENTS

19.1. INTRODUCTION

It is certain that until very recently, the Red Sea has been one of the world's marine areas least affected by man's activities. With the exception of Suez, at the extreme north-west, the few towns situated along the coasts of the Red Sea were small and under-industrialised, and between the towns, only sparse populations of nomads and fisherman inhabited the inhospitable coastline. Industrial impacts were probably limited to the oil field developments of the Gulf of Suez, and the jettisoning of garbage and waste oil by shipping passing through the Red Sea and Suez Canal. It is most unlikely that the low level of artisanal fishing, probably unchanged in effort or technique for hundreds of years, caused significant overfishing, even in the neighbourhood of the coastal towns. In the late 1960s, probably 98% of the total Red Sea coast was in practically virgin condition.

Since that time, the situation has changed very rapidly, as intense commercial and industrial development has taken place along many coastal sections, accelerated by the rapid influx of oil-wealth to the area. Since the re-opening of the Suez Canal in 1976, there has been a resurgence of shipping through the Red Sea, further increased by the opening of trans-Suez and trans-Arabian pipelines, each with oil terminals in the Red Sea. Most countries in the area are endeavouring to expand their sea fisheries, and the populations of rapidly expanding cities such as Jiddah and Yanbu al Bahr have increased five-fold in the last 15 years. These developments are now at a stage when many signs of impact and degradation are becoming common in certain sections of the Red Sea coastline. The rapidly deteriorating environmental situation calls for urgent and effective conservational and environmental management measures to conserve and protect the unique resource of the Red Sea marine environment.

19.2. RATIONALE FOR PROTECTION AND MANAGEMENT

Fortunately, the advantages of protecting environments, including those of the coast and the sea, are increasingly widely appreciated. In the Red Sea context, the main reasons for conservation include the maintenance of fisheries, recreation and tourist resources to which concrete cash values can be set, and the more abstract but important values of education and scientific knowledge, and preserving unique species and ecosystems for our descendants.

19.2.1. Fisheries

Under this heading may be grouped fin and shell fisheries for food, fisheries for products such as mother-of-pearl or coral, and potential future fisheries built around for example the pharmaceutical market. The present fisheries of the Red Sea are briefly described in chapter 17 of this book. It must be remembered that like all coral reef areas, the Red Sea is oligotrophic, of low primary and secondary productivity. The main fishery resources are therefore localised in the extremely limited shallow water shelf areas of reef, mangrove and seagrass, crucial in providing food and habitat for the commercial species.

In general, the countries bordering the Red Sea are taking care to develop their fisheries on a sustainable-yield basis. When the oil finally runs out, the fisheries will represent perhaps the principal remaining renewable resource of the area. The fisheries resource is even more important for the poorer countries without oil reserves. Major fishery development programmes, mostly supported by the FAO

(Food and Agricultural Organisation of the United Nations) are underway in Egypt, The Sudan, Saudi Arabia and North Yemen.

Maintenance of the fishery resource requires control of the water quality, the protection of critical habitats and measures to guard against overfishing. Ideally, each fishery should seek to obtain the maximum sustainable yields, an output as large as possible without running the risk of depleting stocks or causing the extinction of the entire fishery. There should also be minimal environmental damage through direct or indirect ecological perturbations caused by the fishing process or stock removal.

19.2.2. Recreation

The term recreation is used here to distinguish the leisure activities of local people from those of tourists visiting from outside the area. In relation to the Red Sea, it includes sunbathing and picnicking on the beaches, swimming, boating, snorkel and SCUBA diving, fishing and studying wildlife. Until recently few local Red Sea inhabitants pursued such activities, but the number is rapidly growing, especially in the wealthier and more developed countries. Fishing and beach parties are now popular, and many young people are taking up forms of diving. With increasing wealth and awareness of the pleasures to be found in and around the sea, the numbers of people using the Red Sea coast for recreation will increase rapidly.

19.2.3. Tourism

Tourism, both regional and international, is always most important to countries without other natural resources, and with a need for foreign exchange earnings. Most of the coast of the Red Sea is perhaps too hot, or still too remote to attract ordinary tourists, but the coral reefs have an international reputation that attracts divers from all over the world. The largest tourist industry is based in Sinai, including the diving centres of Sharm El Sheikh and the magnificent headland and reefs of Ras Muhammad. By the late 1970s, upwards of 10,000 divers and snorkellers were visiting the Sinai coast every year. Although the use of the area was temporarily checked by political changes, numbers should soon exceed this early level as Egyptian authorities continue to improve the efficiency of their tourist operations.

International tourism has slowly been growing on the mainland coast of Egypt, centred around the port of Hurghada. At the peak of the season as many as 20 boats carrying divers ply daily from the port. There is at least one diving operation at Safaga, and a few independent diving tourists explore as far south as Marsa Alam. The beaches of Suez attract visitors from Cairo, but many are closed by military activities, and the area is not attractive for SCUBA diving.

The Sudan has enormous potential for dive tourism, possessing in the Port Sudan area some of the most spectacular reefs of the whole Red Sea. Dive tourism got off to a good start there in the 1970s, with two or three operators and the construction of a tourist village. Unfortunately, economic difficulties beset the Sudan and tourism was curtailed, the village remaining unoccupied. Recently, tourism has once more increased, and perhaps as many as 1000 divers visit Port Sudan each year.

There has been a slow but steady increase in the numbers of visitors to the North Yemen coast since the country began to open up in the 1970s. Marine activities there still tend to be rather incidental to sightseeing and mountain exploration and there is no specific diving tourism industry. The reefs of the area are not as well developed as those of the central and northern Red Sea.

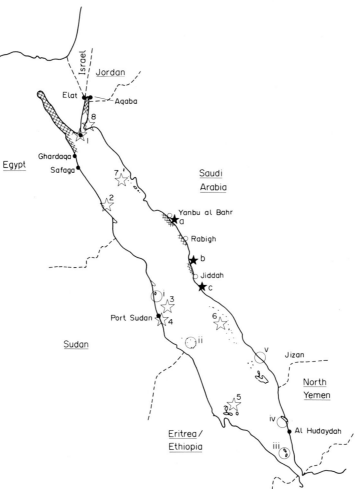

Fig. 19.1. Map of the Red Sea showing areas mentioned in the text where problems of environmental damage have been noted, tourist industries occur or Marine Parks have been proposed.

KEY: Cross hatching = area of pollution or coastal development

    ☆ = Existing or proposed Marine Park
        1. Ras Muhammad and Sharm El Sheikh
        2. Hamata marine zone
        3. Sanganeb Reef
        4. Port Sudan Coast
        5. Dahlak Islands
        6. Farasan Bank
        7. Wajh Bank
        8. Saudi coast of Gulf of Aqaba and environs
        (Several marine parks existed on the west coast of Gulf of Aqaba during Israeli occupation.)

    ○ = Important conservation areas for consideration as protected areas
        i. Taila and Mukawwar Islands
        ii. Suakin Archipelago
        iii. Zuqar Islands
        iv. Isa peninsula
        v. Al Quhmah — Shuqayq coast

    ★ = Possible recreational or education sites
        a. Yanbu al Bahr area
        b. Ras Hatiba,
        c. Shu'aiba

    ● = Towns with some marine tourism industry

Dive tourism is also restricted in Jordan, which has a very small coastline, much of which is occupied by the essential industrial developments. Few tourists travel to Jordan specifically for a coastal holiday, but again the marine activities form a valuable adjunct to a general sightseeing holiday, and five or more hotels operate in Aqaba, catering for international tourists and local holidaymakers.

In the remaining countries bordering the Red Sea international tourism is very restricted and likely to remain so for the foreseeable future. Civil conflict has closed the coasts of Ethiopia and Eritrea to outsiders. The Red Sea coast of Saudi Arabia is considered a holy area and international tourism is not permitted, except for the annual influx of several million Muslim pilgrims for the Haj pilgrimage. In course of time increasing numbers of these pilgrims may be interested in coastal recreation close to the towns which they are visiting.

19.2.4. Education, science and conservation

The intertidal and shallow sublittoral environments of the Red Sea are excellent sites for teaching many aspects of biology and ecology. The coral reefs offer almost unparalleled opportunities for research into the principles of organisation and stability of complex natural ecosystems. In a regional context, education in the natural sciences is becoming as essential for administrators as for biologists, doctors and environmental managers.

Finally, it is perhaps as well that many people regard conservation as an important activity in its own right, reflecting a moral obligation for mankind to attend to the interests of other life forms with which we cohabit this planet. First hand acquaintance with the magnificent reef habitats of the Red Sea is likely to encourage this attitude in people otherwise more swayed by the economic arguments for conservation.

19.3. HUMAN IMPACTS IN THE RED SEA

In almost every part of the world, the intertidal and nearshore marine environment has recently come under pressure from a variety of types of human impact. Ruivo (1972) and Johnston (1976) have reviewed these impacts for the sea in general, while Wood and Johannes (1975) provided an excellent review of the special problems of tropical marine environments. Many types of impact are already obvious in the Red Sea, those concerned with pollution, in the sense of addition of foreign and/or noxious substances have been discussed in Chapter 18. By way of introduction, Table 19.1 lists the main classes of human impact and their causes which could cause problems in the Red Sea.

19.3.1. Sedimentation

Sedimentation is the settling out onto the seabed or shore of significant quantities of silt, sand or other materials put into suspension in the marine environment by various agents including human activities. It is possibly the major cause of damage to marine habitats throughout the tropics. Unfortunately, because sedimentation is not readily perceived as a form of pollution, and because silt and sand are major natural constituents of the seabed and shore, the damage that sedimentation can and is causing in the marine environment is rarely appreciated, by either developers, government officials or the general public.

Sediment occurs naturally in suspension in the sea, especially where a shallow seabed or the shoreline is composed of silt or sand and is stirred up by wave action. Under more sheltered conditions this material settles out onto the seabed again, where it may adversely affect benthic communities. Coral reefs, and to a lesser extent seagrass beds, are not able to tolerate heavy sedimentation. Some coral species such as the brain coral *Platygyra*, *Fungia* and the long-branched Acroporas are able to tolerate some sedimentation, and in shallow sheltered areas where fine material is in the water, specialised coral communities tend to develop. Such communities, dominated by *Acropora* and *Galaxea fascicularis* have been recorded in the Red Sea in shallow inshore areas in, for example, the Dahlak and Farasan archipelagos (Wainwright, 1965; IUCN, 1984). But even those corals able to withstand a little fine sediment cannot survive for long the heavy loads of coarse grain material generated by coastal infilling or dredging, or brought down river into the sea following deforestation and soil erosion on land.

There are of course no permanent rivers joining the Red Sea, and dredging is rarely required, but the practice of land infilling along the coast is becoming more prevalent, and this damages or destroys coral and seagrass communities. The principal example of such damage has occurred along the coast north of Jiddah, where during the late 1970s there was extensive infilling to build a sea front road. Studies sponsored by the Meteorological and Environmental Protection Administration of the Saudi Government have shown that subsequent washing out of fine grained infill material resulted in sedimentation of the adjacent fringing reef (Fig. 19.2). Along some sections of the reef this led to the death of much of the living coral (IUCN, 1984). Following completion of the road sedimentation has decreased and some recolonisation of dead areas is taking place.

19.3.2. Coastal construction work

Coastal construction work can include the construction of harbours, jetties, and residential or industrial sites on the intertidal or immediate supratidal zone. As well as frequently causing sedimentation in adjacent areas, these activities can result in direct destruction of coral communities, seagrass beds, mangrove stands, mud flats, halophyte vegetation or other natural habitats. Such damage is not yet extensive in the Red Sea area, but it is occurring with increasing frequency, mainly around the major towns, and particularly in the regions of Jiddah, Rabigh and Yanbu al Bahr. In addition to major municipal and industrial projects, it is becoming common for individual landowners to construct private jetties and to dredge small boat channels through the fringing reef (Schroeder, 1982). In addition to the direct destruction of coral which is built over or dredged up, solid jetties can cause the area down current of the jetty to become stagnant, and marine life succumbs to the sedimentation and raised temperatures that occur in such enclosed shallow water areas.

TABLE 19.1. Resumé of impacts known or likely to occur in the Red Sea and discussed in the text.

Class of impact	Cause or origin
POLLUTION	
Oil	Production water, pipelines, tankers. Debunkering, waste oil, accidents.
Heavy metals	Industrial, offshore mining, desalination plants.
Chemicals	Industrial discharge.
Sewage	Hotels, municipal and residential waste.
Phosphates	Ore loading, fertiliser plants.
Solid waste	Domestic, municipal, shipping.
OTHER IMPACTS	
Thermal pollution	Desalination plants, power stations.
Sedimentation	Coastal infilling, dredging.
Coastal construction	Harbour works, coastal infilling, Industrial and residential development.
Fishing	Overfishing, dynamiting, spearfishing, aquarium fish trade.
Direct exploitation	Shooting, mangrove felling, egg taking.
Visitor impacts	Coral and shell collecting, anchor and vehicle damage, spearfishing. Souvenir trade.

Resumé of impacts known or likely to occur in the Red Sea and discussed in the text.

Fig. 19.2. The fringing reef, near Jiddah, Saudi Arabia, showing the damage caused by heavy sedimentation resulting from infilling activity associated with the construction of a seafront road. At this point, what ten years previously was a colourful reef with a high density of living coral is now mostly bare reef, coated with a turf of green algae, and only scattered living corals. (Copyright IUCN/TMRU).

Perhaps the most serious aspect of the current expansion of urban regions is the strong tendency for development to occur in strip-like fashion along the coast to either side of each major coastal city. There is an expanding need for ports and docking facilities, and residences with access to the shore are always sought after. Industrialists often prefer to locate major plants along the coastline where cooling water and private docking facilities are available and in order to facilitate waste disposal. However, strip-wise development along the coastline inevitably leads to a slow degradation of the littoral and sublittoral environment. Such strip development is already proceeding rapidly from Jiddah, Yanbu al Bahr, Port Sudan and Aqaba. In the region of Jiddah as much as 100 km of coastline have been or are in the process of development compared with under 10 km less than fifteen years ago. In the Red Sea region, where there are large expanses of arid and unproductive hinterland available, it would seem desirable to direct development away from the coast, protecting the shore and marine environment for recreation, tourism and fisheries.

19.3.3. Fishing and other forms of exploitation

The sustained exploitation of fisheries is a benefit which most marine environmental policies seek to realise. However, uncontrolled fishing, or fishing of certain types, may lead to deterioration or complete loss of the fishery, or cause other environmental damage. In the Red Sea it is unlikely that there has been significant overfishing except in the vicinity of towns and fishing villages, or very possibly of some stocks within the intensively fished Gulf of Suez area. However, throughout the Red Sea and particularly in Egypt, Sudan, Saudi Arabia and North Yemen, intense efforts have been made in the last ten years to modernise and expand the fisheries, and it seems likely that increasingly careful fisheries management will be needed to check that sustainable yields are not exceeded.

Of more concern at present are certain potentially damaging fishing techniques. Fortunately dynamiting reefs for fish is everywhere illegal, and remains uncommon. It has occurred in the northern Gulfs in recent years, and can cause wholesale destruction of the reef structure and community. Large mesh gill-nets are used on a small scale to catch sharks in some areas, but unfortunately also trap and destroy air-breathing animals such as turtles, cetaceans and the rare Dugong.

Suitably managed, spearfishing is an acceptable fishing technique of great selectivity. It is however undesirable in practise for two reasons. Spearfishing is most commonly done by visitors with no personal interest in 'sustainable yields' and is therefore in competition with local fishermen. It also tends to be conducted on the more accessible and attractive reefs most valuable for recreation and tourism. Spearfishing rapidly leads to the elimination of favoured food species such as grouper and snapper, and fish once 'tame' and approachable become extremely wary of divers. Spearfishing is theoretically banned in Sudan, Egypt and Jordan, but it is hard to control fishing by divers from yachts which enter territorial waters without passing through local immigration controls. It is pleasing that dive tour operators have begun to enforce their own spearfishing prohibitions among their visiting clients.

Attempts have been made in Sudan and Saudi Arabia to catch and export reef fish for the aquarium trade. This has caused some local depletion of favoured species and damage to coral colonies. Fish collectors naturally tend to operate on reefs close to major communication routes such as international airports, and these are also the areas most critical for developing tourist industries.

There is also (or has been) direct exploitation of some other natural resources of the Red Sea. These include turtles, nesting birds, shellfish, sea-cucumber, black coral and mangrove wood (*see* for example Hiscock and Al Aiidy, 1972). On the whole, the level of exploitation has been low and sustainable. The collection of mother-of-pearl shell ('trochus') has been quite intensive along the western shore of the Red Sea, but until recently was conducted inefficiently using glass-bottomed boxes to locate the molluscs, and large populations were able to survive on most reefs. Another mollusc resource is the horny operculum of muricid gastropods, which when ground into powder (*dufra*) is sold locally as an aphrodisiac. This trade may have led to serious local stock depletion. The distribution of turtles and the less common of the two mangrove species (*Rhizophora mucronata*) may have been restricted by exploitation for food and building materials respectively.

While the use of such resources in the past has had little or no environmental effect, this may no longer be the case. The coastal populations in some areas are now much greater, access to the coast is much easier, and people are becoming equipped with modern diving gear, guns and engine-powered boats. Few local inhabitants have been educated to have any concern for the long term health of the environment, but most wish to acquire material possessions and are becoming more aware of the economic potential of Red Sea resources. Near the major towns for example, black coral (*Antipathes*), once used traditionally for prayer beads and jewelry, is now practically extinct within divable depths, having been removed by visitors and expatriates with SCUBA gear. The few turtle colonies on the Saudi coast are becoming exploited by foreign labourers from third world countries where turtles have traditionally been used for food (IUCN, 1984). In the same area, the casual shooting of migrant birds for 'fun' rather than food is becoming common. A further serious threat arises from the establishment of many military stations along the coastline and on islands. Soldiers typically make considerable use of bird or turtle eggs, and spend their off-duty time in intensive fishing and shooting within the immediate area.

19.3.4. Visitor Impacts

While recreation and tourism present major reasons for protecting the coastal and marine environment, uncontrolled use by visitors can itself result in serious impacts. In the Red Sea context, extensive collection of corals, shells and other reef animals, spearfishing, damage to corals by swimmers and anchors, destruction of coastal vegetation by trampling and vehicles, and proliferation of waste and garbage, are the types of damage that may be caused by visitors and tourists. Most of these impacts are detectable in areas close to the major coastal towns, and in particular signs of such damage are especially apparent along 100–200 km of coastline in the region of Jiddah. Around Jiddah the most noticeable visitor impact arises from the collection of a wide variety of reef animals (as food and souvenirs) by workers from third world countries, hundreds of whom may arrive at a coastal site and, in line abreast, work their way across the lagoon and reef flat collecting corals, shellfish, urchins, octopus and fish. Extensive loss of the narrow strip of supralittoral halophyte vegetation is also occurring with the proliferation of 4-wheel drive vehicles. At weekends more than 100 vehicles per km of shore may be parked by picnickers and visitors along shorelines well away from any made-up road. Shell collecting, mainly by European and American expatriates is also resulting in the increasing rarity of many species in the Jiddah region.

In countries where diving tourism is more important, such as Egypt, Jordan, Sudan (Fig. 19.3), the main noticeable impacts have been due to anchor damage, coral collection, spearfishing and handline fishing (by the crews of the diving boats!) at the most popular diving sites, and collection of corals, shells and other souvenirs by fishermen and merchants for sale in shops and stalls to tourists. Such damage has not been too great, but has been sufficient to cause comment by more discerning visitors. As described below, moves have been made in each of these countries to try and control these impacts.

In Egypt especially, a threat may have arisen from the sale of dried and inflated puffer fish as souvenirs. In particular many specimens of the puffer *Arothron hispidus* are made into lampshades (Fig. 19.4). This fish is a major predator of the Crown-of-Thorns Starfish (Ormond and Campbell, 1974), and of the needle-spined sea urchin *Diadema setosum*, both of which can cause extensive damage to corals. There is evidence that reef damage near Hurghada by abundant urchins may be related to the elimination of their predators in this way.

19.4. HABITATS AND SPECIES

Having indicated the need for conservation and control, the question arises as to what precisely one should seek to protect. This question is most readily considered in terms of species and of habitats.

Most widely appreciated is the concern to protect endangered or threatened species. In practice there are relatively few such species in the Red Sea, or at least few whose identity is known. Most obvious are

Fig. 19.3. Colony of *Porites solida* on a Sudanese reef showing extensive damage by clumsy snorkel divers. To the top left, a clear imprint of a rubber-soled shoe is visible as a zone of bleached coral tissue. Elsewhere, dark patches of algae with bleached borders represent areas of previous damage which have not regenerated. (*Photo*: S. M. Head).

Fig. 19.4. Preserved pufferfishes (*Arothron hispidus*), many turned into lampshades, for sale in a souvenir shop at Hurghada, Egypt. The pufferfish is a major predator of the Crown-of-thorns starfish (*Acanthaster planci*) and of large sea urchins such as *Diadema setosum*, and population outbreaks of these echinoderms may be associated with a decline in number of the pufferfishes. (*Photo*: R. F. G. Ormond/TMRU).

the Dugong and the species of turtle (*see* Chapter 14), of which the populations of Hawksbill turtles are probably the most significant from the international point of view. Among the birds of the sea and coast none are rare, but there are several which are endemic to the Middle Eastern region and which are nowhere common, so that their Red Sea populations deserve top priority protection; these include the Crab plover (*Dromas ardeola*), White-eyed gull (*Larus leucophthalmus*) and Sooty falcon (*Falco concolor*). Some other bird species which breed in small numbers in different parts of the Red Sea also justify protection at various levels of priority, including, for example, Spoonbill (*Platalea leucorodia*), Brown booby (*Sula leucogaster*) and Goliath heron (*Ardea goliath*).

As regards fish and marine invertebrates our state of knowledge is not sufficiently advanced to determine which species are rare and/or in need of protection. However, because in the Red Sea impacts to the sublittoral environment have to date been very localised, it seems very unlikely that any fish or marine invertebrates have yet been threatened within the region specifically because of man's activities. It makes most sense to emphasise the need to ensure the preservation of representative areas of different littoral and sublittoral habitat, so that within them protection can be given to common and rare species alike and to emphasise the fact that within different marine invertebrate groups typically 10—20% of the fauna is endemic to the Red Sea.

Conservation at the habitat level is in any case essential. Protection of a species from direct harm will be unsuccessful if the habitat on which it depends is lost — for example, protection of the Dugong from hunting would be pointless if sufficiently large areas of the seagrass on which it feeds were not also available. Equally it is the habitats as a whole which are required for recreation and tourism, for science and the fisheries.

As described throughout this volume there are a wide range of habitats, littoral, benthic and oceanic, found within the Red Sea. While all are of interest, conservation and management effort should be directed towards those that are most affected by the impacts described in the previous section, and in particular towards what are now generally termed, following Ray (1976), *critical habitats*. Critical habitats are those which are of much greater ecological significance than surrounding areas, either because they are highly productive and directly or indirectly generate the food on which populations living in the region depend, or because they provide breeding or nursery habitats, or because they harbour a very high diversity of living species.

The principal critical habitats of the Red Sea are the coral reefs, seagrass beds, mangrove stands, creeks and shallow bays and offshore islands. The reefs, seagrass and mangroves are important because they are highly productive, contrasting with the low productivity of surrounding waters (Lewis, 1977). All the habitats mentioned are important nursery and breeding grounds for many marine organisms, invertebrates, fish, seabirds and turtles.

The conservation of critical habitats is also important on economic grounds, since the majority of fish stocks exploited in the Red Sea are dependent on one or more of the critical habitats. Thus handlining for demersal species is the principal fishing method in the Red Sea (Neve and Al Aiidy, 1973), and the majority of species caught are mainly or frequently associated with the reef environment (Wray, 1979).

19.5. MANAGEMENT AND CONSERVATION STRATEGY

19.5.1. General considerations

Two significant features of the marine environment need to be kept in mind when considering developing strategies to protect rare species and critical habitats from the threats described above. Salm (1983) makes the point that many of the problems of impact and over-exploitation of the marine environment relate to its legal status as a 'commons'. Just as a terrestrial common belongs to the whole village or township, the sea and its resources are generally regarded as being held in common by and accessible to the whole population. This means there is little incentive for any one individual or organisation to avoid damage or practise restraint because on the whole it is only others who will benefit from their prudence. It must be galling for a fisherman who reduces his own income to maintain stocks in an area, when less scrupulous colleagues move in and decimate the fish population by continued fishing only to move on when that fishery is finished. By contrast, when a resource is owned by or is exclusive to one person or group, they will take care to leave sufficient stock to maintain it in a healthy state.

Ownership of exclusive fishing rights is common in freshwater fisheries, but relatively rare in marine areas. Johannes (1978, 1981) has described how in parts of the Pacific Ocean, individual villages may own the fishing rights to specific fishing areas. In Bahrain in the Arabian Gulf, individual families effectively own near-shore areas in which they maintain extensive fence-like fishing traps called *haddra* (IUCN, 1983). This exclusive ownership naturally reduces the tendency to overfish, and gives the owners greater legal redress against other parties who may pollute or damage the fishing ground. Restricted fishing rights such as these are not found in the Red Sea, so although avoidance of overfishing is in the interests of the whole community, it is not in the interests of any individual fisherman. Protective measures therefore have to be instigated and fishing controlled by the community as a whole.

The second significant feature of the marine environment concerns the spread of pollutants. Diffusion and water currents will transport pollutants released at one point to affect the marine environment and its resources in other areas. Such pollutants are often non-visible and their insidious spread typically difficult to track. On the land however, most pollutants dumped tend not to affect areas far removed but remain largely in the area of deposition where they may be visually quite obvious. This feature of the marine environment makes it essential to establish regulations to control dumping or waste discharge, and to be effective these regulations must be operative locally, nationally and internationally.

Given these considerations, the management and conservation needs of the area and the impacts to which the Red Sea environment is subject it is possible to develop a management strategy to be applied at regional and local level. This strategy should contain four essential elements:

1. Promotion and enforcement of international agreements and national regulations to maintain water quality.

2. Enactment and enforcement of regulations and standards to control impacts due to developments and other activities throughout coastal and offshore areas.

3. The protection of areas of coastal and marine habitat as marine parks or reserves, or as planning zones within which development would be severely restricted.

4. Full protection of rare or uncommon species and the appropriate regulation of fisheries.

19.5.2. Water quality

Protection of water quality requires the enforcement of discharge standards for all forms of liquid and solid waste, and since wastes can spread so easily, these standards must be internationally or regionally agreed. There must also be prohibition of dumping by ships in international offshore waters. International agreements to protect the marine environment have been promoted through the Regional Seas Project of the United Nations Environment Programme (UNEP). Generally this has made considerable strides particularly when compared with the slow progress of the United Nations Law of the Sea Conference. In the Red Sea the regional organisation is the Environmental Programme for the Red Sea and Gulf of Aden (PERSGA), established under the auspices of ALECSO (the Arab League Educational, Cultural and Scientific Organisation). PERSGA now has its headquarters in Jiddah, where three international conferences have been held since 1974. These have led to the initiation of a regional convention which should effectively outlaw pollution and environmental damage within the region (PERSGA, 1981).

Establishment of this convention is a major step towards preserving the marine life described in this volume, but inevitably its effects are as yet limited. Several Red Sea nations have yet to ratify the treaty, enforcement within international waters remains a problem, and the convention has relatively little force within national waters where pollution control is the responsibility of national authorities. National action to establish and enforce discharge standards is still at an early stage. In Saudi Arabia the Meteorological and Environmental Protection Administration was established in the late 1970s and comprehensive standards based on US and European models were promulgated in 1982. In Egypt a new Water Pollution Prevention Law was enacted in 1983, and it is hoped that the creation of an Egyptian Environment Affairs Agency will lead to measures enforcing and strengthening the legislation.

The need for national and international action is most clearly illustrated by the problem of oil pollution outlined in Chapter 18. There is serious oil pollution in the Gulf of Suez linked to oil production in Egyptian waters. National action is needed to establish and enforce standards. At the same time, ships and tankers discharging waste oil and ballast are responsible for an increasing (although still low) incidence of pollution in other parts of the sea. At present blame for individual spills is very hard to apportion. Regular international surveillance, source analysis of spills and strict enforcement of penalties are required.

19.5.3. Regulation of other impacts

Controls or regulations are required to deal with development and other related human impacts. Permits should be required for any dredging or land fill activity, and should be dependent on the methods and safeguards used. Dredging should only be permitted in certain types of areas, and should involve controls on cut angle and the use of curtains to limit sediment spread. Infilling should only be allowed within urban areas or after rigorous impact evaluation studies. Ideally no new developments should be allowed within 200–500 m of the waters edge, except for essential ports and desalination plants which obviously require direct marine access.

19.5.4. Planning zones and protected areas.

In addition to general development control, key marine areas should be given extra protection by stricter planning controls or blanket development prevention. These areas would be selected as those of outstanding fishery or recreational value, or because of the rare species or critical habitats they contain. In the Red Sea region, urban and industrial expansion are proceeding apace but rather locally, and the critical habitats are distributed patchwork-fashion along the coast. Consequently an appropriate management policy for the region may be to establish a comprehensive zoning scheme under which four or more classes of areas are recognised:

1. Urban and development areas in which commercial and industrial development should be concentrated.

2. Standard areas where normal planning regulations apply and impact assessments should be carried out.

3. Multiple Use Management Areas (IUCN, 1983) within which higher standards would apply and development would be restricted. Traditional uses would be allowed to continue.

4. Marine Parks and Reserves affording complete protection to species and habitats. Some areas would be managed for regulated recreation, others would be strict nature reserves for conservation or scientific study.

A pre-requisite for the establishment of a zoning scheme would be a complete habitat survey to determine which areas were most appropriate for which levels of protection.

19.5.5. Protection of species

Protection of individual species is required for two categories of organisms — those which are rare or uncommon, and those which are of commercial value and may become over-exploited. The protection of these species would apply outside Marine Parks and Reserves as well as within them, and species which were subject to regulated fishing outside protection areas would be afforded complete immunity from exploitation within them.

As discussed in section 19.4 above, there are relatively few Red Sea species that can at this stage be identified as rare or endangered. Certainly the list would include turtles and Dugong (Chapter 14) and Whale shark (*Rhincodon typus*). A somewhat longer list could be drawn up of animals which were uncommon or of restricted distribution, and of particular interest. This would include several breeding birds (Chapter 15), cetaceans (Chapter 14), manta rays and perhaps some of the larger and more 'collectable' gastropod molluscs (Chapter 10). Likewise there are some uncommon plant species along the Red Sea shoreline, especially species of *Euphorbia*.

Among exploited species requiring regulation are many fish species (Chapter 17) and various invertebrates mentioned in section 19.3.4 above. As yet it is unlikely that strict regulation measures would be needed except for populations of crayfish (*Panulirus penicillatus*) near fishing centres. Catch regulation by quota imposition would probably therefore be premature, and would in any case be very hard to enforce. As fisheries develop in the different countries of the Red Sea, regulation could most appropriately take the form of declaring certain areas closed on a permanent or seasonal basis. Other successful control measures include a limit on the numbers of fishing boat licenses issued, and restrictions on the size of fish that may be taken.

19.6. NATIONAL DEVELOPMENTS

19.6.1. Egypt

The northern part of the Egyptian coast and the Gulf of Suez were the first Red Sea areas to become heavily influenced by human activities and so were arguably the first areas to require national conservation legislation. Despite the awareness and experience of numbers of Egyptian marine scientists, progress in this direction was initially slow, for political and military reasons. Following the 1967 war, occupying Israeli authorities established a series of protected marine areas along the western shore of the Gulf of Aqaba including a provisional marine park at Ras Muhammad. The park, staffed by a team of wardens and laid out with trails, signs and other facilities greatly stimulated the tourist influx to Ras Muhammad and Sharm el Sheikh. In the absence of fishing and especially spearfishing activities, the numbers of large fish increased and they became much more tame.

With the return of the area to Egypt, Ras Muhammad's marine life was no longer protected, and the headland was under military occupancy. Some of the small numbers of tourists began to use spearguns, and commercial fishing returned to the area on which large fish stocks had naturally built up. Unfortunately some fishing was conducted by dynamite, destroying reef structure as well as fish. During the late 1970s and early 1980s pressure began to mount, both from Egyptian scientists and organisations and from international bodies, for the formal establishment of Ras Muhammad as a National Park, and for the protection of other marine areas of scientific or touristic value.

At last, in 1983, enabling legislation was ratified by the Egyptian Parliament and President, allowing the creation of natural protectorates or nature reserves. The areas tentatively identified include Ras Muhammad and the Hamata marine zone, south of Marsa Alam. The law specified that areas would be designated and delineated by decree of the Prime Minister, on recommendations of the Egyptian Environmental Affairs Agency. Subsequently a Prime Ministerial decree was issued on 13th November 1983, stating that the Ras Muhammad area was considered a natural protectorate, and emphasising that the following activities were forbidden within the area:

> Hunting, fishing, killing, transporting or disturbing the fauna or flora, or carrying out any activities that could result in their extermination.
> Fishing, taking or transplanting any organisms or organic materials such as shells or reef rocks or soil for any purpose whatsoever.
> Contaminating the soil or water or air of the protectorate in any form or manner.

The order also established a branch of the Environmental Affairs Agency to supervise the operation of the protectorate.

This, the establishment of the first major marine park within the Red Sea proper, must be seen as an event of major national and regional significance. It is to be hoped that the responsibility for the necessary enforcement of this legislation can now be clarified so that activities such as dynamiting, spearfishing and coral and shell collecting, still prevalent in 1983, can be stopped. It is also to be hoped that in time other marine reserves can be established in Egyptian waters to include important areas of critical habitat along the mainland coast of Egypt.

19.6.2. Israel

Although Israel has only a toehold on the Gulf of Aqaba, in the form of the port and holiday resort of Elat, the first formal marine reserve in the Red Sea was set up there in 1960, and administered by Israel's Nature Reserves Authority. However the reserve only occupies some 700 m of coast and is unavoidably close to the city of Elat and a major phosphate loading bay and oil terminal. The corals and marine life of the reserve have deteriorated markedly, as well documented by Fishelson (1973) and Loya (1975, 1976).

The Nature Reserves Authority has however had a greater effect on the conservation of the Gulf of Aqaba than the short coastal reserve at Elat would imply. Following the Israeli occupation of the Sinai peninsula several interdisciplinary research expeditions undertook studies of the whole coast, leading to the adoption of marine reserve legislation. This established a series of zones along the western shore of the Gulf of Aqaba (Fishelson, 1980). Within eleven reserve zones damage to marine life and fishing was completely prohibited, while in the intervening sportsfishing and commercial fishing zones, all corals and non-edible fish and invertebrates were protected. All forms of waste disposal were forbidden. The Israeli scheme, which it is hoped the Egyptian Authorities can develop, is of significance as a model of the type of zoning system needed throughout the Red Sea.

19.6.3. Jordan

Jordan has only 27 km of coastline in the Gulf of Aqaba, and the need for conservation measures there was realised at an early stage. Spearfishing and coral collecting have been banned for some years, and the visitor is reminded of this by beach signs. The ban is enforced by coastguards and personnel from the Marine Science Station, and a tiny reserve area has been established adjacent to the Marine Station. The greatest threats to the area arise from the industrial and municipal pollution, and from the need to make use of the limited coastline for port and industrial developments. Small but significant incidents of pollution by oil, sewage and phosphate dust from the ore port occur frequently (Walker and Ormond, 1982). The port facilities at the northern end of the Gulf have expanded southwards, while the southern section of the coast (adjacent to the Saudi Arabia) has been allocated to industrial development, following a consultants report which recommended that the central section be set aside for recreational and residential use.

A subsequent report (Ormond, 1978) developed this theme and proposed the establishment of a National Marine Park to include the coastline and reefs of the central section. The coastal strip and foreshore would be available for general public use, and residential development should be restricted to the landward side of the coast road, and into the coastal hills. There is now an official local committee considering planning and protection of the marine environment. It is understood that the planning proposals to establish the marine park formally are in the process of being approved, and that funding is being allocated to this task. Hopefully, the National Marine Park will soon be established, but the problem of maintaining adequate water quality in the area will be very difficult. Pollution can spread from the port and industrial area or across the Gulf from Elat, and there will have to be strict observance of appropriate discharge standards. The future of the Jordanian marine environment at Aqaba will be of great interest, since it may represent in microcosm, the problems the whole Red Sea may face within the next hundred years.

19.6.4. Sudan

The 750 km of Sudan's coast include many of the richest marine areas in the Red Sea, and until now the impact of man on the marine environment there has been slight and very localised. Marine conservation within this area therefore acquires a regional as well as a national importance. Local conservation problems were reviewed by Schroeder (1981).

Consideration of the need for active marine conservation began about 1970 when it became clear that diving tourists were causing noticeable damage to popular diving sites near to Port Sudan. An *ad hoc* Sudan Marine Conservation Committee (SMCC) was set up, later receiving official recognition (1979) as a subcommittee of the National Committee for Environment. Following SMCC recommendation in 1975 the existing fishery legislation was modified and strengthened to give a measure of protection to the reefs. Spear guns were completely banned, and coral and shell collecting, and the taking of inedible reef fish prohibited. The Environmental Health Act of 1975 also stipulates that no solid, liquid or gas which may be harmful to man or animal shall be put into the sea.

Subsequently, progress has been made towards the establishment of the country's first marine parks (Schroeder, 1982) when a new Marine Fisheries Act was ratified by the National Cabinet and National Assembly. The Act includes enabling legislation to permit the establishment of marine parks and reserves, and regulates construction near and in the sea. The first of Sudan's Marine National Parks is likely to be established at Sanganeb, the spectacular coral atoll about 30 km north of Port Sudan. The reef was declared closed to fishing several years ago, and IUCN and the World Wildlife Fund have sponsored a feasibility study and the preparation of management plans. At the time of writing the official establishment of the Sanganeb Park is under consideration. Protection and policing of the site should start as soon as possible, since some parts of the reefs are clearly deteriorating under pressure of fishing and souvenir collection.

In addition to the Sanganeb project, a preliminary proposal (Schroeder, 1982) has been submitted for a Port Sudan Marine National Park, incorporating the reefs and shore close to Port Sudan. Further afield Ormond (1980) has identified other areas of great conservation importance. These include the Suakin Archipelago south of Port Sudan, a maze of scattered reefs and islands with some of the richest coral communities in the Red Sea and important turtle and seabird colonies. A second region comprising Taila Islands, Marsa Inkafeil and Mukawwar Island north of Port Sudan, includes well developed mangroves, reefs, seabird colonies and turtle breeding grounds.

19.6.5. Ethiopia and Eritrea

Political events of the last 10–15 years have overtaken attempts at conservation of the marine environment along the coast of Eritrea. As early as 1968 parts of the Dahlak Island group were proposed as a marine park, and legislation was apparently prepared but not implemented (Berhanu, 1975). Recent impacts on Dahlak and the rest of the coast remain unknown, but it is likely that military and other personnel may have had a major effect in some areas. When political conditions allow, a detailed habitats and resources survey should be undertaken to define management parameters. Plans for the Dahlak Marine Park could benefit from further research, since some of the outer islands, not included within the proposed boundaries, are likely to contain better developed critical habitats than the more inshore islands included in the Dahlak Marine Park proposal.

19.6.6. North Yemen

Almost no biological survey work has been carried out along the North Yemen coast. The Yemeni authorities are developing their own marine conservation policy within the framework of PERSGA, and a marine resources and habitat survey is currently being developed by the University of Sana'a.

Preliminary data suggest that two areas may be of particular significance (Ormond, 1980). The southwest side of the Isa peninsula not far from Salif probably has the best coral reefs on the Yemen mainland and is convenient for tourists from Al Hudaydah. The Zuqar islands apparently have a rich sublittoral fauna and are interestingly situated at the extreme south of the Red Sea. This suggests that some atypical species of Indian Ocean origin may occur there, and the islands would be a good site for studies on regional oceanography and monitoring oceanic pollution.

19.6.7. Saudi Arabia

Saudi Arabia has by far the largest coast of any country on the Red Sea (about 1500 km), and this coast is studded with reefs, islands and other habitats of considerable importance. Until about 1970 it was hardly exposed to human impact, but since then commercial and industrial development fuelled by the oil economy have proceeded at unbelievable rates, especially near Jiddah and Yanbu al Bahr. Some of the observed impacts, resulting from sedimentation and infilling, have been described above.

With the support of IUCN, the Meteorological and Environmental Protection Administration has recently undertaken a survey of the whole Saudi Red Sea coast (IUCN, 1984). Intertidal and subtidal habitats have been mapped, and significant areas of critical habitat identified, including reefs, mangrove stands, seagrass beds and seabird and turtle islands. From the survey finding over 70 sites have been identified as candidates for protection, and an interim protection order to prevent development of these areas has been placed on these sites by the Saudi Government. The survey has also led to a proposal for a comprehensive zoning scheme intended to form a management basis for the whole coastline.

Four areas are under consideration for National Parks, two in the Gulf of Aqaba region, one enclosing the coast, reefs and islands of the Wajh bank, and a major one including all the reefs and islands of the outer Farasan Bank, scattered over 250 km of the Red Sea. This last area, if given formal status, would constitute one of the most significant Marine National Parks anywhere in the world, and it would give protection to an area enclosing habitats representative of much of the Red Sea. Development of detailed proposals and regulations for this and the other sites may take some time.

Consideration is also being given to coastal areas near major population centres. These areas are already experiencing substantial impacts, but are of obvious importance for recreation and educational use. Important areas include Ras Hatiba 60 km north of Jiddah, the Shu'aiba area about 75 km south of Jiddah, and sites within or close to Yanbu Industrial City. Another region likely to receive early attention is the coastline between Al Qahmah and Shuqayq about 100 km north of Jizan. This falls within the boundaries of the newly established Asir National Park, although the extension of the Park Provisions to this area has not yet been authorised, and a management policy for the marine areas of the park has yet to be formulated. This and any other Marine National Parks set up fall under the authority of the National Parks Section of the Ministry of Agriculture and Water in Riyadh. Hopefully, as in other countries, regional and local authorities may also take an initiative in establishing parks and other protected areas at provincial and local levels.

19.7. CONCLUDING REMARKS

As this account has shown, the last ten years have seen considerable activity directed at conservation and protection of the environment and marine life of the Red Sea. This progress has very largely been due to the efforts of comparatively few local scientists and administrators, who, with the assistance of expatriate specialists, have spearheaded an informal campaign to increase regional public awareness and to initiate investigation and control measures.

Despite this progress, there can be no doubt that as this chapter and the last have shown, marine environmental conditions in the Red Sea have begun to deteriorate. Indeed it may only be as a result of obvious damage and pollution that authorities and public opinion have come to support the cause of conservation and management. The situation is not yet serious everywhere, but it is becoming serious. Experience elsewhere in the world suggests conditions will get a lot worse before, if ever, they get better. Ultimately all the marine life described in this volume is at stake. The Red Sea is almost completely enclosed, and even if the various Marine Parks proposed are eventually set up, there remains the immense problem of guaranteeing high water quality within the enclosed Red Sea basin in the face of accelerating development and human impact.

Ultimately the choice is one between a dead sea, whose margin is occupied by cities and developments planned without adequate consideration to the environment, or a living sea whose marine life can continue to fascinate the human mind and uplift the spirit. This sea's living resources, carefully husbanded, can continue to contribute towards our basic food and leisure needs. The task of breaking through the technological and especially the economic barriers to this vision of a garden sea is a huge and perhaps impossible one. But it is a vital task, and one to which every reader of this volume can contribute by promoting public awareness of the issues or by collecting and collating personal observations during their own visits to this remarkable marine environment.

REFERENCES

Berhanu, A. (1975). A report on Dahlak Islands Marine Park. *International Conference on Marine Parks and Reserves*, Tokyo, Japan.

Fishelson, L. (1973). Ecology of a coral reef in the Gulf of Aqaba (Red Sea) influenced by pollution. *Oecologia* 12, 55–68.

Fishelson, L. (1980). Marine reserves along the Sinai Peninsula (northern Red Sea). *Helgoländer Meeresunters* 33, 624–40.

Hiscock, K. and Al-Aiidy, H. (1972). *Some notes on the exploitation and conservation of the coral reef community on the Red Sea coast of Saudi Arabia*. Saudi Arabia Project, Marine Science Laboratories, Menai Bridge.

I. U. C. N. (1983) *Ecological study of sites on the coast of Bahrain*. Kuwait Action Plan. Tropical Marine Research Unit, University of York.

I. U. C. N. (1984) *Management of Red Sea coastal resources: Recommendations for protected areas*. Saudi Arabia Marine Conservation Programme, Meteorology and Environmental Protection Administration, Jiddah.

Johannes, R. E. (1978). Traditional marine conservation methods in Oceania and their demise. *Ann. Rev. Ecol. Syst.* 9, 349–64.

Johannes, R. E. (1981). Working with fishermen to improve coastal tropical fisheries and resource management. *Bull. mar. Sci.* 31, 673–80.

Johnston, R. (ed.) (1976). *Marine Pollution*. Academic Press, New York, London & San Francisco.

Lewis, J. B. (1977). Processes of organic production on coral reefs. *Biol. Rev.* 52, 305–47.

Loya, Y. (1975). Possible effects of water pollution on the community structure of Red Sea corals. *Mar. Biol.* 29, 177–85.

Loya, Y. (1976). Recolonisation of Red Sea corals affected by natural catastrophes and man-made perturbations. *Ecology* 57, 278–89.

Neve, P. and Al Aiidy, H. (1973). The Red Sea Fisheries of Saudi Arabia. *Bull. Mar. Res. Centre, Saudi Arabia* No.3.

Ormond, R. F. G. (1978). *A marine park for Jordan. Report on the Feasibility of establishing a Marine Park at Aqaba*. Report prepared for ALECSO (Arab League Educational Cultural and Scientific Organisation).

Ormond, R. F. G. (1980). Management and conservation of Red Sea habitats. *Proc. Symp. Coastal and Marine Environment of the Red Sea, Gulf of Aden and Tropical Western Indian Ocean* 2, 135–62.

Ormond, R. F. G. and Campbell, A. C. (1974). Formation and breakdown of *Acanthaster planci* aggregations in the Red Sea. *Proc. 2nd Int. Coral Reef Symp* 1, 595–619.

P. E. R. S. G. A. (1981). *Final Act of the Jeddah Regional Conference of Plenipotentiaries on the Conservation of the Marine Environment and Coastal Areas of the Red Sea and Gulf of Aden*. Environmental Programme for the Red Sea and Gulf of Aden. (PERSGA).

Ray, G. C. (1976). *Critical Marine Habitats*. IUCN Publications, New Series, No. 37, 15–59.

Ruivo, M. (ed.) (1972). *Marine Pollution and Sea Life*. Fishing News (Books), London.

Salm, R. B. (1983). *Managing Coastal and Marine Protected Areas*. IUCN Commission on National Parks and Protected Areas, IUCN.

Schroeder, J. H. (1981). Man versus reef in The Sudan: Threats, destruction, protection. *Proc. 4th Int. Coral Reef Symp* 1, 253–7.

Schroeder, J. H. (1982). *Aspects of coastal zone management of the Sudanese Red Sea: characteristics, resources, pollution, conservation and research*. Environmental Research Report 3, Institute of Environmental Studies, University of Khartoum, Sudan.

Wainwright, S. A. (1965). Reef communities visited by the Israel South Red Sea Expedition 1962. *Bull. Sea Fish. Res. Stn. Israel* 38, 40–53.

Walker, D. I. and Ormond, R. F. G. (1982). Coral death from sewage and phosphate pollution at Aqaba, Red Sea. *Mar. Poll. Bull.* 13, 21–5.

Wood, E. J. F. and Johannes, R. E. (eds.) (1975). *Tropical Marine Pollution*. Elsevier Oceanography Series, Vol.12, Elsevier Scientific, Amsterdam.

Wray, T. (ed.) (1979) *Commercial Fishes of Saudi Arabia*. White Fish Authority, U.K.

POSTSCRIPT

STEPHEN M. HEAD

Since the completion of the text of this book, the proceedings of an important international symposium have become available (Angel, 1984). The Mabahiss/John Murray International Symposium was held in September 1983 in Alexandria, Egypt. The symposium commemorated the fiftieth aniversary of the 1933–1934 Anglo-Egyptian expedition to the Arabian Sea, during which a number of fundamental advances in our knowledge of the marine environment of the area were made.

The Mabahiss/John Murray symposium covered a wide range of topics in oceanography, geology and biology, and as a whole the Proceedings are a most valuable review of our knowledge of many aspects of the marine environment of the north-western Indian Ocean and its neighbouring waters, including the Red Sea. Of the 28 published papers, about half deal exclusively with the Indian Ocean, Arabian Gulf or Gulf of Oman, and hence are of peripheral interest to the readers of the present book. Five historical papers at the beginning of the Proceedings describe the scientific and national achievements of the Mabahiss expeditions. These historical papers are of particular interest to European and American readers for the light they shed on the question of "technology transfer" within cooperative research programmes run jointly by developed and developing nations. In this regard the Anglo-Egyptian Mabahiss expeditions were a resounding success, laying the foundations for Egypt's present high level of expertise in marine science.

Several other papers in the Proceedings are of direct interest to readers of 'Key Environments — Red Sea'. Girdler (1984) provides a review of the history of formation of the Red Sea and Gulf of Aden complementary to that of Dr. Braithwaite in the present volume. Girdler's account tells essentially the same story as that of our chapter two, (although there are minor differences in the dating applied to some crustal expansion phases), but emphasises in particular the tectonic background. Poisson *et al.* (1984) provide some new observations on water exchange and biogeochemical cycles at the southern end of the Red Sea, and show that summer mid-water inflows of nutrients can balance the loss through outflowing bottom currents.

Two short papers cover recent developments in the deep Red Sea metalliferous sediment mining project. Nawab (1984) provides estimates of the resource size, and the techniques and economics of practical mining. Abu Gidieri (1984) briefly reviews potential impacts of the project; like Dr. Karbe in the present volume he emphasises the importance of a deep-water tailings discharge. Halim (1984) contributes a short review of Red Sea plankton ecology, this does not include data not covered in chapter 5 of the present book, but provides an interesting comparison between the ecology of the Red Sea and Arabian Gulf.

Two papers in the Mabahiss Symposium Proceedings are concerned with coral reefs, and partly supplement chapter 7. Mergner (1984) provides a substantial history of reef studies in the Red Sea, with a very useful set of references, besides reproducing in review form many of his own observations on Red Sea reef ecology. Scheer (1984) provides a revised and updated account of the distribution of coral genera within the whole Indian Ocean province, including the new Red Sea records discussed in chapter 7, tabulated at generic and sub-generic level. The central Red Sea is seen to have a highly diverse fauna even compared with many central Indian Ocean sites, although the problem of comparing areas of different sampling intensity remains.

Dr. Dicks, who contributed the general review of Red Sea pollution which forms Chapter 18 of the present book, includes within the Mabahiss Proceedings a detailed account of the pollution problems of the Ras Budran oil field and oil terminal, in which considerably more information is presented on impacts, recovery and sensitivity of marine communities to oil spillage, besides practical recommendations for amelioration. Finally, Venema (1984) gives a useful review of the fisheries of the north Arabian Sea and adjacent waters, in which the low Red Sea production levels can be seen in the context of the upwelling areas off the south-eastern Arabian peninsula.

REFERENCES

Abu Gideiri, Y. B. (1984). Impacts of mining on central Red Sea environment. *Deep Sea Research Part A*. 31, 823–8.

Angel, M. V. (ed.) (1984). Marine Science of the north-west Indian Ocean and adjacent waters. Proceedings of the Mabahiss/John Murray International Symposium, Egypt, 3–6 September 1983. *Deep-Sea Research Part A*. 31, 571–1035.

Dicks, B. (1984). Oil pollution in the Red Sea — Environmental monitoring of an oilfield in a coral area, Gulf of Suez. *Deep-Sea Research Part A*. 31, 833–55.

Girdler, R. W. (1984). The evolution of the Gulf of Aden and Red Sea in space and time. *Deep-Sea Research Part A*. 31, 747–63.

Halim, Y. (1984). Plankton of the Red Sea and the Arabian Sea. *Deep-Sea Research Part A*. 31, 969–83.

Poisson, A., Morcos, S., Souvermezoglou, E., Papaud, A. and Ivanoff, A. (1984). Some aspects of biogeochemical cycles in the Red Sea with special reference to new observations made in summer 1982. *Deep-Sea Research Part A*. 31, 707–18.

Venema, S. C. (1984). Fishery resources in the north Arabian Sea and adjacent waters. *Deep-Sea Research Part A*. 31, 1001–18.

Geographical Index

The geographical positions of the sites listed here are shown in the maps on the endpapers, with the exception of the ancient or archaeological sites mentioned in Chapter 16, which are marked on separate maps within the chapter.

Subject Index

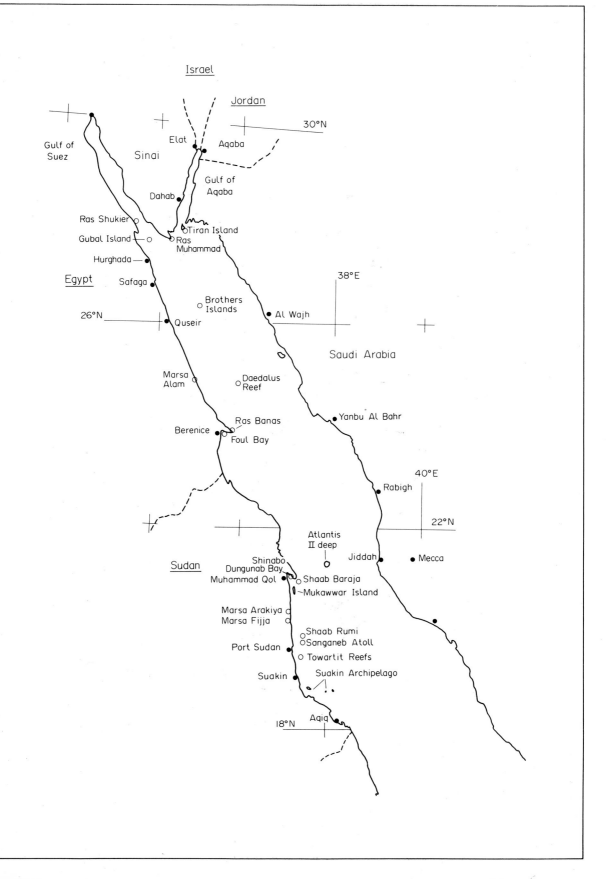